Regulators V

CONTEMPORARY MATHEMATICS

842

Regulators V

Conference
Regulators V
June 3–13, 2024
University of Pisa

Federico Binda
José Ignacio Burgos Gil
James Lewis
Deepam Patel
Gregory Pearlstein
Editors

AMS AMERICAN MATHEMATICAL SOCIETY

2020 *Mathematics Subject Classification.* Primary 19F27, 14C25, 14G45, 14C30, 14D07, 14F42, 19D55.

Library of Congress Cataloging-in-Publication Data

Names: Regulators (Conference) (5th : 2024 : Pisa, Italy) author | Binda, Federico, 1988- editor | Burgos Gil, José I. (José Ignacio), 1962- editor | Lewis, James Dominic, 1953- editor | Patel, Deepam, 1981- editor | Pearlstein, Gregory, 1970- editor

Title: Regulators V / Federico Binda, José Ignacio Burgos Gil, James Lewis, Deepam Patel, Gregory Pearlstein.

Other titles: Regulators 5

Description: Providence, Rhode Island : American Mathematical Society, [2026] | Series: Contemporary mathematics, 0271-4132 ; volume 842 | "The Regulator V conference (3-13 June 2024 in Pisa, Italy)"–page 17.

Identifiers: LCCN 2026013140 | ISBN 9781470479176 paperback | ISBN 9781470485870 ebook

Subjects: LCSH: Regulators (Mathematics) | Algebraic fields | Geometry, Algebraic | AMS: K-theory – K-theory in number theory – Étale cohomology, higher regulators, zeta and L-functions | Algebraic geometry – Cycles and subschemes – Algebraic cycles | Algebraic geometry – Cycles and subschemes – Transcendental methods, Hodge theory | Algebraic geometry – Families, fibrations – Variation of Hodge structures | Algebraic geometry – (Co)homology theory – Motivic cohomology; motivic homotopy theory | K-theory – Higher algebraic K-theory – K-theory and homology; cyclic homology and cohomology | LCGFT: Conference papers and proceedings

Classification: LCC QA247 .R417 2024

LC record available at https://lccn.loc.gov/2026013140

Contemporary Mathematics ISSN: 0271-4132 (print); ISSN: 1098-3627 (online)

DOI: https://doi.org/10.1090/conm/842

---

10 9 8 7 6 5 4 3 2 1 31 30 29 28 27 26

Contents

Preface vii

Scientific program ix

List of participants xi

The A-factor locus of a normal function
DEVIN AKMAN 1

Regulators and derivatives of Vologodsky functions with respect to $\log(p)$
AMNON BESSER 11

A motivic approach to rational p-adic cohomologies
FEDERICO BINDA and ALBERTO VEZZANI 45

Quadratically enriched binomial coefficients over a finite field
CHONGYAO CHEN and KIRSTEN WICKELGREN 77

Knots, primes and class field theory
ALAIN CONNES and CATERINA CONSANI 105

The Grassmannian complex and Goncharov's motivic complex in weight 4
HERBERT GANGL 133

A finite regulator map for the Bloch group
HERBERT GANGL and MAXIM KONTSEVICH 153

Calculating Archimedean height pairing via generalized cross-ratio
ALEXANDER B. GONCHAROV 167

Atypical Hodge loci
PHILLIP GRIFFITHS 171

Motivic irrationality proofs
MATT KERR 187

On singular supports in mixed characteristic
TAKESHI SAITO 213

Picard-Fuchs equations of twisted differential forms associated to Feynman integrals
PIERRE VANHOVE 261

Preface

The concept of "regulator" dates back to the early 1800s, with the work of Dirichlet and the proof of the celebrated class number formula for a number field. Since the 1960s Dirichlet's fundamental discovery has been found potentially to occur elsewhere in number theory, in algebraic geometry, in class field theory, in algebraic K-theory, in the theory of algebraic cycles and motives, and in Hodge theory. In fact, the central conjectures in each of these disciplines (the Birch-Swinnerton-Dyer conjecture, the Lichtenbaum conjecture, the Beilinson-Soulé conjecture, the Bloch-Kato conjecture, the Stark conjecture, etc.) are all outgrowths of Dirichlet's formula and each in its own way involves a regulator. Regulators come in many different forms, according to the context. For instance, the Borel regulator is the higher degree analogue of the Dirichlet regulator, considered as a map on algebraic K-theory in degree one or K_1. On the other hand, in Riemann surface theory, the regulators might involve abelian integrals and Jacobians, extending the ideas of the 19th century analytic number theorists and geometers. Generally speaking, in its current incarnation, a regulator is a map from the algebraic K-theory of an algebraic variety to a suitable cohomology theory such as étale cohomology or Deligne cohomology.

Several conferences have been organized in past decades on the subject, starting from the famous 1998 "Regulators" meeting in Oberwolfach (organised by S. Bloch, M. Kolster, P. Schneider and V. Snaith), that involved a number of participants working on regulators from different perspectives. The Oberwolfach workshop was regarded as a great success, and from it sprang a series of conferences under the same name. Regulators II (organized by J. D. Lewis and V. Snaith) in December of 2005 at the Banff International Research Station; Regulators III, in July of 2010 in the city of Barcelona; Regulators IV, in May 2016 at the IMG-PRG in Paris.

Regulators V was held at the University of Pisa (June 3–13, 2024), as the fifth edition in the Regulators series, bringing together the world's leading experts on regulators, algebraic cycles, and motives. This was also an occasion to celebrate the 80th birthday of Spencer Bloch and the 60th birthday of Rob de Jeu. Once again, the conference covered the different aspects and applications of the complex world of regulators, mixing old and new aspects and approaches.

This volume brings together a selection of contributions that highlight the breadth and vitality of current research on the subject and its arithmetic, geometric, and physical applications.

We would like to express our sincere gratitude to all lecturers, speakers, participants, and administrative personnel for contributing to the success of this conference. Moreover, we would like to thank all authors for their contributions to these proceedings, as well as the expert referees and the staff at AMS.

Federico Binda
José Ignacio Burgos Gil
James D. Lewis
Deepam Patel
Gregory Pearlstein

Scientific program

Invited Speakers

Amina Abdurrahman, IHES: *A topological formula for the central value of symplectic L-functions and Reidemeister torsion.*

Tomoyuki Abe, Kavli IPMU: *Continuity principle in ramification theory.*

Joseph Ayoub, University of Zurich: *Nori connectivity for squared hypersurfaces and classicality of the motivic Galois group.*

Tom Bachmann, University of Mainz: *Algebraizability of vector bundles and motivic homotopy theory.*

Amnon Besser, Ben Gurion University: *Regulators and derivatives of Vologodsky integrals with respect to* $\log(p)$.

Spencer Bloch, University of Chicago: *Generalized Cross-Ratios.*

Francis Brown, Oxford University: *Higher regulators of the ring of integers.*

François Brunault, École Normale Supérieure de Lyon: *Modular regulators and multiple modular values.*

Laura Capuano, Università di Roma III: *Unlikely Intersections and applications to Diophantine Geometry.*

Caterina Consani, Johns Hopkins University: *Primes, knots and the scaling site.*

Rob de Jeu, Vrije Universiteit: K_2 *of elliptic curves over non-Abelian cubic and quartic fields.*

Robin de Jong, University of Leiden: *Heights and periods of limit mixed Hodge structures.*

Javier Fresán, École Polytechnique: *Euler's constant and exponential motives.*

Martin Gallauger, University of Warwick: *The spectrum of Artin motives.*

Herbert Gangl, Durham University: *Depth reduction for multiple polylogarithms.*

Evangelia Gazaki, University of Virginia: *Hyperelliptic Curves mapping to Abelian Surfaces and applications to Beilinson's Conjecture for 0-cycles.*

Thomas Geisser, Rikkyo University: *Brauer and Neron-Severi groups of surfaces over finite fields.*

Alexander Goncharov, Yale University: *Explicit constructions of motivic Galois Groups.*

Souvik Goswami, University of Barcelona: *Framed height pairing.*

Phillip Griffiths, Institute for Advanced Study: *Atypical Hodge Loci.*

Annette Huber-Klawitter, University of Freiburg: *Motives, Periods and Species.*

Matt Kerr, Washington University, St. Louis: *Hypergeometric families and Beilinson's conjectures.*

Matilde Lalín, University of Montreal: *Evaluations of areal Mahler measure.*

Davide Lombardo, Università di Pisa: *Sato-Tate groups of Fermat Jacobians.*

Rita Pardini, Università di Pisa: *Stable I-surfaces of index 2 and generalized spin curves of genus 2.*

Sujatha Ramdorai, University of British Columbia: On the second Iwasawa cohomology and rational points on elliptic curves.

Takeshi Saito, University of Tokyo: *On relative singular support in mixed characteristic.*

Emre Sertöz, Leiden University: *Linear relations between 1-periods.*

Vasudevan Srinivas, University at Buffalo: *Enriched Hodge structures and cycles on analytic thickenings.*

Pierre Vanhove, Institut de Physique théorique: *Picard–Fuchs Equations of Dimensionally Regulated Feynman Integrals.*

Alberto Vezzani, Università Statale di Milano: *Relative rigid cohomology via motivic homotopy theory.*

Kirsten Wickelgren, Duke University: *A quadratically enriched logarithmic zeta function.*

Wei Zhang, MIT: *Height pairings of cycles on modular varieties and L-functions.*

Organizers:

Federico Binda
Università degli Studi di Milano

Gregory Pearlstein
Università di Pisa

James Lewis
University of Alberta, Emeritus

Tamás Szamuely
Università di Pisa

Deepam Patel
Purdue University

Scientific Committee:

Alexander Beilinson
University of Chicago

Marc Levine
University of Duisburg-Essen

José Ignacio Burgos Gil
ICMAT, Madrid

Vincent Maillot
Institut de Mathématiques de Jussieu

Hélène Esnault
Freie Universität Berlin

Shuji Saito
University of Tokyo

List of participants

Ricardo Acuna
Washington University in St. Louis

Devin Akman
Washington University in St. Louis

Ran Azouri
Université Sorbonne Paris Nord

Luca Barbieri Viale
Università degli studi di Milano

Shomrik Bhattacharya
University of Southern Denmark

Giulio Bresciani
Scuola Normale Superiore di Pisa

Zhelun Chen
Leiden University

George Cooper
University of Oxford

Alessandro D'Angelo
KTH Royal Institute of Technology

Erick Miguel Diaz Lopez
University of Miami Marco Franciosi
Università di Pisa

Lorenzo Furio
Università di Pisa

Andrea Gallese
Scuola Normale Superiore in Pisa

Vasily Golyshev
ICPT

Pengju Guan
Vrije Universiteit Amsterdam

Adam Keilthy
Chalmers Technical University

Niklas Kipp
University of Regensburg

Pietro Leonardini
Università di Pisa

Deven Manam
Northwestern University

Nicola Mazzari
University of Padova

Alberto Merici
Università degli studi di Milano

Francesco Naccarato
ETH Zürich

Anubhav Nanavaty
UC Irvine

Alberto Navarro
Universidad Politécnica de Madrid

Gyujin Oh
Columbia University

Luc Pirio
UVSQ (Paris-Saclay)

Fei Ren
Bergische Universität Wuppertal

Diego Rojas
Northwestern University

Francesco Sala
University of Pisa

Luca Speciale
Universität Paderborn

Muhammed Ashar Tafhim
Bergische Universität Wuppertal

Mattia Talpo
University of Pisa

Guoping Tang
University of Chinese Acad. of Sciences

Thu Ha Trieu
École normale supérieure de Lyon

Vijaylaxmi Trivedi
University of Buffalo

Shuddhodan Kadattur Vasudevan
IHES

Runlei Xiao
University of Padova

Yuenian Zhou
Mathématiques Orsay (Paris-Saclay)

Beat Zurbuchen
ETH Zurich

Contemporary Mathematics
Volume **842**, 2026
https://doi.org/10.1090/conm/842/16848

The A-factor locus of a normal function

Devin Akman

ABSTRACT. The A-polynomial of a 3-manifold with toric boundary captures information about families of hyperbolic structures. We present an exposition of the theory of A-polynomials, including a characterization of their irreducible factors via torsion in K_2. Our previous work shows that the torsion locus and hence A-factor locus of the normal function associated with a family of smooth curves is finite, thus settling a conjecture of Antonin Guilloux and Julien Marché. We derive an additional finiteness result for A-factors by introducing the notion of overgenus.

1. Introduction

Let M be a compact connected orientable 3-manifold with boundary $\partial M = T^2$. The primary example to keep in mind will be the exterior of a knot in S^3. There is an invariant $A_M \in \mathbb{C}[x^{\pm}, y^{\pm}]$ called the A-polynomial [**2**] of M.[1] Guilloux and Marché proved [**8**] that an irreducible Laurent polynomial P is a factor of the A-polynomial of some manifold if and only if the Steinberg symbol $\{x, y\}$ is torsion in $K_2(F_P)$, where $F_P := \operatorname{Frac} \mathbb{C}[x^{\pm}, y^{\pm}]/(P)$ is its function field (we call such polynomials "A-factors" and the corresponding curves "A-curves").[2] Following the notation of F. Rodriguez-Villegas [**16**], let $\Delta := \Delta(P)$ be the Newton polygon of P and $\mathcal{T}$ the set of roots of the edge polynomials P_τ. For every $t \in \mathcal{T}$, there is a discrete valuation w on F_P and a nonzero integer n such that $\{x, y\}_w = \pm t^n$. Therefore, the roots of the edge polynomials are roots of unity, and the Laurent polynomials obtained by varying the g interior coefficients of P are *tempered*. Compactifying those curves in the toric variety $\mathbb{P}_\Delta$ yields a smooth projective family of genus g curves $\pi\colon X \to S := \mathbb{A}^g \setminus \Sigma$.

Guilloux and Marché conjectured that the exact locus of the family, which contains the A-factor locus

$$K := \{\, s \in S(\mathbb{C}) \mid \{x, y\} \text{ is torsion in } K_2(\mathbb{C}(X_s)) \,\},$$

2020 *Mathematics Subject Classification.* Primary 14-02; Secondary 57K31, 11C08, 19C99, 14M25, 32G20, 57K32.

We thank Nathan Dunfield for helpful discussions about the A-polynomial.

[1]Computing the A-polynomial requires choosing a basis $\{\mathfrak{L}, \mathfrak{M}\}$ for $\pi_1(\partial M)$. This allows it to distinguish mirror images of chiral knots even though they have diffeomorphic exteriors.

[2]The precise definition of an A-curve is an irreducible curve with an open subset U and a closed immersion $U \hookrightarrow \mathbb{G}_{\mathrm{m}}^2$ whose image is $V(P)$. This arises in the context of curves in toric surfaces.

is finite. We will state our theorem in terms of K since we have not defined the exact locus; however, the proof does pass through finiteness of the exact locus.

THEOREM 1.1. *K is finite.*

PROOF. Associated to π is an integral polarized variation of Hodge structure $\mathbf{V} = R^1\pi_*\mathbb{Z}_X$ of weight and level 1 with Hodge numbers (g, g). Temperedness of the family implies the existence of a higher Chow cycle $\Xi \in CH^2(X, 2)$ which restricts to an integer multiple of $\{x, y\} \in K_2(\mathbb{C}(X_s)) \cong CH^2(\eta_{X_s}, 2)$ on each fiber [**5**]. Evaluating the integral regulator $r\colon CH^2(X, 2) \to H^2_{\mathcal{D}}(X, \mathbb{Z}(2))$ on Ξ produces an admissible normal function $R \in \mathrm{ANF}(\mathbf{V}(2))$ with ∇R of maximal rank. If the fiber X_s is an A-curve, then $R(s) = r(\Xi|_{X_s}) \in H^1(X_s, \mathbb{C}/\mathbb{Z}(2))$ is torsion. Hence, K is contained in the torsion locus $T(R)$. Theorem 1.2 of [**11**] tells us that the $\mathbb{R}$-split locus $Z_{\mathbb{R}}(R) \supset T(R) \supset K$, which equals the exact locus defined in [**8**] by the vanishing of the regulator 1-form in cohomology, is finite. □

If the edge coefficients are algebraic (we can always assume this), then $K \subset S(\overline{\mathbb{Q}})$, and the Bloch-Beilinson conjecture predicts that $K = T(R) \cap S(\overline{\mathbb{Q}}) = T(R)$. The purpose of this article is to provide a coherent exposition of the construction and significance of the A-polynomial and a detailed proof of the K-theoretic criterion for A-factors in service of Theorem 1.1. Additionally, in § 4, we define a measure of complexity for A-factors called the *overgenus* and show that the number of A-factors with bounded positive overgenus is finite.

2. The A-polynomial

2.1. Character varieties. Given a finitely generated group G, the representation variety $R(G) := \mathrm{Hom}(G, \mathrm{SL}_2(\mathbb{C}))$ is a complex affine variety with field of definition $\mathbb{Q}$. Fixing a generating set $S = \{g_1, \dots, g_n\} \subset G$ puts coordinates on $R(G)$; it is the subvariety of $\mathbb{A}^{4n}$ consisting of functions $\rho\colon S \to M_{2\times 2}(\mathbb{C})$ such that $\det(\rho(g_i)) = 1$ for all i and that the images of the generators satisfy the same relations as the generators themselves (these are both clearly polynomial conditions in the entries with integer coefficients). Since conjugate representations are equivalent, we quotient by the conjugation action of $\mathrm{SL}_2(\mathbb{C})$ to obtain the character variety $X(G) := R(G) /\!\!/ \mathrm{SL}(2, \mathbb{C}) = \mathrm{Spec}(\mathbb{C}[R(G)]^{\mathrm{SL}_2(\mathbb{C})})$. Geometrically, the quotient map $R(G) \twoheadrightarrow X(G)$ identifies points whose orbit closures intersect, and each fiber contains a unique closed orbit. C. Procesi showed [**12**] that the ring of invariants is generated by the functions $\{I_{g_{i_1}\cdots g_{i_k}}\}$, where $1 \le i_1 < \cdots < i_k \le n$ and $I_g(\rho) := \chi_\rho(g) = \mathrm{tr}(\rho(g))$. While non-conjugate representations may have the same characters, the irreducible representations are a Zariski-open subset of $X(G)$, and their (closed) orbits are in one-to-one correspondence with their characters [**4**]. This may serve to motivate the name "character variety."

When the fundamental group of a manifold is finitely generated, we define the representation and character varieties of the manifold as those of its fundamental group. These have a close connection to hyperbolic structures as outlined in § 2.2. For now, we will determine the character variety $X(\partial M)$, which is a crucial ingredient in the definition of the A-polynomial. Because $\pi_1(\partial M)$ is the free abelian group on the longitude and meridian cycles $\{\mathfrak{L}, \mathfrak{M}\}$, points of the representation variety are pairs of commuting matrices in $\mathrm{SL}(2, \mathbb{C})$. It is not hard to check that the quotient identifies (A, B) with the orbit of a diagonal representation

$$\rho(\mathfrak{L}) = \begin{bmatrix} x & 0 \\ 0 & x^{-1} \end{bmatrix} \qquad \rho(\mathfrak{M}) = \begin{bmatrix} y & 0 \\ 0 & y^{-1} \end{bmatrix},$$

where $\{x, x^{-1}\}$ and $\{y, y^{-1}\}$ are eigenvalues of A and B, respectively. The only remaining ambiguity comes from simultaneously swapping the eigenvalues; thus, we have a degree two quotient map $q\colon \mathbb{G}_\mathrm{m}^2 \twoheadrightarrow X(\partial M) \cong \mathbb{G}_\mathrm{m}^2 /\!\!/ \mathbb{Z}_2$ induced by the inclusion

$$\mathbb{C}[x + x^{-1}, y + y^{-1}, xy + x^{-1}y^{-1}] \hookrightarrow \mathbb{C}[x^{\pm}, y^{\pm}].$$

2.2. Hyperbolic structures. Let M° be the interior of the manifold in the introduction. A hyperbolic structure on M° is an atlas with charts taking values in $\mathbb{H}^3$ and locally constant transition functions $U_i \cap U_j \to \mathrm{Isom}(\mathbb{H}^3) \cong \mathrm{PSL}_2(\mathbb{C})$. We can also think of it as a $\mathrm{PSL}_2(\mathbb{C})$-bundle with fiber $\mathbb{H}^3$, a flat connection, and a "developing section" coming from the atlas which is transverse to the horizontal foliation. The map

$$\mathrm{hol}\colon \mathrm{Def}^{(\mathrm{PSL}_2(\mathbb{C}), \mathbb{H}^3)}(M^\circ) \to \mathrm{Hom}(\pi_1(M), \mathrm{PSL}_2(\mathbb{C})) /\!\!/ \mathrm{PSL}_2(\mathbb{C})$$

from the space of isotopy classes of hyperbolic structures to the $\mathrm{PSL}_2(\mathbb{C})$-character variety defined by taking the holonomy of the flat bundle is a local homeomorphism [**7**]. Its image comprises precisely the representations which lift to $\mathrm{SL}_2(\mathbb{C})$ [**15**]. It is not injective in general, except on the locus of complete hyperbolic structures, those for which for which the developing section pulls back to a diffeomorphism $\widetilde{M^\circ} \to \mathbb{H}^3$ on the universal cover. When the structure is complete, $M^\circ \cong \mathbb{H}^3/\rho(\pi_1(M))$ is a finite-volume hyperbolic manifold with one cusp, corresponding to the torus boundary component. A rich source of examples is hyperbolic knot complements. The remarkable Mostow-Prasad rigidity theorem states that if $\Gamma_1, \Gamma_2 \subset \mathrm{PSL}_2(\mathbb{C})$ are isomorphic discrete subgroups such that $\mathbb{H}^3/\Gamma_i$ has finite volume, then they are conjugate. As a consequence, there is a unique discrete faithful character (DFC) in the $\mathrm{PSL}_2(\mathbb{C})$-character variety of M, and any isomorphism of fundamental groups $\pi_1(M) \to \pi_1(N)$ is induced by a unique isometry $M^\circ \to N^\circ$. Finite-volume hyperbolic 3-manifolds[3] are uniquely determined by their fundamental groups.

In order to study lifts of representations, we consider the central extension

$$1 \longrightarrow \mathbb{Z}_2 \longrightarrow \mathrm{SL}_2(\mathbb{C}) \longrightarrow \mathrm{PSL}_2(\mathbb{C}) \longrightarrow 1.$$

If a representation $\pi_1(M) \to \mathrm{PSL}_2(\mathbb{C})$ has a lift $\pi_1(M) \to \mathrm{SL}_2(\mathbb{C})$, then the set of lifts is a torsor for the group

$$\mathrm{Hom}(\pi_1(M), \mathbb{Z}_2) \cong \mathrm{Hom}(H_1(M, \mathbb{Z}), \mathbb{Z}_2) \cong H^1(M, \mathbb{Z}_2).$$

Therefore, M has 2^{b_1} DFCs, where b_1 is the first mod 2 Betti number. Alexander duality implies that $b_1 = 1$ for a knot complement. An irreducible component $C_i \subset X(M)$ containing a DFC is called a canonical component, and a famous result of Thurston shows that the dimension of a canonical component equals the number of cusps [**13**]. It is also believed that a DFC is a smooth point of $X(M)$, which would make it contained in a unique canonical component. Not only do we know that canonical components are curves, but Culler and Shalen proved [**2**] that $X(M)$ never has isolated points and that every component is a curve when M contains no closed incompressible surfaces. Hopefully, this overview provides the

[3] And more generally n-manifolds for $n \geq 3$.

reader an idea of the structure of $X(M)$, which we then attempt to capture via the A-polynomial.

2.3. Definition. Now that all the ingredients are in place, the definition of the A-polynomial is fairly straightforward. Let $r\colon X(M) \to X(\partial M)$ be restriction of representations to the boundary. Then

$$D_M := q^{-1}(\overline{\operatorname{im} r}) \setminus \{\text{isolated points}\} \subset \mathbb{G}_{\mathrm{m}}^2$$

is a possibly reducible curve defined by a Laurent polynomial A_M unique up to units.[4,5] An easy argument shows that the field of definition of (A_M) is $\mathbb{Q}$ so that A_M can be normalized to have integer coefficients. How much information about the character variety does the A-polynomial retain? Dunfield bounded the degree of the morphism $r\colon C \to \overline{r(C)}$ by 2^{b_1-1} when C is a canonical component [**6**]. In particular, it is birational when $b_1 = 1$.

The definition leads to a simple procedure for calculating A_M:

(1) Begin with a presentation $G \cong \langle g_1, \dots, g_n \mid R \rangle$ and images $\mathfrak{L}, \mathfrak{M} \in G$ of the generators of $\pi_1(\partial M)$.
(2) Construct the polynomial ring $S := \mathbb{Q}[a_k^{ij}, T_{\mathfrak{L}}, T_{\mathfrak{M}}, T_{\mathfrak{L}\mathfrak{M}}]$ for $1 \leq i, j \leq 2$ and $1 \leq k \leq n$. Define the ideals

$$\begin{aligned} D &:= (\det(A_1) - 1, \dots, \det(A_n) - 1) \\ J &:= (\text{relations among the matrices coming from } R) \\ K &:= (T_{\mathfrak{L}} - I_{\mathfrak{L}}, T_{\mathfrak{M}} - I_{\mathfrak{M}}, T_{\mathfrak{L}\mathfrak{M}} - I_{\mathfrak{L}\mathfrak{M}}) \\ I &:= D + J + K. \end{aligned}$$

Here, the matrices $A_k = (a_k^{ij})$ represent the images of the generators of G, and the polynomials I_g are the trace functions we defined earlier. When defining J, it is more computationally efficient to split the relators in half, as this reduces the degrees of the polynomials.
(3) Eliminate the variables a_k^{ij} from I, and substitute

$$\begin{aligned} T_{\mathfrak{L}} &\to x + x^{-1} \\ T_{\mathfrak{M}} &\to y + y^{-1} \\ T_{\mathfrak{L}\mathfrak{M}} &\to xy + x^{-1}y^{-1} \end{aligned}$$

to get an ideal $L \subset \mathbb{Q}[x^{\pm}, y^{\pm}]$.
(4) If L is principal, then we are done. Otherwise, find the height one minimal primes over L.

When M is a knot exterior, the data in the first step can be obtained from the Wirtinger presentation of the knot diagram. Correct orientation of the longitude and meridian allows one to distinguish between a chiral knot and its mirror image.

3. Algebraic K-theory and topology

In this section, we provide a careful account of the proof of the theorem mentioned in the introduction.

[4]Passing to $\mathbb{G}_{\mathrm{m}}^2$ both lets us bring the tools of toric varieties to bear and remedies the fact that not every height one prime ideal of $\mathbb{C}[X(\partial M)]$ is principal.

[5]Cooper and Long show in [**3**] that the closure of the image cannot be two-dimensional.

THEOREM 3.1 ([**8**]). *An irreducible Laurent polynomial $P \in \mathbb{C}[x^{\pm}, y^{\pm}]$ is an A-factor if and only if $\{x, y\}$ is torsion in $K_2(F_P)$.*

3.1. Milnor K-theory. When Λ is a field, we have an isomorphism $\mathrm{SL}(\Lambda) \cong E(\Lambda)$ between the stable special linear group and the group of elementary matrices. We will use the former notation from now on. Since $\mathrm{SL}(\Lambda)$ is perfect, meaning

$$H_1(\mathrm{SL}(\Lambda)) \cong \mathrm{SL}(\Lambda)^{\mathrm{ab}} = 0,$$

it has a universal central extension

$$1 \longrightarrow K_2(\Lambda) \longrightarrow \mathrm{St}(\Lambda) \longrightarrow \mathrm{SL}(\Lambda) \longrightarrow 1,$$

which is initial in the category of central extensions of $\mathrm{SL}(\Lambda)$. The terms in the sequence satisfy $K_2(\Lambda) \cong H_2(\mathrm{SL}(\Lambda))$ and

$$H_1(\mathrm{St}(\Lambda)) \cong H_2(\mathrm{St}(\Lambda)) = 0.$$

We can therefore take the "Schur multiplier" $H_2(\mathrm{SL}(\Lambda))$ as the definition of the second K-group of Λ. Matsuomoto's theorem [**10**] gives us a very nice way to compute in $K_2(\Lambda)$. It asserts that

$$K_2(\Lambda) \cong \frac{\Lambda^\times \otimes_{\mathbb{Z}} \Lambda^\times}{\langle a \otimes (1-a) \mid a \neq 0, 1 \rangle}.$$

The symbol $\{f, g\}$ is traditionally used to denote $f \otimes g$. The isomorphism is constructed by defining the star product on commuting matrices $U, V \in \mathrm{SL}(\Lambda)$ as follows: let $u, v \in \mathrm{St}(\Lambda)$ be lifts, and set $U \star V := [u, v]$. This commutator lies in $K_2(\Lambda)$ and is independent of the choice of lifts. We have

$$\mathrm{diag}(f_1, \dots, f_n) \star \mathrm{diag}(g_1, \dots, g_n) = \prod \{f_i, g_i\},$$

and specifically

$$\mathrm{diag}(f, f^{-1}, 1) \star \mathrm{diag}(g, 1, g^{-1}) = \{f, g\}.$$

Given a smooth curve C with function field F_P and generic point $\eta_C := \mathrm{Spec}(F_P) \to C$, we have isomorphisms

$$K_2(F_P) \cong CH^2(\eta_C, 2) \cong \varinjlim_{U \subset C} CH^2(U, 2).$$

The localization sequence yields

$$\cancel{CH^1(q,2)} \xrightarrow{\;0\;} CH^2(C, 2) \longrightarrow CH^2(C \setminus \{q\}, 2) \xrightarrow{\text{Tame}} \mathbb{C}^\times$$

for a point $q \in C(\mathbb{C})$. We take the limit over all points to obtain

$$0 \longrightarrow CH^2(C, 2) \longrightarrow K_2(F_P) \xrightarrow{\text{Tame}} \bigoplus_{q \in C(\mathbb{C})} \mathbb{C}^\times.$$

The only obstructions for some multiple of a K_2 symbol to extend (uniquely) to the entire curve are non-torsion tame symbols. This is why temperedness of the family is required to construct the higher Chow cycle in Theorem 1.1. For details, see [**5**].

3.2. Aspherical manifolds. We call topological spaces whose higher homotopy groups are trivial *aspherical*. A smooth manifold with contractible universal cover is aspherical, and the converse follows from Whitehead's theorem. Their salient feature is that singular homology computes the group homology of their fundamental groups functorially. If we assume that the manifold M for which we are computing an A-polynomial is prime, then it is aspherical by a corollary of the sphere theorem. Indeed, this is a harmless assumption to make, as representations gained by gluing on a closed manifold restrict trivially to ∂M and do not change the A-polynomial. Knowing that the torus and figure-eight are aspherical allows us to compute some group homology "for free":

$$H_i(\mathbb{Z}^2) = \begin{cases} \mathbb{Z} & i = 0 \\ \mathbb{Z}^2 & i = 1 \\ \mathbb{Z} & i = 2 \\ 0 & i > 2 \end{cases} \qquad H_i(F_2) = \begin{cases} \mathbb{Z} & i = 0 \\ \mathbb{Z}^2 & i = 1 \\ 0 & i = 2 \\ 0 & i > 2. \end{cases}$$

3.3. The proof. One key ingredient of the proof is the following lemma from [**2**]:

LEMMA 3.2. *Let Λ be a field; $x, y \in \Lambda^\times$; and G the free abelian group on $\{\mathfrak{L}, \mathfrak{M}\}$. Consider the representation $\rho\colon G \to \mathrm{SL}(\Lambda)$ given by*

$$\rho(\mathfrak{L}) = \begin{pmatrix} x & 0 \\ 0 & x^{-1} \end{pmatrix} \qquad \rho(\mathfrak{M}) = \begin{pmatrix} y & 0 \\ 0 & y^{-1} \end{pmatrix}.$$

There exists a generator $[T^2] \in H_2(G)$ such that $\rho_*[T^2] = \{x, y\}^2$.

PROOF. Let $L, M \in \mathrm{St}(\Lambda)$ be lifts of $\rho(\mathfrak{L})$ and $\rho(\mathfrak{M})$, respectively. Denote by F the free group on $\{\mathfrak{L}, \mathfrak{M}\}$. We have a commutative diagram with exact rows

$$\begin{array}{ccccccccc} 1 & \longrightarrow & [F,F] & \longrightarrow & F & \longrightarrow & G & \longrightarrow & 1 \\ & & \downarrow & & \downarrow & & \downarrow{\scriptstyle \rho} & & \\ 1 & \longrightarrow & K_2(\Lambda) & \longrightarrow & \mathrm{St}(\Lambda) & \longrightarrow & \mathrm{SL}(\Lambda) & \longrightarrow & 1. \end{array}$$

The derived subgroup $[F, F]$ is freely generated by the commutators

$$\{\, [\mathfrak{L}^n, \mathfrak{M}^m] \mid n, m \in \mathbb{Z}_{\neq 0}\},$$

and the leftmost vertical map sends $[\mathfrak{L}^n, \mathfrak{M}^m] \mapsto \rho(L^n) \star \rho(M^m)$. The middle map is simply $\mathfrak{L} \mapsto L$, $\mathfrak{M} \mapsto M$. The five-term exact sequence for group homology yields

$$\begin{array}{ccccccccccc} 0 & \longrightarrow & H_2(G) & \longrightarrow & H_1([F,F])_G & \xrightarrow{0} & \mathbb{Z}^2 & \longrightarrow & \mathbb{Z}^2 & \longrightarrow & 0 \\ & & \downarrow{\scriptstyle \rho_*} & & \downarrow & & & & & & \\ 0 & \longrightarrow & H_2(\mathrm{SL}(\Lambda)) & \longrightarrow & K_2(\Lambda) & \longrightarrow & 0 & \longrightarrow & 0 & \longrightarrow & 0. \end{array}$$

The top and bottom maps of the square are isomorphisms. By Theorem 5.4 of [**9**], the commutator $[\mathfrak{L}, \mathfrak{M}]$ generates

$$H_1([F,F])_G \cong \left(\frac{[F,F]}{[[F,F],[F,F]]} \right)_G \cong \frac{[F,F]}{[F,[F,F]]} \cong \mathbb{Z}.$$

Therefore, a generator $[T^2] \in H_2(G)$ is sent to $[\mathfrak{L}, \mathfrak{M}]$. The commutativity of the square implies that

$$\rho_*[T^2] = \rho(\mathfrak{L}) \star \rho(\mathfrak{M}) = \operatorname{diag}(x, x^{-1}) \star \operatorname{diag}(y, y^{-1}) = \{x, y\}^2.$$

□

We will use this lemma to establish the easier direction, namely that if $P \mid A_M$ for some 3-manifold M, then $\{x, y\}$ is torsion in $K_2(F_P)$. Let $Y' := V(P) \subset \mathbb{G}_{\mathrm{m}}^2$. It covers the curve $Y := q(Y') \subset X(\partial M)$. Take a closed point in the generic fiber of the morphism $R(M) \to X(\partial M)$. Its closure $Y'' \subset R(M)$ is another curve dominating Y. Consider all three function fields as subfields of $\overline{F_P}$, and construct the tautological representation $\rho\colon \pi_1(\partial M) \to \pi_1(M) \to \mathrm{SL}(\mathbb{C}(Y'')) \to \mathrm{SL}(\overline{F_P})$. Since $\rho(\mathfrak{L})$ and $\rho(\mathfrak{M})$ are commuting matrices with simple eigenvalues x and y, respectively, ρ is conjugate to the representation in the statement of Lemma 3.2. Therefore, $\rho_*[T^2] = \{x, y\}^2$. Thanks to asphericity, we can think of ρ_* equally well as the pushforward in singular homology along $\partial M \to M \to B\,\mathrm{SL}(\overline{F_P})$. But the inclusion of $[\partial M]$ in $H_2(M)$ is clearly a boundary, implying that $\{x, y\}^2 = 1$. The injectivity of the transfer map $K_2(F_P) \otimes \mathbb{Q} \to K_2(\overline{F_P}) \otimes \mathbb{Q}$ allows us to conclude. The converse requires two additional lemmas.

LEMMA 3.3. *Let Λ be an algebraically closed field. The group $K_2(\Lambda)$ is uniquely divisible, and the map $H_2(\mathrm{SL}_2(\Lambda)) \to H_2(\mathrm{SL}(\Lambda)) \cong K_2(\Lambda)$ is an isomorphism.*

PROOF. Divisibility is clear. To show uniqueness, let $g \in K_2(\Lambda)[n]$. Then $g = \{\xi, f\} = \{\xi^n, f^{1/n}\} = 1$, where ξ is a primitive nth root of unity [**14**]. In *op. cit.*, Suslin produces the short exact sequence

$$0 \longrightarrow I^3 \longrightarrow H_2(\mathrm{SL}_2(\Lambda)) \longrightarrow H_2(\mathrm{SL}(\Lambda)) \longrightarrow 0,$$

where I is the maximal fundamental ideal of the Witt ring $W(\Lambda)$. Because Λ is algebraically closed, we have $W(\Lambda) \cong \mathbb{Z}/(2)$ and $I = 0$. □

LEMMA 3.4. *Let Ω_n^{SO} denote the oriented bordism functor and H_n ordinary homology. The Hurewicz map $\Omega_n^{\mathrm{SO}} \Rightarrow H_n$ given by $[f\colon M \to X] \mapsto f_*[M]$ is an isomorphism for $n \leq 3$.*

PROOF. The homotopy groups of the spectra $M\,\mathrm{SO}$ and $H\mathbb{Z}$ are

$$\pi_n(M\,\mathrm{SO}) \cong \Omega_n^{\mathrm{SO}}(\bullet) = \begin{cases} 0 & n < 0 \\ \mathbb{Z} & n = 0 \\ 0 & 1 \leq n \leq 3 \\ \mathbb{Z} & n = 4 \\ \vdots & \end{cases} \qquad \pi_n(H\mathbb{Z}) = \begin{cases} 0 & n < 0 \\ \mathbb{Z} & n = 0 \\ 0 & n > 0. \end{cases}$$

The Hurewicz map $M\,\mathrm{SO} \to H\mathbb{Z}$ is therefore 3-connected, i.e., $\pi_n(M\,\mathrm{SO}) \to \pi_n(H\mathbb{Z})$ is an isomorphism for $n \leq 3$ and an epimorphism for $n = 4$. Because the suspension spectrum X_+ is (-1)-connected, the map $M\,\mathrm{SO} \wedge X_+ \to H\mathbb{Z} \wedge X_+$ is also 3-connected. In particular, the Hurewicz map

$$\Omega_n^{\mathrm{SO}}(X) \cong \pi_n(M\,\mathrm{SO} \wedge X_+) \to \pi_n(H\mathbb{Z} \wedge X_+) \cong H_n(X)$$

is an isomorphism for $n \leq 3$. One can see this in more down-to-earth terms (if spectral sequences are considered down-to-earth) via the edge maps of the Leray-Serre-Atiyah-Hirzebruch spectral sequence

$$E^2_{pq} = H_p(X, \Omega^{\mathrm{SO}}_q(\bullet)) \Longrightarrow \Omega^{p+q}(X).$$

□

Now, suppose that $\{x, y\}$ is torsion in $K_2(F_P)$. Lemma 3.3 implies that $\{x, y\}$ is trivial in $K_2(\overline{F_P})$. Consider the representation $\rho\colon \pi_1(T^2) \to \mathrm{SL}_2(\overline{F_P})$ given by

$$\rho(\mathfrak{L}) = \begin{pmatrix} x & 0 \\ 0 & x^{-1} \end{pmatrix} \qquad \rho(\mathfrak{M}) = \begin{pmatrix} y & 0 \\ 0 & y^{-1} \end{pmatrix}.$$

By Lemmas 3.2 and 3.3,

$$\rho_*[T^2] = \{x, y\}^2 = 1 \in K_2(\overline{F_P}).$$

Once again, we consider ρ_* as the pushforward along the classifying map $f\colon T^2 \to B\,\mathrm{SL}_2(\overline{F_P})$. Since $f_*[T^2] = 0$, the bordism class $[f]$ is trivial (Lemma 3.4). This implies the existence of a compact oriented 3-manifold M bounding T^2 and extending ρ to a representation $\pi_1(M) \to \mathrm{SL}_2(\overline{F_P})$. Let $\{g_1, \ldots, g_n\}$ generate $\pi_1(M)$, and define $(h^{ij}_k) := \rho(g_k)$. The morphism $\mathbb{C}[R(M)] \twoheadrightarrow \mathbb{C}[h^{ij}_k]$ corresponds to a curve $Y'' \subset R(M)$ whose tautological representation $\pi_1(M) \to \mathrm{SL}_2(\mathbb{C}(Y''))$ restricts to our diagonal representation $\rho\colon \pi_1(T^2) \to \mathrm{SL}_2(F_P)$ and therefore reproduces the component $V(P) \subset \mathbb{G}^2_{\mathrm{m}}$.

4. Finiteness of smooth A-factors

An A-factor P may be smooth or singular in $\mathbb{G}^2_{\mathrm{m}}$. In the former case, it may still acquire singularities along the boundary of its toric variety $\mathbb{P}_\Delta$ (although this cannot happen if the edge polynomials are separable). In our experience, many A-factors "in the wild" are already singular in the interior. This is expected, as Theorem 1.1 shows that smooth ones are rare in some sense. The Fermat curve $x^p + y^p = 1$ is smooth in its toric variety, namely $\mathbb{P}^2$. A quick K_2 calculation

$$\{x, y\}^{p^2} = \{x^p, y^p\} = \{x^p, 1 - x^p\} = 1$$

proves that it is an A-curve, although we do not know which manifold it comes from! We would like to establish a finiteness result for A-factors which appear as smooth irreducible components of some $\overline{D_M} \subset \mathbb{P}_{\Delta(A_M)}$ without reference to a family. We define the genus of a Laurent polynomial to be the number of interior lattice points of its Newton polygon. When $V(P) \subset \mathbb{P}_\Delta$ is smooth, this really is its geometric genus. For a fixed $g > 0$, there are finitely many genus g lattice polygons up to unimodular equivalence [**1**]. Importantly, this is not to say that there are finitely many families containing A-factors of genus g. The coefficients of an A-polynomial are in $\mathbb{Z}$, whereas the coefficients of an A-factor are in $\overline{\mathbb{Q}}$. The orders of torsion in the roots of the edge polynomials are not bounded. To remedy this, we define the overgenus of P as $o(P) := \inf S_P$, where

$$S_P := \{\, \mathrm{genus}(A_M) \mid V(P) \text{ is a smooth irreducible component of } \overline{D_M}\,\}.$$

We refer to an A-factor with $o(P) < \infty$ as smooth. In contrast to the case of A-factors, there are finitely many families containing A-polynomials of genus o. For each of the finitely many lattice polygons, the edge polynomials are products of cyclotomic polynomials,[6] whose degrees are bounded from below: $\deg(\Phi_n) = \varphi(n) \geq \sqrt{n/2}$. Consequently, the edge lengths bound the orders of torsion. Assume that $A_M = PQ$ and that $V(P)$ is smooth in $\overline{D_M}$. Then $\Delta(A_M) = \Delta(P) + \Delta(Q)$. The edges of the Minkowski sum come from the edges of each summand, sorted by polar angle. The edge polynomials are preserved. If an edge of P and an edge of Q are parallel, then they fuse together, and their edge polynomials multiply. Accordingly, $V(P) \subset \mathbb{P}_{\Delta(P)}$ meets the boundary in the same way as it does in $\mathbb{P}_{\Delta(A_M)}$ and is therefore a smooth fiber in the associated family $X \to \mathbb{A}^g \setminus \Sigma$. Since there are finitely many A-curves in this family, we have proven

THEOREM 4.1. *There are finitely many A-factors satisfying* $1 \leq o(P) \leq N$ *for any* $N \in \mathbb{N}$.

References

[1] W. Castryck, *Moving out the edges of a lattice polygon*, Discrete Comput. Geom. **47** (2012), no. 3, 496–518, DOI 10.1007/s00454-011-9376-2. MR2891244

[2] D. Cooper, M. Culler, H. Gillet, D. D. Long, and P. B. Shalen, *Plane curves associated to character varieties of 3-manifolds*, Invent. Math. **118** (1994), no. 1, 47–84, DOI 10.1007/BF01231526. MR1288467

[3] D. Cooper and D. D. Long, *Representation theory and the A-polynomial of a knot*, Chaos Solitons Fractals **9** (1998), no. 4-5, 749–763, DOI 10.1016/S0960-0779(97)00102-1. Knot theory and its applications. MR1628754

[4] M. Culler and P. B. Shalen, *Varieties of group representations and splittings of 3-manifolds*, Ann. of Math. (2) **117** (1983), no. 1, 109–146, DOI 10.2307/2006973. MR683804

[5] C. F. Doran and M. Kerr, *Algebraic K-theory of toric hypersurfaces*, Commun. Number Theory Phys. **5** (2011), no. 2, 397–600, DOI 10.4310/CNTP.2011.v5.n2.a3. MR2851155

[6] N. M. Dunfield, *Cyclic surgery, degrees of maps of character curves, and volume rigidity for hyperbolic manifolds*, Invent. Math. **136** (1999), no. 3, 623–657, DOI 10.1007/s002220050321. MR1695208

[7] W. M. Goldman, *Locally homogeneous geometric manifolds*, Proceedings of the International Congress of Mathematicians. Volume II, Hindustan Book Agency, New Delhi, 2010, pp. 717–744. MR2827816

[8] A. Guilloux and J. Marché, *Volume function and Mahler measure of exact polynomials*, Compos. Math. **157** (2021), no. 4, 809–834, DOI 10.1112/s0010437x21007016. MR4247573

[9] W. Magnus, A. Karrass, and D. Solitar, *Combinatorial group theory*, 2nd ed., Dover Publications, Inc., Mineola, NY, 2004. Presentations of groups in terms of generators and relations. MR2109550

[10] J. Milnor, *Introduction to algebraic K-theory*, Annals of Mathematics Studies, No. 72, Princeton University Press, Princeton, NJ; University of Tokyo Press, Tokyo, 1971. MR349811

[11] RJ Acu na, Devin Akman, and Matt Kerr. *Real regulator maps with finite 0-locus*, 2024.

[12] C. Procesi, *The invariant theory of $n \times n$ matrices*, Advances in Math. **19** (1976), no. 3, 306–381, DOI 10.1016/0001-8708(76)90027-X. MR419491

[13] P. B. Shalen, *Representations of 3-manifold groups*, Handbook of geometric topology, North-Holland, Amsterdam, 2002, pp. 955–1044. MR1886685

[14] A. A. Suslin, *Torsion in K_2 of fields*, K-Theory **1** (1987), no. 1, 5–29, DOI 10.1007/BF00533985. MR899915

[15] W. P. Thurston, *The geometry and topology of three-manifolds. Vol. IV*, American Mathematical Society, Providence, RI, [2022] ©2022. Edited and with a preface by Steven P. Kerckhoff and a chapter by J. W. Milnor. MR4554426

[16] S. D. Ahlgren, G. E. Andrews, and K. Ono (eds.), *Topics in number theory*, Mathematics and its Applications, vol. 467, Kluwer Academic Publishers, Dordrecht, 1999. In honor of B. Gordon and S. Chowla, DOI 10.1007/978-1-4613-0305-3. MR1691307

[6] All statements about polynomials in this section are made up to a constant factor.

Department of Mathematics, Washington University in St. Louis, 1 Brookings Drive, St. Louis, MO 63130, USA
Email address: akman@wustl.edu

Contemporary Mathematics
Volume **842**, 2026
https://doi.org/10.1090/conm/842/16849

Regulators and derivatives of Vologodsky functions with respect to $\log(p)$

Amnon Besser

ABSTRACT. We describe several instances of the following phenomenon: In bad reduction situations the p-adic regulator has a continuous and a discrete component. The continuous component is computed using Vologodsky integrals. These depend on a choice of the branch of the p-adic logarithm, determined by $\log(p)$. They can be differentiated with respect to this parameter and the result is related to the discrete component.

1. Introduction

Let p be a finite prime. Let K be a finite extension of the field $\mathbb{Q}_p$ of p-adic numbers and let X/K be a smooth variety.

The p-adic analogue of the Beilinson regulator into Deligne cohomology is the *syntomic regulator* [**48**]

$$\mathrm{reg}_{\mathrm{syn}} : H^i_{\mathcal{M}}(X, \mathbb{Q}(j)) \to H^i_{\mathrm{syn}}(X, j).$$

The syntomic regulator $\mathrm{reg}_{\mathrm{syn}}$ can sometimes be computed using Coleman or Vologodsky integration. Let us consider a concrete example of the second algebraic K-theory group $K_2(X)$ of a proper curve X. In this case the regulator takes the form

$$\mathrm{reg}_{\mathrm{syn}} : H^2_{\mathcal{M}}(X, \mathbb{Q}(2)) \to H^2_{\mathrm{syn}}(X, 2). \tag{1.1}$$

Elements of the left-hand side of (1.1) are represented by formal sums of symbols $\{f, g\}$ for rational functions f, g on X, such that all their tame symbols are 0 at all points of X. When X has good reduction the group $H^2_{\mathrm{syn}}(X, 2)$ is isomorphic to the first algebraic de Rham cohomology group $H^1_{\mathrm{dR}}(X/K)$. In this case we proved in [**7**] the following result.

THEOREM 1.1. *To $\omega \in H^1_{\mathrm{dR}}(X/K)$ associate $r_\omega = \omega \cup \mathrm{reg}_{\mathrm{syn}}$. Then, for $\omega \in \Omega^1(X)$,*

$$r_\omega(\{f, g\}) = \int_{(f)} \log(g)\omega. \tag{1.2}$$

The expression on the right-hand side of (1.2) is the Coleman integral (see Section 2) of the form $\log(g) \cdot \omega$ evaluated on the divisor (f) of the rational function f. We note that this theorem is a special case of a more general formula for a general $\omega \in H^1_{\mathrm{dR}}(X/K)$.

When X only has semi-stable reduction over K, we have an isomorphism

$$H^2_{\mathrm{syn}}(X,2) \cong H^1_{\mathrm{dR}}(X/K) \oplus B, \tag{1.3}$$

where B is a subset of the weight 2 part of $H^1_{\mathrm{dR}}(X/K)$ (weight with respect to Frobenius, see Section 3). Focusing for the moment on the first summand in (1.3) (though the second component will be our main focus) we proved in [**11**] an almost identical result by replacing Coleman integration with Vologodsky integration.

THEOREM 1.2 ([**11**, Corollary 1.3]). *Under some technical assumptions on the form ω, the first component of the syntomic regulator is computed by* (1.2), *replacing the Coleman integral with a Vologodsky integral.*

See Theorem 3.10 for a more precise statement.

Vologodsky integration [**57**] extend Coleman integration to varieties with bad reduction. Vologodsky integrals have, built into them, a branch of the p-adic logarithm, determined by the branch parameter $\log(p)$. In fact, in Theorem 1.2 the regulator also depends on (essentially) the branch parameter.

One can wonder about the dependency of the Vologodsky integral on the branch of the logarithm. We may even take the derivative of the integral with respect to the branch parameter $\log(p)$. We can ask *what can we say about this derivative.* The goal of this note it to argue that this is an interesting question since the resulting derivative holds valuable arithmetic information.

The fundamental example of this phenomenon is the p-adic logarithm $\log(z)$, which is the Vologodsky integral of the differential form $\mathrm{d}z/z$ on the punctured affine line. Recall that the p-adic logarithm (sometimes called the Iwasawa logarithm) of $z \in K$ with $|z| = 1$ is uniquely determined by requiring that it has the usual power series expansion near 1 and that it satisfies $\log(ab) = \log(a) + \log(b)$. However, to extend it to a homomorphism $\log : K^\times \to K$ one needs to make a choice of $\log(p)$. Once such a choice has been made the extension of the logarithm to $K^\times$ is given as follows: If ν is the p-adic valuation, normalized in such a way that $\nu(p) = 1$, then

$$\log(z) = \log\left(\frac{z}{p^{\nu(z)}}\right) + \nu(z)\log(p).$$

By differentiating with respect to $\log(p)$ we get the fundamental observation of this entire note:

$$\frac{d}{d\log(p)}\log(z) = \nu(z). \tag{1.4}$$

To see how this observation relates to the theory of p-adic regulators, let us consider the simplest such (log syntomic) regulator, namely the one for K_1 of a field K, $H^1_{\mathcal{M}}(K,1)$. This takes the form of the map

$$K^\times \to H^1_{\mathrm{st}}(K, \mathbb{Q}_p(1)) = K \oplus \mathbb{Q}_p$$

where H_{st} is semi-stable Galois cohomology [**47**]. By [**47**, 1.35] this regulator is given by

$$x \mapsto (\log(x), \nu(x)).$$

We immediately observe the main theme of this note: The regulator has 2 components, one "continuous" and one "discrete". The former is computed using Vologodsky integration, The latter is the derivative of the former with respect to $\log(p)$. We will try to convince the reader that the phenomena that we point out is not only interesting on its own, but is potentially useful as a tool, mostly for computing discrete invariants using continuous tools.

To our knowledge, the first place this idea appears in any form is the work of Colmez [**28**, Théorème II.2.18], where the local height at a prime is found as the derivative with respect to $\log(p)$ of the appropriate Green function. We will prove something similar in Proposition 4.5. See Remark 4.6 for the difference between the two approaches.

We will survey a number of examples where a similar picture emerges. In Section 2 we recall in brief the theories of Coleman and Vologodsky, including its dependency on $\log(p)$. Then, in Section 3 We show, under fairly general conditions, that the syntomic regulator for varieties with semi-stable reduction splits into a continuous and discrete components, and that the discrete one is the derivative of the continuous one with respect to $\log(p)$. In Section 4 we explain how to compute the derivative with respect to $\log(p)$ of a Vologodsky integral in some cases, and give an application to local height pairings on curves. Section 5 is mostly based on [**14**] and we explain the theory of p-adic heights associated to adelic line bundles and how the local heights can be computed using derivatives of Vologodsky functions. In Section 6 we do a non-abelian analogue of Section 3, compute the continuous component of the unipotent Albanese map using Vologodsky integrals and the discrete part using their derivatives with respect to $\log(p)$. Finally, in Section 7 we explain the relation with the theory of the toric regulator developed in [**16**].

The presentation is not meant to be complete, and we will sometimes refer to future work for details.

Notation: K is a finite extension of $\mathbb{Q}_p$, unless stated otherwise. In this case, its residue field κ is finite and has $q = p^f$ elements. We denote by K_0 the maximal unramified extension of $\mathbb{Q}_p$ inside K.

I would like to thank the organizers of the Regulators V conference in Pisa, who gave me the opportunity to present this material, both at the conference and in writing. I would also like to thank Steffen Müller and Padma Srinivasan, my coauthors on [**14, 15**], where the inspiration for the ideas presented here grew, and for allowing me to report on some results that will appear in [**15**]. Finally, I would like to thank an anonymous referee for a careful reading of the paper and for pointing out many useful corrections. The author's research is supported by grant number 2958/24 from the Israel Science foundation.

2. Review of Coleman and Vologodsky integration

Coleman integration is a theory of p-adic integration invented by Robert Coleman and developed by him in [**22, 23, 26**]. It is an integration theory (what of? A bit later) on certain "overconvergent" domains with good reduction. Coleman called these: *basic wide open spaces*, but a consistent theory of Dagger spaces, associated to the Dagger algebras of Monsky and Washnitzer [**44**], was later developed by Grosse-Klönne [**30**] and is the right "home" for this theory. If X is a smooth variety over K, then the associated rigid analytic space can often be covered by such domains. These domains can be chosen with good reduction even if X has bad reduction, and therefore one can sometimes glue integrals over different domains. In general, there will be some monodromy for the covering (see [**6**] for a very general theory along those lines). The Tannakian interpretation of Coleman integration [**8**] allows, among other things, an extension of the theory to more general domains without the need for covering, in particular to complete varieties, but still assuming good reduction.

In the bad reduction case covering by overconvergent domains, if they exist, will in general produce some monodromy. Nevertheless, Vologodsky [**57**] defines an integration theory for general varieties with arbitrary reduction.

The integrals appearing in the theories of Coleman and Vologodsky are of the following form: Let X be either an "overconvergent domain" for Coleman integration or a smooth K variety for Vologodsky integration. Let ω be a closed 1-form on X. These would be overconvergent analytic for Coleman or algebraic for Vologodsky. Then the theory provides a primitive, A "function" $F = \int \omega$ on X in an appropriate sense, so that $\mathrm{d}\, F = \omega$. The function F is a locally analytic function on the $\overline{K}$-points, so that the equation $\mathrm{d}\, F = \omega$ makes sense, and it is Galois equivariant. It is a function on X in the Vologodsky case but is only a function on the "underlying affinoid" in Coleman's terminology (or the associated rigid spaces in Grosse-Klönne's terminology). Now, we can further multiply F by a second form η and the theory gives an integral $\int (F\eta)$, again unique up to a constant, and so on (to be more precise, we can only integrate a combination of such expressions which is again closed. The underlying theory was developed in [**8**] as Coleman originally only considered one dimensional spaces, where this issue does not arise). More concretely, the integration theory gives a K-algebra $\mathcal{O}_{\mathrm{Col}}(X) \supset \mathcal{O}(X)$ of locally analytic functions on X such that there is a short exact sequence

$$0 \to K \to \mathcal{O}_{\mathrm{Col}}(X) \xrightarrow{d} \mathcal{O}_{\mathrm{Col}}(X) \cdot \Omega^1(X)^{d=0} \to 0$$

and everything is functorial with respect to morphisms of "spaces". All earlier described integrals are easily obtained as a consequence of this short exact sequence. Definite, or more generally *iterated* integrals, are obtained from this theory just as in first year calculus. For example

$$\int_x^y \omega = F(y) - F(x), \quad F = \int \omega.$$

Both Coleman integration, in its Tannakian interpretation [**8**], and Vologodsky integration [**57**], have very similar constructions. Given the space X one considers the category $\mathrm{Un}(X)$ of unipotent connections on X. Each point $x \in X(K)$ gives a fiber functor ω_x on $\mathrm{Un}(X)$ associating with the connection the fiber of the underlying vector bundle at x. For each two points $x, y \in X(K)$ one has a Tannakian path space $P_{x,y}$ of paths from x to y, whose A-valued points, for any K-algebra A, are the tensor isomorphisms between $\omega_x \otimes A$ and $\omega_y \otimes A$. It turns out that this path space has a canonical (K-linear) automorphism φ (Frobenius) and, in the case of algebraic varieties over K, also a canonical "monodromy" N, which is a derivation of the associated algebra of regular functions. In both cases one proves ([**8**, Corollary 3.2] and [**57**, Theorem B]) the existence of a canonical path $\gamma_{x,y} \in P_{x,y}(K)$. In both cases this path is Frobenius invariant. In the overconvergent case this property completely characterizes it. In the algebraic case this property is not sufficient and one needs an additional property involving monodromy, which we do not describe. The canonical paths are both invariant under composition, and functorial in an obvious sense.

Following [**8**] we can now define Coleman, respectively Vologodsky functions. These are represented by four-tuples $F = (M, \nabla, v = v_x, s)$ where $(M, \nabla) \in \mathrm{Un}(X)$,

$v \in M_x$ and $s : M \to \mathcal{O}_X$ is a morphism from the underlying vector bundle. Here v_x is either chosen at one fixed point $x \in X(K)$ and transported via the canonical path to all other points, or we can avoid a choice of a point by choosing at each point in a way compatible with the γ's. Obviously both ways give the same thing. The function F can be evaluated at a point y,

$$F(y) = s(v_y) = s(\gamma_{x,y} v_x).$$

A morphism $F \to F'$ is a morphism $(M, \nabla) \to (M', \nabla')$ compatible with s, v in the obvious way. If such a morphism exists, the values of F and F' are identical at each point, and we identify them.

One key difference between Coleman and Vologodsky integration, is the dependency of the latter on a branch of the p-adic logarithm. This is manifested by having $\gamma_{x,y}$ defined over the algebra $K[\log(p)]$ of polynomials in a formal variable $\log(p)$ over K. We can fix the branch by evaluating at one particular value for $\log(p)$, but our interest is rather to study the dependency on $\log(p)$. This makes Vologodsky functions have values in $K[\log(p)]$ We will use the notation $\mathrm{d}/\mathrm{d}\log(p) F$ to mean, more precisely, the value of the derivative of the Vologodsky function F with respect to $\log(p)$ evaluated at $\log(p) = 0$. Often, with π a fixed uniformizer for K, it will make more sense to consider the derivative $\mathrm{d}/\mathrm{d}\log(\pi) F$. As noted before, this is not the situation for Coleman integration, so by its very construction we have

PROPOSITION 2.1. *Let X be an overconvergent space and let $F \in \mathcal{O}_{\mathrm{Col}}(X)$. Then,*

$$\frac{\mathrm{d}}{\mathrm{d}\log(\pi)} F = 0$$

as a function on the associated rigid space.

REMARK 2.2. We will later see some instances where Coleman functions appear to have a dependency on $\log(\pi)$. First We will see in Section 4 that in some cases Coleman functions can be extended to X, and at the "new" points there is a dependency on $\log(\pi)$. More interestingly, in Proposition 4.4 we will see that curves with semi-stable reduction can be built by gluing overconvergent domains with good reduction and and Vologodsky functions restrict to Coleman function on these domains and will have non-trivial derivatives with respect to $\log(\pi)$. This does not contradict Proposition 2.1, which should be more precisely interpreted as saying that Coleman integrals are independent of $\log(\pi)$ as long as the constants of integration are independent of it. Indeed, even the integral of the zero differential, which is a constant, can depend on $\log(\pi)$ if we choose the constant to depend on it.

We turn our attention now to Vologodsky integrals, starting with the non-iterated case. These integrals can be described purely using their functoriality properties, which was in fact the way they were defined, before the emergence of the general theory of Vologodsky, by Zarhin and Colmez [**28**, **59**]. The following is well known.

PROPOSITION 2.3. *If X is an abelian variety and let 0 be the identity element. Then, $\int_0^x \omega$ is the logarithm of X to its Lie algebra, paired with the value of the invariant differential ω at 0. This suffices to determine $\int \omega$. If X is a proper* *variety, and* $\alpha : X \to A = \mathrm{Alb}(X)$ *the map to its Albanese variety, then any* $\omega \in \Omega^1(X)$ *is* $\alpha^*\omega'$ *for some invariant differential* ω' *on* A. *and* $\int \omega = \alpha^* \int \omega'$.

This way of obtaining the integral is independent of $\log(p)$. Therefore, we obtain our second result on the derivative of Vologodsky integrals with respect to $\log(p)$.

COROLLARY 2.4 ([**14**, Proposition 9.14]). *Let* ω *be a holomorphic form on a proper variety* X. *Let* $x_0, x \in X(K)$. *Then,*

$$\frac{\mathrm{d}}{\mathrm{d}\log(p)} \int_{x_0}^{x} \omega = 0.$$

Both Zarhin and Colmez extend their (non-iterated) theory to varieties which are non-proper. In this case, just as for the logarithm, there is indeed a dependency on $\log(p)$. These theories coincide with Vologodsky's more general integration theory.

3. Syntomic regulators

We approach syntomic regulators via étale regulators. This has the advantage of being able to treat, for the type of situations we are considering, varieties with arbitrary reduction, without the need for the sophisticated machinery of [**48**]. We stress that for proving some important results about syntomic regulators, as well as for computations, one is forced to get a much closer understanding of their theory.

We assume that X is a proper smooth variety over K. For every finite prime ℓ we have étale regulator maps

$$r_\ell : H^i_{\mathcal{M}}(X, \mathbb{Q}(j)) \to H^i_{\text{ét}}(X, \mathbb{Q}_\ell(j)),$$

where the cohomology on the right-hand side is continuous étale cohomology as defined by Jannsen [**35**]. By a theorem of Jannsen [**51**, Theorem 1.1] The Hochschild-Serre spectral sequence,

$$E_2^{r,s} = H^r(K, H^s_{\text{ét}}(X \otimes \bar{K}, \mathbb{Q}_\ell(j))) \Rightarrow H^{r+s}_{\text{ét}}(X, \mathbb{Q}_\ell(j)), \tag{3.1}$$

degenerates at E_2 Let $H^i_{\mathcal{M}}(X, \mathbb{Q}(j))_0$ be the kernel of the composition

$$H^i_{\mathcal{M}}(X, \mathbb{Q}(j)) \to H^i_{\text{ét}}(X, \mathbb{Q}_\ell(j)) \to H^0(K, H^i_{\text{ét}}(X \otimes \bar{K}, \mathbb{Q}_\ell(j))). \tag{3.2}$$

We get an induced map

$$H^i_{\mathcal{M}}(X, \mathbb{Q}(j))_0 \to H^1(K, H^{i-1}_{\text{ét}}(X \otimes \bar{K}, \mathbb{Q}_\ell(j))). \tag{3.3}$$

At this point we restrict attention to the case $\ell = p$. One important consequence of [**48**] is that the map (3.3) is actually into the subgroup of so called semi-stable classes,

$$\mathrm{reg}_{\mathrm{syn}} : H^i_{\mathcal{M}}(X, \mathbb{Q}(j))_0 \to H^1_{\mathrm{st}}(K, H^{i-1}_{\text{ét}}(X \otimes \bar{K}, \mathbb{Q}_p(j))). \tag{3.4}$$

Here, $V := H^{i-1}_{\text{ét}}(X \otimes \bar{K}, \mathbb{Q}_p(j))$ is regarded as a potentially semi-stable representation of $G = \mathrm{Gal}(\bar{K}/K)$ and $H^1_{\mathrm{st}}(K, V)$ is semi-stable cohomology. An excellent reference for this theory, including Fontaine's rings and functors and semi-stable cohomology, is contained in the first section of [**47**]. In particular, one may find there the interpretation of $H^1_{\mathrm{st}}(K, V)$ as a group of extensions in the category of potentially semi-stable representation and its explicit computation in terms of the complex $C^\bullet_{\mathrm{st}}(V)$ of [**47**, 1.19] which we recall shortly. The map $\mathrm{reg}_{\mathrm{syn}}$ in (3.4) is our syntomic regulator. Notice that so far this map is independent of any choice of branch.

REMARK 3.1. For use in Section 7 we point out that when $\ell \neq p$ the étale regulator lands in the group H^1_g of unramified classes. Furthermore, there is an integral version.

$$\mathrm{reg}_\ell : H^i_{\mathcal{M}}(X, \mathbb{Z}(j))_0 \to H^1_g(K, H^{i-1}_{\text{ét}}(X \otimes \bar{K}, \mathbb{Z}_\ell(j))).$$

We now analyze more carefully the target of the syntomic regulator using the complex $C^\bullet_{\mathrm{st}}(V)$. At this stage, V can be any de Rham representation. Recall the Fontaine functors $\mathrm{D_{st}}$ and DR. For the Galois representation V, $\mathrm{D_{st}}(V)$ is a K_0-vector space, equipped with a linear nilpotent operator N (called monodromy) and a semi-linear (with respect to the unique lift of Frobenius on K_0 operator φ (Frobenius), satisfying the relation

$$N\varphi = p\varphi N \tag{3.5}$$

(such an object will be called a *Frobenius monodromy module*) while $\mathrm{DR}(V)$ is a K vector space with a descending filtration $F^\bullet$, and these two functors are related by an isomorphism,

$$I_\pi : \mathrm{D_{st}}(V) \otimes_{K_0} K \to \mathrm{DR}(V),$$

which depends on a choice of a uniformizer π of K. The dependency of the uniformizer is such that if π' is another uniformizer, then

$$I_{\pi'} = I_\pi \circ \exp(\log(\pi'/\pi)N). \tag{3.6}$$

The collection of data $(D = \mathrm{D_{st}}(V), D_K = \mathrm{DR}(V), I_\pi)$ with the above conditions form a category, called the category of *filtered Frobenius monodromy modules*, $M_{\varphi,N,F^\bullet}(K)$. The subcategory of $M_{\varphi,N,F^\bullet}(K)$ coming from semi-stable Galois representations is the abelian category of *admissible* filtered Frobenius monodromy modules. The category $M_{\varphi,N,F^\bullet}(K)$ has a canonical object $\mathbf{1}$ corresponds to the trivial one dimensional representation and objects $D \in M_{\varphi,N,F^\bullet}(K)$ have Tate twists $D(n)$ corresponding to shifting the filtration by n and multiplying φ by p^{-n}. The Fontaine functors give an obvious functor

$$\mathbb{D} : \mathrm{Rep}_{\mathbb{Q}_p}(\mathrm{Gal}(\overline{K}/K)) \to M_{\varphi,N,F^\bullet}(K). \tag{3.7}$$

For a de Rham representation V The complex $C^\bullet_{\mathrm{st}}(V)$ of [**47**, 1.19] computing $H^\bullet_{\mathrm{st}}(K, V)$ is given as follows:

$$D \xrightarrow{(\varphi-1, N, -I_\pi)} D \oplus D \oplus D_K/F^0 \xrightarrow{N+1-p\varphi+0} D. \tag{3.8}$$

Similarly, we have a complex $C'^\bullet_{\mathrm{st}}(V)$ given by

$$D \xrightarrow{(\varphi-1, N)} D \oplus D \xrightarrow{N+1-p\varphi} D. \tag{3.9}$$

These two complexes sit in a short exact sequence

$$0 \to D_K/F^0[1] \to C^\bullet_{\mathrm{st}}(V) \to C'^\bullet_{\mathrm{st}}(V) \to 0. \tag{3.10}$$

A short exact sequence in $M_{\varphi,N,F^\bullet}(K)$ is a sequence of morphisms, which gives an exact sequence on the D's and a strict exact sequence on the D_K's. As usual, we have a Yoneda $\mathrm{Ext}^1(\mathbb{D}_1, \mathbb{D}_2)$ of isomorphism classes of extensions of $\mathbb{D}_1$ by $\mathbb{D}_2$ forming a group under the Baer sum.

LEMMA 3.2. *The group* $H^1(C^\bullet_{\mathrm{st}}(V))$ *is isomorphic to the Yoneda* $\mathrm{Ext}^1(\mathbf{1},\mathbb{D}(V))$, *where* $\mathbb{D}$ *is the Fontaine functor from* (3.7). *Similarly, the group* $H^1(C'^\bullet_{\mathrm{st}}(V))$ *is isomorphic to the Yoneda* $\mathrm{Ext}^1(\mathbf{1},\mathrm{D}_{\mathrm{st}}(V))$ *in the category of Frobenius monodromy modules.*

PROOF. Let

$$0\to\mathbb{D}(V)\to\mathbb{D}'\to\mathbf{1}\to 0. \tag{3.11}$$

be an extension in $M_{\varphi,N,F^\bullet}(K)$. We can find lifts, $A\in D'$ of $1\in\mathbf{1}$ and $B\in F^0D'_K$ of $1\in\mathbf{1}_K$, the latter since $1\in F^0\mathbf{1}_K$ and the morphisms are strict. We associate with the pair (A,B) the triple

$$(x,y,z)=((\varphi-1)A,NA,B-I_\pi(A))\in D\oplus D\oplus D_K/F^0.$$

We quotient by F^0 to account for the fact that B is only determined up to an element of F^0D_K. This element is clearly in the kernel of $N+1-p\varphi$. When we add to A an element $w\in D$ we add to (x,y,z) the element $(\varphi-1,N,-I_\pi)(w)$. Thus, an extension gives a well-defined element of $H^1(C^\bullet_{\mathrm{st}}(V))$. Conversely, it is easy to construct an extension from such an element. The second statement follows similarly (and more easily). □

PROPOSITION 3.3. *Suppose* $D^{\varphi=1,N=0}=0$. *Then we have a short exact sequence,*

$$0\to D_K/F^0\xrightarrow{\iota}H^1(C^\bullet_{\mathrm{st}}(V))\xrightarrow{\rho}H^1(C'^\bullet_{\mathrm{st}}(V))\to 0, \tag{3.12}$$

given by $\iota(z)=(0,0,z)$ *and* $\rho(x,y,z)=(x,y)$.

PROOF. This is an immediate consequence of the long exact sequence associated with (3.10). □

REMARK 3.4. It is easy to interpret the short exact sequence (3.12) as follows: The projection forgets the D_K data leaving an extension of Frobenius monodromy modules. The injection gives all possible filtration on the trivial extension of Frobenius monodromy modules.

In order to split the short exact sequence (3.12) we now restrict attention to $V:=H^{i-1}_{\text{ét}}(X\otimes\bar{K},\mathbb{Q}_p(j))$ for X with semi-stable reduction. By the semi-stable conjecture of Fontaine, proved by Tsuji [**56**] the associated $D=\mathrm{D}_{\mathrm{st}}(V)$ is isomorphic to Hyodo-Kato cohomology of the special fiber (viewed as a log-scheme) while $D_K=\mathrm{DR}(V)$ is isomorphic to the de Rham cohomology $H^{i-1}_{\mathrm{dR}}(X/K)$, both twisted by j. Consequently, by [**43**] there is a weight decomposition,

$$D=\oplus_k D^k,$$

where the linear Frobenius $\phi=\varphi^f$ operates on D^i with eigenvalues which are Weil numbers of weight i. The relation (3.5) immediately implies that $ND^i\subset D^{i-2}$.

PROPOSITION 3.5. *Suppose either* $D^0=0$ *or that* $N:D^0\to D^{-2}$ *is an isomorphism. Then, the sequence* (3.12) *canonically splits. Furthermore, we have* $H^1(C'^\bullet_{\mathrm{st}}(V))\cong D^{-2}$ *if* $D^0=0$ *or* 0 *if* N *is an isomorphism.*

PROOF. Consider a triple (x,y,z) representing a class in $H^1(C^\bullet_{\mathrm{st}}(V))$. It satisfies the relation $Nx=(p\varphi-1)y$. Decompose

$$x=\sum x_k,\quad x_k\in D^k.$$

It is easy to see, by weight considerations, that $\varphi - 1$ is invertible on D^k when $k \neq 0$. Let then, for $k \neq 0$, $w_k = (\varphi - 1)^{-1} x_k$ Suppose first that $D^0 = 0$. Then, $x_0 = 0$ and we see that by subtracting the differential of $w = \sum w_k$, we can make $x = 0$. This makes $H^1(C^\bullet_{\mathrm{st}}(V)$ isomorphic to H^1 of the complex

$$D^{\varphi=1} \xrightarrow{(N,-I_\pi)} D \oplus D_K/F^0 \xrightarrow{1-p\varphi+0} D$$

and, since $D^{\varphi=1} \subset D^0 = 0$, to just $D^{p\varphi=1} \oplus D_K/F^0$ and the map ι is just the embedding of the first factor, and is clearly canonically split. Suppose now that $N : D^0 \to D^{-2}$ is an isomorphism. Since we are not assuming that $D^0 = 0$ we can only assume that $x \in D^0$, hence $y \in D^{-2}$. By the assumption on N we can now modify by the differential of w' to make $y = 0$, but then $Nx = 0$ so that $x = 0$, again by assumption. This proves the result in the second case. □

REMARK 3.6. In terms of of the proof of Lemma 3.2 the splitting in the case $D^0 = 0$ is given as follows: Given an extension ε as in (3.11) Lift $1 \in \mathbf{1}$ to the unique $A \in D'$ which is Frobenius invariant (this is equivalent to having $x = 0$ in the proof of Proposition 3.5). Then send the extension to $B - I_\pi(A)$ as before. Note that since A is Frobenius invariant we have $\rho(\varepsilon) = NA$

DEFINITION 3.7. Suppose $D^0 = 0$ and let $\text{ß}_\pi : H^1_{\mathrm{st}}(G, V) \to D_K/F^0$ be the canonical splitting of (3.12). For a variety X, with $V = H^{i-1}_{\text{ét}}(X \otimes \bar{K}, \mathbb{Q}_p(j))$, let

$$\mathrm{reg}^{c,\pi}_{\mathrm{syn}} : H^i_{\mathcal{M}}(X, \mathbb{Q}(j))_0 \to D_K/F^0$$
$$\mathrm{reg}^{d}_{\mathrm{syn}} : H^i_{\mathcal{M}}(X, \mathbb{Q}(j))_0 \to D^{-2},$$

called the continuous and discrete component of the syntomic regulator respectively, be the compositions of the syntomic regulator $\mathrm{reg}_{\mathrm{syn}}$ from (3.4) followed by the splitting ß_π and the projection ρ from (3.12) respectively.

THEOREM 3.8. *For any extension* ε *we have*

$$\frac{\mathrm{d}}{\mathrm{d}\log(\pi)} \text{ß}_\pi(\varepsilon) = -I_\pi \circ \rho(\varepsilon).$$

Consequently, we have $\frac{\mathrm{d}}{\mathrm{d}\log(\pi)} \mathrm{reg}^{c,\pi}_{\mathrm{syn}} = -\mathrm{reg}^{d}_{\mathrm{syn}}$

PROOF. Following Remark 3.6 we see that, in the notation there,

$$\frac{\mathrm{d}}{\mathrm{d}\log(\pi)} \text{ß}_\pi(\varepsilon) = \frac{\mathrm{d}}{\mathrm{d}\log(\pi)} B - I_\pi(A) = -\frac{\mathrm{d}}{\mathrm{d}\log(\pi)} I_\pi(A),$$

and (3.6) immediately shows that

$$\frac{\mathrm{d}}{\mathrm{d}\log(\pi)} I_\pi(A) = I_\pi NA.$$

Finally, by Remark 3.6 again, we have $NA = \rho(\varepsilon)$. □

There are two cases where it is known that syntomic regulators in the bad reduction case is given by Vologodsky integration. The first is that of zero-cycles on a proper variety X of dimension n. In this case we have $j = n, i = 2n$ and the regulator takes the form of a map

$$CH^n(X)_0 \to H^1_{\mathrm{st}}(K, H^{2n-1}_{\text{ét}}(X \otimes \bar{K}, \mathbb{Q}_p(n))).$$

Note that in this case D^0 and D^{-2} for V correspond to weights $2n$ and $2n - 2$ of $\mathrm{D}_{\mathrm{st}}(H^{2n-1}_{\text{ét}}(X \otimes \overline{K}, \mathbb{Q}_p))$. Thus, if the monodromy weight conjecture holds for X, then $N : D^0 \to D^{-2}$ is an isomorphism. Therefore, by Proposition 3.5, the discrete component of the regulator is 0. The syntomic regulator is therefore equal to its continuous component

$$\mathrm{reg}_{\mathrm{syn}} : CH^n(X)_0 \to H_{\mathrm{dR}}^{2n-1}(X/K)/F^n.$$

The target of the regulator is Poincare dual to $F^1 H_{\mathrm{dR}}^1(X/K)$, the space of holomorphic 1-forms on X. The following is well known.

PROPOSITION 3.9. *The vector space $CH^n(X)_0$ consists of zero cycles of degree 0 on X, and for such a cycle z and a holomorphic form ω one has*

$$\mathrm{reg}_{\mathrm{syn}}(z)(\omega) = \int_z \omega.$$

As we saw in Corollary 2.4, the Vologodsky integral of a holomorphic form ω on a proper X is independent of the choice of the branch of the logarithm. This is consistent with the syntomic regulator having no discrete part in this case.

The second case is the regulator for K_2 of curves. In [**11**] we proved the following result.

THEOREM 3.10 ([**11**, Corollary 1.3]). *Let X be a proper curve with arbitrary reduction over K. Then, with respect to a fixed uniformizer π, the result of Theorem* 1.1 *continues to hold provided we replace the syntomic regulator with its continuous part with this choice of uniformizer, assume that the form ω is in the kernel of the monodromy operator and replace Coleman integration with Vologodsky integration with respect to the branch of logarithm determined by π.*

This is a special case of a more general result where one does not assume that the form ω is holomorphic, but only that it is in the kernel of monodromy. We will say more about this generalization in Section 7. The monodromy operator N is transported from D_{st} to DR via the comparison isomorphism. We will look at this operator for the special case of curves later in the paper.

4. The derivative of a Vologodsky function with respect to $\log(p)$

In this section we study the universal log variant of Vologodsky integration, as introduced in Section 2, and its derivative with respect to $\log(p)$.

The simplest result on this derivative was already mentioned in the introduction. The universal logarithm log on $\mathbb{G}_m$. We proved in the introduction that

$$\frac{\mathrm{d}}{\mathrm{d}\log(p)} \log(z) = v_p(z).$$

Before continuing it is worth pointing out the obvious functoriality property,

$$\frac{\mathrm{d}}{\mathrm{d}\log(p)}(\alpha^* F) = \alpha^* \left(\frac{\mathrm{d}}{\mathrm{d}\log(p)} F \right), \tag{4.1}$$

valid for any Vologodsky function F on X and any morphism $\alpha : Y \to X$ [**14**, Lemma 9.13]. In particular, for any rational function on X we get

$$\left(\frac{\mathrm{d}}{\mathrm{d}\log(p)} \log(f) \right)(x) = v_p(f(x)).$$

There is one particular version of Coleman integration that has some dependency on the branch of the logarithm. Let X be the generic fiber of a smooth proper $\mathcal{O}_K$-curve $\mathcal{X}$ and let $T = \mathcal{X}_\kappa$. There is a reduction map red : $X \to T$, where we abuse notation to write X for its rigid analytification. the preimages of κ-rational points in T are rigid analytic open discs in X called *residue discs*. Let $S \subset T(\kappa)$ be a finite subset and let us, for each $y \in S$, pick a strictly smaller rigid analytic disc $D_y \subset \operatorname{red}^{-1}(y)$. The space $U = X - \cup_{y \in S} D_y$ is a wide open space, and the underlying affinoid is $A = X - \operatorname{red}^{-1}(S)$. Their difference is a disjoint union of annuli, $e_y = \operatorname{red}^{-1}(y) - D_y$. if ω is a rigid 1-form on U, its Coleman integral, which by standard theory is only defined on A, extends to the annuli e_y. On these annuli ω is given by a convergent Laurent series and integration is done term by term, with the residual term integrating to $(\operatorname{res}_{e_y} \omega) \cdot \log(z)$, with z some local parameter. Note that multiplying the parameter by a constant will change the constant of integration, which is anyhow a degree of freedom. To fix the constants of integration, the Tannakian interpretation of Coleman integration is extended to fiber functors associated with the y's [**8**, Section 5]. The theory easily extend to meromorphic forms on U by successively removing smaller and smaller discs around the singular points, and we have a similar local expansion: If x is a singular point for ω and $\operatorname{res}_x \omega \neq 0$, then the integral has a logarithmic term $(\operatorname{res}_x \omega)\log(z_x)$, for a local parameter z_x at x, and this will clearly depend on $\log(p)$. To get a statement about a concrete Coleman function it is useful to pick an auxiliary point x_0 such that ω is holomorphic on its residue disc. Then, $\int_{x_0}^{x}$ is a Coleman function, and by the above, we see that if $\operatorname{res}_x \omega = 1$ and π is a uniformizer for K,

$$\left(\frac{\mathrm{d}}{\mathrm{d}\log(\pi)} \int_{x_0}^{x'} \omega\right) = v_\pi(z_x(x')) \text{ for } x' \in D_x. \tag{4.2}$$

LEMMA 4.1. *Suppose $D = \sum n_i P_i$ is a divisor on X with $P_i \in A(K)$, with A the underlying affinoid of a wide open space $U \subset X$. Let $x_0 \in A(K)$ be a point whose reduction is disjoint from the reduction of the points P_i and let ω be a meromorphic form on U, regular on the residue disc of x_0, whose residue divisor on U is D. Then, for any $P \in A(K)$ not in the support of D we have*

$$\frac{\mathrm{d}}{\mathrm{d}\log(\pi)} \int_{x_0}^{P} \omega = D \cdot P,$$

the intersection multiplicity of D with P on $\mathcal{X}$, where P and D are considered divisors on $\mathcal{X}$ by extending all the points P_i and P to sections and considering the associated images.

PROOF. By linearity we may assume that $D = P_1$ is a single point (note that the residue divisor of a form on a complete curve is always of degree 0 but here we are considering the residue divisor only on U, allowing further residues at the removed discs). The integral $\int_{x_0}^{x}$ is a Coleman function. Thus, if P_1 and P are in different residue discs both sides of the equality are 0 by intersection theory and Proposition 2.1. On the other hand, if P and P_1 are in the same residue disc, then letting z be a uniformizing parameter at P_1 we have by (4.2).

$$\left(\frac{\mathrm{d}}{\mathrm{d}\log(\pi)} \int_{x_0}^{P} \omega\right) = v_\pi(z(P)) = P_1 \cdot P.$$

□

Results for more general integrals are known only for curves with semi-stable reduction, as this is the only case where the relation between Vologodsky integration and Coleman integration is known, first, in the non-iterated case by [**17**], and then, in general, by [**36**].

We recall the setup of [**12**, Section 4]. We assume that the curve X is the generic fiber of a proper regular flat $\mathcal{O}_K$ scheme $\mathcal{X}$ of relative dimension 1 with semi-stable reduction T which decomposes into smooth irreducible components

$$T = \cup_i T_i. \tag{4.3}$$

For simplicity we will assume that components T_i and T_j intersect at at most one point.

Let $\Gamma(X)$ be the dual graph of T with vertices V and edges E (this is of course an abuse of notation as it really depends on the particular model). The vertices correspond to the components T_v while the edges are ordered pairs of intersecting components (T_v, T_w) oriented from v to w, so that an edge e has tail $e^+ = v$ and head $e^- = w$.

The reduction map red : $X \to T$ allows us to cover X with rigid analytic domains $U_v = \mathrm{red}^{-1} T_v$ which are basic wide open spaces. These then intersect along annuli corresponding bijectively to the unoriented edges of $\Gamma(X)$. An orientation of an annulus fixes a sign for the residue along this annulus, and we match oriented edges with oriented annuli as in [**12**, Definition 4.7]. We use the same notation for the edge and for the associated oriented annulus.

We give a crash course on graph cohomology. All of this material is well known. See [**25**] for the relation between graph cohomology and the cohomology of curves with semi-stable reduction. For us, a graph Γ has a set V of vertices and E of oriented edges. Each edge $e \in E(\Gamma)$ has a tail and head, $e^+, e^- \in V(\Gamma)$ respectively, and for each edge we have the edge with reverse orientation $-e$ such that $(-e)^+ = e^-$ and $(-e)^- = e^+$ and K any field, let K^V denote as usual the vector space of functions from V to K, and K_-^E the set of functions $E \to K$ which are anti-symmetric with respect to orientation-reversing: $f(-e) = -f(e)$. There is a differential $\mathrm{d} : K^V \to K_-^E$, $\mathrm{d} f(e) = f(e^+) - f(e^-)$ and we define graph cohomology by

$$H^1(\Gamma, K) = K_-^E / \mathrm{d}\, K^V.$$

Define a scalar product on K^V by $(f, g) = \sum_{v \in V} f(v) g(v)$ and define a scalar product on K_-^E similarly. The dual operator to d with respect to these, $\mathrm{d}^* : K_-^E \to K^V$, is given by $\mathrm{d}^* g(v) = \sum_{e^+ = v} g(e)$. The space of *harmonic cochains* is defined as $\mathcal{H}(\Gamma, K) = \operatorname{Ker} d^*$.

For a connected Γ the harmonic decomposition

$$K_-^E = \mathrm{d}\, K^V \oplus \mathcal{H}(\Gamma, K) \tag{4.4}$$

follows by observing that the kernel of the *Laplacian*

$$\Delta = \mathrm{d}^* \circ \mathrm{d} : K^V \to K^V, \quad (\Delta f)(v) = \sum_{(v,w) \in E} (f(v) - f(w)) \tag{4.5}$$

is the space of constant functions and its image is its orthogonal complement. We note the following easy observation.

LEMMA 4.2. *The projection $K_-^E \to \mathrm{d}\, K^V$ coming from the harmonic decomposition* (4.4) *is simply*

$$\mathrm{d} \circ \Delta^{-1} \circ \mathrm{d}^* .$$

One source of harmonic cochains is the monodromy operator N. For a meromorphic form on X with no singularities on the annuli it is defined as

$$N\omega(e) := \operatorname{res}_e(\omega),$$

where the residue of an analytic form on an oriented annulus is defined in [**27**, Lemma 2.1]. The residue theorem for wide open spaces [**27**, Proposition 4.3] immediately shows that

$$\mathrm{d}^* N\omega(v) = \operatorname{res}_{U_v} \omega, \tag{4.6}$$

where the right-hand side means the sum of the residues of ω on U_v. Recall that a meromorphic form on a curve is said to be of the *second kind* if all its residues are 0. For such a form ω we immediately get from (4.6) that $N\omega$ is harmonic. The monodromy operator as defined here is essentially the same as the one coming from Fontaine's theory. Indeed, by [**25**] the space $\mathcal{H}(\Gamma, K)$ embeds into $H^1_{\mathrm{dR}}(X/K)$ and the Fontaine monodromy operator is just the composition of N as defined here with this embedding.

The results of [**17**] are for a Vologodsky integral with respect to a fixed branch, so at this point we fix such a branch. To state the main result of [**17**] we point out that as the domains U_v are basic wide open spaces with good reduction, forms on them can be Coleman integrated. As discussed earlier in this section, these integrals extend, once a branch of the logarithm is fixed, to the annuli e.

THEOREM 4.3 ([**17**]). *Let ω be a meromorphic form on X and let F_ω be a Vologodsky integral of ω. Then, for each vertex v of $\Gamma(X)$ there exist Coleman integrals F^v_ω of $\omega|_{U_v}$ such that F_ω equals F^v_ω on $U_v(K)$. For an oriented edge e let $c_\omega(e) = F^{e^-}_\omega|_e - F^{e^+}_\omega|_e \in K$. Then c_ω is a harmonic cochain on $\Gamma(X)$ with values in K, $c_\omega \in \mathcal{H}(\Gamma(X), K)$.*

It is important to observe that F_ω is *not* a function on the annuli. It is only a function on K-rational points and the annuli do not have any such points. This is why the single valued function F_ω can be defined on the U_v by Coleman functions that do not match on the intersections. Obviously, one can extend F_ω to the annuli by extending the base field, but this will force a change in the semi-stable model and no contradiction arises (see [**17**]).

Theorem 4.3 gives a way of recovering Vologodsky integrals on curves, provided one knows how to compute Coleman integrals (For an alternative description of the relation between the theories, at least for holomorphic forms, see [**37**]. In both approaches one also recovers a description of the behavior of Vologodsky integrals on annuli, that was first observed by Stoll [**55**]). Namely, to compute F_ω first choose Coleman integrals G^v for ω on the corresponding U_v. There is a choice of a constant of integration for each v and one makes such a choice arbitrarily. Then one computes the cochain $c(e) = G^{e^-} - G^{e^+}$. This cochain has a unique decomposition, by (4.4),

$$c = c_\omega + \mathrm{d}\,\gamma, \quad c_\omega \text{ harmonic}, \gamma \in K^V,$$

where γ is uniquely determined by $\mathrm{d}\,\gamma$ up to an additive constant. The required F^v_ω computing a Vologodsky integral of ω are then clearly given by

$$F^v_\omega = G^v - \gamma(v).$$

Note that by the discussion above we have

$$\gamma = \Delta^{-1}\,\mathrm{d}^*\,c.$$

To give a flavor of the type of results one might prove using this type of analysis, let us begin by reproving Corollary 2.4 (The reader might find it useful to look first at the very simple case of a Tate elliptic curves considered in Section 3 of [**17**]). The key to understanding how the description of Vologodsky integration affects the dependency on $\log(p)$ is to analyze what happens at an annulus. Suppose an annulus e is mapped, via a coordinate z, to the annulus $|\pi| < |z| < 1$ and a form ω is given, whose local expansion on e is

$$\omega = \sum_k a_k z^k \frac{\mathrm{d}\, z}{z},$$

so that $\operatorname{res}_e \omega = a_0$. The integral of ω on the domain which connects to the inner side of e is given by integrating term by term

$$F_1 = \sum_{k \neq 0} \frac{a_k}{k} z^k + a_0 \log(z) + C_1$$

for some constant C_1. To get the integral on the domain connecting to the outer side we first make the change of variables $z = \pi/w$,

$$\omega = -\sum_k a_k \left(\frac{\pi}{w}\right)^k \frac{dw}{w},$$

giving an integral

$$F_2 = \sum_{k \neq 0} \frac{a_k \pi^k}{k} w^{-k} - a_0 \log(w) + C_2.$$

Substituting back $z = \pi/w$ we get

$$F_1 - F_2 = C_1 - C_2 + a_0 \log(\pi) = C_1 - C_2 + \operatorname{res}_e \omega \log(\pi).$$

Suppose that F is a Vologodsky integral of ω with respect to a certain branch $\log(\pi)$. On each U_v it is represented by a Coleman integral with a choice of constant of integration, and the chain of differences $c_\omega(e)$ is harmonic. It is better to write here $c_{\omega,\log(\pi)}$ to stress the dependency on $\log(\pi)$ According to the above computation, for the same choices of constants of integration, with respect to a different branch of the logarithm $\log'(\pi)$, the chain of differences is

$$c_{\omega,\log'(\pi)} = c_{\omega,\log(\pi)} + (\log'(\pi) - \log(\pi)) N\omega. \tag{4.7}$$

Since $N\omega$ is harmonic, this chain is harmonic as well, hence the same constants of integration give again the Vologodsky integral. Therefore, the integral is independent of the choice of the branch.

The advantage of this proof is that it applies verbatim for integrals of forms of the second kind, which are therefore also independent of the branch of the logarithm. For general meromorphic forms we have the following easy generalization.

PROPOSITION 4.4. *Let ω be a meromorphic form on X with no singularities on the annuli. Let F_ω be a Vologodsky integral of ω. Then, the function $\mathrm{d}/\mathrm{d}\log(\pi) F_\omega$ is constant on each domain U_v except near the logarithmic singularities of ω, and associating this constant with the vertex v one gets a function on the dual graph which satisfies*

$$\Delta \frac{\mathrm{d}}{\mathrm{d}\log(\pi)} F_\omega(v) = \operatorname{res}_{U_v} \omega.$$

PROOF. The same analysis as before together with Lemma 4.2 gives that the derivative is

$$\Delta^{-1} d^* N\omega$$

and we are done by (4.6). □

Note that the above result determines the derivative only up to an additive constant. This is unavoidable since one could in theory have a different constant of integration associated with F_ω for each branch of the logarithm. To remedy this problem, we can make the integral independent of $\log(\pi)$ at some point, e.g., by setting it to 0 at that point. However, there are cases where we have the constant built in. For example, if f is a rational function on X and $\omega = \mathrm{d}\log(f)$, then it is natural to pick $F_\omega = \log(f)$ and this might have a constant of integration which is a multiple of $\log(\pi)$. In fact, we clearly have

$$\frac{\mathrm{d}}{\mathrm{d}\log(\pi)} \log(f)(v) = \mathrm{ord}_{T_v} f.$$

Note that this is consistent with [**17**, Lemma 2.1].

As a particular case, and as a prelude to the discussion in Section 5, let us use Proposition 4.4 to recover local height pairings on curves. In [**12**] we extended the Coleman-Gross p-adic height pairing on curves [**24**] using Vologodsky integration instead of Coleman integration at primes above p, and proved its compatibility with the Nekovář p-adic height pairing. When X is a smooth complete curve over a number field F, and D, E are two divisors of degree 0 on X, their p-adic height pairing $h(D, E)$ is a sum of local terms corresponding to the different finite completions of F. If K is such a completion, with respect to a prime above p, the local term for K is the trace down to $\mathbb{Q}_p$ (we do not discuss this trace here, see Section 5), of the expression

$$(D, E) = \int_E \omega_D.$$

Here, the integral is a Vologodsky integral evaluated at the divisor E while the form ω_D is a specially chosen meromorphic form of the third kind whose residue divisor, $\sum_x \mathrm{res}_x(\omega)\cdot x$, is D (recall that a meromorphic 1-form on a curve is of the *third kind* if it only has simple poles and all of its residues are integers). The determination of a canonical such ω_D is one of the key elements in the construction, but fortunately for us, we do not need to discuss it here, as we will only be interested in

$$\frac{\mathrm{d}}{\mathrm{d}\log(\pi)}(D, E) = \frac{\mathrm{d}}{\mathrm{d}\log(\pi)} \int_E \omega_D,$$

and, since ω_D is well-defined up to the addition of a holomorphic form, this quantity is independent of the precise choice by Corollary 2.4.

PROPOSITION 4.5. *The expression*

$$\frac{\mathrm{d}}{\mathrm{d}\log(\pi)}(D, E)$$

equals, up to a fixed multiple, to the local height pairing as defined in the proof of [**24**, Proposition 1.2].

PROOF. For simplicity, we assume that D and E are formal sums of K-rational points and skip the formal argument to show that this is sufficient. The relevant, "rational part", of the local height pairing is defined in [**24**] as the intersection product on a regular model $\mathcal{X}$ of X, of divisors extending the divisors D and E, to $\mathbb{Q}$-divisors which have a zero intersection with each component of the special fiber. In other words, we extend the points in D and E to sections and then add rational multiples of the components of the special fiber to satisfy the requirement. Clearly, it suffices to do this for only one of the divisors. So let D and E denote also the divisor on $\mathcal{X}$ obtained by extending the points in D and E to sections over $\mathcal{O}_K$ and let $\tilde{D} = D + \sum_v a_v T_v$ be the required extension, so that we need to compute

$$\tilde{D} \cdot E = D \cdot E + \sum_v a_v E \cdot T_v. \tag{4.8}$$

We may decompose D and E into their components inside each U_v, $D = \sum_v D_v$ and similarly for E. We further fix a K-rational point x_v in each U_v, which does not share a residue disc with any of the points in the support of either D or E (extending K if needed). We clearly have $D \cdot E = \sum_v D_v \cdot E_v$. Apply now Lemma 4.1 to U_v, the form ω_D, $x_0 = x_v$ and P a point in the support of E_v. We get

$$\frac{\mathrm{d}}{\mathrm{d}\log(\pi)} \int_{x_v}^{P} \omega_D = D_v \cdot P.$$

Now sum with coefficients over all those P to obtain

$$\frac{\mathrm{d}}{\mathrm{d}\log(\pi)} \int_{E_v - (\deg E_v) x_v} \omega_D = D_v \cdot E_v.$$

Summing now over all v we get, noting that $\sum_v \deg E_v = \deg E = 0$,

$$\frac{\mathrm{d}}{\mathrm{d}\log(\pi)} \int_{E} \omega_D = D \cdot E + \frac{\mathrm{d}}{\mathrm{d}\log(\pi)} \int_{\sum_v (\deg E_v) x_v} \omega_D.$$

In view of (4.8), to prove the proposition we need to show that

$$\frac{\mathrm{d}}{\mathrm{d}\log(\pi)} \int_{\sum_v (\deg E_v) x_v} \omega_D = \sum_v a_v D \cdot T_v.$$

Pick one of the x_v to be x_0 and let

$$f(v) = \frac{\mathrm{d}}{\mathrm{d}\log(\pi)} \int_{x_0}^{x_v} \omega_D$$

so that

$$\frac{\mathrm{d}}{\mathrm{d}\log(\pi)} \int_{\sum_v (\deg E_v) x_v} \omega_D = \sum_v (\deg E_v) f(v) = \sum_v f(v) E \cdot T_v.$$

The result will be proved if we show that $f(v) - a_v$ is a constant function on V, which is equivalent to being in the kernel of Δ. By Proposition 4.4 we know that $\Delta f = (v \longmapsto \deg D_v)$ and so the result follows from the fact that $\Delta(v \longmapsto a_v) = (v \longmapsto \deg D_v)$. This is well known (see [**1, 2**] for the relation between graph Laplacians and intersection multiplicities) but we review this simple argument. The condition on $\tilde{D}$ is that for every $w \in V$ we have

$$0 = (D + \sum_v a_v T_v) \cdot T_w = \deg D_w + \sum_v a_v T_v \cdot T_w.$$

The sum on the right is clearly only over v for which there exist an edge (v, w), and over $v = w$. Our assumptions about the reduction imply that for $v \neq w$ in the sum we have $T_v \cdot T_w = 1$. Furthermore, since the special fiber is just $\sum_v T_v$ and its intersection with each T_w is 0, we see that

$$T_w \cdot T_w = - \sum_{(v,w)\in E} 1$$

so that we have

$$0 = \deg D_w + \sum_{(v,w)\in E} a_v - a_w,$$

which, by (4.5), is equivalent to $\Delta(v \longmapsto a_v) = (v \longmapsto \deg D_v)$ □

REMARK 4.6. As mentioned in the introduction, a similar observation has been made by Colmez [**28**], in the proof of Théorème II.2.18. The method is quite different. Colmez simply notes that the derivative satisfies the basic list of properties which uniquely characterize the height pairing. Here on the other hand, we directly show that the result is the same as the construction of the local height pairing in terms of intersection theory.

In order to treat iterated integrals we need the work of Katz and Litt [**36**]. We will present it in an equivalent way (the equivalence with the original formulation was proved by Katz in a private communication). The new formulation is a direct generalization of [**17**].

To describe the result, choose for simplicity a K rational point x_v in each domain U_v. Let (M, ∇) be a unipotent connection on X. We have "monodromy matrices" corresponding to the following "loops" associated with each edge $e = (v, w)$: Start with x_v, use Vologodsky parallel transport to go to x_w, use Coleman parallel transport to go to the side of the annulus e on the U_w side, cross to the U_v side of e and finally Coleman parallel transport back to x_v.

This gives an operator on the fiber at x_v of M. Up to conjugacy it is independent of the choice of x_v. Denote it by $A(e)$.

THEOREM 4.7 (Reinterpreted Katz-Litt). *The association $e \to A(e)$ is harmonic in the sense that for each $v \in V$, $\sum_{e^+=v} \log(A(e)) = 0$.*

In [**15**] we use this result to derive a formula for the derivative with respect to $\log \pi$ of certain iterated Vologodsky integrals.

THEOREM 4.8 ([**15**]). *Suppose ω and η are two forms of the second kind on X. Let*

$$F = \int (\omega \int \eta)$$

with respect to the universal branch. The function

$$\frac{\mathrm{d}}{\mathrm{d}\log(\pi)} F$$

is constant on each domain U_v except near the singularities of ω or η, and associating this constant with the vertex v one gets a function on the dual graph which satisfies

$$\Delta \frac{\mathrm{d}}{\mathrm{d}\log(\pi)} F_\omega(v) = \frac{1}{2} \sum_{e^+=v} c_\eta(e) \operatorname{res}_e \omega - c_\omega(e) \operatorname{res}_e \eta - \sum_{e^+=v} \langle \int \eta, \int \omega \rangle_e.$$

*Here, c_ω and c_η are the harmonic cochains associated with ω and η in Theorem 4.3 (with respect to the canonical branch) and the last term is the local index on an annulus (see [**7**]).*

5. p-adic heights

In this section we report on work with Müller and Srinivasan [**14**]. In this work we describe a p-adic analogue to the theory of heights associated to line bundles with adelic metrics [**60**]. Our main interest is in applications to the problem of finding rational point on curves through the method known as "Quadratic Chabauty" [**3**–**5**], but the theory of heights we develop is of independent interest. For analogy, let us first quickly review the classical theory.

Let F be a number field with a system of absolute values $|\ |_\nu$ satisfying the product formula. Let X be a smooth complete variety over F.

DEFINITION 5.1. An adelic line bundle on X is a line bundle $\mathcal{L}$ over X together with a collection of norms, $\|\ \|_\nu$ on the line bundle $\mathcal{L} \otimes F_\nu$ (which means norms on all the fibers of the line bundle), for each place, finite or infinite, ν of F. This collection needs to satisfy certain conditions, which guarantee, among other things, that the associated height

$$h_{\hat{L}} : X(F) \to \mathbb{R}, \quad h_{\hat{L}}(x) = \sum_\nu \log \|w\|_\nu$$

is a finite sum, where w is any vector in the fiber of L over x.

From the product formula it is immediate to see that the height indeed depends only on x and not on the particular choice of w. We avoid discussing conditions on the norms for ease of presentation, since we will mostly be concerned with the p-adic theory, and since canonical norms that we want to study satisfy these automatically.

Following Tate's definition of canonical heights for Abelian varieties, one can attempt to obtain canonical heights in various situations. The following scenario, which we dub "the dynamical situation", is considered by Zhang [**60**](see also [**19**]). Suppose that $f\colon X \to X$ is an endomorphism X and the line bundle $\mathcal{L}$ is such that there exists an integer $d > 1$ and an isomorphism

$$\beta\colon \mathcal{L}^{\otimes d} \to f^*\mathcal{L}. \tag{5.1}$$

Zhang constructs, at each place ν, a canonical norm on $\mathcal{L}_v$ for which β is an isometry (we will say that the metric is *compatible* with the dynamical situation), by starting with any norm $\|\cdot\|_\nu$ and repeatedly replacing $\|\cdot\|_\nu$ with $(\beta^{-1}f^*\|\cdot\|_\nu)^{\frac{1}{d}}$ and taking a limit. The resulting local metrics turn out to give an adelic metric. In the case where X is an abelian variety, $f : X \to X$ is the multiplication by 2 map and $\mathcal{L}$ is a symmetric line bundle on X, one gets the Tate canonical height associated with $\mathcal{L}$.

The p-adic analogue of the above picture is considered in [**14**]. The p-adic analogue of the system of absolute values satisfying the product formula is a continuous idèle class character

$$\chi = \sum_\nu \chi_\nu \colon \mathbb{A}_F^\times / F^\times \to \mathbb{Q}_p.$$

The local components χ_ν are the analogues of the logs of the absolute values and the fact that the character vanishes on $F^\times$ is the analogue of the product formula. Note the important difference from the classical case: The primes above ∞ play no role as the sum is only over finite primes.

For $\nu \mid p$, there is a decomposition

$$F_\nu^\times \xrightarrow{\chi_\nu} \mathbb{Q}_p, \qquad F_\nu^\times \xrightarrow{\log_\nu} K_\nu \xrightarrow{t_\nu} \mathbb{Q}_p \tag{5.2}$$

for some $\mathbb{Q}_p$-linear map t_ν, which is usually called the trace map, and some choice of a branch $\log_\nu$ of the p-adic logarithm. In analogy with the classical case we can define local norms, or rather, logs of local norms, on a line bundle $\mathcal{L}$, to be functions ℓ_ν on the total space of $\mathcal{L}$ minus the zero section, which satisfy the relation

$$\ell_\nu(\alpha w) = \chi_\nu(\alpha) + \ell_\nu(w).$$

However, at all places ν not above p one has the property $\chi_\nu(\alpha) = \nu(\alpha)\chi(\pi_\nu)$, where π_ν is a uniformizer at ν and we write ν also for the associated valuation, so that it makes sense to divide the function by $\chi_\nu(\pi_\nu)$. With this in mind we make the following.

DEFINITION 5.2. Let K be a finite extension of $\mathbb{Q}_q$, for some finite prime q. Let X be a variety over K and let $\mathcal{L}$ be a line bundle over X and denote by 0 the 0-section. Let $\mathcal{L}^\times$ denote the total space of $\mathcal{L}$ minus the image of 0. Then, a p-adic metric on $\mathcal{L}$ is:

- For $q = p$ a *log function*

  $$\log : \mathcal{L}^\times \to \overline{K}$$

  A function satisfying $\log(\alpha w) = \log_\nu(\alpha) + \log(w)$, $\alpha \in K^\times$, $w \in \mathcal{L}_x$, where $\log_\nu$ is the branch of the p-adic logarithm determined by (5.2).
- For $q \neq p$ a *valuation*

  $$\mathrm{val} : \mathcal{L}^\times \to \mathbb{Q}$$

  satisfying $\mathrm{val}(\alpha w) = \nu(\alpha) + \mathrm{val}(w)$ where ν is the valuation on $K^\times$, normalized so that $\nu(\pi_\nu) = 1$.

Let F be a number field, X a variety over F and $\mathcal{L}$ a line bundle over X. An adelic metric on $\mathcal{L}$ is a collection of metrics for the completions of $\mathcal{L}$ over all finite places of F, satisfying a certain technical condition (that the valuation is almost always a "model valuation"). A line bundle $\mathcal{L}$ over X together with an adelic metric will be called an adelically metrized line bundle, usually denoted $\hat{\mathcal{L}}$. The p-adic height $h_{\hat{\mathcal{L}}} : X(F) \to \mathbb{Q}_p$ associated to the adelically metrized line bundle $\hat{\mathcal{L}}$ is given by the formula

$$h_{\hat{L}}(x) = \sum_{\nu \mid p} t_\nu \log_\nu(w) + \sum_{\nu \nmid p} \mathrm{val}_\nu(w) \cdot \chi_\nu(\pi_\nu)$$

where w is any vector in the fiber of $\mathcal{L}$ over x and t_ν is the trace map from (5.2).

As in the classical case, it is obvious that the definition of the height is independent of the choice of w.

REMARK 5.3. One can define valuations that take values in other groups. This will be useful later on (see Proposition 5.6). For the global theory and the construction of heights, only $\mathbb{Q}$-valued valuations are useful.

The part of the theory which involves finite primes different from p, is in many respects joint between the p-adic and the classical case. Indeed, given a valuation val_ν, where ν is a place above a finite prime $q \neq p$, such that the associated absolute value is $|\ |_\nu = c^{\nu(\ )}$ for some $c \in (0,1)$, the function

$$\|\ \|_\nu = c^{\mathrm{val}_\nu(\ )}$$

is a norm in the classical sense. Clearly, it is not the most general type of norm. However, it turns out that in many situations the norms one considered are in fact of this form. Conversely, if we obtain a norm taking values in $c^{\mathbb{Q}}$, then taking the log with respect to c gives us a valuation.

Let us turn now to the situation above p. As we mentioned for the classical theory, defining heights using metrics can be done without any serious restrictions on the type of metric used. However, in practice one imposes some restrictions on the norms. In the p-adic situation, this is both a restriction and a good way of constructing metrics. Recall that if K is a finite extension of $\mathbb{Q}_p$, and $\mathcal{L}/X$ is a line bundle over a smooth variety over K, then a p-adic metric on $\mathcal{L}$ consists of a log function on the total space of $\mathcal{L}$ minus the zero section. It is natural to ask that this log function is a Vologodsky function. We in fact ask for a far stronger condition: that log is at most a second iterated Vologodsky integral. Let us call such log functions *admissible* (note, in [**9**] we called these log functions and what we now call log functions we called there Pseudo log functions). The key result concerning these functions, allows their characterization as well as construction using the notion of curvature. This is described in the following Theorem.

THEOREM 5.4 ([**9**, Proposition 4.4]). *Suppose X is proper and let*

$$\cup : H^1_{\mathrm{dR}}(X/K) \otimes \Omega^1(X) \to H^2_{\mathrm{dR}}(X/K)$$

be the cup product map. Suppose $\mathrm{ch}_1(\mathcal{L})$, the first Chern class of $\mathcal{L}$, is in the image of $\cup$. Then one can associate to each admissible log function on $\mathcal{L}$ a curvature form, *$\alpha \in H^1_{\mathrm{dR}}(X/K) \otimes \Omega^1(X)$, such that $\cup(\alpha) = \mathrm{ch}_1(\mathcal{L})$ Conversely, any curvature form α cupping to $\mathrm{ch}_1(\mathcal{L})$ is the curvature of an admissible log function on $\mathcal{L}$, unique up to an integral of $\omega \in \Omega^1(X)$.*

We will not explain the construction of the curvature or the proof of the theorem. An explicit description of a log function with a given curvature is not so easy in general. However, for the case of curves, a fairly easy description can be given as follows.

THEOREM 5.5 ([**14**, Section 3]). *Let $\mathcal{L}$ be a line bundle over a complete curve X over K and let α be a curvature form cupping to $\mathrm{ch}_1(\mathcal{L})$. Fix a rational section s of $\mathcal{L}$. To determine a log function on $\mathcal{L}$ it suffices to determine* $\log(s)$. *Let $\alpha = \sum_i \omega_i \otimes [\eta_i] \in \Omega^1(X) \otimes H^1_{\mathrm{dR}}(X)$ for some holomorphic forms $\omega_i \in \Omega^1(C)$ and forms of the second kind η_i with corresponding cohomology classes $[\eta_i] \in H^1_{\mathrm{dR}}(C)$. Then, the formula*

$$\mathrm{d}\log_{\mathcal{L}}(s) = \sum \omega_i \int \eta_i + \gamma, \tag{5.3}$$

determines a log function on $\mathcal{L}$ with curvature α. Here, γ is any meromorphic form with the property that the right-hand side has only simple poles on X and its residue divisor is exactly $\mathrm{div}(s)$.

We observe that the degree of freedom in choosing γ is exactly adding a holomorphic form, so the degree of freedom in the log function is exactly adding the integral of a holomorphic form, which is exactly the degree of freedom in Theorem 5.4.

If we try to copy the construction of canonical norms in Dynamical situations to the p-adic setting, we immediately observe a problem: The limiting procedure defining the norms will not converge p-adically. Thus, another idea is required for finding canonical metrics.

For $q \neq p$, one can hope that a canonical norm obtained through the limiting procedure over $\mathbb{R}$ actually takes values in some $c^{\mathbb{Q}}$, hence a valuation may be obtained from it. This will not work for $q = p$, where the log function is truly p-adic.

Instead of using limiting methods, one may use the theory of the curvature, Theorem 5.4. Consider then K a finite extension of $\mathbb{Q}_p$, a complete smooth variety X over K, a line bundle $\mathcal{L}$ over X, an endomorphism $f : X \to X$ and an isomorphism (5.1). If the log function $\log_{\mathcal{L}}$ with curvature α makes β an isometry, then we have

$$f^*\alpha = \mathrm{d}\,\alpha \tag{5.4}$$

This is consistent with the fact that $f^*\mathrm{ch}_1(\mathcal{L}) = \mathrm{d}\,\mathrm{ch}_1(\mathcal{L})$, which needs to hold since α cups to $\mathrm{ch}_1(\mathcal{L})$. Conversely, suppose that (5.4) holds with α a curvature form cupping to $\mathrm{ch}_1(\mathcal{L})$. Then, for any log function on $\mathcal{L}$ with curvature α, which exists by Theorem 5.4, The two log functions, $\mathrm{d}\log_{\mathcal{L}}$ and $\beta^{-1}f^*\log_{\mathcal{L}}$, on $\mathcal{L}^{\otimes d}$, have the same curvature and therefore, again by Theorem 5.4, differ by the integral of a holomorphic form on X. At this point one can study how this form is changed when $\log_{\mathcal{L}}$ is changed by the integral of a holomorphic form, and, depending on the action of f^* on $\Omega^1(X)$, it's often possible to make β an isometry as required.

One case of the above was key to [**14**]. There we construct canonical metrics on a symmetric line bundle $\mathcal{L}$ on an abelian variety X. In this case f is the multiplication by 2 map and $d = 4$. For further details see [**14**, 4.2.1].

At this point we apply the general theme of this work. Up to now we defined the p-adic metric above p with a fixed choice of a branch, dictated by the idèle class character χ. Now, instead, we will consider the metric as a Vologodsky function with respect to the universal branch of the logarithm, and differentiate with respect to $\log(p)$. As we will see, this gives valuations, in the more general sense of Remark 5.3, i.e., with values in more general groups than the additive group of the rationals.

PROPOSITION 5.6. *Let $\log_{\mathcal{L}}$ be a log function with the universal branch of the logarithm. Then $\mathrm{d}/\mathrm{d}\log(\pi)\log_{\mathcal{L}}$ is a $\overline{K}$-valued valuation on $\mathcal{L}$. We call it the* Vologodsky valuation associated with the log function.

PROOF. This is immediate from (1.4). □

PROPOSITION 5.7. *The Vologodsky valuation associated with a log function on $\mathcal{L}$ depends only on the associated curvature α. Thus, we may denote it val_α. This valuation is defined for every curvature form α cupping to $\mathrm{ch}_1(\mathcal{L})$.*

PROOF. This is clear from Corollary 2.4. □

The following is obvious.

PROPOSITION 5.8. *Let $\mathcal{L}$ be a line bundle on X and let α be a curvature form cupping to* $\mathrm{ch}_1(\mathcal{L})$. *If $g: Y \to X$ is a morphism, then* $g^* \mathrm{val}_\alpha = \mathrm{val}_{g^*\alpha}$.

COROLLARY 5.9. *In the dynamical situation, suppose α is a curvature form cupping to* $\mathrm{ch}_1(\mathcal{L})$ *and satisfying* $f^*\alpha = \mathrm{d}\,\alpha$. *Then β is an isometry (up to a constant) for the valuation* val_α.

PROOF. The two valuations one obtains on $\mathcal{L}^{\otimes d}$ have the same curvature. □

Thus, in dynamical situation, there are (at least) two ways to construct a compatible valuation locally at a place above the prime q. One method is via limiting procedures using norms over $\mathbb{R}$, hoping that the resulting norm has values in some $c^{\mathbb{Q}}$. The other is via q-adic analysis and curvature forms. The former method has several advantages

- It does not need to assume the existence of a well-behaved curvature form
- The latter method may not produce $\mathbb{Q}$-valued valuations. The former method may produce norms which are not coming from valuations, but these may still be used in the theory of heights while non-$\mathbb{Q}$-valued valuations are useless there.

We note though that for any line bundle $\mathcal{L}$ on an abelian variety X the Vologodsky valuation associated with any curvature form for $\mathcal{L}$ is $\mathbb{Q}$-valued. In fact, it does not depend on the curvature but only on the line bundle itself and is the canonical valuation so may be used for the computation of the canonical height. This is proved in [**14**, Theorem 9.24].

The method of Vologodsky valuations has one very important advantage. If a valuation is constructed in a dynamical situation for the line bundle $\mathcal{L}$ over X using the limiting procedure, then there would be some way to control, compute or approximate it over X. However, if we pull back via $g: Y \to X$, there will be no obvious way to compute the pulled back valuation over Y. On the other hand, using Vologodsky valuations, the pulled back valuation is associated with the pulled back curvature, and is therefore, at least in principle, computable. This is especially true if Y is a curve, as Theorem 5.5 provides an explicit description of the log function in terms of the curvature, and Theorem 4.8 provides a way of computing the associated valuation. In [**15**] we use this to explicitly compute the pullback to a curve of the canonical valuation on a symmetric line bundle on the Jacobian of the curve, with applications to Quadratic Chabauty.

6. The unipotent Albanese map

The classical Albanese map between a variety and its Albanese variety has a non-abelian analogue, the *Unipotent Albanese map* first appearing in its Hodge incarnation in the work of Hain [**33**] and later becoming a cornerstone of Kim's non-abelian Chabauty's method [**38, 39**].

Our Goal is to discuss the case of curves with semi-stable reduction, but doing so in a rather informal manner. In particular, we do not claim any relation with the étale theory.

Let us discuss things first is some generality. Suppose X is a variety over a field K, which for the moment can be any field. We assume that we can define for $x_0 \in X(K)$ some sort of fundamental group $G = \pi_1(X, x_0)$, possibly with some extra structure, and, given another rational point $x \in X(K)$, a fundamental path space $\pi_1(X, x_0, x)$, again with some extra structure, which is a right torsor for G with compatible structures (using the theory of Tannakian categories one can often streamline the meaning of these compatible structures). Here, by torsor we mean a principal homogeneous space in the most naive sense of the term.

REMARK 6.1. Extra structures on (unipotent or general) fundamental groups have a very long history. The Hodge theory of the fundamental group was probably first studied by Chen [**20**], Morgan [**45**] and more recently by Hain [**32**]. The étale fundamental group was invented by Groethendieck and is studied in detail in [**46**]. The algebraic de Rham fundamental group with its filtration is studied in [**58**]. The p-adic Hodge theory was studies by a great number of authors.

DEFINITION 6.2. Let $\mathrm{Alb}(G)$ be the set of isomorphism classes of G-torsors with compatible structure and define the *abstract Albanese map* as the map

$$\alpha_X : X(K) \to \mathrm{Alb}(\pi_1(X, x_0))$$

sending $x \in X(K)$ to the isomorphism class of the torsor $\pi_1(X, x_0, x)$.

Perhaps the most obvious example of an abstract Albanese map is the one for which π_1 is the étale fundamental group (respectively étale path space) of $\overline{X} := X \otimes_K \overline{K}$, which is a group (respectively a torsor) with an action of $\mathrm{Gal}(\overline{K}/K)$. In this case we clearly have

$$\mathrm{Alb}(\pi_1(X, x_0)) = H^1(\mathrm{Gal}(\overline{K}/K), \pi_1(\overline{X}, x_0)),$$

the non-abelian cohomology of $\mathrm{Gal}(\overline{K}/K)$ with coefficients in $\pi_1(\overline{X}, x_0)$.

In the second example K is a finite extension of $\mathbb{Q}_p$, X is a proper variety over K with good reduction and the fundamental group is the unipotent de Rham fundamental group, $\pi_1^{\mathrm{dR}}(X, x_0)$, the pro-algebraic group associated by Tannakian duality to the category of unipotent integrable connections on X with the fiber functor of taking the fiber at x_0. This has a descending filtration By subgroups F^i [**58**] and an action of a Frobenius automorphism (we take the linear Frobenius and not the semi-linear one for convenience), and path spaces have the same structure.

For the proof of Theorem 6.4, let us recall that that function on an affine group scheme G over a field K can be given by "matrix coefficients". These are defined by triple (V, v, s), where V is a finite dimensional representation of G, $v \in V$ and $s \in V^*$. Such a triple defines a function on G by the formula

$$(V, v, s)(g) = s(gv).$$

There are obvious relations between such triple that produce the same function. In particular, if $T : V \to V'$ is a G-map, then the triple (V, v, T^*s') and (V', Tv, s') produce the same function. If G is the fundamental group of a neutral Tannakian category $\mathbb{T}$, with fiber functor ω, then these triple are given in terms of $\mathbb{T}, \omega$ as (V, v, s) where $V \in \mathbb{T}$, $v \in \omega(V)$ and $s \in \omega(V)^*$. This allows a direct construction of the underlying Hopf algebra. In the standard definition one uses in the construction $\omega(V) \otimes \omega(V)^*$ but it is easy to see that simple tensors suffice.

LEMMA 6.3. *Let $\delta : H \to G$ be a homomorphism of affine group schemes over K. A function $f : G \to K$ is $\delta(H)$ left-invariant if and only if it can be represented by a triple (V, v, s) where s is an H-homomorphism $V \to K$, with K having the trivial H-action.*

PROOF. Clearly if such a representation exists, the function is left-invariant. Conversely, suppose f is H invariant from the left and is represented by a triple (V, v, s). By replacing V with a sub G-module we may assume that v generates V as a G-module. But then the equation $s(\delta(h)gv) = s(gv)$ for every $g \in G$ and $h \in H$ implies that $s(\delta(h)w) = s(w)$ for every $w \in V$. $\square$

In [**10**] we proved the following result.

THEOREM 6.4. *We have a canonical bijection of sets*

$$\mathrm{Alb}(\pi_1^{\mathrm{dR}}(X, x_0)) \cong F^0 \backslash \pi_1^{\mathrm{dR}}(X, x_0)(K) \tag{6.1}$$

Furthermore, there exists a correspondence between regular functions on $\pi_1^{\mathrm{dR}}(X, x_0)$, *left-invariant under* F^0, *and Coleman functions on* X *such that if the function* f *corresponds to the Coleman function* F *on* X, *then*

$$f(\alpha_X(x)) = F(x), \quad x \in X(K).$$

PROOF. By [**8**, Corollary 3.2] each torsor P for $G = \pi_1^{\mathrm{dR}}(X, x_0)$ has a canonical Frobenius invariant element γ_ϕ (In Section 2 we called it $\gamma_{x,y}$, a path connecting the two points, but the proof of the theorem, which is an immediate consequence of [**8**, Theorem 3.1] clearly applies to any path space for G). Pick any $\gamma_{\mathrm{dR}} \in F^0 P$. Then $\gamma_\phi = \gamma_{\mathrm{dR}} g$ for a unique $g \in G$. Any other such choices of γ_{dR} will be of the form $\gamma_{\mathrm{dR}} h$ for some $h \in F^0 G$, replacing g by hg. This establishes the bijection (6.1). Suppose now that $f : G \to K$ is a regular function that is invariant from the left under F^0. By [**31**, Remark 3.8] the subgroup $F^0 G$ is the image of H, the fundamental group of the neutral Tannakian category of unipotent vector bundles, under the map induced by the functor that forgets the connection on bundles with connections. Let $((M, \nabla), v, s)$ be a representation for f whose existence is guaranteed by Lemma 6.3. Thus, (M, ∇) is a unipotent connection on X, which is the same thing as a representation of G, $v \in M_{x_0} =$ the underlying vector space for the representation, and $s : M_{x_0} \longrightarrow K$ is a functional, which is an H-homomorphism. This immediately translates to it being the fiber at x_0 of a map of vector bundles $M \longrightarrow \mathcal{O}_X$, which we continue to call s. With the new interpretation of s, the triple $((M, \nabla), v, s)$ is now a Coleman function F. Its value at x is by definition

$$F(x) = s(\gamma_\phi v).$$

The value of f on $\alpha(x) = g$ is $s(gv)$. The result now follows because

$$s(\gamma_\phi v) = s(\gamma_{\mathrm{dR}}(gv)) = \gamma_{\mathrm{dR}}(s(gv) = s(gv).$$

The last two equalities follow because γ_{dR}, being in F^0, commutes with arbitrary morphisms of vector bundles, in particular s, and on the trivial line bundle it is just the identity. $\square$

Next we consider a variety X with semi-stable reduction over K. We then *should* have a similar picture with a fundamental group $G = \pi_1(X, x_0)$ in the category of filtered (φ, N)-modules, and a corresponding Albanese map. To be more precise, There is a triple $G = (G_{\mathrm{cr}}, G_{\mathrm{dR}}, I_\pi)$ consisting of an affine group scheme G_{cr} over K_0 with Frobenius and monodromy operators, an affine group scheme G_{dR} over K with a filtration, and a comparison isomorphism

$$I_{\mathrm{HK},\pi} : G_{\mathrm{cr}} \otimes_{K_0} K \to G_{\mathrm{dR}}$$

depending on the choice of a uniformizer π. The dependency on π is such that if we change to π' we have, exactly as in (3.6),

$$I_{\pi'} = I_\pi \circ \exp(\log(\pi'/\pi)N). \tag{6.2}$$

Note that N (and any multiple of it), being a topologically nilpotent derivation of a Hopf algebra, can be exponentiated to an automorphism of the algebra, hence of the associated affine group scheme.

REMARK 6.5. We intentionally wrote *should have.* There have been many approaches to the unipotent fundamental group in the log scheme or semi-stable case (a probably not inclusive list includes [**29**, **36**, **40**, **41**, **53**, **57**]). Not all of them state all required results. For example, to the best of our knowledge, the dependency on the uniformizer of the Hyodo-Kato isomorphism is only hinted to in the introduction to [**41**]. Also, while it is often claimed that all of these approaches are the same, this seems to never be actually proved, excluding some results in [**21**]. In particular, to actually make all of this into actual theorems, the theory should be compared with the theory of Vologodsky.

The Albanese set $\mathrm{Alb}(G)$ is the set of isomorphism classes of G-torsors with compatible structures $P = (P_{\mathrm{cr}}, P_{\mathrm{dR}}, I_\pi)$ in the obvious sense.

PROPOSITION 6.6. *There exists a bijection*

$$P \mapsto (P_{c,\pi}, P_d) : \mathrm{Alb}(G) \to (F^0\backslash G_{\mathrm{dR}}) \times \mathrm{Nil}(G_{\mathrm{cr}})$$

where $\mathrm{Nil}(G_{\mathrm{cr}})$ *classifies compatible* N *on the trivial torsor* G_{cr} *with its Frobenius.*

PROOF. Let P be a G-torsor. By [**57**] the torsor P_{cr} has a canonical path γ_ϕ. This path is fixed by Frobenius. As a torsor with a Frobenius action, P_{cr} is therefore trivial and P_{dR} is trivialized by $I_\pi(\gamma_\phi)$. The set $\mathrm{Alb}(G)$ is thus the product of the set of possible filtrations on G_{dR} as a G_{dR}-torsor, with $\mathrm{Nil}(G_{\mathrm{cr}})$. The former is clearly classified by $F^0\backslash G_{\mathrm{dR}}$: One can declare an $x \in G_{\mathrm{dR}}$ to be in F^0 and then the i-th part of the filtration is forced to be $(F^i G_{\mathrm{dR}}) \cdot x$. An x' defines the same filtration if and only if it is gx, with $g \in F^0 G_{\mathrm{dR}}$. □

Note that since the Vologodsky path is not killed by monodromy, there is still some wiggling room for the monodromy, and $\mathrm{Nil}(G_{\mathrm{cr}})$ is not trivial. Also note that the map to $F^0\backslash G_{\mathrm{dR}}$ depends on the choice of the uniformizer π, since the Vologodsky path needs to be imported via I_π to G_{dR} in order to define it.

The map in Proposition 6.6 is explicitly given as follows

$$P_{c,\pi} = \gamma_{\mathrm{dR}}^{-1} I_\pi(\gamma_\phi), \quad P_d = \gamma_\phi^{-1} N(\gamma_\phi). \tag{6.3}$$

The element $N(\gamma_\phi)$ belongs to $\mathrm{Lie}(G_{\mathrm{cr}})$ and characterizes P_d. Vologodsky theory puts some restrictions on this element, which we will not attempt to describe here.

REMARK 6.7. The product decomposition of the Albanese was already observed by Betts.

DEFINITION 6.8. The compositions of the Albanese map $X(K) \to \mathrm{Alb}(G)$ with the projections on the factors $F^0\backslash G_{\mathrm{dR}}$ and $\mathrm{Nil}(G_{\mathrm{cr}})$ will be called the continuous and discrete components of the semi-stable Albanese map respectively, denoted $\alpha_{c,\pi}$ and α_d respectively.

THEOREM 6.9. *For each choice of uniformizer π, the continuous component of the semi-stable Albanese map is computed using Vologodsky integrals: There exists a correspondence between regular functions on G_{dR}, left-invariant under F^0, and Vologodsky functions with respect to the branch of the logarithm determined by $\log(\pi) = 0$, such that if the regular function $f : G_{\mathrm{dR}} \to K$ corresponds to the Vologodsky function F on X, then, with $\alpha_{c,\pi}$ as in Definition 6.8,*

$$f(\alpha_{c,\pi}(x)) = F(x), \quad x \in X(K).$$

The discrete component of the semi-stable Albanese is the logarithmic derivative, with respect to $\log(\pi)$ of the continuous component:

$$P_{c,\pi}^{-1} \frac{\mathrm{d}}{\mathrm{d}\log(\pi)} P_{c,\pi} = I_\pi(P_d).$$

PROOF. The statement about the continuous component is word for word the proof in the good reduction case. The second statement is obtained by deriving (6.3) with respect to $\log(\pi)$ using (6.2). □

The similarity between the construction in this section and the discussion of syntomic regulators in Section 3, and in particular Theorem 3.8, should be obvious. Unlike the syntomic regulators case, we have not attempted to spell out what this result means in specific situations.

7. Toric regulators

In this section we recall the main results of our paper [**16**] with Wayne Raskind and relate it to the main theme of this paper.

In [**16**] we defined (conditionally since some reasonable conjectures are assumed) a new type of regulator maps. It applied to varieties X over a p-adic field K with so called "totally degenerate" reduction. This means that X is smooth projective, has strictly semi-stable reduction, and the components of its reduction have all of their cohomologies generated by cycles (see [**52**, Section 2] for a precise definition). The target of the toric regulator is a "toric intermediate Jacobian" which is a p-adic torus $\mathrm{CoKer} : T^0 \to T^{-1} \otimes K^\times$, where T^* are finitely generated $\mathbb{Z}$-modules whose definition we recall shortly. Note that T^0 and T^{-1} may not be of the same rank.

Some simple examples of toric regulators include:

- The identity map $H^1_{\mathcal{M}}(K,1) = K^\times \to K^\times$;
- For E/K a Tate elliptic curve $\mathbb{G}_m/q^{\mathbb{Z}}$ the identity map

$$H^2_{\mathcal{M}}(E,1)_0 \cong E(K) \to K^\times/q^{\mathbb{Z}};$$

- More generally, for a Mumford curve X with Jacobian J, the Abel map $X(K) \to J(K)$, where $J(K)$ is viewed as a p-adic torus via its p-adic uniformization [**42**];
- The rigid analytic regulator of Pál [**49**]. For X a Mumford curve with dual graph Γ, it is a map

$$\mathrm{reg}_P : K_2(X) \to \mathcal{H}(\Gamma, K^\times)$$

the right-hand side being harmonic cochains with values in $K^\times$.

We now recall the construction of the toric regulator. The special fiber Y decomposes as a union $Y = \cup_{i=1}^n Y_i$. For a subset $I \subset \{1, \dots, n\}$ we let

$$Y_I = \bigcap_{i \in I} Y_i.$$

Fix integers $k, r \geq 0$. For each finite prime ℓ, let

$$M_\ell = H^k_{\text{ét}}(X \otimes_K \bar{K}, \mathbb{Z}_\ell).$$

THEOREM 7.1 ([**52**]). *Ignoring finite torsion and cotorsion there are finitely generated $\mathbb{Z}$-modules T^i, $i \in \mathbb{Z}$ (depending on k and r) such that for each ℓ the Galois module $M_\ell(r)$ has an increasing filtration $U^\bullet$ with $\operatorname{gr}_U M_\ell(r) = \oplus_i T^i \otimes \mathbb{Z}_\ell(-i)$.*

PROOF. See [**52**, Corollaries 1 and 6]. The proof uses the Rapoport-Zink weight spectral sequence [**50**] for $\ell \neq p$ and result by Mokrane, Hyodo and Tsuji [**34, 43, 56**] when $\ell = p$. In particular, when $\ell \neq p$, the filtration U is a reindexed version of the weight filtration induced by the Rapoport-Zink weight spectral sequence. □

Recall that the residue field of K is denoted κ. Put

- $\bar{Y}_I := Y_I \otimes \bar{\kappa}$
- $\bar{Y}^{(m)} = \bigcup_{|I|=m} \bar{Y}_I$
- $C_j^{i,k} = CH^{i+j-k}(\bar{Y}^{(2k-i+1)})$, $k \geq \max(0, i)$
- $C_j^i = \bigoplus_k C_j^{i,k}$

For every index s let $I_s = I - \{i_s\}$ and consider the obvious inclusion $\rho_s : Y_I \to Y_{I_s}$. Let

$$\theta_{i,m} = \sum_{s=1}^{m+1} (-1)^{s-1} \rho_s^* : CH^i(\bar{Y}^{(m)}) \to CH^i(\bar{Y}^{(m+1)}),$$

$$\delta_{i,m} = \sum_{s=1}^{m+1} (-1)^s \rho_{s*} : CH^i(\bar{Y}^{(m+1)}) \to CH^{i+1}(\bar{Y}^{(m)}),$$

$$d' = \bigoplus_{k \geq \max(0,i)} \theta_{i+j-k, 2k-i+1},$$

$$d'' = \bigoplus_{k \geq \max(0,i)} \theta_{i+j-k, 2k-i}.$$

Finally, let

$$d_j^i = d' + d'' : C_j^i \to C_j^{i+1},$$

set

$$T_j^i := H^i(C_j^\bullet),$$

and renumber for ease of notation:

$$T^i = T_{i+r}^{k-2r-2i}.$$

The monodromy map,

$$N : T^i \to T^{i-1}, \tag{7.1}$$

is given by "identity on identical components".

There is, in general, a non-trivial action of $\operatorname{Gal}(\overline{K}/K)$ on the T's. However, this will factor through a finite quotient and so, for simplicity, we will assume that this action is in fact trivial, which may be achieved after a finite field extension.

The quotient $M'_\ell = U^0 M_\ell(r)/U^{-2}M_\ell(r)$ sits in a short exact sequence

$$0 \to T^{-1} \otimes \mathbb{Z}_\ell(1) \to M'_\ell \to T^0 \otimes \mathbb{Z}_\ell \to 0.$$

Recall the Bloch-Kato group of geometric cohomology classes H^1_g [**18**, Section 3] which in particular satisfies

$$H^1_g(K, \mathbb{Z}_\ell(1)) \cong K^{\times(\ell)},$$

where the right-hand side is the ℓ-completion of the multiplicative group of K. It has a completed valuation map

$$\mathrm{val} : K^{\times(\ell)} \to \mathbb{Z}_\ell.$$

We have a boundary map

$$\tilde{N}_\ell : T^0 \otimes \mathbb{Z}_\ell \to H^1_g(K, T^{-1} \otimes \mathbb{Z}_\ell(1)) \cong T^{-1} \otimes K^{\times(\ell)}.$$

LEMMA 7.2 ([**16**, Corollary 1.7]). *The composition* $T^0 \otimes \mathbb{Z}_\ell \xrightarrow{\tilde{N}_\ell} T^{-1} \otimes K^{\times(\ell)} \xrightarrow{\mathrm{val}} T^{-1} \otimes \mathbb{Z}_\ell$ *equals* $N \otimes \mathbb{Z}_\ell$*, where* N *is the monodromy map* (7.1).

COROLLARY 7.3. [**16**, Corollary 1.9] *There exists a homomorphism* $T^0 \xrightarrow{\tilde{N}} T^{-1} \otimes K^\times$ *such that* $\tilde{N}_\ell = \tilde{N} \otimes \mathbb{Z}_\ell$ *for each* ℓ.

DEFINITION 7.4. [**16**, Definition 1.10] We define the toric intermediate Jacobian

$$H^{k+1}_{\mathcal{T}}(X, \mathbb{Z}(r)) := \mathrm{CoKer}\, \tilde{N}.$$

Here, and in what follows, observe a shift of indexing $k \to k+1$ which matches [**16**] but defers from Section 3.

We next define the toric regulator completed at ℓ. The étale regulator map,

$$\mathrm{reg}_\ell : H^{k+1}_{\mathcal{M}}(X, \mathbb{Z}(r))_0 \to H^1_g(K, M_\ell(r))$$

(see Remark 3.1), factors, up to possibly multiplying by some integer to avoid finite torsion and cotorsion issues, via $H^1_g(K, U^0 M_\ell(r))$ (recall that $U^\bullet$ is the filtration defined in Theorem 7.1) because $H^0(K, \mathbb{Q}_\ell(j)) = H^1_g(K, \mathbb{Q}_\ell(j)) = 0$ for $j < 0$ [**18**, Example 3.9]. Applying the projection $U^0 \to M'_\ell$ we get, again after possibly multiplying by some integer,

$$\mathrm{reg}''_\ell : H^{k+1}_{\mathcal{M}}(X, \mathbb{Z}(r))_0 \to H^1_g(K, M'_\ell) \cong \mathrm{CoKer}(T^0 \otimes \mathbb{Z}_\ell \xrightarrow{\tilde{N}_\ell} T^{-1} \otimes K^{\times(\ell)}).$$

The idea is now to construct the toric regulator in such a way that the above regulator at the prime ℓ is its ℓ-completion. For this, however, we need to recall the theory of the Sreekantan regulator [**54**]. Sreekantan's motivation was to obtain analogues of Beilinson's conjectures in the function field case. He defines a "Deligne style cohomology", $H^{k+1}_{\mathcal{D}}(X, \mathbb{Q}(r))$. We do not recall the general definition which involves similar ingredients to our definition of the lattices T^i. In fact, assuming completely degenerate reduction we have, for most ranges of indices, an isomorphism [**16**, (2.2)],

$$H^{k+1}_{\mathcal{D}}(X, \mathbb{Q}(r)) \cong \mathrm{CoKer}(T^0 \otimes \mathbb{Q} \to T^{-1} \otimes \mathbb{Q}).$$

Sreekantan defines his Deligne cohomology in the relevant range as higher Chow groups of the special fiber

$$H^{k+1}_{\mathcal{D}}(X, \mathbb{Q}(r)) \cong CH^{r-1}(Y, 2r-k-2) \otimes \mathbb{Q}.$$

Assuming standard conjectures this is equivalent to a definition using the Consani complex [**16**, Proposition 2.3]. Sreekantan defines the "regulator",

$$r_{\mathcal{D}} : H^{k+1}_{\mathcal{M}}(X, \mathbb{Z}(r)) \to H^{k+1}_{\mathcal{D}}(X, \mathbb{Q}(r)),$$

as the boundary map in higher Chow groups.

$$H^{k+1}_{\mathcal{M}}(X, \mathbb{Z}(r)) \cong CH^r(X, 2r-k-1) \xrightarrow{\partial} CH^{r-1}(Y, 2r-k-2) \otimes \mathbb{Q}.$$

In [**16**] we made the following conjectures.

CONJECTURE 1. [**16**, Conjecture 1] *For each prime ℓ the valuation of the toric regulator at ℓ is the Sreekantan regulator tensored with $\mathbb{Q}_\ell$.*

We show that this conjecture suffices to prove the existence of the toric regulator.

THEOREM 7.5. [**16**, Theorem 1.14] *Assuming Conjecture* 1, *There exists a map, called the* toric regulator,

$$H^{k+1}_{\mathcal{M}}(X, \mathbb{Z}(r)) \xrightarrow{\text{reg}_t} H^{k+1}_{\mathcal{T}}(X, \mathbb{Z}(r)), \tag{7.2}$$

giving the toric regulator at ℓ for each ℓ by completion.

The toric regulator, when it exists, is related with the syntomic regulator of Section 3. The relation is roughly that the syntomic regulator is the logarithm of the toric regulator. One reason to expect such a relation, is the case of zero cycles. In this case, the toric regulator is just the Albanese map into an appropriate p-adic uniformization of the Albanese, while the syntomic regulator is the Albanese map composed with the logarithm of the Albanese variety, which is compatible with the logarithm on the torus uniformizing the Albanese. In general, we have the following.

THEOREM 7.6 ([**16**, Theorem 4.6]). *Let X be a variety with totally degenerate reduction over K, k, r positive integers, and let T^{-1} and T^0 be as in Theorem 7.1. Let $V = H^k_{\acute{e}t}(X \otimes_K \bar{K}, \mathbb{Q}_p(r))$. Then the toric regulator at p exists, there is a canonical map*

$$H^1_{\text{st}}(K, V) \to \text{DR}(V)/(F^0 + T^0 \otimes \mathbb{Q}_p)$$

and we have a commuting diagram

$$\begin{array}{ccccc} H^{k+1}_{\mathcal{M}}(X, \mathbb{Z}(r))_0 & \xrightarrow{\text{reg}_{\text{syn}}} & H^1_{\text{st}}(K, V) & \longrightarrow & \text{DR}(V)/(F^0 + T^0 \otimes \mathbb{Q}_p) \\ \downarrow & & & & \downarrow \\ H^{k+1}_{\mathcal{T}}(X, \mathbb{Z}(r)) & & \xrightarrow{\quad\log\quad} & & T^{-1} \otimes K/T^0 \otimes \mathbb{Q}_p. \end{array}$$

An obvious expectation from the result above is that the discrete part of the syntomic regulator, which is the derivative with respect to $\log(\pi)$ of the continuous part, should be related to the Sreekantan regulator. Let us briefly indicate how this can be seen in the case of K_2 of curves. As mentioned after Theorem 3.10, the expression for the cup product of the (continuous part of the) regulator with a holomorphic form ω given there is a special case of a more general expression valid for any form of the second kind. The expression we need is found at Proposition 3.7 in [**11**] which is the same as the expression for the regulator by the main theorem [**11**, Theorem 1.2]. This expression is

$$\sum_v \langle \log(f), F_\omega; \log(g)\rangle_{(U_v-Z)^\dagger} - \sum_e c_\omega(e)\cdot \langle \log(f), \log(g)\rangle_e.$$

Here, Z is a subset containing all the singularities of f and g. The last term in the expression is the local (double) index, which was mentioned in Theorem 4.8. The first term in this expression involves the triple index, which we do not explain here. In [**16**] we pointed out that in the totally degenerate reduction case, when all the components of the reduction are projective lines, this term vanishes because all those triple indices vanish by [**13**, Proposition 8.4]. This led to a formula for the regulator, which we proved is the logarithm of the Pál regulator, consistent with Theorem 7.6. Without assuming completely degenerate reduction, it is still true that this first term is independent of the choice of branch. Recall that the formula applies to cup products with forms which are in the kernel of N. By (4.7) we see that for such forms $c_\omega(e)$ is independent of the branch. By Theorem 3.8 we see that

$$\operatorname{reg}^d_{\mathrm{syn}}(\{f,g\})\cup\omega = \sum_e c_\omega(e)\cdot \frac{\mathrm{d}}{\mathrm{d}\log(\pi)}\langle \log(f), \log(g)\rangle_e.$$

By Definition 3.7, $\operatorname{reg}^d_{\mathrm{syn}}$ lands in D^{-2}, which, in our case, is the weight 2 subspace of $H^1_{\mathrm{dR}}(X/K)$ (more precisely, it is a K_0-subspace). This subspace is mapped isomorphically by N on the weight 0 space, which is isomorphic to $\mathcal{H}(\Gamma,K)$. In addition, it is also dual to the weight 0 subspace via the cup product pairing. By [**12**, Corollary 5.3] the pairing

$$\mathcal{H}(\Gamma,K)\times\mathcal{H}(\Gamma,K)\to K,\quad (x,y)\mapsto x\cup N^{-1}y,$$

is none other than the pointwise scalar product $\sum_e x(e)\cdot y(e)$. Thus, assuming that the main result of [**11**] holds without restrictions on ω, or that the cochains c_ω suffice to cover all of $\mathcal{H}(\Gamma,K)$, as is the case for example for Mumford curves, we get the conjectural formula

$$N\operatorname{reg}^d_{\mathrm{syn}}(\{f,g\}) = \left(e\mapsto \frac{\mathrm{d}}{\mathrm{d}\log(\pi)}\langle \log(f), \log(g)\rangle_e\right).$$

Let us see how this matches with the conjecture that $\operatorname{reg}^d_{\mathrm{syn}}$ factors via the Sreekantan regulator. The map

$$K_2(X)\to\mathcal{H}(\Gamma,\mathbb{Z})$$

is derived from the boundary map in K-theory of the integral model $\mathcal{X}$, which in this case is given by tame symbols at the irreducible components of the reduction,

$$\{f,g\}\mapsto (v\mapsto h_v = t_{T_v}(f,g)),$$

followed by the map that send a collection

$$(v\mapsto h_v\in\kappa(T_v)^\times) \text{ such that } \sum_v \operatorname{div}(h_v)=0 \text{ on } \mathcal{X}$$

to

$$e\mapsto \operatorname{ord}_e\left(h_{e^+}\right),$$

where here e is viewed as the intersection point of the components T_{e^+} and T_{e^-} and the condition implies that $\operatorname{ord}_e h_{e^+} = -\operatorname{ord}_e h_{e^-}$. To see how the Sreekantan regulator is related with $\operatorname{reg}^d_{\mathrm{syn}}$ it suffices to prove the following.

LEMMA 7.7. *For any two rational functions f, g on X and any annulus e we have*

$$\frac{\mathrm{d}}{\mathrm{d}\log(\pi)}\langle\log(f),\log(g)\rangle_e = \mathrm{ord}_e\left(h_{e^+}\right).$$

PROOF. It is enough to consider the case when $\mathrm{ord}_{T_v}(f) = 1$ while $\mathrm{ord}_{T_v}(g) = 0$, with $v = e^+$. Then locally on U_v near e we can write $f = \pi f_0$ and

$$\begin{aligned}\langle\log(f),\log(g)\rangle_e &= \langle\log(f_0),\log(g)\rangle_e + \langle\log(\pi),\log(g)\rangle \\ &\langle\log(f_0),\log(g)\rangle_e + \log(\pi)\,\mathrm{res}_e\,\mathrm{d}\log(g),\end{aligned}$$

so that

$$\frac{d}{\mathrm{d}\log(p)}\langle\log(f),\log(g)\rangle_e = \mathrm{res}_e\,\mathrm{d}\log(g) = \mathrm{ord}_e(g|_{T_v}),$$

where, for this last equation it suffices to consider g which is a local equation for T_{e^-} and use [**17**, Lemma 2.1]. □

References

[1] M. Baker, *Specialization of linear systems from curves to graphs*, Algebra Number Theory **2** (2008), no. 6, 613–653, DOI 10.2140/ant.2008.2.613. With an appendix by Brian Conrad. MR2448666

[2] M. Baker and S. Norine, *Riemann-Roch and Abel-Jacobi theory on a finite graph*, Adv. Math. **215** (2007), no. 2, 766–788, DOI 10.1016/j.aim.2007.04.012. MR2355607

[3] J. S. Balakrishnan and N. Dogra, *Quadratic Chabauty and rational points, I: p-adic heights*, Duke Math. J. **167** (2018), no. 11, 1981–2038, DOI 10.1215/00127094-2018-0013. With an appendix by J. Steffen Müller. MR3843370

[4] J. S. Balakrishnan and N. Dogra, *Quadratic Chabauty and rational points II: Generalised height functions on Selmer varieties*, Int. Math. Res. Not. IMRN **15** (2021), 11923–12008, DOI 10.1093/imrn/rnz362. MR4294137

[5] J. S. Balakrishnan, A. Besser, and J. S. Müller, *Quadratic Chabauty: p-adic heights and integral points on hyperelliptic curves*, J. Reine Angew. Math. **720** (2016), 51–79, DOI 10.1515/crelle-2014-0048. MR3565969

[6] V. G. Berkovich, *Integration of one-forms on p-adic analytic spaces*, Annals of Mathematics Studies, vol. 162, Princeton University Press, Princeton, NJ, 2007, DOI 10.1515/9781400837151. MR2263704

[7] A. Besser, *Syntomic regulators and p-adic integration. II. K_2 of curves.* part B, Proceedings of the Conference on p-adic Aspects of the Theory of Automorphic Representations (Jerusalem, 1998), Israel J. Math. **120** (2000), no. part B, 335–359, DOI 10.1007/BF02834844. MR1809627

[8] A. Besser, *Coleman integration using the Tannakian formalism*, Math. Ann. **322** (2002), no. 1, 19–48, DOI 10.1007/s002080100263. MR1883387

[9] A. Besser, *p-adic Arakelov theory*, J. Number Theory **111** (2005), no. 2, 318–371, DOI 10.1016/j.jnt.2004.11.010. MR2130113

[10] A. Besser, *Heidelberg lectures on Coleman integration*, The arithmetic of fundamental groups—PIA 2010, Contrib. Math. Comput. Sci., vol. 2, Springer, Heidelberg, 2012, pp. 3–52, DOI 10.1007/978-3-642-23905-2_1. MR3220512

[11] A. Besser, *The syntomic regulator for K_2 of curves with arbitrary reduction*, Arithmetic L-functions and differential geometric methods, Progr. Math., vol. 338, Birkhäuser/Springer, Cham, [2021] ©2021, pp. 75–89, DOI 10.1007/978-3-030-65203-6_3. MR4311239

[12] A. Besser, *p-adic heights and Vologodsky integration*, J. Number Theory **239** (2022), 273–297, DOI 10.1016/j.jnt.2021.12.005. MR4434496

[13] A. Besser and R. de Jeu, *The syntomic regulator and K_4 of curves*, Pacific J. Math. **260** (2012), no. 2, 305–380, DOI 10.2140/pjm.2012.260.305. MR3001797

[14] A. Besser, J. S. Müller, and P. Srinivasan, *p-adic adelic metrics and quadratic Chabauty I*, J. Reine Angew. Math. **828** (2025), 223–305, DOI 10.1515/crelle-2025-0063. MR4979241

[15] A. Besser, J. S. Müller, and P. Srinivasan, *p-adic adelic metrics and Quadratic Chabauty II*, Work in progress, 2023.

[16] A. Besser and W. Raskind, *Toric regulators*, Arithmetic L-functions and differential geometric methods, Progr. Math., vol. 338, Birkhäuser/Springer, Cham, [2021] ©2021, pp. 91–120, DOI 10.1007/978-3-030-65203-6_4. MR4311240

[17] A. Besser and S. L. Zerbes, *Vologodsky integration on curves with semi-stable reduction*, Israel J. Math. **253** (2023), no. 2, 761–770, DOI 10.1007/s11856-022-2377-4. MR4575425

[18] S. Bloch and K. Kato, *L-functions and Tamagawa numbers of motives*, The Grothendieck Festschrift, Vol. I, Progr. Math., vol. 86, Birkhäuser Boston, Boston, MA, 1990, pp. 333–400. MR1086888

[19] G. S. Call and J. H. Silverman, *Canonical heights on varieties with morphisms*, Compositio Math. **89** (1993), no. 2, 163–205. MR1255693

[20] K. T. Chen, *Iterated path integrals*, Bull. Amer. Math. Soc. **83** (1977), no. 5, 831–879, DOI 10.1090/S0002-9904-1977-14320-6. MR454968

[21] B. Chiarellotto, V. Di Proietto, and A. Shiho, *Comparison of relatively unipotent log de Rham fundamental groups*, Mem. Amer. Math. Soc. **288** (2023), no. 1430, v+111, DOI 10.1090/memo/1430. MR4627087

[22] R. F. Coleman, *Torsion points on curves and p-adic abelian integrals*, Ann. of Math. (2) **121** (1985), no. 1, 111–168, DOI 10.2307/1971194. MR782557

[23] R. Coleman and E. de Shalit, *p-adic regulators on curves and special values of p-adic L-functions*, Invent. Math. **93** (1988), no. 2, 239–266, DOI 10.1007/BF01394332. MR948100

[24] R. F. Coleman and B. H. Gross, *p-adic heights on curves*, Algebraic number theory, Adv. Stud. Pure Math., vol. 17, Academic Press, Boston, MA, 1989, pp. 73–81, DOI 10.2969/aspm/01710073. MR1097610

[25] R. Coleman and A. Iovita, *The Frobenius and monodromy operators for curves and abelian varieties*, Duke Math. J. **97** (1999), no. 1, 171–215, DOI 10.1215/S0012-7094-99-09708-9. MR1682268

[26] R. F. Coleman, *Dilogarithms, regulators and p-adic L-functions*, Invent. Math. **69** (1982), no. 2, 171–208, DOI 10.1007/BF01399500. MR674400

[27] R. F. Coleman, *Reciprocity laws on curves*, Compositio Math. **72** (1989), no. 2, 205–235. MR1030142

[28] P. Colmez, *Intégration sur les variétés p-adiques* (French, with English and French summaries), Astérisque **248** (1998), viii+155. MR1645429

[29] F. Déglise and W. Nizioł, *On p-adic absolute Hodge cohomology and syntomic coefficients. I*, Comment. Math. Helv. **93** (2018), no. 1, 71–131, DOI 10.4171/CMH/430. MR3777126

[30] E. Grosse-Klönne, *Rigid analytic spaces with overconvergent structure sheaf*, J. Reine Angew. Math. **519** (2000), 73–95, DOI 10.1515/crll.2000.018. MR1739729

[31] M. Hadian, *Motivic fundamental groups and integral points*, Duke Math. J. **160** (2011), no. 3, 503–565, DOI 10.1215/00127094-1444296. MR2852368

[32] R. M. Hain, *The geometry of the mixed Hodge structure on the fundamental group*, Algebraic geometry, Bowdoin, 1985 (Brunswick, Maine, 1985), Proc. Sympos. Pure Math., vol. 46, Amer. Math. Soc., Providence, RI, 1987, pp. 247–282, DOI 10.1090/pspum/046.2/927984. MR927984

[33] R. M. Hain, *Higher Albanese manifolds*, Hodge theory (Sant Cugat, 1985), Lecture Notes in Math., vol. 1246, Springer, Berlin, 1987, pp. 84–91, DOI 10.1007/BFb0077531. MR894044

[34] O. Hyodo, *A note on p-adic étale cohomology in the semistable reduction case*, Invent. Math. **91** (1988), no. 3, 543–557, DOI 10.1007/BF01388786. MR928497

[35] U. Jannsen, *Continuous étale cohomology*, Math. Ann. **280** (1988), no. 2, 207–245, DOI 10.1007/BF01456052. MR929536

[36] E. Katz and D. Litt, *p-adic iterated integration on semi-stable curves*, Preprint, `arXiv:2202.05340`, 2022.

[37] E. Katz, J. Rabinoff, and D. Zureick-Brown, *Uniform bounds for the number of rational points on curves of small Mordell-Weil rank*, Duke Math. J. **165** (2016), no. 16, 3189–3240, DOI 10.1215/00127094-3673558. MR3566201

[38] M. Kim, *The motivic fundamental group of* $\mathbf{P}^1\backslash\{0,1,\infty\}$ *and the theorem of Siegel*, Invent. Math. **161** (2005), no. 3, 629–656, DOI 10.1007/s00222-004-0433-9. MR2181717

[39] M. Kim, *The unipotent Albanese map and Selmer varieties for curves*, Publ. Res. Inst. Math. Sci. **45** (2009), no. 1, 89–133, DOI 10.2977/prims/1234361156. MR2512779

[40] M. Kim and R. M. Hain, *A de Rham-Witt approach to crystalline rational homotopy theory*, Compos. Math. **140** (2004), no. 5, 1245–1276, DOI 10.1112/S0010437X04000442. MR2081156

[41] M. Kim and R. M. Hain, *The Hyodo-Kato theorem for rational homotopy types*, Math. Res. Lett. **12** (2005), no. 2-3, 155–169, DOI 10.4310/MRL.2005.v12.n2.a2. MR2150873

[42] Yu. Manin and V. Drinfeld, *Periods of p-adic Schottky groups*, J. Reine Angew. Math. **262(263)** (1973), 239–247, DOI 10.1515/crll.1973.262-263.239. MR396582

[43] A. Mokrane, *La suite spectrale des poids en cohomologie de Hyodo-Kato* (French), Duke Math. J. **72** (1993), no. 2, 301–337, DOI 10.1215/S0012-7094-93-07211-0. MR1248675

[44] P. Monsky and G. Washnitzer, *Formal cohomology. I*, Ann. of Math. (2) **88** (1968), 181–217, DOI 10.2307/1970571. MR248141

[45] J. W. Morgan, *The algebraic topology of smooth algebraic varieties*, Inst. Hautes Études Sci. Publ. Math. **48** (1978), 137–204. MR516917

[46] J. P. Murre, *Lectures on an introduction to Grothendieck's theory of the fundamental group*, Tata Institute of Fundamental Research Lectures on Mathematics, No. 40, Tata Institute of Fundamental Research, Bombay, 1967. Notes by S. Anantharaman. MR302650

[47] J. Nekovář, *On p-adic height pairings*, Séminaire de Théorie des Nombres, Paris, 1990–91, Progr. Math., vol. 108, Birkhäuser Boston, Boston, MA, 1993, pp. 127–202, DOI 10.1007/s10107-005-0696-y. MR1263527

[48] J. Nekovář and W. Nizioł, *Syntomic cohomology and p-adic regulators for varieties over p-adic fields*, Algebra Number Theory **10** (2016), no. 8, 1695–1790, DOI 10.2140/ant.2016.10.1695. With appendices by Laurent Berger and Frédéric Déglise. MR3556797

[49] A. Pál, *A rigid analytical regulator for the K_2 of Mumford curves*, Publ. Res. Inst. Math. Sci. **46** (2010), no. 2, 219–253, DOI 10.2977/PRIMS/8. MR2722778

[50] M. Rapoport and Th. Zink, *Über die lokale Zetafunktion von Shimuravarietäten. Monodromiefiltration und verschwindende Zyklen in ungleicher Charakteristik* (German), Invent. Math. **68** (1982), no. 1, 21–101, DOI 10.1007/BF01394268. MR666636

[51] W. Raskind, *Higher l-adic Abel-Jacobi mappings and filtrations on Chow groups*, Duke Math. J. **78** (1995), no. 1, 33–57, DOI 10.1215/S0012-7094-95-07803-X. MR1328751

[52] W. Raskind and X. Xarles, *On the étale cohomology of algebraic varieties with totally degenerate reduction over p-adic fields*, J. Math. Sci. Univ. Tokyo **14** (2007), no. 2, 261–291. MR2351367

[53] A. Shiho, *Crystalline fundamental groups and p-adic Hodge theory*, The arithmetic and geometry of algebraic cycles (Banff, AB, 1998), CRM Proc. Lecture Notes, vol. 24, Amer. Math. Soc., Providence, RI, 2000, pp. 381–398, DOI 10.1090/crmp/024/19. MR1738868

[54] R. Sreekantan, *Non-Archimedean regulator maps and special values of L-functions*, Cycles, motives and Shimura varieties, Tata Inst. Fund. Res. Stud. Math., vol. 21, Tata Inst. Fund. Res., Mumbai, 2010, pp. 469–492. MR2906033

[55] M. Stoll, *Uniform bounds for the number of rational points on hyperelliptic curves of small Mordell-Weil rank*, J. Eur. Math. Soc. (JEMS) **21** (2019), no. 3, 923–956, DOI 10.4171/JEMS/857. MR3908770

[56] T. Tsuji, *p-adic étale cohomology and crystalline cohomology in the semi-stable reduction case*, Invent. Math. **137** (1999), no. 2, 233–411, DOI 10.1007/s002220050330. MR1705837

[57] V. Vologodsky, *Hodge structure on the fundamental group and its application to p-adic integration* (English, with English and Russian summaries), Mosc. Math. J. **3** (2003), no. 1, 205–247, 260, DOI 10.17323/1609-4514-2003-3-1-205-247. MR1996809

[58] Z. Wojtkowiak, *Cosimplicial objects in algebraic geometry*, Algebraic K-theory and algebraic topology (Lake Louise, AB, 1991), NATO Adv. Sci. Inst. Ser. C: Math. Phys. Sci., vol. 407, Kluwer Acad. Publ., Dordrecht, 1993, pp. 287–327, DOI 10.1007/978-94-017-0695-7_15. MR1367304

[59] Yu. G. Zarhin, *p-adic abelian integrals and commutative Lie groups*, J. Math. Sci. **81** (1996), no. 3, 2744–2750, DOI 10.1007/BF02362339. Algebraic geometry, 4. MR1420227

[60] S. Zhang, *Small points and adelic metrics*, J. Algebraic Geom. **4** (1995), no. 2, 281–300. MR1311351

Ben-Gurion University of the Negev
Email address: bessera@math.bgu.ac.il

Contemporary Mathematics
Volume **842**, 2026
https://doi.org/10.1090/conm/842/16850

A motivic approach to rational p-adic cohomologies

Federico Binda and Alberto Vezzani

ABSTRACT. We survey over some recent applications of motivic homotopy theory in the definition and the study of p-adic cohomology theories. In particular, we revisit the proof of the p-adic weight-monodromy conjecture for smooth projective hypersurfaces in light of the motivic definition of nearby cycles and monodromy operators.

1. Introduction

1.1. From the special fiber to the generic fiber. One of the first motivations for the introduction of analytic geometry over non-archimedean fields is the attempt of defining a p-adic Weil cohomology theory (as opposed to the ℓ-adic étale cohomology with $\ell \neq p$) for varieties over a finite field k of characterstic p, and the study of the associated zeta functions.

Following ideas of Dwork, Monsky and Washnitzer [**47**] introduced a way to attach to an affine, smooth variety over k a vector space $H^i_{MW}(X/K)$ over $K = \operatorname{Frac} W(k)$. This can be thought as an "overconvergent" de Rham cohomology of a (dagger rigid analytic) generic fiber of a weakly complete smooth lift of X over $W(k)$.

We may interpret Monsky-Washnitzer's plan was as a way to define a "functor"

$$\operatorname{Sm}/k \rightsquigarrow \operatorname{FSm}/\mathcal{O}_K \rightsquigarrow \operatorname{RigSm}/K \rightsquigarrow \operatorname{RigSm}^\dagger/K \tag{1.1}$$

with K being the p-adic non-archimedean field $W(k)[1/p]$, and where Sm [resp. FSm resp. RigSm resp. RigSm†] stands for the category of smooth varieties [resp. formal schemes, resp. rigid analytic variety, resp. dagger varieties] over the the base.

Nonetheless, of all the arrows sketched above, only the middle one is an actual functor. As a matter of fact, the existence of some formal lift is only granted locally, for affine varieties X. It is not even enough to work locally (that is, considering the Zariski topos): indeed, even if affine smooth lifts are unique up to isomorphism, their "weak completions" depend on a choice of coordinates for X. The same ambiguity exists when trying to lift maps between affine smooth varieties.

It was only thanks to Berthelot's introduction of the overconvergent site [**13**] that Monsky–Washnitzer cohomology could be turned into a well-defined functorial cohomology theory, extendable to arbitrary varieties over k and with the usual

The authors are partially supported by the PRIN 20222B24AY "The arithmetic of motives and L-functions".

properties of a Weil cohomology theory (a Künneth formula, finite dimensionality etc.).

We propose an alternative way to "straighten up" the squiggly arrows of (1.1) based on motivic homotopy theory. Indeed, by results of Elkik [**28**] and Arabia [**3**], we know that potential lifts of maps are unique, *up to homotopy.* Such results suggest that one should consider a more flexible environment, in which not only can one work Zariski-locally (or even étale locally) but also *up to homotopy.* More or less by definition, this environment is the so-called category of (derived, étale, rational) motives that we will denote by DA or RigDA or $\mathrm{RigDA}^\dagger$ in the algebraic resp. analytic resp. dagger setting.

The aim of Section 3, in which we collect results from [**29, 44, 59**] is to sketch how one can prove the following results.

PROPOSITION 1.2. *There is a functor*

$$\xi\colon \mathrm{Sm}/k \to \mathrm{DA}(k) \to \mathrm{RigDA}^\dagger(K)$$

Whenever X/k admits a weak formal smooth lift $\mathfrak{X}^\dagger$ over $W(k)$, then $\xi(X) \simeq \mathsf{M}(\mathfrak{X}^\dagger_K)$.

We can use this functor to define Monsky–Washnitzer/rigid cohomology and prove even new features of them. In what follows, we let Fr/S be the category of proper frames over S in the sense of Berthelot.

THEOREM 1.3. *Let S be a scheme over k. There is a functor*

$$\xi\colon \mathrm{Sm}/S \to \mathrm{DA}(S) \to \lim_{(T,\bar{T},\mathfrak{P})\in \mathrm{Fr}/S} \mathrm{RigDA}^\dagger(]T[^\dagger_{\mathfrak{P}})$$

which, composed to an overconvergent relative de Rham functor gives rise to a cohomology theory:

$$R\Gamma_{\mathrm{rig}}\colon :\ \mathrm{Sm}/S \to \mathrm{DA}(S) \to \lim_{(T,\bar{T},\mathfrak{P})\in \mathrm{Fr}/S} \mathrm{QCoh}(]T[^\dagger_{\mathfrak{P}})^{\mathrm{op}}.$$

computing (relative) rigid cohomology $R\Gamma_{\mathrm{rig}}(X/(T,\bar{T},\mathfrak{P}))$.

In particular, if $S = \operatorname{Spec} k$ we get an absolute Monsky–Washnitzer realization:

$$R\Gamma_{\mathrm{rig}}\colon \mathrm{Sm}/k \to \mathcal{D}(K)^{\mathrm{op}}.$$

COROLLARY 1.4. *Let S be a scheme over k.*

(1) *If $S = \operatorname{Spec} k$, then (absolute) Monsky–Washnitzer cohomology is well-defined, functorial, has étale and h-descent, and can be extended to arbitrary varieties by putting $R\Gamma_{\mathrm{rig}}(X/K) = R\Gamma_{\mathrm{rig}}(\mathsf{M}_k(X))$. It has the structure of a φ-module, and is finite dimensional whenever X is a smooth quasi-compact rigid variety, or the analytification of an algebraic variety.*
(2) *Relative rigid cohomology has étale descent and h-descent. If X/S is smooth and proper, then $R^i\Gamma_{\mathrm{rig}}(X/(T,\bar{T},\mathfrak{P}))$ is a vector bundle over $]T[^\dagger_{\mathfrak{P}}$ and is equipped with a canonical structure of an overconvergent F-isocrystal.*

We note that point (1) was essentially established by Berthelot and Chiarellotto–Tsuzuki, so that motivic methods help re-visit classical resuls. On the other hand, point (2) answers positively to an open conjecture by Berthelot which was previously known only in the (log-)liftable case, and for a special class of frames [**13, 55, 57**].

1.2. From generic fibers to special fibers. One of the classical ways to define an analytic variety over a non-archimedean field arises when one has a generically smooth family X_t of varieties over a curve C over a field k of characteristic 0. When restricting around an infinitesimal "punctured neighborhood" of a point $0 \in C$, one ends up with a formal model of a smooth rigid analytic variety over $k((t))$, having a (potentially singular) special fiber X_0 over k. By standard resolution of singularities, one may even suppose that such special fiber is semi-stable (at least after an extension of scalars).

By classic work by Clemens [**21**], Schmid [**53**], Steenbrink [**56**], Saito [**52**] et al. when $k = \mathbb{C}$, the limit Hodge cohomology of the elements of the family $\lim_{t\to 0} H(X_t)$ can be expressed in terms of a nearby cycle functor ΨX, introduced either via passing to a universal covering and using the functor $\iota^* j_*$, or as an explicit bi-complex of Hodge modules depending on the geometry of X_0. It comes equipped with a monodromy operator.

In the arithmetic situation, by replacing the punctured complex disc and the non-archimedean field $\mathbb{C}((t))$ with a local field K, this last procedure has been translated to the study of ℓ-adic cohomology by Rapoport–Zink [**51**], interpreting the Weil-Deligne cohomology of the generic fiber in terms of ℓ-adic cohomology of a bi-complex ΨX, equipped with a monodromy operator. Similarly, this procedure was adapted to the p-adic case, with the introduction of a so called Hyodo–Kato cohomology, by work of Hyodo–Kato [**37**], Mokrane [**46**], Nakkajima [**49**], Große-Klönne [**34**], Colmez–Niziol [**22**] and Ertl–Yamada [**30**]. Once again, Hyodo–Kato cohomology of the (rigid analytic) generic fiber can be described as rigid cohomology of a bi-complex ΨX, and comes equipped with a monodromy operator.

In all three circumstances, the strategy is then to consider a "nearby cycle functor"

$$\Psi\colon \operatorname{RigSm}/K \rightsquigarrow \operatorname{Sm}/k + \text{monodromy}$$

and then define/study a "limit" cohomology for X/K via a cohomology theory for ΨX. Once again, this functor is not really well-defined, as typically the nearby cycle functor is given by some bi-complex of objects in the category of *coefficients* (Hodge modules, ℓ-adic sheaves or φ-modules) albeit defined using the geometry of the special fiber.

Also in this setting, working with a motivic category allows one to straighten up this arrow again into a well-defined functor, that we will discuss in Section 4 based on results of [**14, 15**]:

THEOREM 1.5. *Let C be an algebraically closed non-archimedean field, with valuation group equal to $\mathbb{Q}$ and residue field k. There is an equivalence of categories*

$$\operatorname{RigDA}(C) \simeq \operatorname{DA}(k)_N$$

where the category on the right hand side is the category of locally nilpotent "monodromy maps" $M \xrightarrow{N} M(-1)$ with M in $\operatorname{DA}(k)$.

In particular, if X is a quasi-compact rigid variety or an algebraic variety over a local field K, then up to replacing K with a finite extension, X is canonically associated to a motive $\Psi X \in \operatorname{DA}(k)$ equipped with a monodromy operator.

This provides a geometric (yet, motivic) universal way to perceive and define the various functors Ψ on each realization, and even prove new facts about the associated "limit cohomologies".

COROLLARY 1.6. *Hyodo–Kato cohomology for rigid analytic varieties over C can be defined as $R\Gamma_{\mathrm{rig}}(\Psi X)$ without recourse to log-geometry. It is canonically equipped with a weight filtration and fits in a Clemens-Schmid chain complex.*

Once again, while the first point of the previous corollary admits some variations which were already known, the second point is new, and answers positively to a conjecture by Flach–Morin [**31**].

1.3. On the p-adic weight-monodromy conjecture. As a matter of fact, this "motivic" intepretation of the Hyodo–Kato cohomology also sheds some new light on the proof of the p-adic weight monodromy conjecture for smooth projective hypersurfaces, obtained in [**15**] based on Scholze's strategy [**54**]. The most precise statement goes as follows:

THEOREM 1.7 ([**15**]). *Let K be a finite extension of $\mathbb{Q}_p$ and Y be a smooth scheme-theoretic complete intersection inside a projective smooth toric variety over K. Then the p-adic weight-monodromy conjecture holds for Y.*

Indeed, one of the technical ingredients of the main proof of [**15**], is the compatibility of the motivic tilting equivalence $\mathrm{RigDA}(C) \simeq \mathrm{RigDA}(C^\flat)$ from [**60**] with the Hyodo–Kato realization functors based on log-geometry. We will show that this check can alternatively be done based on the identification of Equation (1.5) and some detection principles developed in [**14**].

We will revisit the strategy of the above theorem in Section 4.4, by making some generalizations of certain intermediate results in light of the equivalence of Equation (1.5). As such, this section contains some original results (see the proof of Equation (4.32)).

2. Properties of motivic categories

In this section, we list all the "formal" properties of motivic categories that we use for our explicit, cohomological applications later on. We will survey them based on [**9, 10, 14, 44**].

2.1. Basic definitions. Motivic sheaves are defined by means of a universal property, that we now recall.

DEFINITION 2.1. We will consider the following ∞-categories of motivic sheaves.

(1) Let S be a scheme locally of finite Krull dimension. We write $\mathrm{DA}(S) = \mathrm{DA}_{\text{ét}}(S;\mathbb{Q})$ for the ∞-category of hypercomplete étale motives over S, with rational coefficients as defined in [**6**, Section 3] by putting $\Lambda = \mathbb{Q}$. It is constructed from the full sub-∞-category $\mathrm{DA}^{\mathrm{eff}}_{\text{ét}}(S;\mathbb{Q})$ of
$$\mathrm{Shv}_{\text{ét}}(\mathrm{Sm}/S;\mathcal{D}(\mathbb{Q}))$$
spanned by those objects which are local with respect to the collection of maps $\mathbb{Q}(\mathbb{A}^1_Y) \to \mathbb{Q}(Y)$, by formally inverting the tensor product operation with the Tate twist $\mathbb{1}(1)$.

(2) Let S be a rigid analytic space locally of finite Krull dimension, in the sense of [**32**, Definition II.2.2.18]. We write $\mathrm{RigDA}(S) = \mathrm{RigDA}_{\text{ét}}(S;\mathbb{Q})$ for the ∞-category of hypercomplete rigid analytic motives over S with respect to the étale topology, as defined in [**10**, Definition 2.1.15]. Its construction is analogous to (1), by replacing Sm/S with RigSm/S, and the affine line $\mathbb{A}^1$ by the closed unit disk $\mathbb{B}^1$.

These are stable, compactly generated monoidal ∞-categories, underlying a six-functor formalism, see [**4**, **5**, **10**].

REMARK 2.2. In the setting of (2), we will mostly be interested in the case where $S = \mathrm{Spa}(K, \mathcal{O}_K) = \mathrm{Spf}(\mathcal{O}_K)^{\mathrm{rig}}$, for K a complete non-archimedean field with a perfect residue field. Here, for a quasi-compact and quasi-separated formal scheme $\mathfrak{X}$, we write $\mathfrak{X}^{\mathrm{rig}}$ or $\mathfrak{X}_\eta$ for the Raynaud generic fiber. We will simply write $\mathrm{RigDA}(K)$ for $\mathrm{RigDA}(\mathrm{Spf}(\mathcal{O}_K)^{\mathrm{rig}})$. Note that in this case by [**10**, Lemma 2.4.18] the non-hypercomplete version of the construction of RigDA agrees with the one considered above, and the reader can simply ignore the difference between the two.

2.2. Weil cohomology theories in algebraic and analytic geometry. We follow the exposition of Ayoub [**9**, §1], and quickly recall the notion of mixed Weil cohomology theory, as introduced by Cisinski and Déglise [**19**, §2], specialized to our setting of $\mathbb{Q}$-linear stable ∞-categories.

We begin by recalling the following basic definition.

DEFINITION 2.3. Let k be a field (resp. let K be a complete non-archimedean local field). A monoidal *realization functor* or simply a realization is a morphism

$$R\Gamma^*\colon \mathrm{DA}(k) \to \mathscr{D}^{\otimes}$$

(resp., $R\Gamma^*\colon \mathrm{RigDA}(K) \to \mathscr{D}^{\otimes}$) in $\mathrm{CAlg}(\mathrm{Pr}^L)$, where $\mathscr{D}^{\otimes}$ is presentable monoidal stable ∞-category. We will mostly be interested in the case where

$$\mathscr{D}^{\otimes} \in \mathrm{Mod}_{\mathrm{Mod}_{\mathbb{Q}}}(\mathrm{Pr}^L)$$

is $\mathbb{Q}$-linear. A "classical" example is given by $\mathscr{D}^{\otimes} = \mathrm{Mod}_\Lambda$, for $\Lambda \in \mathrm{CAlg}(\mathrm{Mod}_{\mathbb{Q}})^{\mathrm{cn}}$ (i.e., a commutative connective dg $\mathbb{Q}$-algebra).

REMARK 2.4. To give a realization functor for $\mathrm{DA}(k)$ is thus equivalent to give a symmetric monoidal functor $\mathrm{Sm}/k \to \mathscr{D}^{\otimes}$ satisfying étale descent, $\mathbb{A}^1$-invariance and Tate stability. Similarly, to give a realization functor for $\mathrm{RigDA}(K)$ is equivalent to give a symmetric monoidal functor $\mathrm{RigSm}/K \to \mathscr{D}^{\otimes}$ satisfying rig-étale descent, $\mathbb{B}^1$-invariance and Tate stability.

Realizations that are Mod_Λ-valued are classically introduced in terms of cohomology theories for algebraic (resp. analytic) varieties in the following sense.

DEFINITION 2.5 ([**9**, Definition 1.1 and 2.1]). Let k be a field (resp. let K be a complete non-archimedean local field), and let $\Lambda \in \mathrm{CAlg}(\mathrm{Mod}_{\mathbb{Q}})^{\mathrm{cn}}$.

A *mixed Weil cohomology theory* for algebraic k-varieties (resp. for rigid analytic K-varieties) is a presheaf Γ_W of commutative Λ-algebras on Sm/k (resp. on RigSm/K) satisfying the following properties:

(1) For any $X \in \mathrm{Sm}/k$ (resp. $X \in \mathrm{RigSm}/K$) the natural morphism
$$\Gamma_W(X) \to \Gamma_W(X \times \mathbb{A}^1)(\text{resp. } \Gamma_W(X) \to \Gamma_W(X \times \mathbb{B}^1))$$
is an equivalence;

(2) The $\Gamma_W(\mathrm{pt})$-module $\mathrm{cofib}(\Gamma_W(\mathrm{pt}) \to \Gamma_W(\mathbb{P}^1))$ is invertible;

(3) For every $X, Y \in \mathrm{Sm}/k$ (resp. for every $X, Y \in \mathrm{RigSm}/K^{\mathrm{qcqs}}$) there is an equivalence
$$\Gamma_W(X) \otimes_{\Gamma_W(\mathrm{pt})} \Gamma_W(Y) \simeq \Gamma_W(X \times Y);$$

(4) $\Gamma_W(-)$ has étale descent.

Note that Γ_W is automatically an object of $\mathrm{DA}^{\mathrm{eff}}(k; \Lambda)$ (resp. of $\mathrm{RigDA}^{\mathrm{eff}}(K; \Lambda)$).

REMARK 2.6. Any mixed Weil cohomology theory for algebraic or for analytic varieties gives rise naturally to a commutative algebra object $\mathbf{\Gamma}_W$ in $\mathrm{DA}(k;\Lambda)$ (resp. in $\mathrm{RigDA}(K;\Lambda)$), satisfying some extra assumptions. See [**9**, Proposition 1.9, 2.5] and [**19**, 2.1.5]. In fact, the categories of *Weil spectra* and that of Λ-valued Weil cohomology theories [**9**, Definition 1.2] are equivalent. See [**9**, Theorems 1.16, 2.12]

REMARK 2.7. Given any monoidal realization functor $R\Gamma^*$ as in Equation (2.3), write $R\Gamma_*$ for its right adjoint. We denote $R\Gamma_*(\mathbb{1})$ by E_Γ. Since $R\Gamma^*$ is monoidal, $R\Gamma_*$ is lax monoidal. In particular, it carries commutative algebra objects to commutative algebra objects, so that $E_\Gamma \in \mathrm{CAlg}(\mathrm{DA}(k))$ (resp. in $\mathrm{CAlg}(\mathrm{RigDA}(K))$).

The assignment $R\Gamma^* \mapsto E_\Gamma$ determines in fact an equivalence of categories between the category of algebraic (resp. analytic) Mod_Λ-valued realization functors and the category of mixed Weil spectra in $\mathrm{DA}(k)$ (resp. in $\mathrm{RigDA}(K)$). See [**9**, Theorems 1.21, 2.16].

We list a series of examples of classical realization functors. In what follows, we keep the convention that k is a field and K is a complete non-archimedean local field.

EXAMPLE 2.8 (ℓ-adic realizations). Let ℓ be a prime invertible in k (resp. in K). The classical ℓ-adic étale cohomology is an example of mixed Weil cohomology theory: for every $X \in \mathrm{Sm}/k$ (resp. $X \in \mathrm{RigSm}/K$), write $R_\ell(X) \in \mathcal{D}_{\text{proét}}(k, \mathbb{Q}_\ell)$ for the object computing the ℓ-adic étale homology of X, see e.g. [**6**, Théorème 6.9] in the algebraic context. The assignment $X \mapsto R_\ell(X)$ satisfies all the assumptions of Equation (2.5) and gives rise to a monoidal functor $\mathrm{DA}(k) \to \mathcal{D}_{\text{proét}}(k, \mathbb{Q}_\ell)$ (resp. $\mathrm{RigDA}(K) \to \mathcal{D}_{\text{proét}}(k, \mathbb{Q}_\ell)$).

One can describe more accurately the ℓ-adic étale realization (both in the algebraic and in the analytic setting), and over a general base, by means of the so-called rigidity theorem: in this language, the étale realization arises as the left adjoint to the canonical functor induced by the inclusion of the small étale site into the big étale site. See [**10**, §2.10] and [**11**, Theorem 3.2] in the analytic setting and the already mentioned [**6**] in the algebraic setting for the compatibility of the ℓ-adic realization with the six-functor formalism.

EXAMPLE 2.9 (Betti and Hodge realization). Let $k = \mathbb{C}$. For a finite type $\mathbb{C}$-scheme X, let X^{an} be the set of $\mathbb{C}$-points $X(\mathbb{C})$, equipped with the classical complex analytic topology. The functor $X \mapsto X^{\mathrm{an}}$ can be promoted to a symmetric monoidal functor $\mathsf{Bti}\colon \mathrm{DA}(\mathbb{C}) \to \mathrm{DA}^{\mathrm{an}}(\mathbb{C}) \simeq \mathrm{Mod}_{\mathbb{Q}}$, where $\mathrm{DA}^{\mathrm{an}}(\mathbb{C})$ is the (stabilization of the) category of $\mathbb{D}^1_{\mathbb{C}}$-local sheaves on $\mathbb{C}$-smooth analytic spaces. By [**7**, §2] it is equivalent to the classical derived ∞-category of $\mathbb{Q}$-modules. For any smooth $\mathbb{C}$-scheme X, the homology groups $H_*(\mathsf{Bti}(\mathsf{M}(X)))$ compute the singular homology of $X(\mathbb{C})$.

The Betti realization can be further enriched to obtain a mixed Hodge realization. Let $\mathrm{MHS}^p_{\mathbb{Q}}$ be the Abelian category of polarisable mixed Hodge structures and let $\mathcal{D}^b(\mathrm{MHS}^p_{\mathbb{Q}})$ be its bounded derived stable ∞-category. The presheaf defined by sending $X \in \mathrm{Sm}/\mathbb{C}$ to the object $R\Gamma^*_{\mathrm{Hodge}}(X) \in \mathcal{D}^b(\mathrm{MHS}^p_{\mathbb{Q}})$ computing the singular cohomology of X equipped with its natural mixed Hodge structure is a mixed Weil cohomology theory. As such, it can be promoted to a symmetric monoidal (cohomological) realization functor

$$\mathrm{DA}(\mathbb{C}) \to \mathcal{D}^b(\mathrm{MHS}^p_{\mathbb{Q}})^{\mathrm{op}}.$$

Restricting to the subcategory of compact objects $\mathrm{DA}(\mathbb{C})_\omega$ and pre-composing with the duality functor $M \mapsto M^\vee$ gives the homological Hodge realization functor $\mathrm{DA}(\mathbb{C})_\omega \to \mathcal{D}^b(\mathrm{MHS}^p_{\mathbb{Q}})$. Since $\mathrm{DA}(\mathbb{C}) \simeq \mathrm{Ind}(\mathrm{DA}(\mathbb{C})_\omega)$, this can be further extended to a symmetric monoidal functor $\mathrm{DA}(\mathbb{C}) \to \mathrm{Ind}\,\mathcal{D}^b(\mathrm{MHS}^p_{\mathbb{Q}})$ taking values in the Ind-derived ∞-category. As for the ℓ-adic realization, it turns out that also the Hodge realization can be defined in the relative setting, in a way that is compatible with the six-functor formalism. See [**58**, Theorem 4.4].

EXAMPLE 2.10 (de Rham realization). Let k be a field of characteristic zero. The presheaf $X \mapsto R\Gamma(X, \Omega_{X/k})$ is another example of mixed Weil cohomology, giving rise to a monoidal (cohomological) realization functor $\mathrm{dR}\colon \mathrm{DA}(k) \to \mathcal{D}(k)^{\mathrm{op}}$. We denote by $R\Gamma_{\mathrm{dR},k}$ the covariant realization functor obtain by pre-composing the restriction of dR to the subcategory of compact objects with the duality functor, as in Equation (2.9), further extended to a unique monoidal, compact-preserving, colimit-preserving functor $R\Gamma_{\mathrm{dR},k}\colon \mathrm{DA}(k) \to \mathcal{D}(k)$.

2.3. The six functor formalism. For any morphism $f\colon S \to T$ of schemes (resp. of rigid analytic spaces), we have natural adjunctions

$$\mathrm{DA}(T) \underset{f_*}{\overset{f^*}{\rightleftarrows}} \mathrm{DA}(S) \quad (\text{resp. } \mathrm{RigDA}(T) \underset{f_*}{\overset{f^*}{\rightleftarrows}} \mathrm{RigDA}(S)\ )$$

In fact, the assignment $\mathrm{Sch} \ni S \mapsto \mathrm{DA}(S)$ (resp. $\mathrm{RigSpc} \ni S \mapsto \mathrm{RigDA}(S)$) naturally extends into a functor

$$\mathrm{DA}(-)\colon \mathrm{Sch}^{\mathrm{op}} \longrightarrow \mathrm{CAlg}(\mathrm{Pr}^L) \quad (\text{resp. } \mathrm{RigDA}(-)\colon \mathrm{RigSpc}^{\mathrm{op}} \longrightarrow \mathrm{CAlg}(\mathrm{Pr}^L))$$

More is true: the presheaves $\mathrm{DA}(-)$ and $\mathrm{RigDA}(-)$ underlie a six-functor formalism, in the following sense. For any morphism $f\colon S \to T$ of schemes (resp. of rigid analytic spaces) locally of finite type there are adjunctions

$$\mathrm{DA}(S) \underset{f^!}{\overset{f_!}{\rightleftarrows}} \mathrm{DA}(T) \quad (\text{resp. } \mathrm{RigDA}(S) \underset{f^!}{\overset{f_!}{\rightleftarrows}} \mathrm{RigDA}(T)\ )$$

satisfying the following list of properties (see [**4**, **5**, **10**, **20**]:

(i) (Functoriality). If f and g are composable morphisms, then $(g \circ f)^* \simeq f^* \circ g^*$ and $(g \circ f)_* \simeq g_* \circ f_*$. If both f and g are locally of finite type, then $(g \circ f)_! \simeq g_! \circ f_!$ and $(g \circ f)^! \simeq f^! \circ g^!$.

(ii) (Proper maps). There is a natural transformation $f_! \to f_*$. If f is proper, then $f_* \simeq f_!$.

(iii) (Smooth morphisms, Relative purity). If f is smooth, the functor f^* admits a left adjoint $f_\sharp$. If f is smooth of relative dimension d, then $f^! \simeq f^*(-)(d)[2d]$ and $f_! \simeq f_\sharp(-)(-d)[-2d]$. In particular, if f is étale, then $f^* \simeq f^!$.

(iv) (Absolute purity). If $\iota\colon S \to T$ is a regular closed immersion of regular schemes of pure codimension d, then $\iota^! \Bbbk_T \simeq \Bbbk_S(-d)[-2d]$.

(v) (Localization). Let $\iota\colon Z \to X$ be a closed immersion of schemes (resp. a closed immersion of rigid analytic spaces [**10**, Definition 1.1.14]) and let

$j\colon U = X - Z \to X$ be the open complement. Then there are fiber sequences of endofunctors

$$j_\sharp j^* \to \mathrm{id}_X \to \iota_*\iota^*, \quad \iota_*\iota^! \to \mathrm{id}_X \to j_*j^*$$

of $\mathrm{DA}(X)$ (resp. of $\mathrm{RigDA}(X)$).

(vi) (Base-change theorems). Let

$$\begin{array}{ccc} Y' & \xrightarrow{g'} & Y \\ \downarrow{\scriptstyle f'} & & \downarrow{\scriptstyle f} \\ X' & \xrightarrow{g} & X \end{array}$$

be a cartesian morphism of schemes (resp. of rigid analytic spaces).

(a) If f is smooth, the natural transformations

$$(f')_\sharp \circ (g')^* \longrightarrow g^* \circ f_\sharp, \quad f^* \circ g_* \longrightarrow (g')_* \circ (f')^*$$

are equivalences.

(b) If f is proper then the natural transformation

$$g^* \circ f_* \longrightarrow (f')_* \circ (g')^* \tag{2.11}$$

is an equivalence. If g is moreover smooth, then also the natural transformation

$$g_\sharp \circ (f')_* \longrightarrow f_* \circ (g')_\sharp$$

is an equivalence. The equivalence (2.11) holds also when f is a quasi-compact, and quasi-separated morphism of rigid analyitic spaces.

(c) If f is locally of finite type, then the natural transformation

$$g^* \circ f_! \longrightarrow (f')_! \circ (g')^*$$

is an equivalence

(vii) (Projection formulas). If $f\colon Y \to X$ is locally of finite type, there is an equivalence

$$M \otimes f_! N \simeq f_!(f^*M \otimes N)$$

for every $M \in \mathrm{DA}(X)$ (resp. in $\mathrm{RigDA}(X)$) and $N \in \mathrm{DA}(Y)$ (resp. in $\mathrm{RigDA}(Y)$). In particular, when f is proper, there is an equivalence $f_*(f^*M \otimes N) \simeq M \otimes f_*N$.

(viii) (Duality). If $f\colon Y \to X$ is a smooth and proper morphism of schemes or of rigid analytic spaces. Then $f_\sharp \mathbb{1}_Y$ is strongly dualizable in $\mathrm{DA}(X)$ (resp. in $\mathrm{RigDA}(X)$), with dual given by $f_*\mathbb{1}_Y$. More generally, if $M \in \mathrm{DA}(Y)$ (resp. in $\mathrm{RigDA}(Y)$) is dualizable, then $f_\sharp M$ also is, with dual $f_*M^\vee$.

If $p\colon X \to S$ is a morphism locally of finite type of schemes (resp. of rigid analytic spaces), we call the object $p_*p^*\mathbb{1}_S = p_*\mathbb{1}_X$ the *cohomological motive of* X (relative to S), denoted $\mathsf{M}^{\mathrm{coh}}_S(X)$, and we call the object $p_!p^!\mathbb{1}_S$ the *homological motive of* X (relative to S), denoted $\mathsf{M}_S(X)$. For a smooth morphism $X \to S$, this agrees with the image of X under the Yoneda functor, as mentioned above. When S is the spectrum (resp. the adic spectrum) of field we will omit the subscript S and simply write the homological [resp. cohomological] motive of X as $\mathsf{M}(X)$ [resp. $\mathsf{M}^{\mathrm{coh}}(X)$].

Similarly, we sometimes denote by $\mathsf{M}_S(X)^{\mathrm{c}} = p_*p^!\mathbb{1}_S$ the *motive with compact support of* X. It agrees with the homological motive of X when p is proper.

REMARK 2.12 (Localization sequences for motives over a field). Let k be a field (resp. let K be a complete non-archimedean local field). Let $X \in \mathrm{Sch}/k$ (resp. $X \in \mathrm{RigSpc}^{\mathrm{qcqs}}/K$) and let $\iota\colon Z \to X$ be a closed immersion of schemes (resp. a closed immersion of rigid analytic spaces). The localization sequences of (v)(v) takes the (possibly more familiar) form of a fiber sequence in $\mathrm{DA}(k)$ (resp. in $\mathrm{RigDA}(K)$)

$$\mathsf{M}(Z)^{\mathrm{c}} \xrightarrow{\iota_*} \mathsf{M}(X)^{\mathrm{c}} \xrightarrow{j^*} \mathsf{M}(U)^{\mathrm{c}}$$

If X and Z are both smooth, then combining purity and duality we have the fiber sequence

$$\mathsf{M}(U) \xrightarrow{j_*} \mathsf{M}(X) \xrightarrow{\iota^!} \mathsf{M}(Z)(c)[2c]$$

if c is the codimension of Z in X.

2.4. Frobenius data. We start by recording the following fact about algebraic (resp. analytic) motives. Recall that a morphism $e\colon S' \to S$ of schemes is a universal homeomorphism if for any $Y \to S$, the base change $Y \times_S S' \to Y$ is a homeomorphism. It is well-known that this is equivalent to require that e is integral, radicial and surjective. For example, the canonical closed immersion $X_{\mathrm{red}} \to X$ is a universal homeomorphism, as it is any closed embedding $Z \to X$ defined by a nilpotent sheaf of ideals. In positive characteristic, both the absolute and the relative Frobenius are examples of universal homeomorphisms.

This notion has an exact analogue in the analytic world: we say that a morphism of rigid analytic spaces $S' \to S$ is a universal homomorphism if it is quasi-compact, quasi-separated, surjective and radicial. See [**10**, 2.9.1].

n Universal homeomorphisms of schemes induce equivalences of étale sites [**1**, Exposé VIII, Théorème 1.1]. This fact has the following motivic generalization, sometimes known as the "semi-separatedness property" of the stable motivic categories.

THEOREM 2.13 ([**6**, Theorem 1.2], resp. [**10**, Corollary 2.9.11]). *Let $e : X' \to X$ be a universal homeomorphism of schemes (resp. of rigid analytic spaces with X locally of finite Krull dimension). Then, the base change functor*

$$e^*\colon \mathrm{DA}(X) \to \mathrm{DA}(X') \qquad (\textit{resp. } e^* : \mathrm{RigDA}(X) \to \mathrm{RigDA}(X'))$$

is an equivalence. In particular, $\mathrm{DA}(X) \simeq \mathrm{DA}(X^{\mathrm{Perf}})$.

We obtain immediately an equivalence between the associated cohomological motives of X, X' as follows.

COROLLARY 2.14. *Let $X \to X'$ be a universal homeomorphism of schemes over S. Then the map $\mathsf{M}^{coh}_S(X') \to \mathsf{M}^{coh}_S(X)$ is invertible.*

REMARK 2.15. Equation (2.13) is stated for the $\mathbb{Q}$-linear categories $\mathrm{DA}(S)$ and $\mathrm{RigDA}(S)$, but holds more generally for the stable motivic homotopy categories $\mathrm{SH}(S)$ and $\mathrm{RigSH}(S)$ for S any scheme (or rigid analytic space locally of finite Krull dimension) of characteristic $p > 0$, after inverting the morphisms $\cdot p\colon E \to E$. In particular, for any *oriented* motivic spectrum $\mathbf{E} \in \mathrm{CAlg}(\mathrm{SH}(S))$ representing a generalized cohomology theory on S-schemes, the semi-separatedness of $\mathrm{SH}(S)[\frac{1}{p}]$ implies that for any universal homeomorphism $f\colon Y \to X$ of S-schemes, the induced maps $f^*\colon \mathbf{E}^{*,*}(X)[\frac{1}{p}] \to \mathbf{E}^{*,*}(Y)[\frac{1}{p}]$ and $f_*\colon \mathbf{E}^{\mathrm{BM}}_{*,*}(Y)[\frac{1}{p}] \to \mathbf{E}^{\mathrm{BM}}_{*,*}(X)[\frac{1}{p}]$ are isomorphisms. When $\mathbf{E} = \mathrm{KGL}$ is the spectrum representing algebraic K-theory, we obtain in particular that the canonical map $K(S)[\frac{1}{p}] \to K(S^{\mathrm{perf}})[\frac{1}{p}]$ is an equivalence. See [**41**] for more general results in this direction. Note that this statement would be false before inverting p, as one can check by considering $K_1(k)$ for an imperfect field k.

We can specialize the above Theorem to construct a canonical "Frobenius enrichment" for motives over schemes or analytic varieties defined over a field of positive characteristic p. We first observe that for S any scheme (resp. a rigid analytic space locally of finite Krull dimension) of characteristic p and X smooth over S, the relative Frobenius defines a natural transformation $X \to X \times_{\varphi} S = \varphi_S^* X$ of S-schemes (resp. of analyitic spaces over S). In particular, there is a natural transformation $\varphi^*\colon \mathrm{id} \to \varphi_S^*$ of endofunctors of $\mathrm{DA}(S)$ (resp. of $\mathrm{RigDA}(S)$), which is in fact an equivalence in light of Equation (2.13). If we denote by $\mathrm{DA}(S)^{\varphi^*}$ (resp. by $\mathrm{RigDA}(S)^{\varphi^*}$) the category of *homotopy fixed points of* φ^*, i.e., the equalizer of φ^* and the identity functor id, computed in $\mathrm{CAlg}(\mathrm{Pr}^L_\omega)$, we obtain the following Corollary.

COROLLARY 2.16 ([**44**, Corollary 2.26]). *Let S be a scheme (resp. a rigid analytic space locally of finite Krull dimension) of characteristic p and φ be the absolute Frobenius map $\varphi\colon S \to S$. There is a canonical functor*

$$\mathrm{DA}(S) \to \mathrm{DA}(S)^{\varphi^*} \qquad \left(\textit{resp. } \mathrm{RigDA}(S) \to \mathrm{RigDA}(S)^{\varphi^*}\right)$$

which informally sends $\mathsf{M}_S(X)$ to $(\mathsf{M}_S(X), \varphi_S\colon \mathsf{M}_S(X) \xrightarrow{\sim} \mathsf{M}_S(X \times_{\varphi} S))$ where φ_S is the relative Frobenius map.

2.5. The spreading out. The functor $S \mapsto \mathrm{RigDA}(S)$ has the property of converting limits of certain cofiltered inverse systems of rigid analytic spaces into filtered colimits of presentable ∞-categories.

THEOREM 2.17 ([**10**, Theorem 2.8.15]). *Let $(X_i)_{i\in I}$ be a cofiltered inverse system of rigid analytic spaces, with quasi-compact and quasi-separated transition maps, and admitting a weak limit X_∞ in the sense of Huber (see* [**36**, Definition 2.4.2]*). We further assume that X_∞ and the X_i's are locally of finite Krull dimension.*

Then, the obvious functor

$$\varinjlim_i \mathrm{RigDA}(X_i) \to \mathrm{RigDA}(X_\infty), \tag{2.18}$$

where the colimit is taken in Pr^L, is an equivalence.

REMARK 2.19. We like to call this a "spreading out" theorem as it ultimately says that any compact motive over X_∞ has a model over some X_i. For example, by putting X_i to be $\mathrm{Spa}\, L$ with L ranging over finite extensions of $\mathbb{Q}_p$ in $\mathbb{C}_p$, we deduce

$$\mathrm{RigDA}(\mathbb{C}_p) \simeq \varinjlim_{[L:K]<\infty} \mathrm{RigDA}(L).$$

We remark that the analogous statement would not hold if we simply considered the big étale sites over $\mathbb{C}_p$ and over the L's (for example, a map $\mathrm{Spa}\,\mathbb{C}_p \to \mathbb{B}^1_{\mathbb{C}_p}$ hitting an element of $\mathbb{C}_p$ which is trascendental over $\mathbb{Q}_p$ doesn't admit any model over any L).

Also, by putting X_i to be the affinoid open neighborhoods of a K-rational point $\operatorname{Spa} K = x \in S(K)$ in some rigid variety S over K, we deduce

$$\mathrm{RigDA}(x) \simeq \varinjlim_{x \in U \subset S} \mathrm{RigDA}(U).$$

In this case, the explicit quasi-inverse of the canonical pull-back functor x^* is given by Π^* i.e. the pull-back along the structural morphism $\Pi\colon U \to \operatorname{Spa} K$. Informally, this says that any compact motive M over S becomes constant around each rational point, i.e. for any sufficiently small open neighborhood $j\colon U \hookrightarrow S$ of x, one has $j^*M \simeq \Pi^* x^* M$.

We note that the "spreading out" formula in this last case (a rational point x as the limit of its open neighborhoods) follows immediately from the localization triangle (Section (v)(v)) and plays an important role in proving finiteness results of de Rham cohomologies (see [**44**, Theorem 4.46], and even [**63**] for an algebraic analogue).

The following is basically a restatement of the spreading out theorem, in terms on cohomological motives.

PROPOSITION 2.20. *Let $(\pi_i\colon X_i \to S)_{i \in I}$ be a cofiltered inverse system of rigid analytic spaces locally of finite Krull dimension which are qcqs over S admitting a weak limit $\pi_\infty\colon X_\infty \to S$ over S which is locally of finite Krull dimension. Let $f_{ij}\colon X_i \to X_j$ be the natural transition maps, with $i, j \in I \cup \{\infty\}$. Then for any $M \in \mathrm{RigDA}(X_0)$, the canonical map in* $\mathrm{RigDA}(S)$

$$\varinjlim \pi_{i_*} f_{i0}^* M \to \pi_{\infty *} f_{\infty 0}^* M$$

is invertible in $\mathrm{RigDA}(S)$.

PROOF. We let C be a compact object in $\mathrm{RigDA}(S)$. By [**10**, Proposition 2.4.22], the functors f_{ij}^* and π_i^* are compact-preserving. By Equation (2.17), we know that $\mathrm{RigDA}(X_\infty)$ is equivalent to $\varinjlim \mathrm{RigDA}(X_i)$ in $\mathrm{CAlg}(\mathrm{Pr}^L_\omega)$. As such, for any two objects A, B in $\mathrm{RigDA}(X_0)$ with A compact, the mapping complex $\operatorname{map}(f_{\infty 0}^* A, f_{\infty 0}^* B)$ can be computed as $\varinjlim \operatorname{map}(f_{i0}^* A, f_{i0}^* B)$. We then deduce the following sequence of canonical equivalences:

$$\begin{aligned}\operatorname{map}(C, \varinjlim \pi_{i_*} f_{i0}^* M) &\simeq \varinjlim \operatorname{map}(f_{i0}^* \pi_0^* C, f_{i0}^* M)\\ &\simeq \operatorname{map}(\pi_\infty^* C, f_{\infty 0}^* M)\\ &\simeq \operatorname{map}(C, \pi_{\infty *} f_{\infty 0}^* M).\end{aligned}$$

As $\mathrm{RigDA}(S)$ is compactly generated (see [**10**, Proposition 2.4.22]), this is enough to prove our claim. □

By putting $M = \mathbb{1}$ and $S = \operatorname{Spa} K$ we deduce:

COROLLARY 2.21. *Let $X_\infty \sim \varprojlim X_i$ be a qcqs weak limit of qcqs adic spaces over K, with finite Krull dimension. The canonical map $\varinjlim \mathsf{M}^{\mathrm{coh}}(X_i) \to \mathsf{M}^{\mathrm{coh}}(X_\infty)$ is an equivalence in* $\mathrm{RigDA}(K)$.

The previous result is particularly useful when one knows that the maps in the direct system induce equivalences in the homotopy category, e.g. if they are Frobenius lifts (see Section 2.4). If they are finite, we also deduce a special relation between the motive of the limit and each motive of the elements in the tower.

COROLLARY 2.22. *Let (X_i, f_{ij}) be a projective system of qcqs smooth rigid varieties over K, whose transition maps f_{ij} are finite surjective, and let X_∞ be a qcqs adic space such that $X_\infty \sim \varprojlim X_i$. The motive $\mathsf{M}^{\mathrm{coh}}(X_0)$ is a direct summand of the motive $\mathsf{M}^{\mathrm{coh}}(X_\infty)$.*

PROOF. We may and do assume that all the spaces involved are connected. We recall that, based on the equivalence between motives with and without transfers $\mathrm{RigDA}(K) \simeq \mathrm{RigDM}(K)$, the finite transition maps induce motivic maps $f_{ij} \colon \mathsf{M}(X_i) \to \mathsf{M}(X_j)$ as well as transpose correspondences $f_{ij}^T \colon \mathsf{M}(X_j) \to \mathsf{M}(X_i)$ such that $f_{ij} \circ f_{ij}^T = \deg(f_{ij}) \cdot \mathrm{id}$. We then deduce that the motive $\mathsf{M}(X_0)$ is a direct summand of each $\mathsf{M}(X_i)$, and dually that the motive $\mathsf{M}^{\mathrm{coh}}(X_0)$ is a direct summand of each $\mathsf{M}^{\mathrm{coh}}(X_i)$. We obtain the claim passing to the colimit $\varinjlim \mathsf{M}^{\mathrm{coh}}(X_i) \simeq \mathsf{M}^{\mathrm{coh}}(X_\infty)$. □

As an extra corollary of the spreading out result, we get the existence of "motivic tubular neighborhoods" for hypersurfaces defined as the zero-locus of a smooth map (see [**15**, Corollary 4.5]).

COROLLARY 2.23. *Let $f \colon X \to S$ be a smooth map of smooth rigid varieties over K and $s \in S(K)$ a rational point. There exists a final family of open neighborhoods U of s for which $\mathsf{M}^{\mathrm{coh}}(X_U) \simeq \mathsf{M}^{\mathrm{coh}}(X_s)$ in $\mathrm{RigDA}(K)$.*

PROOF. The point s has a final family of open neighborhoods U which are all isomorphic to poly-discs. In light of Equation (2.19), we may and do suppose that $\mathsf{M}_U(X_U) \simeq \mathsf{M}_U(X_s \times U)$. This implies that $\mathsf{M}_K(X_U) \simeq \mathsf{M}_K(X_s \times U) = \mathsf{M}_K(X_s)$ and hence that $\mathsf{M}^{\mathrm{coh}}(X_U) = \mathsf{M}^{\mathrm{coh}}(X_s)$. □

3. From characteristic p to mixed characteristic

The aim of this section is to prove Equation (1.3) that is, to use homotopy methods to give a new alternative definition of relative rigid cohomology, and even deduce new results about it. For the absolute case, see [**59, 62**].

3.1. Formal motives vs algebraic motives. We first show how Elkik and Arabia's solutions to the (local) "lifting problems" in the Monsky–Washnitzer approach to rigid cohomology can be condensed into suitable motivic equivalences.

The first result is due to Ayoub (see e.g., [**8**, Corollary 1.4.29]), and is a consequence of the localization theorem.

THEOREM 3.1 ([**10**, Theorem 3.1.10]). *Let $\mathfrak{S}$ be a formal scheme. The functors*

$$\sigma^* : \mathrm{FDA}(\mathfrak{S}) \rightleftarrows \mathrm{DA}(\mathfrak{S}_\sigma) : \sigma_*. \tag{3.2}$$

induced by taking special fibers, are equivalences of ∞-categories.

Indeed, for a formal scheme $\mathfrak{S}$, the complement of $\mathfrak{S}_\sigma$ in $\mathfrak{S}$ is empty and so by the localization triangle we obtain that the natural transformation $\mathrm{id} \to \sigma_*\sigma^*$ is invertible.

There is also a log-generalization of the equivalence provided by eq. (3.1) above, based on a modification of the argument to the logarithmic setting. For S a log scheme (resp. $\mathfrak{S}$ a log formal scheme), we let $\mathrm{logDA}^{\mathrm{eff}}(S)$ (resp. $\mathrm{logFDA}^{\mathrm{eff}}(\mathfrak{S})$) be the full monoidal ∞-subcategory of $\mathbb{A}^1$-invariant objects in the category

$$\mathrm{Shv}_{s\acute{e}t}(\mathrm{Sm}/S; \mathcal{D}(\mathbb{Q}))(\mathrm{resp.}\mathrm{Shv}_{s\acute{e}t}(\mathrm{Sm}/\mathfrak{S}; \mathcal{D}(\mathbb{Q})))$$

of strict-étale hypersheaves on the category of log smooth log schemes over S (resp. log formal schemes over $\mathfrak{S}$). This is the exact generalization to the log setting of Equation (2.1). Formally inverting the Tate twist $\mathbb{K}(1)$ we obtain the non-effective version $\operatorname{logFDA}(S)$ (resp. $\operatorname{logFDA}(\mathfrak{S})$). See [**15**, Section 2.2] for an extended discussion, [**40**] for basic results about fine log structures and [**42**, Appendix A] for non-necessarily fine log structures on formal schemes.

THEOREM 3.3 ([**15**, Theorem 2.22]). *Let $\mathfrak{S}$ is a quasi-coherent integral log formal scheme. Assume that at least one of the following holds:*

(1) *The log structure $M_{\mathfrak{S}}$ is fine;*
(2) *The underlying scheme $\underline{\mathfrak{S}}$ has finite topological Krull dimension.*

Then the functors

$$\sigma^* : \operatorname{logFDA}(\mathfrak{S}) \rightleftarrows \operatorname{logDA}(\mathfrak{S}_\sigma) : \sigma_*. \tag{3.4}$$

induced by taking special fibers, are equivalences of ∞-categories.

3.2. Dagger motives vs analytic motives. The possibility to "lift" varieties over $\mathfrak{S}_\sigma$ to (formal) motives over $\mathfrak{S}$ and hence to (analytic) motives over $\mathfrak{S}_{\mathbb{Q}_p}$ is not yet enough to produce a motivic p-adic cohomology theory "à la de Rham", which is finite dimensional on objects of geometric origin (for example on a smooth quasi-compact rigid analytic variety X over K). The problem being, that the de Rham complex of the closed disc $\mathbb{B}^1$ has non-trivial higher cohomology.

This obstacle is classically overcome as follows. Any smooth rigid analytic variety X can be embedded, locally on X, as a strict open subvariety of a larger smooth variety X'. This choice is called a "dagger variety structure on X" and denoted by $X^\dagger$. One can then replace the de Rham complex $\Omega_{X/K}$ by its subcomplex $\Omega_{X^\dagger/K}$ given by those differential forms that are defined over some strict neighborhood of X inside X'. One can make a similar construction relatively over a (dagger) base S.

Just like the case of formal lifts, the choice of dagger variety structures are only local, and canonical up to homotopy. It comes as no surprise that this admits a motivic restatement.

THEOREM 3.5 ([**59**, Theorem 4.23]). *Let $\mathcal{S}^\dagger$ be a dagger variety. The functors*

$$l^* : \operatorname{RigDA}(\mathcal{S}^\dagger) \rightleftarrows \operatorname{RigDA}(\mathcal{S}) : l_*. \tag{3.6}$$

induced by taking the underlying rigid analytic variety, are equivalences of ∞-categories.

We can then introduce a suitable generalization of the overconvergent site above a scheme S/k as follows.

DEFINITION 3.7. Let S be a scheme over a perfect field k of characteristic p. The category Fr/S of generalized proper frames over S is given by triples $(\mathfrak{T}, f\colon \mathfrak{T}_\sigma \to S, \mathfrak{T}^\dagger_{\mathbb{Q}_p})$ where $\mathfrak{T}$ is a formal scheme over $W(k)$ (not necessarily adic over it), f is a morphism over k and $\mathfrak{T}^\dagger_{\mathbb{Q}_p}$ is a dagger structure on the $\mathbb{Q}_p$-generic fiber of $\mathfrak{T}$.

Our notation for the name is justified by the following remark: any proper frame over S in the sense of Berthelot ($T \subset \bar{T} \subset \mathfrak{P}, f\colon T \to S$) given by an open inclusion into a closed subscheme of a proper formal scheme $\mathfrak{P}$ induces a generalized proper frame by taking $(\mathfrak{U}, f,]T[_{\mathfrak{P}} \Subset]\bar{T}[_{\mathfrak{P}})$, where $\mathfrak{U}$ is the open formal subscheme of $\mathfrak{P}^{\wedge \bar{T}}$ whose special fiber is T, and whose $\mathbb{Q}_p$-generic fiber is the tube $]T[_{\mathfrak{P}}$.

We can then mix together Equation (3.1) and Equation (3.5) to finally create a Monsky–Washnitzer relative realization (with a Frobenius structure!).

THEOREM 3.8 ([**29**, Proposition 3.19]). *Let S be a scheme over k. There is a canonical functor*

$$\mathcal{MW}_S \colon \mathrm{DA}(S) \to \left(\lim_{(\mathfrak{T}, f, \mathfrak{T}^\dagger_{\mathbb{Q}_p}) \in \mathrm{Fr}/S} \mathrm{RigDA}^\dagger(\mathfrak{T}^\dagger_{\mathbb{Q}_p}) \right)^{\varphi}.$$

By taking $S = \operatorname{Spec} k$, we may in particular obtain the "absolute" realization

$$\mathcal{MW}_k \colon \mathrm{DA}(k) \to \mathrm{RigDA}^\dagger(W(k)[1/p])^{\varphi}.$$

3.3. de Rham and rigid cohomology. As we have finally completed the journey from characteristic p to characteristic 0 (plus a dagger structure), we have all the ingredients to define a "de Rham-like" cohomology theory.

PROPOSITION 3.9 ([**44**, Corollary 4.39],[**29**, Corollary 3.20]). *The functor*

$$\mathcal{X}^\dagger \mapsto \underline{\Omega}_{\mathcal{X}^\dagger/\mathcal{S}^\dagger}$$

induces a motivic realization functor

$$\mathrm{RigDA}(\mathcal{S}^\dagger) \to \mathcal{D}(\mathcal{S}^\dagger)^{op}$$

where on the right hand side we consider the category of overconvergent solid modules on $\mathcal{S}^\dagger$. In particular, we deduce (absolute and relative) rigid cohomology functors

$$R\Gamma_{\mathrm{rig}} \colon \mathrm{DA}(k)^{op} \to \mathcal{D}(W(k)[1/p])^{\varphi} \qquad R\Gamma_{\mathrm{rig}}(-/S) \colon \mathrm{DA}(S)^{op} \to \left(\lim \mathcal{D}(\mathfrak{T}^\dagger_{\mathbb{Q}_p})\right)^{\varphi}.$$

We may then prove specific properties of the realization of smooth and proper varieties by using the fact that their associated motives are dualizable: one uses crucially the explicit description of dualizable solid modules as (classical) perfect complexes, see [**2**].

COROLLARY 3.10 ([**44**, Theorem 4.46],[**29**, Corollary 3.20]). *If $M \in \mathrm{DA}(S)$ is dualizable, then $R\Gamma_{\mathrm{rig}}(M/S)$ is a pefect complex on each frame. Moroever, its cohomology groups $R^i\Gamma_{\mathrm{rig}}(M/S)$ have a canonical structure of overconvergent F-isocrystals.*

REMARK 3.11. In case $M = \mathsf{M}_S(X)$ for some smooth and proper variety X/S, the overconvergent sheaves underlying the $R^i\Gamma(M/S)$'s can be compared to Berthelot's rigid cohomology groups. The result above answers positively to a conjecture by Berthelot, which was previously known only in the liftable case, or considering variants of it (see in particular [**17, 55**]).

4. Motivic nearby cycles and applications

4.1. Analytic motives vs algebraic motives with monodromy. In the classical literature (see [**38**]) monodromy operators are certain nilpotent endomorphisms between cohomology groups $H^*(X)$ (e.g. de Rham cohomology, ℓ-adic cohomology), where X arises as a fiber of a family over a disk, or X is a variety defined over a local field. They come from the action of a generator of a fundamental group $\pi^1(\Delta^*) \simeq \mathbb{Z}$ or a topological generator of tame inertia groups (isomorphic to $\widehat{\mathbb{Z}}$). If one equips the groups $H^*(X)$ with suitable extra structures (in the de Rham case: a mixed Hodge module structure, in the ℓ-adic case: the structure of a representation over $\mathrm{Gal}(k)$) then it makes sense to define (positive and negative) twists of them, and it turns out that the monodromy maps are actually morphisms of the form

$$N\colon H^*(X) \to H^*(X)(-1)$$

hence their nilpotence (for weight reasons).

We can then define a category of monodromy operators as follows:

DEFINITION 4.1. Let $\mathscr{C}$ be stable ∞-category, and $(-1) : \mathscr{C} \to \mathscr{C}$ be an exact functor. We let $(-n)$ be the iterated functor $(-1)^n$.

(1) We let $\mathscr{C}_{lax}^{(-1)}$ to be the ∞-category whose objects are pairs (X, f) with $X \in \mathscr{C}$ and $f\colon X \to X(-1)$. Mapping complexes are defined in the obvsious way, i.e.

$$\mathrm{map}((X,f),(Y,g))) = \mathrm{fib}(\mathrm{map}(X,Y) \xrightarrow{f^*(-1)-g_*} \mathrm{map}(X,Y(-1)).$$

(2) We define $\mathscr{C}_N$ as the full sub-∞-category of $\mathscr{C}_{lax}^{(-1)}$ spanned by objects (X, f) with ind-nilpotent morphism f, that is, such that

$$\mathrm{colim}(X \to X(-1) \to X(-2) \to \cdots) = 0.$$

The following facts are proved in [**14**, Section 2].

REMARK 4.2 (See [**14**, Remark 2.20]). If $\mathscr{C}$ is compactly generated, then also $\mathscr{C}_{lax}^{(-1)}$ is. Its compact objects are given by pairs (X, f) with X compact, and f nilpotent.

REMARK 4.3 (See [**14**, Remark 2.23]). The category $\mathscr{C}_{lax}^{(-1)}$ underlies a structure of a symmetric monoidal category, as soon as $\mathscr{C}$ does and the functor (-1) is lax-monoidal. Informally, one has $(X, f) \otimes (Y, g) = (X \otimes Y, f + g)$.

REMARK 4.4. We note that there is a natural adjunction

$$(N = 0)\colon \mathscr{C} \rightleftarrows \mathscr{C}_N\colon \mathrm{fib}(N)$$

which, in good situations, satisfies the hypotheses of a Barr–Beck–Lurie functor, thus identifiying the category $\mathscr{C}_N$ with a category of modules $\mathrm{Mod}_{\mathrm{fib}(N)(\mathbb{1})}\mathscr{C} = \mathrm{Mod}_{\mathbb{1}\oplus(-1)[-1]}(\mathscr{C})$.

Let K be a complete non-archimedean field with a perfect residue field k. Consider the generic fiber functor

$$\xi\colon \mathrm{DA}(k) \simeq \mathrm{FDA}(\mathcal{O}_K) \xrightarrow{(-)^{\mathrm{rig}}} \mathrm{RigDA}(K)$$

where the first equivalence is provided by Equation (3.1) and $(-)^{\mathrm{rig}}$ is the functor induced by sending a formal scheme to its associated rigid space (see [**10**, Notation 3.8.14]). The main theorem of [**14**] states the following.

THEOREM 4.5 ([**14**, Corollary 4.14]). *Let K be as above and assume that it has a value group of (rational) rank 1. Let $\mathrm{RigDA}(K)^{\mathrm{gr}}$ be the essential image of the functor $\xi\colon \mathrm{DA}(k) \to \mathrm{RigDA}(K)$. Depending on a choice of a pseudo-uniformizer $\varpi \in K$, the functor above exhibits an equivalence between $\mathrm{DA}(k)_N$ and $\mathrm{RigDA}(K)^{\mathrm{gr}}$.*

DEFINITION 4.6. Notation-wise, we shall write the image in $\mathrm{DA}(k)_N$ of an object $M \in \mathrm{RigDA}(K)^{\mathrm{gr}}$ via the previous equivalence as $(\Psi M, N_M)$.

PROOF. We can highlight the main points of the proof as follows.

Step 1: We note that the functor $\xi\colon \mathrm{DA}(k) \to \mathrm{RigDA}(K)^{\mathrm{gr}}$ satisfies the hypotheses of the Barr–Beck–Lurie theorem. Indeed, it is a left adjoint which sends a set of compact generators to a set of compact generators (more or less by definition), which implies that it preserves colimits and its right adjoint preserves filtered colimits. The projection formula $\chi(\xi M \otimes N) \simeq M \otimes \chi N$ follows formally from the fact that all compact objects in $\mathrm{DA}(k)$ are dualizable.

Step 2: Because of Step 1, we deduce that ξ exhibits a a monoidal equivalence $\mathrm{RigDA}(K)^{\mathrm{gr}} \simeq \mathrm{Mod}_{\chi\mathbb{1}}(\mathrm{DA}(k))$. Thanks to the computations of [**10**], we have a concrete description of the algebra $\chi\mathbb{1}$ as the split square-zero extension $\mathbb{1} \oplus (-1)[-1]$. (This identification depends on a choice of a pseudo-uniformizer.) In particular, we can alternatively identify $\mathrm{Mod}_{\chi\mathbb{1}}(\mathrm{DA}(k))$ with $\mathscr{C}_N$, as shown above. □

REMARK 4.7. We note that under the equivalence $\mathrm{DA}(k)_N \simeq \mathrm{RigDA}(K)^{\mathrm{gr}}$ of Equation (4.5), the adjunction

$$\xi\colon \mathrm{DA}(k) \rightleftarrows \mathrm{RigDA}(K)^{\mathrm{gr}}\colon \chi$$

is identified with the adjunction of Equation (4.4):

$$(N=0)\colon \mathrm{DA}(k) \rightleftarrows \mathrm{DA}_N(K)\colon \mathrm{fib}(N).$$

REMARK 4.8. Step 1 of the proof above can be generalized over a base, giving rise to an equivalence between the essential image of $\xi\colon \mathrm{FDA}(\mathfrak{S}) \to \mathrm{RigDA}(\mathfrak{S}_\eta)$ and $\mathrm{Mod}_{\chi\mathbb{1}}(\mathrm{DA}(\mathfrak{S}_\sigma)$. In this setting, the projection formula is not automatic, and is the core result of [**10**, Section 3.6].

The previous theorem is only mildly useful if one doesn't have a good handle of the categories $\mathrm{RigDA}(K)^{\mathrm{gr}}$. Fortunately, de Jong's alteration of singularities and some explicit homotopies allow one to conclude the following.

PROPOSITION 4.9. [**14**, Remarks 4.6, 4.7] *Let K be a non-archimedean field.*

1. *If X/K is a quasi-compact smooth rigid variety having a semi-stable formal model (more generally, a pluri-nodal model in the sense of* [**12**, Definition 1.1]*) then* $\mathsf{M}(X) \in \mathrm{RigDA}(K)^{\mathrm{gr}}$.
2. *If K is algebraically closed, then* $\mathrm{RigDA}(K) = \mathrm{RigDA}(K)^{\mathrm{gr}}$. *In particular, for any $M \in \mathrm{RigDA}(K)$ which is compact, there exists a finite extension K'/K such that $M_{K'} \in \mathrm{RigDA}(K')^{\mathrm{gr}}$.*

REMARK 4.10. Passing to finite extensions of the base field in order to have an alteration from a variety admitting a semi-stable formal model is a standard technique which is used to bootstrap results from the semi-stable case to the general case. Similarly, we will focus on the categories $\mathrm{RigDA}(K)^{\mathrm{gr}}$, as any compact motive $M \in \mathrm{RigDA}(K)$ lies in them, up to some finite field extension (use the previous proposition and the spreading-out).

4.2. Revisiting the tilting equivalence. Equation (4.5) has the following straightforward corollary.

COROLLARY 4.11. *Let K be a perfectoid field with rational rank $\mathbb{Q}$. Then there is an equivalence*

$$\mathrm{RigDA}(K) \simeq \mathrm{RigDA}(K^\flat)$$

PROOF. If K is algebraically closed, then both categories are equivalent to $\mathrm{DA}_N(k)$ where k is their common residue field. For the general case, observe that this equivalence is equivariant with the action of use $\mathrm{Gal}(K) \simeq \mathrm{Gal}(K^\flat)$. □

The result above is a special case of an equivalence which holds in much greater generality, and whose proof is based on the spreading out theorem, and some explicit computations of mapping spaces:

THEOREM 4.12. *Let P be a perfectoid space over K. There is an equivalence $\mathrm{RigDA}(P) \simeq \mathrm{RigDA}(P^\flat)$ which is compatible with pull-back functors. In particular, under the equivalence $\mathrm{RigDA}(K) \simeq \mathrm{RigDA}(K^\flat)$, the cohomological motive $\mathsf{M}^{\mathrm{coh}}(P)$ of a perfectoid space P/K corresponds to $\mathsf{M}^{\mathrm{coh}}(P^\flat)$.*

PROOF. The first part is [**44**, Theorem 5.13]. For the last point, it suffices to point out that the identification $\mathrm{RigDA}(P) \simeq \mathrm{RigDA}(P^\flat)$ is also compatible with the functors f_*, as it is compatible with their left adjoint functors f^*. □

We note that there is no analogue of the previous result for *homological* motives: if $f\colon P \to K$ is perfectoid, there is no functor $f_!$ as f is not locally of topologically finite type, in general.

REMARK 4.13. The equivalence of Equation (4.11) is a special case of the one of Equation (4.12). In order to see this, it suffices to show that the tilting equivalence is compatible with the functors $\mathrm{DA}(k) \to \mathrm{RigDA}(K)$ and $\mathrm{DA}(k) \to \mathrm{RigDA}(K^\flat)$, as shown in [**61**].

REMARK 4.14. In this remark, we review one of the main results of [**15**] in which the tilting equivalence is shown to be compatible with the theory of log-motives on the log point.

Let C be the completion of an algebraic closure of a discrete valuation field, let $\mathcal{O}_C$ be its valuation ring and k its residue field. We can canonically equip $\mathrm{Spf}(\mathcal{O}_C)$ with a logarithmic structure, denoted $\mathcal{O}_C^\times$, by considering the inclusion of the submonoid of non-zero elements $\mathcal{O}_C - \{0\} \subset \mathcal{O}_C$. The residue field k also acquires a logarithmic structure as the so-called log point $k^0 = (k, k^* \oplus_{\mathcal{O}_C^*} \mathcal{O}_C - \{0\})$. Note that this is simply the induced log structure for the embedding $\mathrm{Spec}(k) \to \mathrm{Spf}(\mathcal{O}_C)$: in terms of maps of monoids, it is the log structure associated to the pre-log structure $\mathbb{Q}_{\geq 0} \to k, 0 \neq m \mapsto 0$, where $\mathbb{Q}_{\geq 0}$ is the valuation group of C.

We can consider the category $\mathrm{Sm}^{\mathrm{ss}}_{k^0}$ of *semistable* log schemes $X = (\underline{X}, \mathsf{M}_X)$ over k^0: it is the full subcategory of all log smooth log schemes over k^0 consisting of those X such that, Zariski locally on $\underline{X}$, there exist a strict étale map $X \to \mathrm{Spec}(k[X_0, \ldots, X_n]/(X_1 \cdots X_r))$, equipped with the standard log structure induced by $\mathbb{N}^r \to k[X_0, \ldots, X_n]/(X_1 \cdots X_r)$, $e_i \mapsto X_i$, for some $1 \leq r \leq n$. See [**15**, Definition 2.11(1), (2)] for more details.

Write $\log \mathrm{DA}^{\mathrm{ss}}(k^0)$ for the category of semistable log motives over k^0: this is the full subcategory of $\log \mathrm{DA}(k^0)$ spanned by motives of semistable log varieties over k^0. The equivalence of Equation (3.3) restricts in particular to an equivalence

$$\log \mathrm{DA}^{\mathrm{ss}}(k^0) \simeq \log \mathrm{FDA}^{\mathrm{ss}}(\mathcal{O}_C^\times),$$

where the right-hand side is defined in the obvious way. By construction, the log structures on semistable formal schemes over $\mathcal{O}_C$ are in particular *vertical*, i.e., the sheaf of monoids $\mathsf{M}_{\mathfrak{X}}$ on a semistable formal scheme $\mathfrak{X} \to \mathrm{Spf}(\mathcal{O}_C)$ becomes a sheaf of groups after inverting a pseudo-uniformizer ϖ. In particular there is a well-defined functor

$$\mathrm{Sm}^{\mathrm{ss}}_{\mathcal{O}_C^\times} \to \mathrm{RigSm}_C, \quad \mathfrak{X} = (\underline{\mathfrak{X}}, \mathsf{M}_{\mathfrak{X}}) \mapsto \underline{\mathfrak{X}}^{\mathrm{rig}}$$

where $\underline{\mathfrak{X}}^{\mathrm{rig}}$ is the rigid generic fiber. This passes to the category of motives, providing a functor

$$\mathcal{MW}_k^{\log} \colon \log \mathrm{DA}^{\mathrm{ss}}(k^0) \simeq \log \mathrm{FDA}^{\mathrm{ss}}(\mathcal{O}_C^\times) \longrightarrow \mathrm{RigDA}(C)$$

that we can call the log Monsky–Washnitzer functor. This is compatible with the tilting equivalence, in the sense that if C is perfectoid field with tilt $C^\flat$, then there is a commutative diagram

$$\begin{array}{ccccc} & & \log \mathrm{FDA}^{\mathrm{ss}}(\mathcal{O}_C^\times) & \longrightarrow & \mathrm{RigDA}(C) \\ & \overset{\simeq}{\nearrow} & & & \Big| \wr \\ \log \mathrm{DA}^{\mathrm{ss}}(k^0) & & & & \Big| \simeq \\ & \underset{\simeq}{\searrow} & & & \Big| \\ & & \log \mathrm{FDA}^{\mathrm{ss}}(\mathcal{O}_{C^\flat}^\times) & \longrightarrow & \mathrm{RigDA}(C^\flat) \end{array}$$

see [**15**, Proposition 2.31]. In fact, the horizontal arrows in the previous diagram are actually localizations with respect to the rig-étale topology.

REMARK 4.15. This can be made even more explicit: one can consider the category of log motives $\log \mathrm{DA}^{\mathrm{ss}}_{\mathrm{div}}(k^0)$ for the *dividing* étale topology [**16**, Section 3.3], i.e., the topology generated by strict étale maps together with universally surjective proper log étale monomorphisms (informally, locally given by blow-ups associated to subdivisions of fans). Then by [**50**, Theorem 3.3] and Equation (4.5), there are equivalences

$$\log \mathrm{DA}^{\mathrm{ss}}_{\mathrm{div}}(k^0) \simeq \mathrm{DA}(k)_N \simeq \mathrm{RigDA}(C)$$

This allows one to consider log motives as algebraic avatars of analytic motives. In equi-characteristic zero, Park further shows that the motivic nearby cycle functor Ψ can be computed via the functor $\Psi^{\log}$. See [**50**, Definition 2.2].

4.3. A quick detection principle. As another corollary of Equation (4.5), we point out that one can immediately extend (monoidal) functors $F \colon \mathrm{DA}(k) \to \mathscr{D}^{\otimes}$ to functors from $\mathrm{RigDA}(K)^{\mathrm{gr}}$. More precisely let $F \colon \mathrm{DA}(k) \to \mathscr{D}^{\otimes}$ be a colimit-preserving monoidal functor between monoidal stable presentable ∞-categories. Let $u = F(\Bbbk(-1)) \in \mathscr{D}$, and write $\mathscr{D}_N$ for the category of Equation (4.1)-(2) with respect to the exact endofunctor $(-1) := - \otimes u$. The choice of a pseudo-uniformizer $\varpi \in \mathcal{O}_K$ gives rise to a monoidal extension

$$\widehat{F} \colon \mathrm{RigDA}(K)^{\mathrm{gr}} \to \mathscr{D}_N \tag{4.16}$$

endowed with equivalences $\widehat{F} \circ \xi \simeq (N = 0) \circ F$ and $\pi \circ \widehat{F} \simeq F \circ \Psi$. Informally, it sends the motive M to $(F(\Psi M), F(N_{\Psi M}))$.

We can be even more precise than that, by classifying these possible extensions explicitly.

PROPOSITION 4.17. *Let $F\colon \mathrm{DA}(k) \to \mathscr{D}^{\otimes}$ be a functor of monoidal presentable categories. There is an equivalence*

$$\mathrm{Map}_{\mathrm{DA}(k)^{\otimes}/}(\mathrm{RigDA}^{\mathrm{gr}}(K)^{\otimes}, \mathscr{D}^{\otimes}) \simeq \mathrm{Map}_{\mathscr{D}}(F(\mathbb{1}(-1)[-1]), \mathbb{1}).$$

In particular, in case $\mathscr{D} = \mathcal{D}(\mathcal{A})$ for some monoidal abelian category $\mathcal{A}$, there is an equivalence between $\pi_0 \mathrm{Map}_{\mathrm{DA}(k)^{\otimes}/}(\mathrm{RigDA}^{\mathrm{gr}}(K)^{\otimes}, \mathscr{D}^{\otimes})$ and $\mathrm{Ext}^1(F(\mathbb{1}(-1)), \mathbb{1})$ under which the functor $F \circ \Psi$ corresponds to 0.

PROOF. For the first claim, we let G be the right adjoint to F and we note the following identifications (using Equation (4.5) and the Barr–Beck–Lurie theorem [**45**, Proposition 5.29], see the proof of [**14**, Proposition 4.52])

$$\begin{aligned}
\mathrm{Map}_{\mathrm{DA}(k)^{\otimes}/}(\mathrm{RigDA}^{\mathrm{gr}}(K)^{\otimes}, \mathscr{D}^{\otimes}) &\simeq \mathrm{Map}_{\mathrm{DA}(k)^{\otimes}/}(\mathrm{Mod}_{\mathbb{1}\oplus\mathbb{1}(-1)[-1]}\,\mathrm{DA}(k)^{\otimes}, \mathscr{D}^{\otimes})\\
&\simeq \mathrm{Map}_{\mathrm{CAlg}(\mathrm{DA}(k))}(\mathbb{1} \oplus \mathbb{1}(-1)[-1], G\mathbb{1})\\
&\simeq \mathrm{Map}_{\mathrm{DA}(k)}(\mathbb{1}(-1)[-1], G\mathbb{1})\\
&\simeq \mathrm{Map}_{\mathscr{D}}(F\mathbb{1}(-1)[-1], \mathbb{1}).
\end{aligned}$$

For the last sentence, we might consider the case $\mathscr{D} = \mathrm{DA}(k)$ and remark that the base change along the augmentation $\mathbb{1} \oplus \mathbb{1}(-1)[-1] \to \mathbb{1}$, induced by the zero map $\mathbb{1}(-1)[-1] \to \mathbb{1}$, corresponds to the functor Ψ. □

EXAMPLE 4.18. Write $K_0 = W(k)[1/p]$. If $\mathcal{A}$ is the category of (φ, N)-modules and $F(\mathbb{1}(-1)) = \mathbb{1}(-1)$, then $\mathrm{Ext}^1(\mathbb{1}(-1), \mathbb{1}) = K_0$ having as K_0-basis the "Kummer extension", i.e. the φ-module $\mathbb{1} \oplus \mathbb{1}(-1)$ with monodromy $\left(\begin{smallmatrix} 0 & 1 \\ 0 & 0 \end{smallmatrix}\right)$.

In general, the object in $\mathrm{Ext}^1(F(\mathbb{1}(-1)), \mathbb{1})$ associated to F can be read out of the image of the *Kummer motive* $\mathcal{K} \in \mathrm{RigDA}^{\mathrm{gr}}(K)$, (see [**14**, §3.1]) which is a direct summand of the motive of the Tate curve $\mathbb{G}_m^{\mathrm{an}}/\varpi^{\mathbb{Z}}$ (responsible for its H^1-cohomology groups). Note that this curve is (the analytification of) an elliptic curve with nodal reduction. Testing the image of this motive via classical realization functors, one can then prove the following result (see [**14**, §3.2, §4.8, §4.9]):

COROLLARY 4.19. *Let $\ell \neq p$ be a prime. The realizations*

$$\begin{gathered}
\mathrm{RigDA}^{\mathrm{gr}}(K) \simeq \mathrm{DA}_N(k) \to \mathcal{D}_{\varphi,N}(\mathbb{Q}_\ell)\\
\mathrm{RigDA}^{\mathrm{gr}}(K) \simeq \mathrm{DA}_N(k) \to \mathcal{D}_{\varphi,N}(K_0)\\
\mathrm{RigDA}^{\mathrm{gr}}(\mathbb{C}(\!(t)\!)) \simeq \mathrm{DA}_N(\mathbb{C}) \to (\mathrm{Ind}\,\mathcal{D}^b(\mathrm{MHS}^p_{\mathbb{Q}}))_N
\end{gathered}$$

compute the Weil-Deligne ℓ-adic representation attached to the ℓ-adic cohomology, resp. Hyodo-Kato cohomology resp. limit Hodge cohomology.

COROLLARY 4.20. *If K is perfectoid, the tilting equivalence $\mathrm{RigDA}^{\mathrm{gr}}(K) \simeq \mathrm{RigDA}^{\mathrm{gr}}(K^\flat)$ is compatible with the ℓ-adic and the Hyodo-Kato realizations.*

PROOF. This is because the tilting equivalence for $\mathrm{RigDA}^{\mathrm{gr}}(K)$ can be obtained via the equivalences

$$\mathrm{RigDA}^{\mathrm{gr}}(K) \simeq \mathrm{DA}_N(k) \simeq \mathrm{RigDA}^{\mathrm{gr}}(K^\flat).$$

□

4.4. The weight-monodromy conjecture. One of the first applications of the theory of perfectoid spaces and the almost purity theorem is Scholze's proof of the weight-monodromy conjecture for the ℓ-adic cohomology of smooth proper varieties obtained as the intersection of smooth hypersurfaces in projective smooth toric varieties [**54**]. As formulated by Deligne [**26**] (see [**38**, Conjecture 3.9]), the conjecture predicts in general that for any smooth projective variety Y defined over a local field K, with residue field k of characteristic $p > 0$, the i-th ℓ-adic cohomology groups $H^i(Y_{\overline{K}}, \mathbb{Q}_\ell)$ for $\ell \neq p$ have the property that the action of Frobenius on the j-th graded quotient with respect to the monodromy filtration $\mathrm{gr}_j^M H^i(Y_{\overline{K}}, \mathbb{Q}_\ell)$ is pure of weight $i+j$. This conjecture admits a p-adic variant, as predicted by Fontaine and Jannsen (see [**39**, p. 347]), where the ℓ-adic cohomology groups are replaced by the Hyodo-Kato cohomology groups $H^i_{HK}(Y)$, which are equipped with their canonical structure of (φ, N)-modules over $W(k)[\frac{1}{p}]$.

In [**15**], we show that a part of Scholze's proof is of "motivic nature". As such, it can be extend almost verbatim to the case of an arbitrary realization, such as the one induced by the p-adic Hyodo–Kato cohomology. Recall the following definition.

DEFINITION 4.21. A (smooth) toric variety over a field L is a (smooth) separated L-scheme X containing a dense open subset U isomorphic to a split torus $T = \mathbb{G}^r_{m,L}$ such that the tautological action of T on U by multiplication extends to X (i.e., T acts on X).

REMARK 4.22. Every abstract normal separated toric variety can be realized as the toric variety associated to a fan Σ. More precisely, let N be a free abelian group of finite rank, and write $N_\mathbb{R}$ for $N \otimes \mathbb{R}$. A fan Σ in $N_\mathbb{R}$ is a finite collection of strongly convex rational polyhedral cone $\sigma \subset N_\mathbb{R}$ subject to the condition of being closed under taking faces (i.e., for each $\sigma \in \Sigma$ and each face F of σ we have $F \in \Sigma$), and such that for each $\sigma_1, \sigma_2 \in \Sigma$, we have that $\sigma_1 \cap \sigma_2$ is a face of both (and so is also in Σ).

Writing $M = \mathrm{Hom}(N, \mathbb{Z})$, we can associate to each $\sigma \in \Sigma, \sigma \subset N_\mathbb{R}$ the affine scheme $U_\sigma = \mathrm{Spec}(L[\sigma^\vee \cap M])$. Since σ is rational, it can be obtained as the cone spanned by a finite subset $S \subseteq N$. By [**23**, Theorem 1.3.12], the scheme U_σ is a non-singular L-scheme if the generators of σ form a part of a $\mathbb{Z}$-basis of N (this is in particular independent of the field L). The various U_σ for $\sigma \in \Sigma$ can be glued together to define the toric variety $X_{\Sigma,L}$. By Sumihiro's theorem [**23**, Corollary 3.1.8], for any normal separated toric variety X, there exists a fan Σ such that $X \simeq X_\Sigma$. In fact, Σ can be taken as a fan in $N_\mathbb{R}$ for N the character group of the torus T acting on X.

REMARK 4.23. A basic example of a toric variety is given by the projective space $\mathbb{P}^n_K$: this is obtained by the usual gluing of the affine schemes $\mathrm{Spec}(K[\mathbb{N}^n]) \simeq \mathbb{A}^n_K$. The reader unfamiliar with the general theory of toric varieties can safely stick to the case $X_\Sigma = \mathbb{P}^n$ in all statements below.

REMARK 4.24. The multiplication by $p\colon Q \to Q$ on any monoid Q induces an endomorphism on the corresponding affine toric variety $\mathrm{Spec}(\mathbb{Z}[Q])$. More generally, for any fan Σ, the multiplication by p map induces a map $\varphi\colon X_\Sigma \to X_\Sigma$.

If K is a perfectoid field and $X_\Sigma = X_{\Sigma,K}$ is a toric variery over K associated to a fan Σ, we can construct a perfectoid space $X^{\mathrm{an}}_{\Sigma_\infty}$ by gluing together the affinoid spaces $\mathrm{Spa}(K[\sigma^\vee \cap M[1/p])$. Note that this is equivalent to the weak limit $X^{\mathrm{an}}_{\Sigma_\infty} \sim$

$\varprojlim_{\varphi} X_{\Sigma}^{\mathrm{an}}$, where X_{Σ}^{an} is the analytification of X_{Σ}. We will omit the superscript an if clear from the context.

We can use a version of the spreading out theorem to control the cohomological motives associated to the perfectoid covers of toric varieties as follows.

COROLLARY 4.25. *Let $X_{\Sigma\infty}$ be the perfectoid cover of $X_{\Sigma} = X_{\Sigma,K}$ obtained as $X_{\Sigma\infty} \sim \varprojlim_{\varphi} X_{\Sigma}$ where φ is the induced by the multiplication by p map as in Equation* (4.24). *The canonical projection induces an equivalence $\mathsf{M}^{coh}(X_{\Sigma\infty}) \simeq \mathsf{M}^{coh}(X_{\Sigma})$.*

PROOF. In light of Equation (2.20), it suffices to show that φ induces an equivalence on the associated (cohomological) motives. Note that X_{Σ} (and the map φ as well) has a smooth proper model $\mathcal{X}_{\Sigma}$ over $\mathcal{O}_K$ so that the motivic map φ is the image under ξ of the motivic map on $\mathsf{M}^{\mathrm{coh}}(\mathcal{X}_{\Sigma,k})$ induced by $\bar{\varphi}\colon \mathcal{X}_{\Sigma,k} \to \mathcal{X}_{\Sigma,k}$. The latter is a universal homeomorphism (by looking at local coordinates, it is in fact a relative Frobenius), and hence invertible in $\mathrm{DA}(k)$ (see Equation (2.14)). □

NOTATION 4.26. From now on, we fix a smooth toric variety X_{Σ} obtained from some fan Σ of a given Euclidean space $N_{\mathbb{R}}$. Let K be a finite extension of $\mathbb{Q}_p$, and let ϖ be a uniformizer of K. Let Z be a smooth projective K-scheme, which is a scheme-theoretic complete intersection inside $X_{\Sigma,K}$. Write X_{Σ}^{an} (resp. Z^{an}) for the analytification of X_{Σ} (resp. of Z).

As a special case of Equation (2.23), we have the following.

COROLLARY 4.27 ([**15**, Corollary 4.5]). *Let $Z \subset X_{\Sigma}$ be as above. There is an open neighborhood $Y \subseteq X_{\Sigma}^{\mathrm{an}}$ of Z^{an} for which $\mathsf{M}^{coh}(Z^{\mathrm{an}}) \simeq \mathsf{M}^{coh}(Y)$.*

Our next goal is to revisit the proof of the weight-monodromy conjecture for Z, isolating the "motivic" part of the argument. Since the statement of the conjecture is insensitive to purely ramified extensions, we may and do replace K by the completion of $K(\varpi^{1/p^{\infty}})$. In particular, we can assume that Z is defined over a perfectoid field. Write $K^{\flat}$ for its tilt. Similarly, write $X_{\Sigma}^{\flat}$ for the smooth toric variety defined over $K^{\flat}$ by the same fan used to construct X_{Σ}. We also write $X_{\Sigma,\infty}^{\mathrm{an}}$ (resp. $X_{\Sigma,\infty}^{\mathrm{an}\flat}$) for the perfectoid cover of X_{Σ} (resp. of $X_{\Sigma}^{\flat}$) considered in Equation (4.24).

The following approximation result is due to Scholze and is one of the essential geometric arguments for the proof.

THEOREM 4.28 ([**54**, Proposition 8.7, Corollary 8.8]). *Let K be a perfectoid field. Let Z be a hypersurface in a smooth proper toric variety X_{Σ} over K and Y a tubular neighborhood of Z^{an} in X_{Σ}^{an}. There exists a hypersurface $Z^{\flat}$ in $X_{\Sigma}^{\flat}$ and an open neighborhood $Y^{\flat}$ of $Z^{\flat\,\mathrm{an}}$ in $X_{\Sigma}^{\flat\,\mathrm{an}}$ such that the open Y_{∞} corresponds to $Y^{\flat}$ under the homeomorphism $|X_{\Sigma\infty}^{\mathrm{an}}| \simeq |X_{\Sigma\infty}^{\mathrm{an}\,\flat}| \simeq |X_{\Sigma}^{\mathrm{an}\,\flat}|$. For any chosen dense subfield of $K^{\flat}$, we may even suppose that the hypersurface $Z^{\flat}$ admits a model over it.*

More generally, if $Z = Y_1 \cap \ldots \cap Y_c$ is a codimension c complete intersection in X_{Σ} and Y is a tubular neighborhood of Z^{an} in X_{Σ}^{an}, there exists a closed subvariety $Z^{\flat}$ in $X_{\Sigma}^{\flat}$ of codimension c and an open neighborhood $Y^{\flat}$ of $Z^{\flat\,\mathrm{an}}$ in $X_{\Sigma}^{\flat\,\mathrm{an}}$ such that the open Y_{∞} corresponds to $Y^{\flat}$. Again, for any chosen dense subfield of $K^{\flat}$, we may suppose that the hypersurface $Z^{\flat}$ admits a model over it.

Construction 4.29. Consider again the situation of Equation (4.26). Replacing again K by the completion of $K(\varpi^{1/p^\infty})$, assume that K is a perfectoid field. Using Equation (4.28), and choosing some motivic tubular neighborhood Y of Z^{an} as in Equation (4.27), we obtain a diagram of adic spaces:

$$\begin{array}{ccccccc}
X^{\mathrm{an}}_{\Sigma} & \longleftarrow & X^{\mathrm{an}}_{\Sigma\infty} & \longleftrightarrow & X^{\flat}_{\Sigma\infty} & \longrightarrow & X^{\flat\,\mathrm{an}}_{\Sigma} \\
\uparrow & & \uparrow & & \uparrow & & \uparrow \\
Y & \longleftarrow & Y_{\infty} & \longleftrightarrow & Y^{\flat}_{\infty} & \longrightarrow & Y^{\flat} \\
\uparrow & & & & & & \uparrow \\
Z^{\mathrm{an}} & & & & & & Z^{\flat\,\mathrm{an}}
\end{array}$$

where in the middle the arrows indicate a tilting equivalence of two perfectoid spaces over K and $K^\flat$. Denote by $(-)^\sharp$ the equivalence $\mathrm{RigDA}(K^\flat) \simeq \mathrm{RigDA}(K)$ of Equation (4.11). The diagram above induces the following diagram of *motives* over K:

$$\begin{array}{ccccccc}
\mathsf{M}^{\mathrm{coh}}(X^{\mathrm{an}}_{\Sigma}) & \xrightarrow{\sim} & \mathsf{M}^{\mathrm{coh}}(X^{\mathrm{an}}_{\Sigma\infty}) & \overset{\sim}{\text{——}} & \mathsf{M}^{\mathrm{coh}}(X^{\flat}_{\infty})^\sharp & \xleftarrow{\sim} & \mathsf{M}^{\mathrm{coh}}(X^{\flat\,\mathrm{an}}_{\Sigma})^\sharp \\
\downarrow & & \downarrow & & \downarrow & & \downarrow \\
\mathsf{M}^{\mathrm{coh}}(Y) & \xrightarrow{\oplus} & \mathsf{M}^{\mathrm{coh}}(Y_{\infty}) & \overset{\sim}{\text{——}} & \mathsf{M}^{\mathrm{coh}}(Y^{\flat}_{\infty})^\sharp & \xleftarrow{\sim} & \mathsf{M}^{\mathrm{coh}}(Y^{\flat})^\sharp \\
\downarrow{\scriptstyle\sim} & & & & & & \downarrow \\
\mathsf{M}^{\mathrm{coh}}(Z^{\mathrm{an}}) & & & & & & \mathsf{M}^{\mathrm{coh}}(Z^{\flat\,\mathrm{an}})^\sharp
\end{array}$$

where the symbol $\oplus$ denotes an inclusion of a direct factor (by Equation (2.22), but this is not important), the equivalences in the middle follow from Equation (4.12), the one in the bottom left follows from the choice of Y (see Equation (4.27)), the one on the top left from Equation (4.25) and the ones on the right from Equation (2.14).

In particular, we have a map $\mathsf{M}^{\mathrm{coh}}(Z^{\mathrm{an}}) \to \mathsf{M}^{\mathrm{coh}}(Z^{\flat\,\mathrm{an}})^\sharp$. We may even use [**25**, Theorem 4.1] to find a smooth alteration $T^\flat$ of $Z^\flat$ therefore giving rise to a map $\mathsf{M}^{\mathrm{coh}}(Z^{\mathrm{an}}) \to \mathsf{M}^{\mathrm{coh}}(T^{\flat\,\mathrm{an}})^\sharp$. We also note that this map is compatible with the diagonal maps $Z \to Z \times Z$ resp. $T^\flat \to T^\flat \times T^\flat$ (as it is given by a composition of geometric maps, tilting equivalences and inverses thereof).

We are going to show that $\mathsf{M}^{\mathrm{coh}}(Z^{\mathrm{an}})$ is a *direct summand* of $\mathsf{M}(T^\flat)^\sharp$. This was observed by Scholze at the level of ℓ-adic cohomology groups, but in fact the proof only relies on a form of Poincaré duality, hence on motivic results, as we now explain.

Construction 4.30. Let X be an algebraic [resp. analytic] variety over K. The associated cohomological motive $M := p_*\mathbb{1} = \mathsf{M}^{\mathrm{coh}}(X) \in \mathrm{DA}(K)$ [resp. in $\mathrm{RigDA}(K)$] inherits a ring structure. If X is smooth, proper of dimension d, then there is a natural duality pairing

$$M \otimes M \to M \to \mathbb{1}(-d)[-2d]$$

which exhibits M as its self-dual, up to twist and shift: $M \simeq \underline{\mathrm{Hom}}(M, \mathbb{1}(-d)[-2d])$.

Assume now M' is another ring object, equipped with a similar duality pairing giving rise to $M' \simeq \underline{\mathrm{Hom}}(M', \mathbb{1}(-d)[-2d])$, for the same d as before. Any ring map

$f\colon M \to M'$ then gives also rise to a dual map

$$f^\vee\colon M' \simeq \underline{\mathrm{Hom}}(M', \mathbb{1}(-d)[-2d]) \xrightarrow{f^*} \underline{\mathrm{Hom}}(M, \mathbb{1}(-d)[-2d]) \simeq M.$$

Note that the composition $f^\vee \circ f\colon M \to M$ corresponds to a pairing $M \otimes M \to \mathbb{1}(-d)[-2d]$ obtained as follows:

$$\begin{array}{ccccc}
M \otimes M & \xrightarrow{\mu} & M & \longrightarrow & \mathbb{1}(-d)[-2d] \\
\downarrow{\scriptstyle f\otimes \mathrm{id}} & & & & \\
M' \otimes M & & \downarrow{\scriptstyle f} & & \vdots\ {\scriptstyle \sim} \\
\downarrow{\scriptstyle \mathrm{id}\otimes f} & & & & \downarrow \\
M' \otimes M' & \xrightarrow{\mu} & M' & \longrightarrow & \mathbb{1}(-d)[-2d]
\end{array}$$

In case there is an invertible automorphism (the dotted arrow above) of $\mathbb{1}(-d)[-2d]$ making the diagram commute, then we can deduce that the map $f^\vee \circ f$ is equivalent to the map corresponding to the duality pairing (i.e. the identity map of M) and hence that M is a direct summand of M'. Luckily, there is an easy criterion to fill in the missing edge, as the next proposition shows.

PROPOSITION 4.31. *Let X be a connected smooth and proper algebraic variety over K of dimension d, with semi-stable reduction. Then*

$$\mathrm{Hom}(\mathsf{M}^{\mathrm{coh}}(X^{\mathrm{an}}), \mathbb{1}(-d)[-2d]) \simeq \mathbb{Q}.$$

In particular, all non-zero maps $\mathsf{M}^{\mathrm{coh}}(X) \to \mathbb{1}(-d)[-2d]$ are equivalent with each other, up to a scalar multiplication.

PROOF. By duality, we may alternatively prove that $\mathrm{Hom}(\mathbb{1}(d)[2d], \mathsf{M}(X)) \simeq \mathbb{Q}$. This group is, by adjunction, the same as $\mathrm{Hom}(\mathbb{1}(d)[2d], \chi\mathsf{M}(X))$. The object on the left is pure with respect to the Chow-weight structure on $\mathrm{RigDA}^{\mathrm{gr}}(K)$ (see [**14**, §4.3]). For the second, there is an explicit description of its weight complex in terms of the irreducible components D_i of its special fiber as shown in [**14**, Example 4.72]. More precisely (we exhibit the complex starting from degree 2, we let D_I be $\bigcap_{i\in I} D_i$ and omit the notation $\mathsf{M}(-)$):

$$\begin{aligned} w(\chi\mathsf{M}(X)) = \cdots \to \bigoplus D_{ij}(-1)[-2] \to \bigoplus D_i(-1)[-2] \to \bigoplus D_i \\ \to \bigoplus D_{ij}(1)[2] \to \cdots \end{aligned}$$

We remark that $\mathrm{Hom}(\mathbb{1}(d)[2d], \bigoplus D_i) \simeq \bigoplus \mathbb{Q}$ and $\mathrm{Hom}(\mathbb{1}(d)[2d], \bigoplus D_{ij}(1)[2]) \simeq \bigoplus \mathbb{Q}$. The terms in positive degree $\mathrm{Hom}(\mathbb{1}(d)[2d], \bigoplus D_I(1)[2])$ are on the other hand equal to 0 for $|I| > 2$. This follows from the identification (as a consequence of e.g. [**64**, Corollary 2])

$$\begin{aligned} \mathrm{Hom}_{\mathrm{DA}(k)}(\mathbb{1}(d)[2d], \mathsf{M}(D_I)(1)[2]) &\simeq \mathrm{Hom}_{\mathrm{DA}(k)}(\mathsf{M}(D_I), \mathbb{1}(1-d+r)[2(1-d+r)]) \\ &\simeq \mathrm{CH}^{1-d+r}(D_I)_{\mathbb{Q}} \end{aligned}$$

where $r = \dim(D_I) = d - |I| + 1$. The exponent $1-d+r$ is negative for $|I| > 2$, so the term is zero. We deduce that $\mathrm{Hom}(\mathbb{1}(d)[2d], \mathsf{M}(X))$ is the kernel of the natural "difference" map $\bigoplus_i \mathbb{Q} \to \bigoplus_{i\neq j} \mathbb{Q}$ which is given by the diagonal embedding $\mathbb{Q} \to \bigoplus_i \mathbb{Q}$. □

COROLLARY 4.32. *Let Z be a smooth scheme-theoretic complete intersection in a smooth projective toric variety X_Σ. The weight-monodromy conjecture holds for the Hyodo–Kato cohomology of Z.*

PROOF. As before, we may and do replace K with a perfectoid extension, and even assume that Z has semi-stable reduction (as we may take any finite extension of K). We let $T^\flat$ be as in Equation (4.29). We may and do suppose that it is defined over some finite extension L of the field $k((p^\flat)) \subseteq K^\flat$ and that $\mathsf{M}^{\mathrm{coh}}(T^\flat) \in \mathrm{RigDA}^{\mathrm{gr}}(L)$.

We then consider the Hyodo–Kato realization functor

$$\mathrm{RigDA}^{\mathrm{gr}}(K) \to \mathcal{D}_{\varphi,N}(W(k)[1/p])$$

and observe that:

(1) The composite realization

$$\mathrm{RigDA}^{\mathrm{gr}}(L) \to \mathrm{RigDA}^{\mathrm{gr}}(K^\flat) \simeq \mathrm{RigDA}^{\mathrm{gr}}(K) \to \mathcal{D}_{\varphi,N}(W(k)[1/p])$$

must agree with the Hyodo–Kato realization functor defined by Lazda–Pal [**43**, Theorem 2.52] by means of Equation (4.19).

(2) The map of motives

$$M := \mathsf{M}^{\mathrm{coh}}(Z) \to M' := \mathsf{M}^{\mathrm{coh}}(T^\flat)^\sharp$$

is a map of ring objects (since the multiplication is induced by the diagonal map, see the end of Equation (4.29)) and the map $M \to M' \to \mathbb{1}(-d)[-2d]$ fits in the following diagram

$$\begin{array}{ccccc} \mathsf{M}^{\mathrm{coh}}(X_\Sigma^{\mathrm{an}}) & \longrightarrow & M & & \\ \downarrow{\scriptstyle\sim} & & \downarrow & \searrow & \\ \mathsf{M}^{\mathrm{coh}}(X_\Sigma^{\flat\mathrm{an}})^\sharp & \longrightarrow & M' & \longrightarrow & \mathbb{1}(-d)[-2d] \end{array} \tag{4.33}$$

(3) The lower horizontal map is the (untilt of the) analytification of the algebraic map

$$\mathsf{M}^{\mathrm{coh}}(X_\Sigma^\flat) \to \mathsf{M}^{\mathrm{coh}}(T^\flat) \to \mathbb{1}(-d)[-2d]$$

in $\mathrm{DA}(L)$. All the motives above are dualizable, and the analytification functor on motives is monoidal. We then deduce a diagram (where we denote by $\{n\}$ the operation $(n)[2n]$, for brevity):

$$\begin{array}{ccc} \mathrm{Hom}_{\mathrm{DA}(L)}(\mathbb{1}\{-d\}, \mathsf{M}^{\mathrm{coh}}(X_\Sigma^\flat)) \simeq CH^d(X_\Sigma^\flat)_\mathbb{Q} & \longrightarrow & \mathrm{Hom}_{\mathrm{RigDA}(L)}(\mathbb{1}\{-d\}, \mathsf{M}^{\mathrm{coh}}(X_\Sigma^{\mathrm{an}\,\flat})) \\ \downarrow{\scriptstyle \iota^!} & & \downarrow \\ \mathrm{Hom}_{\mathrm{DA}(L)}(\mathbb{1}\{-d\}, \mathsf{M}^{\mathrm{coh}}(T^\flat)) \simeq CH^d(T^\flat)_\mathbb{Q} & \longrightarrow & \mathrm{Hom}_{\mathrm{RigDA}(L)}(\mathbb{1}\{-d\}, \mathsf{M}^{\mathrm{coh}}(T^{\mathrm{an}\,\flat})) \\ \downarrow & & \downarrow \\ \mathrm{Hom}_{\mathrm{DA}(L)}(\mathbb{1}\{-d\}, \mathbb{1}\{-d\}) & \longrightarrow & \mathrm{Hom}_{\mathrm{RigDA}(L)}(\mathbb{1}\{-d\}, \mathbb{1}\{-d\}) \\ \downarrow{\scriptstyle\sim} & & \downarrow{\scriptstyle\sim} \\ \mathbb{Q} & \xrightarrow{\sim} & \mathbb{Q} \end{array}$$

where ι is the inclusion $T^\flat \subset X^\flat_\Sigma$. Note that the left column is a non-zero map. Indeed, the composition

$$\mathrm{CH}^d(X^\flat_\Sigma) \xrightarrow{\iota^!} \mathrm{CH}^d(T^\flat) \xrightarrow{\deg} \mathbb{Z}$$

is always non-zero (it suffices to notice that $\deg(h^!(c_1(\mathcal{L})^d)) \neq 0$ where $\mathcal{L}$ is an ample line bundle on $X^\flat_\Sigma$ by [**33**, Lemma 12.1]).

As a consequence, we deduce that the composite map in the right column is non-zero as well. In particular, the lower horizontal map of the diagram (4.33) is non-zero, hence so is the diagonal map.

We then deduce by Equation (4.31) that $\mathsf{M}^{\mathrm{coh}}(X)$ is a direct summand of $\mathsf{M}^{\mathrm{coh}}(T^\flat)^\sharp$, and so are its Hyodo–Kato cohomology groups (that are (φ, N)-modules). We can then conclude that $H^*_{HK}(X)$ satisfies the weight-monodromy conjecture, since $H^*_{HK}(T^\flat)$ does: this is a result by Crew [**24**] and Lazda–Pal [**43**, Theorem 5.58], that builds on a proof of Deligne [**27**]. □

REMARK 4.34. In the previous proof, one can test the "zero-ness" of the map $M \to \mathbb{1}(-d)[-2d]$ under any given realization functor. In particular, using the notation above, if one shows that the induced map $H^{2d}(X) \to H^{2d}(M') \simeq \mathbb{1}(d)$ is non-zero for some cohomological realization, then one can deduce that $M = \mathsf{M}^{\mathrm{coh}}(X)$ is a direct summand of M'. This is the approach taken in [**54**] (and in [**15**]).

4.5. On the Clemens-Schmid sequence. The motivic nearby cycle can also be used to give a uniform treatment (in the ℓ-adic, p-adic and in the complex analytic setting) of the so-called local invariant cycle theorem.

Let us begin by casting the problem in the complex analytic case. Let $f\colon X \to S$ be a proper morphism between smooth complex analytic spaces, where S has dimension one. Localizing around an analytic neighborhood of $0 \in S(\mathbb{C})$ and passing to a finite cover, we may replace S by an open disc Δ, suppose that the central fiber $Y = f^{-1}(0)$ is a normal crossing divisor in X, and that f is smooth away from zero. Write X_t for the (smooth) fiber of f at a point $t \neq 0$. Since the restriction of f to Δ^* is topologically a locally trivial bundle, a generator of the fundamental group $\pi_1(\Delta^*) \simeq \mathbb{Z}$ acts on the vector space $H^q(X_t, \mathbb{Q})$ by an operator N, which is in fact nilpotent. As such, it induces a canonical filtration M, for which N^r restricts to an isomorphism $\mathrm{gr}^M_r H^q(X_t, \mathbb{Q}) \to \mathrm{gr}^M_{-r} H^q(X_t, \mathbb{Q})$ The morphism N can be promoted to a morphism of mixed Hodge structures when the vector space $H^q(X_t, \mathbb{Q})$ is equipped with Steenbrink's limit Hodge structure $H^n_{\lim}(X, \mathbb{Q}(m))$ from [**56**]. The kernel of the monodromy operator captures some cohomological information of the singular space Y. More precisely, one has the following result:

THEOREM 4.35 (Local invariant cycle theorem [**21**, 3.7]). *Assume that X is a Kähler manifold. The sequence*

$$H^n(Y, \mathbb{Q}(m)) \xrightarrow{\mathrm{sp}} H^n_{\lim}(X, \mathbb{Q}(m)) \xrightarrow{N} H^n_{\lim}(X, \mathbb{Q}(m-1)) \tag{4.36}$$

where sp *is the canonical specialization morphism, is exact in the category of mixed Hodge structures.*

The specialization morphism arises as follows. Let $\overline{X}^*$ be the base change of $X^* = X \times_\Delta \Delta^*$ to a universal cover $\overline{\Delta^*} \to \Delta^*$. Then for $t \neq 0$ we have $H^*(\overline{X}^*, \mathbb{Q}) \simeq H^*(X_t, \mathbb{Q})$ as $\mathbb{Q}$-vector spaces, and there is a natural morphism

$H^n(Y, \mathbb{Q}(m)) \to H^n(\overline{X}^*, \mathbb{Q}(m))$ by localization. The main content of the local invariant cycle theorem is that this is indeed a morphism of mixed Hodge structures when $H^*(X_t, \mathbb{Q})$ is equipped with the limit Hodge structure $H^n_{\lim}(X, \mathbb{Q}(m))$, and that the kernel of the monodromy map N has the correct weights.

Using Poincaré duality one can in fact complete the exact sequence (4.36) to the left and to the right and obtain a long exact sequence:

THEOREM 4.37 (Clemens-Schmid sequence [**48**], [**21**]). *Assume that X is a Kähler manifold. Then there are (two) long exact sequences*

$$\cdots \to H^n(Y, \mathbb{Q}(m)) \xrightarrow{\text{sp}} H^n_{\lim}(X, \mathbb{Q}(m)) \xrightarrow{N} H^n_{\lim}(X, \mathbb{Q}(m-1)) \to$$
$$\to H_{2d-n}(Y, \mathbb{Q}(m-d-1)) \longrightarrow H^{n+2}(Y, \mathbb{Q}(m-1)) \xrightarrow{\text{sp}} H^{n+2}_{\lim}(X, \mathbb{Q}(m-1)) \to \dots$$

of mixed Hodge structures.

The local invariant cycle theorem is then part of the exactness of Clemens-Schmid sequence. Note that by duality the exactness at the spot $H^n_{\lim}(X, \mathbb{Q}(m))$ is essentially enough to prove exactness at the remaining terms (see the remark at [**21**, p. 229]). In turn, this is a consequence of the complex analytic version of the weight-monodromy conjecture.

THEOREM 4.38 (Monodromy-Weight, Hodge version [**35**, Théorème 5.2]). *In the above assumptions, for each $r \geq 0$ and each n, the morphism $N^r : H^n_{\lim}(X, \mathbb{Q}) \to H^n_{\lim}(X, \mathbb{Q}(-r))$ induces an isomorphism of pure Hodge structures*

$$\mathrm{gr}^W_{n+r} H^n_{\lim}(X, \mathbb{Q}) \xrightarrow{\simeq} \mathrm{gr}^W_{n-r} H^n_{\lim}(X, \mathbb{Q})$$

where W is the weight filtration of the Hodge structure $H^n_{\lim}(X, \mathbb{Q})$. Equivalently, the weight and the monodromy filtrations agree (up to a shift by n).

For the proof of the exactness of the Clemens-Schmid sequence, it is enough to know that the kernel of the monodromy N on $H^i_{\lim}(X)$ has weights $\leq i$. But this is precisely guaranteed by the concidence of the weight filtration and of the monodromy filtration on $H^i_{\lim}(X)$.

REMARK 4.39. Equation (4.35) is due to Deligne (see [**26**]) in the algebraic setting, and to Clemens in the stated generality. The most general approach using the theory of mixed Hodge modules is due to Saito [**52**] (see [**38**] for an extensive discussion). Note that the assumptions are optimal: as remarked by Nakkajima [**49**] (see also the discussion in the original paper of Clemens [**21**]), the statement is false for arbitrary proper families of complex analytic spaces. From the motivic point of view, this implies that a generalization of the local invariant cycle theorem cannot hold for arbitrary objects of $\mathrm{RigDA}(\mathbb{C}((t)))^{\mathrm{gr}}$.

REMARK 4.40. Following the approach of Deligne in the proof of the ℓ-adic weight-monodromy conjecture (see the proof of [**27**, Théorème 1.8.4]), having an estimate for the weights of the kernel of the monodromy operator on H^i, together with a product formula, is sufficent to establish the full conjecture. For this reason, we expect that the exactness of the Clemens-Schmid sequence for the variety X and its self-products is in fact equivalent to the weight-monodromy conjecture.

From our perspective, all of the above arises more naturally as realization of the monodromy action on the motivic nearby cycle. To explain this, let us consider the following general setting.

CONSTRUCTION 4.41. Let K be a complete non-archimedean field with a perfect residue field and a value group of (rational) rank 1. Let k be its residue field and denote by $\mathcal{O}_K$ the valuation ring. Let us now write $\mathrm{DA}(\mathcal{O}_K)^{\mathrm{gr}}$ [resp. $\mathrm{DA}(K)^{\mathrm{gr}}$] for the full subcategory of $\mathrm{DA}(\mathcal{O}_K)$ [resp. $\mathrm{DA}(K)^{\mathrm{gr}}$] defined as the inverse image of $\mathrm{RigDA}(K)^{\mathrm{gr}}$ along the composite

$$\mathrm{DA}(\mathcal{O}_K) \xrightarrow{j^*} \mathrm{DA}(K) \xrightarrow{\mathrm{An}^*} \mathrm{RigDA}(K).$$

[resp. An^*]. This category in particular contains the motive of any regular scheme $\mathcal{X}$, flat over $\mathcal{O}_K$ and with strict semi-stable reduction over k. We then have:

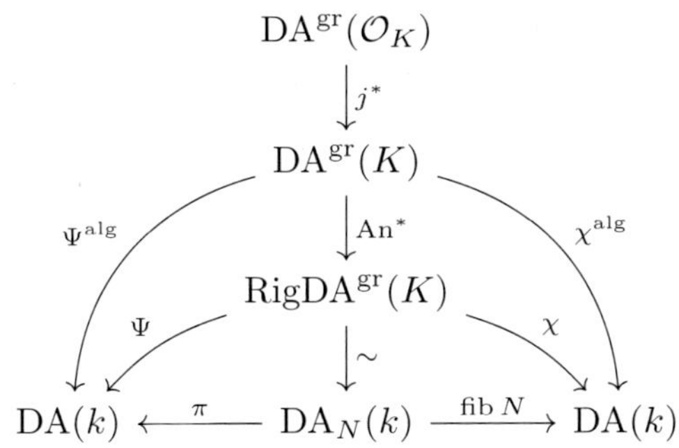

where χ^{alg} is the motivic algebraic specialization functor $\iota^* j_*$, and χ is the motivic analytic specialization functor, right adjoint to the generic fiber functor ξ. The commutations on the left are by definition, while on the right they follow from [**10**, Corollary 3.8.18 and Theorem 3.8.19], and Equation (4.7).

Combining the localization sequence of Item (v)-(v) with the previous discussion we obtain the following commutative diagram of natural transformation of functors from $\mathrm{DA}(\mathcal{O}_K)^{\mathrm{gr}}$ to $\mathrm{DA}(k)$

(4.42)

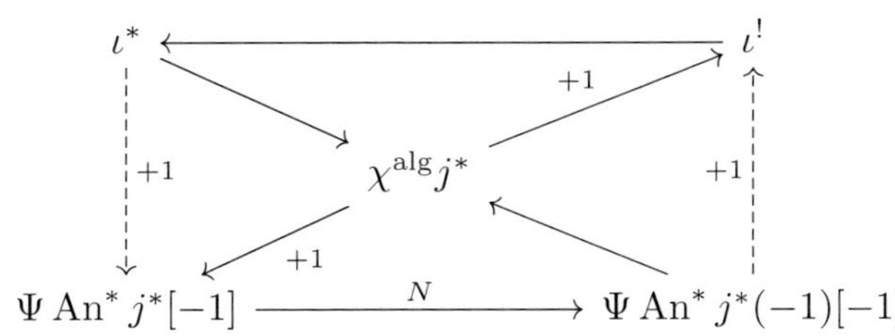

We can compose the square above with any realization functor $R\Gamma\colon \mathrm{DA}(k) \to \mathscr{D}$ to obtain a Clemens–Schimd style complex. More precisely, we have the following result.

COROLLARY 4.43 ([**14**, Remark 4.68, Corollary 4.70]). *Let* $R\Gamma\colon \mathrm{DA}(k) \to \mathscr{D}$ *be a monoidal functor of presentable stable* ∞*-categories, and assume that* $\mathscr{D}$ *is endowed with a t-structure compatible with tensor products and with heart* $\mathcal{A}$. *Let* $\widehat{R\Gamma}$ *be the canonical extension of* (4.16), *following Equation* (4.5).

Let $\mathcal{X}$ *be a flat, proper, regular scheme of essential relative dimension* d *over* $\mathcal{O}_K$ *which is generically smooth and with polystable special fiber* Y *over* k *(so that*

$\mathrm{An}^* j^* \mathsf{M}(\mathcal{X}) = \mathsf{M}(\mathcal{X}_K^{\mathrm{An}}) \in \mathrm{RigDA}(K)^{\mathrm{gr}}$ *by Equation* (4.9)). *Then there are (two) chain complexes in* $\mathcal{A}$*:*

$$\cdots \to \mathrm{H}^*(Y) \to \widehat{\mathrm{H}}^*(\mathcal{X}_K^{\mathrm{An}}) \xrightarrow{N} \widehat{\mathrm{H}}^*(\mathcal{X}_K^{\mathrm{An}})(-1) \to \mathrm{H}_{2d-*}(Y)(-d-1) \to \mathrm{H}^{*+2}(Y) \to \cdots$$

where we let $\mathrm{H}_*(M)$ *resp.* $\widehat{\mathrm{H}}_*(M)$ *be the objects* $\mathrm{H}_* R\Gamma(M)$ *resp.* $\mathrm{H}_* \pi\widehat{R\Gamma}(M) \cong \mathrm{H}_* R\Gamma(\Psi M)$ *and we let* $\mathrm{H}^*(M)$ *resp.* $\widehat{\mathrm{H}}^*(M)$ *be their duals.*

PROOF. The Corollary follows immediately by applying the realization functor to the square (4.42), together with absolute purity. □

When $\widehat{R\Gamma}$ is the canonical extension of the Hodge realization

$$R\Gamma_{\mathrm{Hodge}} \colon \mathrm{DA}(\mathbb{C}) \to \mathrm{Ind}\,\mathcal{D}^b(\mathrm{MHS}^p_{\mathbb{Q}})$$

of Equation (2.9) we obtain exactly the sequence in Equation (4.37).

REMARK 4.44. When $\widehat{R\Gamma}$ is the Hyodo-Kato realization functor, we obtain in particular that for every $\mathcal{X}/\mathcal{O}_K$ as above a *complex* of φ-modules

$$\ldots \to H^n_{\mathrm{rig}}(Y) \xrightarrow{\mathrm{sp}} H^n_{HK}(\mathcal{X}_K^{\mathrm{An}}) \xrightarrow{N} H^n_{HK}(\mathcal{X}_K^{\mathrm{An}})(-1) \to \ldots$$

that is exact when the p-adic weight-monodromy conjecture holds. Exactness at the spot $H^n_{HK}(\mathcal{X}_K^{\mathrm{An}})$ is the p-adic analogue of the local invariant cycle theorem. The existence of such sequence (as well as the exactness statement) was originally conjectured by Chiarellotto [**18**] in the semi-stable setting, and then conjecturally extended to the analogue of the Clemens-Schmid sequence by Flach and Morin [**31**]. To our knowledge, not even the morphisms appearing in the p-adic Clemens-Schmid complex had been explicitly defined in this generality prior to [**14**].

REMARK 4.45. In more classical terms, for *proper* $\mathcal{X}$, the identifications of [**37**] allow to describe the map sp as follows. First, there is a morphism (constructed by Chiarellotto in [**18**])

$$H^n_{\mathrm{rig}}(Y/K_0) \to H^n((Y,N)/(W(k),\mathbb{N})) \otimes K_0$$

where (Y,N) is the proper semistable log scheme over the log point k^0 obtained by equipping the special fiber of $\mathcal{X}$ with its canonical log structure, and

$$H^n((Y,N)/(W(k),\mathbb{N}))$$

is the log crystalline cohomology group relative to the base $(W(k),L)$, with $1 \mapsto 0$. The Hyodo-Kato isomorphism [**37**, Proposition 4.13] further gives an identification

$$H^n((Y,N)/(W(k),\mathbb{N})) \otimes K_0 \simeq H^n((Y,N)/(W(k)^\times) \otimes K_0$$

where the right-hand side denotes the crystalline cohomology of (Y,N) computed relative to the standard log structure on $W(k)$ given by $1 \mapsto p$. This is in turn isomorphic to the de Rham cohomology of the (analytic) generic fiber $H^n_{\mathrm{dR}}(\mathcal{X}_K^{\mathrm{An}})$ by [**37**, Theorem 5.1] (in this case clearly agrees with algebraic de Rham cohomology, since $\mathcal{X}_K$ is smooth and proper over K). Our method completely bypasses the reference to the log crystalline cohomology groups (relative to different log bases), and constructs directly the map sp between $H^n_{\mathrm{rig}}(Y)$ and $H^n_{HK}(\mathcal{X}_K^{\mathrm{An}})$.

In log-terms, using the comparison isomorphisms above, we can interpret the Clemens–Schmid complex as connecting the rigid cohomology and the log-crystalline cohomology (after inverting p) of the *singular* special fiber Y.

Acknowledgments

We thank the organizers and the participants of the conference "Regulators V" for their hospitality and the useful interactions, and an anonymous referee for their comments and corrections to an early version of this paper.

References

[1] *Théorie des topos et cohomologie étale des schémas. Tome 2*, Lecture Notes in Mathematics, Vol. 270, Springer-Verlag, Berlin, 1972. Séminaire de Géométrie Algébrique du Bois-Marie 1963–1964 (SGA 4), Dirigé par M. Artin, A. Grothendieck et J. L. Verdier. Avec la collaboration de N. Bourbaki, P. Deligne et B. Saint-Donat.

[2] G. Andreychev, *Pseudocoherent and perfect complexes and vector bundles on analytic adic spaces.* `arXiv:2105.12591 [math.AG]`, 2021.

[3] A. Arabia, *Relèvements des algèbres lisses et de leurs morphismes* (French, with English and French summaries), Comment. Math. Helv. **76** (2001), no. 4, 607–639, DOI 10.1007/s00014-001-8322-y. MR1881700

[4] J. Ayoub, *Les six opérations de Grothendieck et le formalisme des cycles évanescents dans le monde motivique. I* (French, with English and French summaries), Astérisque **314** (2007), x+466 pp. (2008). MR2423375

[5] J. Ayoub, *Les six opérations de Grothendieck et le formalisme des cycles évanescents dans le monde motivique. II* (French, with English and French summaries), Astérisque **315** (2007), vi+364 pp. (2008). MR2438151

[6] J. Ayoub, *La réalisation étale et les opérations de Grothendieck* (French, with English and French summaries), Ann. Sci. Éc. Norm. Supér. (4) **47** (2014), no. 1, 1–145, DOI 10.24033/asens.2210. MR3205601

[7] J. Ayoub, *L'algèbre de Hopf et le groupe de Galois motiviques d'un corps de caractéristique nulle, I* (French, with French summary), J. Reine Angew. Math. **693** (2014), 1–149, DOI 10.1515/crelle-2012-0089. MR3259031

[8] J. Ayoub, *Motifs des variétés analytiques rigides* (French, with English and French summaries), Mém. Soc. Math. Fr. (N.S.) **140-141** (2015), vi+386, DOI 10.24033/msmf.449. MR3381140

[9] J. Ayoub, *Weil cohomology theories and their motivic Hopf algebroids*, Indag. Math. (N.S.) **37** (2026), no. 1, 195–249, DOI 10.1016/j.indag.2024.09.009. MR5011370

[10] J. Ayoub, M. Gallauer, and A. Vezzani, *The six-functor formalism for rigid analytic motives*, Forum Math. Sigma **10** (2022), Paper No. e61, 182, DOI 10.1017/fms.2022.55. MR4466640

[11] F. Bambozzi and A. Vezzani, *Rigidity for rigid analytic motives*, J. Inst. Math. Jussieu **20** (2021), no. 4, 1341–1369, DOI 10.1017/S1474748019000501. MR4293798

[12] V. G. Berkovich, *Smooth p-adic analytic spaces are locally contractible*, Invent. Math. **137** (1999), no. 1, 1–84, DOI 10.1007/s002220050323. MR1702143

[13] P. Berthelot, *Géométrie rigide et cohomologie des variétés algébriques de caractéristique p* (French, with English summary), Mém. Soc. Math. France (N.S.) **23** (1986), 3, 7–32. Introductions aux cohomologies p-adiques (Luminy, 1984). MR865810

[14] F. Binda, M. Gallauer, and A. Vezzani, *Motivic monodromy and p-adic cohomology theories.* J. Eur. Math. Soc. (JEMS), Online First, 2025. DOI: 10.4171/JEMS/1634.

[15] F. Binda, H. Kato, and A. Vezzani, *On the p-adic weight-monodromy conjecture for complete intersections in toric varieties*, Invent. Math. **241** (2025), no. 2, 559–603, DOI 10.1007/s00222-025-01344-x. MR4929650

[16] F. Binda, D. Park, and P. A. Østvær, *Triangulated categories of logarithmic motives over a field* (English, with English and French summaries), Astérisque **433** (2022), ix+267, DOI 10.24033/ast. MR4439182

[17] D. Caro, *Sur la préservation de la surconvergence par l'image directe d'un morphisme propre et lisse* (French, with English and French summaries), Ann. Sci. Éc. Norm. Supér. (4) **48** (2015), no. 1, 131–169, DOI 10.24033/asens.2240. MR3335840

[18] B. Chiarellotto, *Rigid cohomology and invariant cycles for a semistable log scheme*, Duke Math. J. **97** (1999), no. 1, 155–169, DOI 10.1215/S0012-7094-99-09707-7. MR1682272

[19] D.-C. Cisinski and F. Déglise, *Mixed Weil cohomologies*, Adv. Math. **230** (2012), no. 1, 55–130, DOI 10.1016/j.aim.2011.10.021. MR2900540

[20] D.-C. Cisinski and F. Déglise, *Triangulated categories of mixed motives*, Springer Monographs in Mathematics, Springer, Cham, [2019] ©2019, DOI 10.1007/978-3-030-33242-6. MR3971240

[21] C. H. Clemens, *Degeneration of Kähler manifolds*, Duke Math. J. **44** (1977), no. 2, 215–290. MR444662

[22] P. Colmez and W. Nizioł, *On p-adic comparison theorems for rigid analytic varieties, I*, Münster J. Math. **13** (2020), no. 2, 445–507. MR4130689

[23] D. A. Cox, J. B. Little, and H. K. Schenck, *Toric varieties*, Graduate Studies in Mathematics, vol. 124, American Mathematical Society, Providence, RI, 2011, DOI 10.1090/gsm/124. MR2810322

[24] R. Crew, *Finiteness theorems for the cohomology of an overconvergent isocrystal on a curve* (English, with English and French summaries), Ann. Sci. École Norm. Sup. (4) **31** (1998), no. 6, 717–763, DOI 10.1016/S0012-9593(99)80001-9. MR1664230

[25] A. J. de Jong, *Smoothness, semi-stability and alterations*, Inst. Hautes Études Sci. Publ. Math. **83** (1996), 51–93. MR1423020

[26] P. Deligne, *Théorie de Hodge. I* (French), Actes du Congrès International des Mathématiciens (Nice, 1970), Gauthier-Villars Éditeur, Paris, 1971, pp. 425–430. MR441965

[27] P. Deligne, *La conjecture de Weil. II* (French), Inst. Hautes Études Sci. Publ. Math. **52** (1980), 137–252. MR0601520

[28] R. Elkik, *Solutions d'équations à coefficients dans un anneau hensélien* (French), Ann. Sci. École Norm. Sup. (4) **6** (1973), 553–603 (1974). MR345966

[29] V. Ertl and A. Vezzani, *Berthelot's conjecture via homotopy theory*, Duke Math. J. **174** (2025), no. 18, 3561–3614, DOI 10.1215/00127094-2025-0021. MR5000780

[30] V. Ertl and K. Yamada, *Rigid analytic reconstruction of Hyodo–Kato theory.* Preprint arXiv:1907.10964, 2019.

[31] M. Flach and B. Morin, *Weil-étale cohomology and zeta-values of proper regular arithmetic schemes*, Doc. Math. **23** (2018), 1425–1560. MR3874942

[32] K. Fujiwara and F. Kato, *Foundations of rigid geometry. I*, EMS Monographs in Mathematics, European Mathematical Society (EMS), Zürich, 2018. MR3752648

[33] W. Fulton, *Intersection theory*, 2nd ed., Ergebnisse der Mathematik und ihrer Grenzgebiete. 3. Folge. A Series of Modern Surveys in Mathematics [Results in Mathematics and Related Areas. 3rd Series. A Series of Modern Surveys in Mathematics], vol. 2, Springer-Verlag, Berlin, 1998, DOI 10.1007/978-1-4612-1700-8. MR1644323

[34] E. Grosse-Klönne, *Frobenius and monodromy operators in rigid analysis, and Drinfel′d's symmetric space*, J. Algebraic Geom. **14** (2005), no. 3, 391–437, DOI 10.1090/S1056-3911-05-00402-9. MR2129006

[35] F. Guillén and V. Navarro Aznar, *Sur le théorème local des cycles invariants* (French), Duke Math. J. **61** (1990), no. 1, 133–155, DOI 10.1215/S0012-7094-90-06107-1. MR1068383

[36] R. Huber, *Étale cohomology of rigid analytic varieties and adic spaces*, Aspects of Mathematics, E30, Friedr. Vieweg & Sohn, Braunschweig, 1996, DOI 10.1007/978-3-663-09991-8. MR1734903

[37] O. Hyodo and K. Kato, *Semi-stable reduction and crystalline cohomology with logarithmic poles*, Astérisque **223** (1994), 221–268. Périodes p-adiques (Bures-sur-Yvette, 1988). MR1293974

[38] L. Illusie, *Autour du théorème de monodromie locale* (French), Astérisque **223** (1994), 9–57. Périodes p-adiques (Bures-sur-Yvette, 1988). MR1293970

[39] U. Jannsen, *On the l-adic cohomology of varieties over number fields and its Galois cohomology*, Galois groups over $\mathbf{Q}$ (Berkeley, CA, 1987), Math. Sci. Res. Inst. Publ., vol. 16, Springer, New York, 1989, pp. 315–360, DOI 10.1007/978-1-4613-9649-9_5. MR1012170

[40] K. Kato, *Logarithmic structures of Fontaine-Illusie*, Algebraic analysis, geometry, and number theory (Baltimore, MD, 1988), Johns Hopkins Univ. Press, Baltimore, MD, 1989, pp. 191–224. MR1463703

[41] S. Kelly and M. Morrow, *K-theory of valuation rings*, Compos. Math. **157** (2021), no. 6, 1121–1142, DOI 10.1112/s0010437x21007119. MR4264079

[42] T. Koshikawa, *Logarithmic prismatic cohomology, I.* Preprint arXiv:2007.14037, 2022.

[43] C. Lazda and A. Pál, *Rigid cohomology over Laurent series fields*, Algebra and Applications, vol. 21, Springer, [Cham], 2016, DOI 10.1007/978-3-319-30951-4. MR3495307

[44] A.-C. Le Bras and A. Vezzani, *The de Rham–Fargues–Fontaine cohomology*, Algebra Number Theory **17** (2023), no. 12, 2097–2150, DOI 10.2140/ant.2023.17.2097. MR4650391

[45] A. Mathew, N. Naumann, and J. Noel, *Nilpotence and descent in equivariant stable homotopy theory*, Adv. Math. **305** (2017), 994–1084, DOI 10.1016/j.aim.2016.09.027. MR3570153

[46] A. Mokrane, *La suite spectrale des poids en cohomologie de Hyodo-Kato* (French), Duke Math. J. **72** (1993), no. 2, 301–337, DOI 10.1215/S0012-7094-93-07211-0. MR1248675

[47] P. Monsky and G. Washnitzer, *Formal cohomology. I*, Ann. of Math. (2) **88** (1968), 181–217, DOI 10.2307/1970571. MR248141

[48] D. R. Morrison, *The Clemens-Schmid exact sequence and applications*, Topics in transcendental algebraic geometry (Princeton, N.J., 1981/1982), Ann. of Math. Stud., vol. 106, Princeton Univ. Press, Princeton, NJ, 1984, pp. 101–119. MR756848

[49] Y. Nakkajima, *Signs in weight spectral sequences, monodromy-weight conjectures, log Hodge symmetry and degenerations of surfaces*, Rend. Sem. Mat. Univ. Padova **116** (2006), 71–185. MR2287344

[50] D. Park, *Log motivic nearby cycles*. `https://arxiv.org/abs/2405.14083`, 2024.

[51] M. Rapoport and Th. Zink, *Über die lokale Zetafunktion von Shimuravarietäten. Monodromiefiltration und verschwindende Zyklen in ungleicher Charakteristik* (German), Invent. Math. **68** (1982), no. 1, 21–101, DOI 10.1007/BF01394268. MR666636

[52] M. Saito, *Modules de Hodge polarisables* (French), Publ. Res. Inst. Math. Sci. **24** (1988), no. 6, 849–995 (1989), DOI 10.2977/prims/1195173930. MR1000123

[53] W. Schmid, *Variation of Hodge structure: the singularities of the period mapping*, Invent. Math. **22** (1973), 211–319, DOI 10.1007/BF01389674. MR382272

[54] P. Scholze, *Perfectoid spaces*, Publ. Math. Inst. Hautes Études Sci. **116** (2012), 245–313, DOI 10.1007/s10240-012-0042-x. MR3090258

[55] A. Shiho, *Relative log convergent cohomology and relative rigid cohomology III*. arXiv:0805.3229 [math.NT], 2008.

[56] J. Steenbrink, *Limits of Hodge structures*, Invent. Math. **31** (1975/76), no. 3, 229–257, DOI 10.1007/BF01403146. MR429885

[57] N. Tsuzuki, *On base change theorem and coherence in rigid cohomology*, Doc. Math. **Extra Vol.** (2003), 891–918. Kazuya Kato's fiftieth birthday. MR2046617

[58] S. Tubach, *On the Nori and Hodge realisations of Voevodsky motives*, Compos. Math. **161** (2025), no. 9, 2155–2201, DOI 10.1112/S0010437X25102431. MR4971596

[59] A. Vezzani, *The Monsky-Washnitzer and the overconvergent realizations*, Int. Math. Res. Not. IMRN **11** (2018), 3443–3489, DOI 10.1093/imrn/rnw335. MR3810223

[60] A. Vezzani, *A motivic version of the theorem of Fontaine and Wintenberger*, Compos. Math. **155** (2019), no. 1, 38–88, DOI 10.1112/s0010437x18007595. MR3880025

[61] A. Vezzani, *Rigid cohomology via the tilting equivalence*, J. Pure Appl. Algebra **223** (2019), no. 2, 818–843, DOI 10.1016/j.jpaa.2018.05.001. MR3850572

[62] A. Vezzani, *The relative (de-)perfectoidification functor and motivic p-adic cohomologies*, Perfectoid spaces, Infosys Sci. Found. Ser., Springer, Singapore, [2022] ©2022, pp. 15–36, DOI 10.1007/978-981-16-7121-0_2. MR4433440

[63] A. Vezzani, *A motivic proof of the finiteness of the relative de Rham cohomology*, Arch. Math. (Basel) **123** (2024), no. 4, 393–398, DOI 10.1007/s00013-024-02024-7. MR4806532

[64] V. Voevodsky, *Motivic cohomology groups are isomorphic to higher Chow groups in any characteristic*, Int. Math. Res. Not. **7** (2002), 351–355, DOI 10.1155/S107379280210403X. MR1883180

Dipartimento di Matematica "F. Enriques" - Università degli Studi di Milano (Italy)
Email address: `federico.binda@unimi.it`

Dipartimento di Matematica "F. Enriques" - Università degli Studi di Milano (Italy)
Email address: `alberto.vezzani@unimi.it`

Contemporary Mathematics
Volume **842**, 2026
https://doi.org/10.1090/conm/842/16851

Quadratically enriched binomial coefficients over a finite field

Chongyao Chen and Kirsten Wickelgren

Abstract. We compute an analogue of Pascal's triangle enriched in bilinear forms over a finite field. This gives an arithmetically meaningful count of the ways to choose j ring homomorphisms into an algebraic closure from an étale extension of degree n. We also compute a quadratic twist. These (twisted) enriched binomial coefficients are defined in joint work of Brugallé and the second-named author, building on work of Serre. Such binomial coefficients support curve counting results over non-algebraically closed fields, using $\mathbb{A}^1$-homotopy theory.

1. Introduction

We consider combinatorics enriched in bilinear forms, in the sense that an integer n is replaced by the class of a symmetric, non-degenerate bilinear form on a vector space of dimension n. The resulting binomial coefficients arose in [**1**] in the context of curve counting over non-algebraically closed fields: to count curves on surfaces, one is led to certain degeneration formulas in which curves of lower degrees are glued together. To perform such gluings, one chooses closed points. However, over non-algebraically closed fields, the points have different residue fields. To obtain a count retaining arithmetic information, it is effective to replace integer valued counts with ones enriched in bilinear forms. The counts now take values in the Grothendieck–Witt group of the base field, defined to be the group completion of isomorphism classes of symmetric, non-degenerate bilinear forms. See for example [**8**] [**11**] [**15**] [**14**] [**13**] [**3**]. In [**1**], Erwan Brugallé and the second named author obtained a wall-crossing formula for $\mathbb{A}^1$ Gromov–Witten invariants using combinatorial identities enriched in bilinear forms. Here we systematically compute the analogue of Pascal's triangle over a finite field of odd characteristic and a quadratic twist, arising in the enumerative context of [**1**],[**7**].

Let $k = \mathbb{F}_q$ be a finite field with odd characteristic. Let $\mathrm{GW}(k)$ denote the Grothendieck–Witt group of k and let $u \in \mathrm{GW}(\mathbb{F}_q)$ denote the class of the bilinear form $\mathbb{F}_q \times \mathbb{F}_q \to \mathbb{F}_q$ sending (x, y) to μxy, where μ is a non-square in $\mathbb{F}_q^*$. There is a ring structure on $\mathrm{GW}(k)$ induced from the tensor product of forms, and the unit, denoted by 1, is represented by the bilinear form $\mathbb{F}_q \times \mathbb{F}_q \to \mathbb{F}_q$ sending (x, y) to

2020 *Mathematics Subject Classification.* Primary 05A10, 11E81, 14F42.

The first author was partially supported by National Science Foundation Awards DMS-2304981 and DMS-2405191.

The second author was partially supported by National Science Foundation Awards DMS-2103838 and DMS-2405191.

xy. For an étale k-algebra L, the paper [**1**] defines $\binom{L/k}{j}$ in $\mathrm{GW}(k)$ to be the trace form of the étale k-algebra associated via the Galois correspondence to the set of subsets of size j of the set of k-maps of L into $\overline{k}$. Here, $\overline{k}$ denotes the algebraic closure of k.

For a quadratic extension Q of k and $[L : k] = 2j$, this set has a twisted action, where $\mathrm{Gal}(\overline{Q}/k)$ acts by swapping two complementary subsets. This defines a quadratically twisted binomial coefficient $\binom{L[Q]/k}{j}$ in $\mathrm{GW}(k)$. We give these definitions in detail in Section 3.

In this paper, we show the following closed formula for the quadratically enriched binomial coefficients over $\mathbb{F}_q$.

THEOREM 1.1. *Let q be an odd prime power and let j be a non-negative integer. Let L/k be the finite extension of $\mathbb{F}_q$ of degree n. Then*

$$\binom{L/k}{j} = \binom{n}{j} - (1-u)\cdot\binom{\frac{n-2}{2}}{\frac{j-1}{2}} \in \mathrm{GW}(\mathbb{F}_q),$$

where u is the non-square class in $\mathrm{GW}(k)$ and our convention is that $\binom{a}{b} := 0$, if either a or b is not in $\mathbb{Z}$.

Theorem 1.1 gives an enrichment of Pascal's triangle in $\mathrm{GW}(\mathbb{F}_q)$. The first few rows $n = 0, 1, \dots 8$ of the (untwisted, $\mathrm{GW}(\mathbb{F}_q)$-enriched) Pascal's triangle $\binom{\mathbb{F}_{q^n}/\mathbb{F}_q}{j}$ are the following:

$$\begin{array}{c}
1\\
1 \quad 1\\
1 \quad 1+u \quad 1\\
1 \quad 3 \quad 3 \quad 1\\
1 \quad 3+u \quad 6 \quad 3+u \quad 1\\
1 \quad 5 \quad 10 \quad 10 \quad 5 \quad 1\\
1 \quad 5+u \quad 15 \quad 20 \quad 15 \quad 5+u \quad 1\\
1 \quad 7 \quad 21 \quad 35 \quad 35 \quad 21 \quad 7 \quad 1\\
1 \quad 7+u \quad 28 \quad 55+u \quad 70 \quad 55+u \quad 28 \quad 7+u \quad 1
\end{array}.$$

We also compute the twisted quadratically enriched binomial coefficients over a finite field:

THEOREM 1.2. *Let q be an odd prime power and let j be a non-negative integer. Let L/k be the finite field extension of $\mathbb{F}_q$ of degree $2j$, and let Q/k the degree 2 field extension of $\mathbb{F}_q$. Then*

$$\binom{L[Q]/k}{j} = \frac{1}{2}\binom{2j}{j}\cdot(1+u) \in \mathrm{GW}(\mathbb{F}_q).$$

EXAMPLE 1.3. The first few terms of the sequence $\binom{L[Q]/k}{j}$ are

$$2, 5+u, 20, 69+u, 252, 924, 3432, 12869+u, \dots.$$

1.1. Summary of the proof. The idea of the proofs is first to rephrase the problems into purely combinatorial forms. In Section 4.1 and 5.1, we show that the proofs of these two theorems reduce to calculations of the parities of the number of orbits of necklaces of even cardinality under different cyclic group actions.

1.1.1. *Untwisted case.* The proof of Theorem 1.1 in the case when n is odd is very straightforward, as shown in Proposition 4.2. In Section 4.3, we prove Theorem 1.1 when j is odd. The proof involves a direct calculation using Möbius inversion and Lucas's theorem. The remaining task is to show that when both n and j are even, the enumeration is always even, which is established combinatorially.

In Section 4.4, we use a C_2-action, referred to as flipping, on the set of cyclic orbits of necklaces. Since we are concerned only with the parity of the enumeration, it suffices to count the C_2-fixed points. Next, in Section 4.5, we introduce the concept of symmetry axes for the orbits of necklaces, which allows us to rewrite the set of C_2-fixed (symmetric) orbits as a non-disjoint union of two subsets: the sets of symmetric orbits with a symmetry axis passing through two beads (we call this a symmetry axis of type 1) and those with a symmetry axis passing between two pairs of beads (type 2). The enumeration methods for these two subsets are different.

In Section 4.5, we also decompose each cyclic orbit of necklaces into two smaller necklace orbits. This decomposition exhibits special properties when the orbit is symmetric, as different types of symmetry axes yield different properties. This decomposition induces a map whose codomain is the symmetric product of the sets of two smaller cyclic orbits of necklaces. The enumeration is carried out by summing the fibers of this map over the codomain. At the end of Section 4.5, we provide the enumeration of symmetric orbits with a type 1 symmetry axis when both n and j are even.

Using this approach, in Section 4.6, we enumerate the symmetric orbits with a type 2 symmetry axis for the case where $n \equiv 2 \mod 4$ and j is even. Combining this result with the results in Section 4.5, we prove Theorem 1.1 for the case where $n \equiv 2 \mod 4$ and j is even by calculating the parity of the enumeration using Kummer's theorem.

In Section 4.7, we consider the case where $n \equiv 0 \mod 4$. For a symmetric orbit with a type 2 symmetry axis, we show how to reduce it to a symmetric orbit with a type 1 symmetry axis. Since $n - 2 \equiv 0 \mod 4$, we can now utilize the results from Section 4.5 and Section 4.6 to conclude the proof of Theorem 1.1 for the case where $n \equiv 0 \mod 4$ and j is even. This completes the proof of the untwisted case.

1.1.2. *Twisted case.* We prove Theorem 1.2 by reducing to an untwisted enumeration that we have studied in detail previously. In Section 5.2, we first provide a description of the twisted orbits in terms of untwisted orbits. Next, we construct a C_2-action, called swapping, on the set of twisted orbits and reduce the problem to the enumeration of the swapping fixed points. Then, by studying the swapping of fixed twisted orbits, we reduce the problem to an enumeration of untwisted orbits under certain conditions.

In Section 5.3, we discuss another way to encode the information of an untwisted cyclic orbit of necklaces. This is the marked cyclic equivalence class of partitions. The condition for the untwisted orbits we need to enumerate can be rephrased in terms of the properties of the partitions, which allows us to calculate the parity of the enumeration using partition theory. This concludes the proof of Theorem 1.2.

More explicit forms of the untwisted and twisted binomial coefficients over a finite field are presented in Section 4.8 and Section 5.5, respectively.

2. Background

The Möbius inversion formula [**6**, 16.4] states that for two functions $f, g : \mathbb{Z}_{>0} \to \mathbb{C}$ such that $g(n) = \sum_{d|n} f(n)$ we have that

$$f(n) = \sum_{d|n} \mu(d) g(\frac{n}{d})$$

where μ is the Möbius function

$$\mu(j) = \begin{cases} 1 & \text{if } j = 1, \\ 0 & \text{if } j \text{ has a squared prime factor,} \\ (-1)^v & \text{if } j \text{ is a product of } v \text{ distinct prime numbers.} \end{cases} \tag{2.1}$$

2.1. Lucas and Kummer's theorems. We recall the classical results of Lucas and Kummer. For a prime p, let $\nu_p : \mathbb{Z} \to \mathbb{Z}_{\geq 0}$ denote the p-adic valuation map. Lucas' classical theorem calculates the mod p residue class for binomial coefficients.

THEOREM 2.1. *[Lucas's theorem* **[12]**] *For $x, y \in \mathbb{Z}_{\geq 0}$, and p a prime*

$$\binom{x}{y} \equiv \prod_i \binom{x_i}{y_i} \mod p$$

where x_i, y_i are the coefficients of the p-adic expansions of x and y, $x = \sum_i x_i p^i$, $y = \sum_i y_i p^i$.

Define the p-adic carrier function for an integer x as $S_p(x) = \sum_i x_i$, where $x = \sum_i x_i p^i$ is the p-adic expansion. Kummer's classical theorem calculates the p-adic valuation of binomial coefficients.

THEOREM 2.2 (Kummer's theorem **[9]**). *The p-adic valuation of $\binom{n}{m}$ is*

$$\nu_p \binom{n}{m} = \frac{S_p(m) + S_p(n-m) - S_p(n)}{p-1}.$$

Some easy but useful corollaries that are relevant to our purpose are the following.

COROLLARY 2.3. *For $n, j \in \mathbb{Z}$, we have $\nu_2 \binom{n}{j} = \nu_2 \binom{2n}{2j} = \nu_2 \binom{2n+1}{2j+1}$.*

COROLLARY 2.4. *For $n, j \in 2\mathbb{Z}$ we have*

$$\nu_2 \binom{n}{j} = \nu_2 \binom{n+1}{j}.$$

COROLLARY 2.5. *For $j \in \mathbb{Z}_{>0}$, we have $\nu_2 \binom{2j}{j} \geq 1$, and the equality holds if and only if $j = 2^m$, $m \in \mathbb{Z}_{>0}$.*

3. Quadratically enriched binomial coefficients over a field

The paper **[1]** defined quadratically enriched (twisted) binomial coefficients over a base scheme. We recall the definition in the case where the base scheme is a field.

Let k be a field, and fix a separable closure k^s of k. For a finite separable extension $k \subseteq L$, let $\mathrm{Emb}_k(L, k^s)$ denote the set of ring homomorphisms of L into k^s over k

$$\mathrm{Emb}_k(L, k^s) = \left\{ f : \begin{array}{ccc} L & \xrightarrow{\quad f \quad} & k^s \\ & \nwarrow \qquad \nearrow & \\ & k & \end{array} \right\}.$$

Here it is not necessary to assume that L is a field; the same definition will carry over verbatim. But note that in this case f is not injective. We keep the notation Emb because when L is a field, $\mathrm{Emb}_k(L, k^s)$ the set of embeddings of L into the separable closure.

Recall that a finite étale k-algebra $k \to E$ has an associated trace map

$$\mathrm{Tr}_{E/k} : E \to k$$

which takes an element e to the trace of the matrix associated to multiplication by e, viewed as an endomorphism of the finite dimensional k-vector space E. This trace map determines an element of $\mathrm{GW}(k)$ denoted $\mathrm{Tr}_{E/k}\langle 1\rangle$ and defined to be the class of the bilinear form

$$E \times E \to E \to k$$

where the first map is multiplication on E and the second is $\mathrm{Tr}_{E/k}$.

Recall that the Galois correspondence

$$E \mapsto \mathrm{Emb}_k(E, k^s)$$

gives an equivalence of categories from finite étale k-algebras to finite sets equipped with a $\mathrm{Gal}(k^s/k)$-action. Under this correspondence, a field F corresponds to a set with a transitive action. A set with a transitive action can be encoded by the stabilizer $H \leq \mathrm{Gal}(k^s/k)$ of any element, and F is recovered as the fixed field $F \cong (k^s)^H$ of the stabilizer.

To enrich binomial coefficients, we introduce the following notation. Let $\mathrm{Emb}_k^j(L, k^s)$ denote the set of subsets of size j of $\mathrm{Emb}_k(L, k^s)$. So $\mathrm{Emb}_k^j(L, k^s)$ is the subset of the power set of $\mathrm{Emb}_k(L, k^s)$ consisting of those subsets of size exactly j. This set is equipped with a $\mathrm{Gal}(k^s/k)$-action coming from the action on $\mathrm{Emb}_k(L, k^s)$. Let $E_k^j(L, k^s)$ denote the finite étale k-algebra that is associated to $\mathrm{Emb}_k^j(L, k^s)$ under the Galois correspondence.

Definition 3.1 ([**1**, Definition 4.1]). For a finite étale k-algebra L, let

$$\binom{L/k}{j} \in \mathrm{GW}(k)$$

denote the class of the trace form of $E_k^j(L, k^s)$.

Here, $E_k^j(L, k^s)$ corresponds to the finite set $\mathrm{Emb}_k^j(L, k^s)$ with $\mathrm{Gal}(k^s/k)$-action induced from the canonical action on $\mathrm{Emb}_k(L, k^s)$.

For $j = 2, 3$, these forms appeared in [**5**, 30.12-30.14].

Remark 3.2. By definition $\binom{L/k}{j} = \mathrm{Tr}_{E_k^j(L,k^s)/k}\langle 1\rangle$. We have $\dim_k E_k^j(L, k^s) = |\mathrm{Emb}_k^j(L, k^s)| = \binom{n}{j}$. For a group G acting on a set S, let $\mathrm{Orb}(G, S)$ denote the set of orbits. The set $\mathrm{Emb}_k^j(L, k^s)$ can be decomposed into Galois orbits $\mathrm{Emb}_k^j(L, k^s) = \amalg_{i\in I}\mathfrak{O}_i$ with

$$I = \mathrm{Orb}(\mathrm{Gal}(k^s/k), \mathrm{Emb}_k^j(L, k^s))$$

being a finite set. Let $H_i \leq \mathrm{Gal}(k^s/k)$ be the stabilizer subgroup of $\mathfrak{O}_i$. Then by the Galois correspondence $E_k^j(L, k^s) = \prod_{i\in I}(k^s)^{H_i}$, and

$$\binom{L/k}{j} = \mathrm{Tr}_{E_k^j(L,k^s)/k}\langle 1\rangle = \sum_{i\in I}\mathrm{Tr}_{(k^s)^{H_i}/k}\langle 1\rangle . \tag{3.1}$$

Note that any element of $\mathrm{Gal}(k^s/k)$ which stabilizes every element of $\mathrm{Emb}_k(L, k^s)$ also stabilizes every element of $\mathrm{Emb}_k^j(L, k^s)$. It follows that $(k^s)^{H_i}$ embeds as a subfield of L.

EXAMPLE 3.3. $\binom{L/k}{1} = \binom{L/k}{[L:k]-1} = \operatorname{Tr}_{L/k}\langle 1\rangle$. For example, suppose $d \in \mathbb{Q}^*$ is not a square. Then $\binom{\mathbb{Q}[\sqrt{d}]/\mathbb{Q}}{1} = \langle 2\rangle + \langle 2d\rangle$, where $\langle d\rangle$ denotes the class of the bilinear form $\mathbb{Q}\times\mathbb{Q}\to\mathbb{Q}$ given $(x,y)\mapsto dxy$. For more on trace forms, see [**2**].

In [**1**], it was important to twist such quadratically enriched binomial coefficients by a degree 2-field extension $k \subset Q$ as follows. There is a canonical isomorphism $\operatorname{Gal}(Q/k) \cong C_2$, where C_2 denotes the cyclic group of order 2. Let

$$q_Q : \operatorname{Gal}(k^s/k) \to \operatorname{Gal}(Q/k) \cong C_2$$

denote the corresponding quotient map.

For a set S of size $2j$, the set of subsets of S of size j has an action of C_2 given by taking a subset to its complement. If a group G acts on S, the set of subsets of size j inherits an action of $G\times C_2$ because taking a set to its complement commutes with any automorphism of S.

DEFINITION 3.4 ([**1**, Definition 4.3]). Let L be a finite étale k-algebra and j such that $\dim_k L = 2j$. Define

$$\binom{L[Q]/k}{j} \in \mathrm{GW}(k)$$

to be the class of the trace form of the étale k-algebra corresponding to $\mathrm{Emb}^j_k(L,k^s)$ with the $\operatorname{Gal}(k^s/k)$-action given by the homomorphism

$$\operatorname{Gal}(k^s/k) \overset{(1,q_Q)}{\to} \operatorname{Gal}(k^s/k)\times C_2$$

and the canonical action of $\operatorname{Gal}(k^s/k)\times C_2$ on $\mathrm{Emb}^j_k(L,k^s)$.

EXAMPLE 3.5. Suppose $a,d \in \mathbb{Q}^*$ are relatively prime non-squares. Let $L = \mathbb{Q}[\sqrt{d}]$ and $Q = \mathbb{Q}[\sqrt{a}]$. Then $\binom{L[Q]/\mathbb{Q}}{1} = \langle 2\rangle + \langle 2da\rangle$.

4. Untwisted case - Proof of Theorem 1.1

Let $k = \mathbb{F}_q$ be a finite field of odd characteristic. We have

$$\mathrm{GW}(\mathbb{F}_q) \cong \frac{\mathbb{Z}[u]}{(u^2-1,2-2u)} \cong \frac{\mathbb{Z}\cdot 1 \oplus \mathbb{Z}\cdot u}{(2-2u)} \cong \mathbb{Z}\times \mathbb{F}_q^*/(\mathbb{F}_q^*)^2,$$

where the first isomorphism is an isomorphism of rings, the second and third only respect group structures, and the third isomorphism is the product of the rank and the discriminant. Note that $\mathbb{F}_q^*/(\mathbb{F}_q^*)^2 \cong \mathbb{Z}/2\mathbb{Z}$. Here as above, u denotes the class of the bilinear form $k\times k\to k$ given by $(x,y)\mapsto \mu xy$ where μ in $\mathbb{F}_q$ is a non-square. See for example [**10**, Theorem 3.5, Corollary 3.6]. In particular, the class of any element in $\mathrm{GW}(\mathbb{F}_q)$ is determined by the rank and the discriminant.

Let L/k be a finite extension of degree n. Then $\operatorname{Gal}(L/k)\cong C_n$, where C_n is the cyclic group of order n that is generated by the Frobenius automorphism $\varphi : x \to x^q$. See for example [**4**, Section 14.3]. It is a classical fact[1] that for $[L:k]=n$, the class of the trace form of L over k in $\mathrm{GW}(k)$ is given:

$$\operatorname{Tr}_{L/k}\langle 1\rangle = \epsilon(n) := \begin{cases} n-1+u, & n\equiv 0 \mod 2 \\ n, & n \equiv 1 \mod 2 \end{cases} \tag{4.1}$$

Indeed, the rank of $\operatorname{Tr}_{L/k}\langle 1\rangle$ is n, so it suffices to compute the discriminant of $\operatorname{Tr}_{L/k}\langle 1\rangle$. The discriminant of the trace form is the discriminant of the minimal

[1] A proof can be found in Lemma 58 of the first ArXiv version of [**8**].

polynomial of a generator. Since the Fröbenius acts as an n cycle on the roots of a minimal polynomial for a generator of $k \subseteq L$, we have that $\mathrm{Tr}_{L/k}\langle 1\rangle$ has square discriminant if and only if an n cycle has even sign, which occurs if and only if n is odd.

As in Remark 3.2, let

$$I = \mathrm{Orb}(\mathrm{Gal}(k^s/k), \mathrm{Emb}_k^j(L, k^s))$$

denote the set of orbits of the action of $\mathrm{Gal}(k^s/k)$ on $\mathrm{Emb}_k^j(L, k^s)$, and let

$$\mathrm{Emb}_k^j(L, k^s) = \amalg_{i\in I}\mathfrak{O}_i$$

denote the decomposition of $\mathrm{Emb}_k^j(L, k^s)$ into orbits. Letting $d_i = |\mathfrak{O}_i|$ denote the size of the ith orbit, we see that

$$\binom{L/k}{j} = \sum_{i\in I} \epsilon(d_i)$$

by Equation (3.1).

For a group G acting on a set S, let $\mathrm{Orb}_{even}(G, S) \subseteq \mathrm{Orb}(G, S)$ denote the subset consisting of those orbits with an even cardinality. Define $\mathrm{Orb}_{odd}(G, S)$ similarly.

The set $\mathrm{Orb}(\mathrm{Gal}(k^s/k), \mathrm{Emb}_k^j(L, k^s))$ decomposes

$$\mathrm{Orb}(\mathrm{Gal}(k^s/k), \mathrm{Emb}_k^j(L, k^s)) = GO_{odd}(L/k, j) \amalg GO_{even}(L/k, j),$$

where we use the abbreviations

$$GO_{odd}(L/k, j) := \mathrm{Orb}_{odd}(\mathrm{Gal}(k^s/k), \mathrm{Emb}_k^j(L, k^s)),$$
$$GO_{even}(L/k, j) := \mathrm{Orb}_{even}(\mathrm{Gal}(k^s/k), \mathrm{Emb}_k^j(L, k^s)).$$

Let $\Delta(n, j)$ denote the difference $\Delta(n, j) := \binom{L/k}{j} - \binom{n}{j}$. The following lemma follows immediately from (4.1).

LEMMA 4.1. $\Delta(n, j) = (u - 1) \cdot |GO_{even}(L/k, j)|$.

Therefore, to determine the value of $\Delta(n, j)$, one only needs to determine the mod 2 residue class of $|GO_{even}(L/k, j)|$.

PROPOSITION 4.2 (Theorem 1.1: case for n odd). *$\binom{L/k}{j} = \binom{n}{j}$ for n odd.*

PROOF. As in Remark 3.2, $(k^s)^{H_i}$ is a subfield of L. Thus, we have that $d_i | n$. Thus, if n is odd, $GO_{even}(L/k, j) = \emptyset$. □

4.1. Necklace interpretation. Let $\mathrm{Neck}(n, j)$ denote the set of necklaces consisting of j blue beads and $n - j$ red beads, with one bead designated as the bead at the top of the necklace. One could alternately view this set as the set of necklaces with j blue beads and $n - j$ red beads, lying on the plane, beads equally spaced on the unit circle with the top bead at $(0, 1)$. It will be useful to consider the natural action of the dihedral group $D_n = C_n \rtimes C_2 = \langle r, f : r^n = 1, f^2 = 1, rfr = f\rangle$ on $\mathrm{Neck}(n, j)$ in which r acts by rotating the necklace one bead counterclockwise and f acts by flipping the necklace over the vertical line through the top bead.

As above, let $k \subseteq L$ be the degree n field extension $\mathbb{F}_q \subseteq \mathbb{F}_{q^n}$. Fix an embedding $p \in \mathrm{Emb}_k(L, k^s)$. Then

$$\mathrm{Emb}_k(L, k^s) = \{p, \varphi p, \varphi^2 p \dots, \varphi^{n-1} p\},$$

where φ denotes the Fröbenius $\varphi\colon k^s \to k^s$, $\varphi(x) = x^q$ of k. It follows that if we identify $\mathrm{Emb}_k^j(L, k^s)$ with the set $\mathrm{Neck}(n, j)$, then the action of $\mathrm{Gal}(L/k) \cong C_n$ on $\mathrm{Emb}_k^j(L, k^s)$ is identified with the induced C_n-action on $\mathrm{Neck}(n, j)$. The induced C_n-action comes from $C_n \to D_n$ which sends the generator to $r \in D_n$. In other words, we have constructed an isomorphism of C_n-sets

$$\mathrm{Emb}_k^j(L, k^s) \cong \mathrm{Neck}(n, j).$$

In particular, $GO_{odd}(L/k, j)$ can be identified with the subset of orbits of $\mathrm{Neck}(n, j)$ under the rotation action of $C_n = \langle r : r^n = 1\rangle$ consisting of those orbits with odd cardinality. A similar statement holds for $GO_{even}(L/k, j)$ as well, giving a canonical bijection

$$GO_{even}(L/k, j) \cong \mathrm{Orb}_{even}(C_n, \mathrm{Neck}(n, j)).$$

Moreover, Lemma 4.1 says that

$$\binom{L/k}{j} = \binom{n}{j} + (u-1)|\,\mathrm{Orb}_{even}(C_n, \mathrm{Neck}(n, j))|.$$

4.2. Möbius inversion. Let $N(n, j)$ denote the cardinality of the set of necklaces in $\mathrm{Neck}(n, j)$ whose stabilizer under the C_n-action by rotation is trivial

$$N(n, j) := |\{l \in \mathrm{Neck}(n, j) : \mathbf{Stab}_{C_n}(l) = \{1\}\}|.$$

By the necklace interpretation of Section 4.1, we have

$$N(n, j) = |\{l \in \mathrm{Emb}_k^j(L, k^s) : \mathbf{Stab}_{\mathrm{Gal}(L/k)}(l) = \{1\}\}|.$$

Note that that the numbers $N(n, j)$ and $|\,\mathrm{Emb}_k^j(L, k^s)| = \binom{n}{j}$ only depend on n and j.

Let $I = \mathrm{Orb}(C_n, \mathrm{Neck}(n, j))$ index the orbits of the action of C_n on $\mathrm{Neck}(n, j)$ and, for i in I, let d_i denote the cardinality of the corresponding orbit. Let $\rho := \frac{j}{n}$ denote the fraction of beads which are blue. Given an orbit $i \in I$ of cardinality d_i, we can take d_i adjacent beads in the necklace and form a new necklace with d_i beads, ρd_i of which are blue, and which has a trivial stabilizer under the rotation action of C_{d_i}. This process can be run in reverse, creating a necklace with n beads from one with d beads for $d|n$. It follows that

$$|\{i \in I, d_i = d\}| = N(d, \rho d).$$

By the necklace interpretation, I is in canonical bijection with an indexing set for the orbits of the $\mathrm{Gal}(L/k)$ action on $\mathrm{Emb}_k^j(L, k^s)$. Combining with the above, we have

$$\binom{n}{j} = |\,\mathrm{Emb}_k^j(L, k^s)| = \sum_{d|n} d\ |\{i \in I, d_i = d\}| = \sum_{d|n} d\ N(d, \rho d), \quad \rho := \frac{j}{n},$$

under the convention $N(d, \rho d) = 0$ if $\rho d \notin \mathbb{Z}$.

We can now apply the Möbius inversion formula, which yields

$$|N(d, b)| = \frac{1}{d} \sum_{j|d} \mu(j) \binom{\frac{d}{j}}{\frac{b}{j}},$$

where $\mu(j)$ is the Möbius function recalled in Equation (2.1).

By definition, $|GO_{even}(L/k,j)|$ is the total number of orbits of even size, thus

$$|GO_{even}(L/k,j)| = \sum_{d|n,2|d} \frac{1}{d} \sum_{j|d} \mu(j) \binom{\frac{d}{j}}{\frac{\rho d}{j}}. \tag{4.2}$$

4.3. Proof of Theorem 1.1 when $\nu_2(n) \geq 1$ and $\nu_2(j) = 0$. Define a partial order $\prec$ on $\mathbb{Q}_{\geq 0}$ as follows.

DEFINITION 4.3. For $x, y \in \mathbb{Q}_{\geq 0}$ written as 2-adic numbers $x = \sum_i x_i \cdot 2^i$, $y = \sum_i y_i \cdot 2^i$, then $x \prec y$ if and only if $x, y \in \mathbb{Z}_{\geq 0}$ and $x_i \leq y_i$ for all i.

PROPOSITION 4.4. *For $\nu_2(n) \geq 1$ and $\nu_2(j) = 0$ we have*

$$\Delta(n,j) = \begin{cases} -1+u, & \frac{j-1}{2} \prec \frac{n-2}{2} \\ 0, & \text{else.} \end{cases}$$

PROOF. As above, define $\rho = j/n$. From (4.2), we have

$$\begin{aligned} |GO_{even}(L/k,j)| &= \sum_{d|n,2|d} \frac{1}{d} \sum_{l|d} \mu(l) \binom{\frac{d}{l}}{\frac{\rho d}{l}}. \\ &= \sum_{d|n,2|d} \frac{1}{d} \sum_{l|d} \mu(l) \frac{1}{\rho} \binom{\frac{d}{l}-1}{\frac{\rho d}{l}-1}. \\ &= \frac{1}{j} \sum_{d|n,2|d} \frac{n}{d} \sum_{l|d} \mu(l) \binom{\frac{d}{l}-1}{\frac{\rho d}{l}-1}. \end{aligned}$$

By Lemma 4.1, $\Delta(n,j)$ only depends on the mod 2 residue of $|GO_{even}(L/k,j)|$. As j is odd by our assumption,

$$|GO_{even}(L/k,j)| \equiv \sum_{d|n,2|d} \frac{n}{d} \sum_{l|d} \mu(l) \binom{\frac{d}{l}-1}{\frac{\rho d}{l}-1} \mod 2.$$

Further, as $\mu(l) \equiv 1 \mod 2$ if and only if l is square free, we have

$$\begin{aligned} |GO_{even}(L/k,j)| &\equiv \sum_{d|n,2|d} \frac{n}{d} \sum_{\substack{l|d \\ l \text{ square free}}} \binom{\frac{d}{l}-1}{\frac{\rho d}{l}-1} \\ &\equiv \sum_{(2^{\nu_2(n)}d)|n} \sum_{\substack{l|(2^{\nu_2(n)}d) \\ l \text{ square free}}} \binom{\frac{2^{\nu_2(n)}d}{l}-1}{\frac{\rho 2^{\nu_2(n)}d}{l}-1} \mod 2. \end{aligned} \tag{4.3}$$

Where in the second line we used the fact that only when $\frac{n}{d}$ is even, the summand has a nontrivial mod 2 contribution. By Corollary 2.3, we have mod 2 equalities

$$\binom{x-1}{y-1} \equiv \binom{2x-2}{2y-2} \equiv \binom{2x-1}{2y-1} \mod 2, \quad \forall x, y \in \mathbb{Z}_{\geq 0}.$$

Thus, the mod 2 residue of each of the summands of (4.3) only depends on $\frac{d}{l}$ and $\frac{\rho d}{l}$.

Notice the sum in (4.3) is over the set P of pairs (d,l) with $2^{\nu_2(n)}d$ dividing n and l square free dividing $2^{\nu_2(n)}d$. Since the summand only depends on $\frac{d}{l}$, we introduce an equivalence relation on P by declaring $(d,l) \sim (d',l')$ when $\frac{d}{l} = \frac{d'}{l'}$.

Now, we need to identify the parity of cardinality of the $\sim$ classes, since those with even cardinality will not contribute to (4.3). Suppose (d,l) in P with l odd. Then we can divide both l and d by l, obtaining a new pair (d',l') with the same ratio $\frac{d}{l}=\frac{d'}{l'}$ and $l'=1$. The number of pairs with this given ratio is then equal to 2^m where m is the number of distinct odd prime factors of $n/(d')$. It follows that all these equivalence classes have even cardinality, except for the class of $(n,1)$. Now suppose (d,l) in P with l even. Then we can divide both l and d by $l/2$, obtaining an equivalence pair $(d',2)$. The number of pairs equivalent to $(d',2)$ is then 2^m where m is the number of odd prime factors of n/d'. Thus, all these equivalence classes have even cardinality, except for the class of $(n,2)$. However, for $(d,l)=(n,2)$, the binomial coefficient $\binom{\frac{2^{\nu_2(n)}d}{l}-1}{\frac{\rho 2^{\nu_2(n)}d}{l}-1}=\binom{\frac{n}{2}-1}{\frac{j}{2}-1}$, which is 0 because $\frac{j}{2}$ is not an integer and by convention (see Theorem 2.1), binomial coefficients with fractional lower terms are 0. Thus,

$$|GO_{even}(L/k,j)|\equiv\binom{n-1}{j-1}\mod 2.$$

By Lucas's theorem (Theorem 2.1), we have $|GO_{even}(L/k,j)|$ is odd if and only if $j-1\prec n-1$. Finally, because n is even and j is odd, it follows that $j-1\prec n-1$ if and only if $j-1\prec n-2$. Then since both $j-1$ and $n-2$ are even, we have $j-1\prec n-2$ if and only if $\frac{j-1}{2}\prec\frac{n-2}{2}$, which gives the desired formula. $\square$

By Lemma 4.1, Proposition 4.4 implies Theorem 1.1 for the case $\nu_2(n)\geq 1$ and $\nu_2(j)=0$.

4.4. Combinatorics of symmetric orbits. We will mainly work on the necklace interpretation for the rest of the paper. Note that C_2 generated by the flip f acts on the set $\mathrm{Orb}(C_n,\mathrm{Neck}(n,j)$. For a set $S\subset\mathrm{Orb}(C_n,\mathrm{Neck}(n,j))$, let S^f denote the subset of S that is fixed by the flipping action. If l is a necklace with orbit $[l]\in\mathrm{Orb}(C_n,\mathrm{Neck}(n,j))^f$, then $fl=r^m l$ for some $m\in\mathbb{Z}$. This implies that l has an axis of symmetry: the axis rotated counterclockwise by $\frac{m\pi}{n}$ from the vertical axis. As we will see in a moment, a symmetry axis will allow us to decompose a necklace. Therefore, it will be important to consider orbits under rotation of pairs consisting of a necklace and a symmetry axis.

Consider the 2-dimensional faithful representation $\Phi: D_n\hookrightarrow O(2,\mathbb{R})$,

$$\Phi(r)=\begin{pmatrix}\cos(\frac{2\pi}{n}) & -\sin(\frac{2\pi}{n})\\ \sin(\frac{2\pi}{n}) & \cos(\frac{2\pi}{n})\end{pmatrix},\quad \Phi(f)=\begin{pmatrix}-1 & 0\\ 0 & 1\end{pmatrix}.$$

Fix $x\in\mathbb{R}^2, x=(0,1)$, then $l\in\mathrm{Neck}(n,j)$ can be represented by a subset of size j in $\Phi(D_n)\cdot x$. If $[l]\in\mathrm{Orb}(C_n,\mathrm{Neck}(n,j))^f$, then for every representative $l\in\mathrm{Neck}(n,j)$ there exists $\sigma\in\mathbb{P}^1(\mathbb{R})$ such that l is invariant under the reflection $f_\sigma\in O(2,\mathbb{R})$ with respect to σ. Any such linear space σ is called a symmetry axis of l. A symmetry axis $[(l,\sigma)]$ for $[l]\in\mathrm{Orb}(C_n,\mathrm{Neck}(n,j))^f$ is the C_n-orbit of the pair (l,σ).

DEFINITION 4.5. For any $[l]\in\mathrm{Orb}(C_n,\mathrm{Neck}(n,j))$, define the period of $[l]$ as $\pi([l]):=|C_n\cdot l|$, which is well defined as $|C_n\cdot l|$ is independent from the choice of representative of $[l]$.

DEFINITION 4.6. For two symmetry axes $[(l_1,\sigma_1)]$ and $[(l_2,\sigma_2)]$ of

$$[l]\in\mathrm{Orb}(C_n,\mathrm{Neck}(n,j))^f,$$

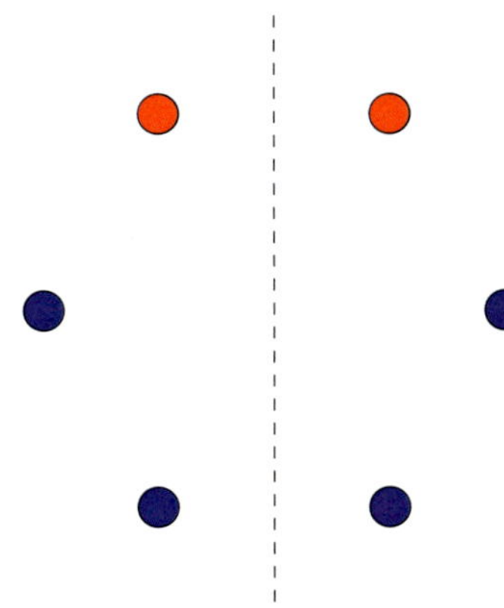

FIGURE 1. An element in $\mathrm{Orb}(C_6, \mathrm{Neck}(6,4))^f$ with a unique symmetry axis.

define their distance as

$$d([(l_1,\sigma_1)],[(l_2,\sigma_2)]) = \min\{|m| : r^m \cdot \sigma_1 = \sigma_2, (l_i,\sigma_i) \in [(l_i,\sigma_i)], i = 1,2\},$$

where m is allowed to be half-integers in the sense that $\Phi(r)^{1/2}$ is the counterclockwise rotation by $\frac{2\pi i}{2n}$.

LEMMA 4.7. *Let* $[l]$ *be in* $\mathrm{Orb}(C_n, \mathrm{Neck}(n,j))^f$. *If* $\frac{n}{\pi([l])}$ *is odd, then there is a unique symmetry axis. Otherwise, there are two distinct symmetry axes* $[(l_1,\sigma_1)]$ *and* $[(l_2,\sigma_2)]$ *with* $d([(l_1,\sigma_1)],[(l_2,\sigma_2)]) = \frac{\pi([l])}{2}$.

PROOF. Let $[(l_1,\sigma_1)]$ and $[(l_2,\sigma_2)]$ be two symmetry axes of $[l]$. As the composition $f_{\sigma_2} \circ f_{\sigma_1} \in \Phi(C_n)$ is r raised to the power of $2d([(l_1,\sigma_1)],[(l_2,\sigma_2)])$, we have

$$\pi([l]) \mid 2d([(l_1,\sigma_1)],[(l_2,\sigma_2)]).$$

On the other hand, by definition $d([(l_1,\sigma_1)],[(l_2,\sigma_2)]) < \pi([l])$. Thus, $d([(l_1,\sigma_1)], [(l_2,\sigma_2)]) = \frac{\pi([l])}{2}$ or 0. Moreover, in the first case, we must have that $\frac{n}{\pi([l])}$ is even. □

For example, Figure 2 shows an element $[l] \in \mathrm{Orb}(C_6, \mathrm{Neck}(6,4))^f$ with two different symmetry axes. In this example, $\pi([l]) = 3$, from which $\frac{6}{\pi([l])} = 2$, and the distance between the two axes of symmetry is $\frac{3}{2}$.

DEFINITION 4.8. A symmetry axis $[(l,\sigma)]$ for $[l] \in \mathrm{Orb}(C_n, \mathrm{Neck}(n,j))^f$ is called *type 1* if σ does not intersect $\Phi(D_n) \cdot x$. It is called *type 2* if σ intersects at least one element of $\Phi(D_n) \cdot x$.

For $[l] \in \mathrm{Orb}(C_n, \mathrm{Neck}(n,j))^f$, if n is odd, then the only symmetry axis of $[l]$ is of type 2. If n is even, then $[l]$ has two symmetry axes, one for each type. As noted above, an example for n is even shown in Figure 2.

For $S \subset \mathrm{Orb}(C_n, \mathrm{Neck}(n,j))^f$, denote S_i as the subset of S that has a symmetry axis of type i, for $i = 1,2$. We have

$$\mathrm{Orb}(C_n, \mathrm{Neck}(n,j))^f = \mathrm{Orb}(C_n, \mathrm{Neck}(n,j))^f_1 \bigcup \mathrm{Orb}(C_n, \mathrm{Neck}(n,j))^f_2.$$

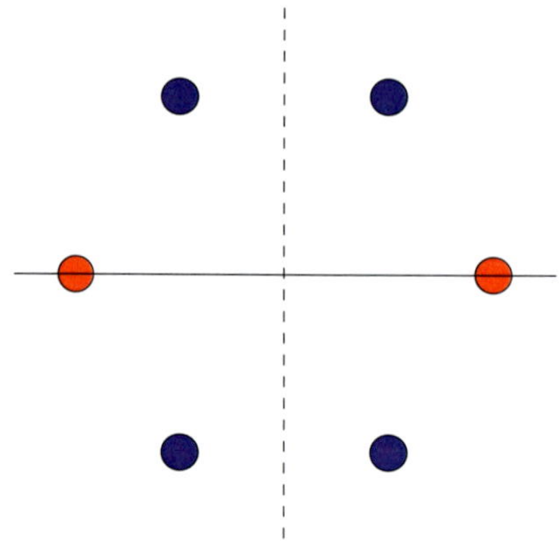

FIGURE 2. An element in $\mathrm{Orb}(C_6, \mathrm{Neck}(6,4))^f$ with two symmetry axes. The dashed line is a symmetry axis of type 1 and the solid line is of type 2.

LEMMA 4.9. *For* $\nu_2(n) \geq 1$,

$$\mathrm{Orb}(C_n, \mathrm{Neck}(n,j))_1^f \cap \mathrm{Orb}(C_n, \mathrm{Neck}(n,j))_2^f = \mathrm{Orb}_{odd}(C_n, \mathrm{Neck}(n,j))^f.$$

In particular,

$$\mathrm{Orb}_{even}(C_n, \mathrm{Neck}(n,j))_1^f \cap \mathrm{Orb}_{even}(C_n, \mathrm{Neck}(n,j))_2^f = \emptyset.$$

PROOF. If $[l] \in \mathrm{Orb}(C_n, \mathrm{Neck}(n,j))^f$ has two axes $[(l_1, \sigma_1)], [(l_2, \sigma_2)]$ of different type, then $d([(l_1,\sigma_1)], [(l_2,\sigma_2)])$ is an half integer. By Lemma 4.7, we have $\pi([l]) = 2d([(l_1,\sigma_1)], [(l_2,\sigma_2)])$ is odd. □

Consider again the necklace in Figure 2. This necklace has period 3, and is contained in both sides of the first equation in Lemma 4.9. An immediate corollary of Lemma 4.9 is the following.

COROLLARY 4.10. *For* $\nu_2(n) \geq 1$, *we have*

$$\begin{aligned}|\mathrm{Orb}_{even}(C_n, \mathrm{Neck}(n,j))| &\equiv |\mathrm{Orb}_{even}(C_n, \mathrm{Neck}(n,j))^f| \mod 2\\ &= |\mathrm{Orb}_{even}(C_n, \mathrm{Neck}(n,j))_1^f|\\ &\quad + |\mathrm{Orb}_{even}(C_n, \mathrm{Neck}(n,j))_2^f|.\end{aligned}$$

4.5. Combinatorics of type 1 symmetric orbits. Now assume n is even. By choosing every other bead of a representative necklace l, any

$$[l] \in \mathrm{Orb}(C_n, \mathrm{Neck}(n,j))$$

can be uniquely decomposed into an unordered pair of necklace orbits (under the rotation action) with $\frac{n}{2}$ beads. This constructs a well-defined map

(4.4)

$$\begin{aligned}\phi : \mathrm{Orb}(C_n, \mathrm{Neck}(n,j)) &\to \bigcup_{j_1+j_2=j} \mathrm{Sym}(\mathrm{Orb}(C_{\frac{n}{2}}, \mathrm{Neck}(\frac{n}{2}, j_1)), \mathrm{Orb}(C_{\frac{n}{2}}, \mathrm{Neck}(\frac{n}{2}, j_2)))\\ [l] &\mapsto ([l_1],[l_2]),\end{aligned}$$

where Sym denotes the symmetric product.

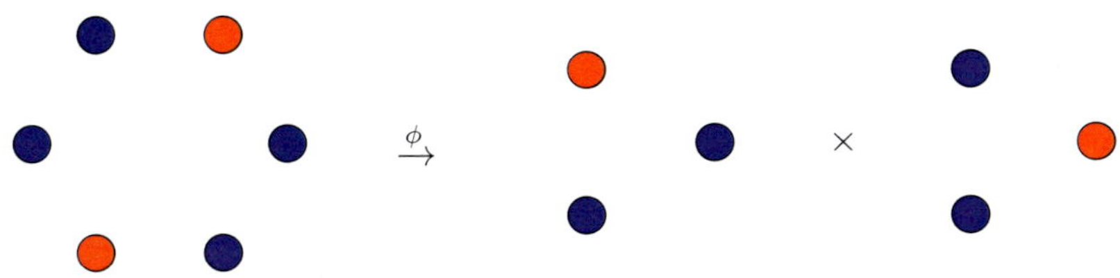

FIGURE 3. An element in $\mathrm{Orb}(C_6, \mathrm{Neck}(6,4))^f$ decomposes to a symmetric product in $\mathrm{Orb}(C_3, \mathrm{Neck}(3,2)^f$ under ϕ.

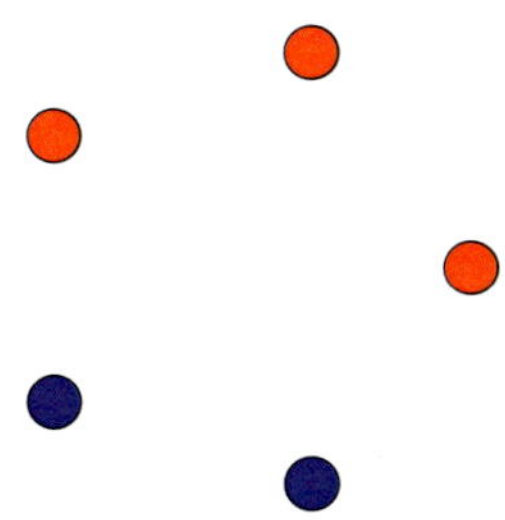

FIGURE 4. An element in $\mathrm{Orb}(C_5, \mathrm{Neck}(5,2))^f$.

LEMMA 4.11. *For $[l] \in \mathrm{Orb}(C_n, \mathrm{Neck}(n,j))_1^f$, we have $\phi([l]) = ([l_1], f\cdot[l_1])$, where $l_1 \in \mathrm{Neck}(\frac{n}{2}, \frac{j}{2})$.*

PROOF. For $[l] \in \mathrm{Orb}(C_n, \mathrm{Neck}(n,j))_1^f$, let $[(l,\sigma)]$ denote a symmetry axis of type 1. The action of f on $\mathrm{Orb}(C_n, \mathrm{Neck}(n,j))_1^f$ can be induced by the reflection f_σ. Since f_σ exchanges the two smaller necklaces, we must have $[l_2] = f\cdot[l_1]$. □

In particular, when restricted to $\mathrm{Orb}(C_n, \mathrm{Neck}(n,j))_1^f$, ϕ surjects to

$$\mathrm{Sym}(\mathrm{Orb}(C_{\frac{n}{2}}, \mathrm{Neck}(\frac{n}{2}, \frac{j}{2})), f\cdot \mathrm{Orb}(C_{\frac{n}{2}}, \mathrm{Neck}(\frac{n}{2}, \frac{j}{2})).$$

Therefore, we have

$$\mathrm{Orb}(C_n, \mathrm{Neck}(n,j))_1^f = \bigcup_{([l], f\cdot[l]))\in \mathrm{Sym}(\mathrm{Orb}(C_{\frac{n}{2}}, \mathrm{Neck}(\frac{n}{2},\frac{j}{2})), f\cdot \mathrm{Orb}(C_{\frac{n}{2}}, \mathrm{Neck}(\frac{n}{2},\frac{j}{2}))} \phi^{-1}([l], f\cdot[l]).$$

LEMMA 4.12. *Consider the restriction of ϕ to $\mathrm{Orb}(C_n, \mathrm{Neck}(n,j))_1^f$. Then*
(1) $|\phi^{-1}([l], f\cdot[l])| = \pi([l])$, *if* $[l] \notin \mathrm{Orb}(C_{\frac{n}{2}}, \mathrm{Neck}(\frac{n}{2}, \frac{j}{2}))^f$,
(2) $|\phi^{-1}([l],[l])| = \frac{\pi([l])+1}{2}$ *if* $[l] \in \mathrm{Orb}(C_{\frac{n}{2}}, \mathrm{Neck}(\frac{n}{2}, \frac{j}{2}))^f$ *and* $\pi([l]) \equiv 1 \mod 2$,
(3) $|\phi^{-1}([l],[l])| = \frac{\pi([l])}{2}$ *if* $[l] \in \mathrm{Orb}(C_{\frac{n}{2}}, \mathrm{Neck}(\frac{n}{2}, \frac{j}{2}))^f$ *and* $\pi([l]) \equiv 0 \mod 2$.

PROOF. $|\phi^{-1}([l], f\cdot[l])|$ is the number of elements in $\mathrm{Orb}(C_n, \mathrm{Neck}(n,j))_1^f$ that can be formed by interweaving $[l]$ and $f\cdot[l]$. The above statement can be seen by considering a fixed representative of $[l]$, and counting how many different ways one can insert $f\cdot l$. □

For example, the element shown in Figure 4 has period 5, which can generate 3 distinct elements by (2) of Lemma 4.12. These three elements are shown in Figure 5.

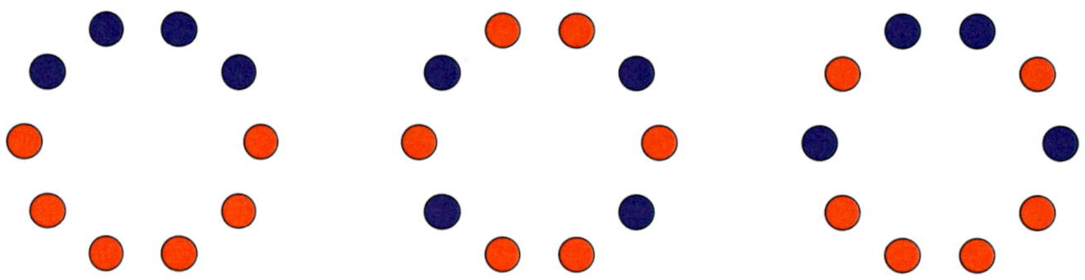

FIGURE 5. The three distinct elements in $\mathrm{Orb}(C_{10}, \mathrm{Neck}(10,4))^f$ that can be generated by the same element in $\mathrm{Orb}(C_5, \mathrm{Neck}(5,2))^f$.

LEMMA 4.13. *Take* $[l] \in \mathrm{Orb}(C_n, \mathrm{Neck}(n,j))_1^f$, *then we have* $\phi([l]) = ([l_1],[l_1])$ *for some* $[l_1] \in \mathrm{Orb}(C_{\frac{n}{2}}, \mathrm{Neck}(\frac{n}{2}, \frac{j}{2}))^f$. *If* $[l_1]$ *satisfies* $\pi([l_1]) \equiv 1 \mod 2$, *then* $\pi([l]) = \pi([l_1])$. *Otherwise,* $\pi([l]) = 2\pi([l_1])$.

PROOF. The exceptional case is the following. View l_1 as a subset of $\Phi(D_n)\cdot x$. Then flip l_1 with respect to $\sigma_0 \in \mathbb{P}^1(\mathbb{R})$, where σ_0 is perpendicular to a symmetry axis of l_1. Then $[l]$ is the C_n-orbit of $l_1 \cup f_{\sigma_0} \cdot l_1$. □

For example, the three elements shown in Figure 5 have period $10, 5, 10$ respectively, where the middle one is the exceptional case in Lemma 4.13.

Therefore, for $[l] \in \mathrm{Orb}_{odd}(C_{\frac{n}{2}}, \mathrm{Neck}(\frac{n}{2}, \frac{j}{2}))^f$, we have

$$\left|\phi^{-1}([l],[l]) \cap \mathrm{Orb}_{even}(C_n, \mathrm{Neck}(n,j))\right| = \frac{\pi([l])-1}{2}. \tag{4.5}$$

PROPOSITION 4.14. *For* $\nu_2(n) \geq 1$,

$$|\,\mathrm{Orb}_{even}(C_n, \mathrm{Neck}(n,j))_1^f| = \frac{1}{2}\left(\binom{\frac{n}{2}}{\frac{j}{2}} - |\,\mathrm{Orb}_{odd}(C_{\frac{n}{2}}, \mathrm{Neck}(\frac{n}{2}, \frac{j}{2}))^f|\right).$$

PROOF. By Lemma 4.12 and (4.5), we have

$$\begin{aligned}
|\,\mathrm{Orb}_{even}(C_n, \mathrm{Neck}(n,j))_1^f| =& \frac{1}{2} \sum_{[l]\notin \mathrm{Orb}(C_{\frac{n}{2}}, \mathrm{Neck}(\frac{n}{2},\frac{j}{2}))^f} \pi([l]) \\
&+ \sum_{[l]\in \mathrm{Orb}_{even}(C_{\frac{n}{2}}, \mathrm{Neck}(\frac{n}{2},\frac{j}{2}))^f} \frac{\pi([l])}{2} \\
&+ \sum_{[l]\in \mathrm{Orb}_{odd}(C_{\frac{n}{2}}, \mathrm{Neck}(\frac{n}{2},\frac{j}{2}))^f} \frac{\pi([l])-1}{2} \\
=& \frac{1}{2}\Bigg(\sum_{[l]\in \mathrm{Orb}(C_{\frac{n}{2}}, \mathrm{Neck}(\frac{n}{2},\frac{j}{2}))} \pi([l]) \\
&- |\,\mathrm{Orb}_{odd}(C_{\frac{n}{2}}, \mathrm{Neck}(\frac{n}{2}, \frac{j}{2}))^f| \Bigg)
\end{aligned}$$

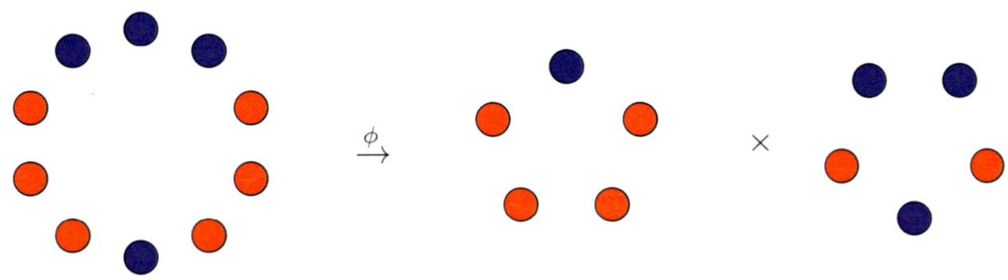

FIGURE 6. An element in $\mathrm{Orb}(C_{10}, \mathrm{Neck}(10,4))_2^f$ decomposes under ϕ to a symmetric product of two elements in $\mathrm{Orb}(C_5, \mathrm{Neck}(5,1))^f$ and $\mathrm{Orb}(C_5, \mathrm{Neck}(5,3))^f$, repsectively.

By definition $\sum_{[l]\in \mathrm{Orb}(C_{\frac{n}{2}}, \mathrm{Neck}(\frac{n}{2},\frac{j}{2}))} \pi([l]) = |\mathrm{Neck}(\frac{n}{2},\frac{j}{2})| = \binom{\frac{n}{2}}{\frac{j}{2}}$. So

$$|\,\mathrm{Orb}_{even}(C_n, \mathrm{Neck}(n,j))_1^f| = \frac{1}{2}\left(\sum_{d|\frac{n}{2}} dN(d,\rho d) - |\,\mathrm{Orb}_{odd}(C_{\frac{n}{2}}, \mathrm{Neck}(\frac{n}{2},\frac{j}{2}))^f|\right)$$

□

LEMMA 4.15. *For any $n, j \in \mathbb{Z}_{\geq 0}$, we have*

$$|\,\mathrm{Orb}_{odd}(C_n, \mathrm{Neck}(n,j))^f| = \begin{cases} \binom{\frac{n}{2^{\nu_2(j)+1}} - \frac{1}{2}}{\frac{j}{2^{\nu_2(j)+1}}}, & \textit{when } \nu_2(j) > \nu_2(n), \\ \binom{\frac{n}{2^{\nu_2(j)+1}} - \frac{1}{2}}{\frac{j}{2^{\nu_2(j)+1}} - \frac{1}{2}}, & \textit{when } \nu_2(j) = \nu_2(n), \\ 0, & \textit{when } \nu_2(j) < \nu_2(n). \end{cases}$$

PROOF. Notice that for any $n, j \in 2\mathbb{Z}$, we have

$$\mathrm{Orb}_{odd}(C_n, \mathrm{Neck}(n,j))^f = \mathrm{Orb}_{odd}(C_{\frac{n}{2}}, \mathrm{Neck}(\frac{n}{2},\frac{j}{2}))^f.$$

Then, the problem reduces to the case when $\nu_2(n)\cdot\nu_2(j) = 0$. In this case, the counting corresponds to the number of ways to assign exactly half of the blue beads to one half of the necklace when the total number of blue beads is even, or half minus one when the number is odd. □

4.6. Proof of Theorem 1.1 when $\nu_2(n) = 1$ and $\nu_2(j) \geq 1$. Consider the restriction of the map ϕ in (4.4) to $\mathrm{Orb}(C_n, \mathrm{Neck}(n,j))_2^f$. As $n \equiv 2 \mod 4$, the image is contained in

$$\bigcup_{j_1+j_2=j} \mathrm{Sym}(\mathrm{Orb}(C_{\frac{n}{2}}, \mathrm{Neck}(\frac{n}{2}, j_1))^f, \mathrm{Orb}(C_{\frac{n}{2}}, \mathrm{Neck}(\frac{n}{2}, j_2))^f).$$

An example of the morphism ϕ in this case is shown in Figure 6.

Moreover, since $\frac{n}{2}$ is odd, and thus any symmetry axis for $[l]$ only passes through one bead, it is of type two. It follows that the restriction of ϕ to

$$\mathrm{Orb}(C_n, \mathrm{Neck}(n,j))_2^f$$

is bijective, and we have

(4.6)

$$|\operatorname{Orb}(C_n, \operatorname{Neck}(n,j))^f_2| = \sum_{j_1+j_2=j, j_1<j_2} |\operatorname{Orb}(C_{\frac{n}{2}}, \operatorname{Neck}(\frac{n}{2}, j_1))^f| \cdot |\operatorname{Orb}(C_{\frac{n}{2}}, \operatorname{Neck}(\frac{n}{2}, j_2))^f|$$
$$+ \frac{1}{2}|\operatorname{Orb}(C_{\frac{n}{2}}, \operatorname{Neck}(\frac{n}{2}, \frac{j}{2}))^f| \cdot (|\operatorname{Orb}(C_{\frac{n}{2}}, \operatorname{Neck}(\frac{n}{2}, \frac{j}{2}))^f| + 1).$$

Next, we have

Lemma 4.16. *Let* $[l] \in \operatorname{Orb}(C_n, \operatorname{Neck}(n,j))^f_2$. *Then*

(1) $\pi([l]) = \pi([l_1])$, *if* $\phi([l]) = ([l_1],[l_1])$,
(2) $\pi([l]) = 2\operatorname{lcm}(\pi([l_1]), \pi([l_2]))$, *if* $\phi([l]) = ([l_1],[l_2])$, $[l_1] \neq [l_2]$.

Proof. l has a stabilizer r^k with $k \equiv 1 \mod 2$ under the C_n action if and only if $\phi([l]) = ([l_1],[l_1])$. Therefore, when $\phi([l]) = ([l_1],[l_1])$, the period of $[l]$ is exactly that of $[l_1]$. For the other case, the reason why there is a multiplier 2 in front of $\operatorname{lcm}(\pi([l_1]), \pi([l_2]))$ is because any consecutive $\pi([l])$ number of beads consists of $\frac{\operatorname{lcm}(\pi([l_1]),\pi([l_2]))}{\pi([l_2])}$ copies of $[l_1]$ and $\frac{\operatorname{lcm}(\pi([l_1]),\pi([l_2]))}{\pi([l_1])}$ copies of $[l_2]$. □

Figure 6 provides an example of the second case. Since $\frac{n}{2}$ is odd, we have

$$|\operatorname{Orb}_{odd}(C_n, \operatorname{Neck}(n,j))^f| = |\operatorname{Orb}_{odd}(C_{\frac{n}{2}}, \operatorname{Neck}(\frac{n}{2}, \frac{j}{2}))^f|.$$

Then, as a simple consequence,

(4.7)

$$|\operatorname{Orb}_{even}(C_n, \operatorname{Neck}(n,j))^f_2| = |\operatorname{Orb}(C_n, \operatorname{Neck}(n,j))^f_2| - |\operatorname{Orb}_{odd}(C_n, \operatorname{Neck}(n,j))^f|$$
$$= |\operatorname{Orb}(C_n, \operatorname{Neck}(n,j))^f_2| - |\operatorname{Orb}_{odd}(C_{\frac{n}{2}}, \operatorname{Neck}(\frac{n}{2}, \frac{j}{2}))^f|.$$

Proposition 4.17.

$$|\operatorname{Orb}_{even}(C_n, \operatorname{Neck}(n,j))^f| = \begin{cases} \binom{\frac{n}{2}}{\frac{j}{2}} - \binom{\frac{n-2}{4}}{\frac{j-2}{4}}, & j \equiv 2 \mod 4, \\ \binom{\frac{n}{2}}{\frac{j}{2}} - \binom{\frac{n-2}{4}}{\frac{j}{4}}, & j \equiv 0 \mod 4. \end{cases}$$

Proof. Since $\frac{n}{2}$ is odd, we have

$$\operatorname{Orb}_{odd}(C_{\frac{n}{2}}, \operatorname{Neck}(\frac{n}{2}, \frac{j}{2}))^f = \operatorname{Orb}(C_{\frac{n}{2}}, \operatorname{Neck}(\frac{n}{2}, \frac{j}{2}))^f.$$

Therefore, combining (4.6), (4.7), and Proposition 4.14, and noting Lemma 4.9, we have

$$|\operatorname{Orb}_{even}(C_n, \operatorname{Neck}(n,j))^f|$$
$$= |\operatorname{Orb}_{even}(C_n, \operatorname{Neck}(n,j))^f_1| + |\operatorname{Orb}_{even}(C_n, \operatorname{Neck}(n,j))^f_2|$$
$$= \frac{1}{2} \sum_{j_1+j_2=j, j_1<j_2} |\operatorname{Orb}(C_{\frac{n}{2}}, \operatorname{Neck}(\frac{n}{2}, j_1))^f| \cdot |\operatorname{Orb}(C_{\frac{n}{2}}, \operatorname{Neck}(\frac{n}{2}, j_2))^f|$$
$$+ \frac{1}{2}\left(\binom{\frac{n}{2}}{\frac{j}{2}} - |\operatorname{Orb}(C_{\frac{n}{2}}, \operatorname{Neck}(\frac{n}{2}, \frac{j}{2}))^f|\right).$$

Since j is even, we have $j_1 \equiv j_2 \mod 2$. As $\frac{n}{2}$ is odd, we further have $\nu_2(j_i) > \nu_2(n)$, $i = 1, 2$ or $\nu_2(j_i) = \nu_2(n)$, $i = 1, 2$. Thus by Lemma 4.15, we have

$$\begin{aligned}
&\frac{1}{2} \sum_{j_1+j_2=j, j_1<j_2} |\operatorname{Orb}(C_{\frac{n}{2}}, \operatorname{Neck}(\frac{n}{2}, j_1))^f| \cdot |\operatorname{Orb}(C_{\frac{n}{2}}, \operatorname{Neck}(\frac{n}{2}, j_2))^f| \\
=&\frac{1}{2} \sum_{i=0}^{\frac{j}{2}-1} \binom{\frac{n-2}{4}}{i} \binom{\frac{n-2}{4}}{\frac{j}{2}-1-i} + \frac{1}{2} \sum_{i=0}^{\frac{j}{2}} \binom{\frac{n-2}{4}}{i} \binom{\frac{n-2}{4}}{\frac{j}{2}-i} \\
=&\frac{1}{2} \left(\binom{\frac{n}{2}-1}{\frac{j}{2}-1} + \binom{\frac{n}{2}-1}{\frac{j}{2}} \right) = \frac{1}{2} \binom{\frac{n}{2}}{\frac{j}{2}},
\end{aligned}$$

where in the last line we have used the Pascal identity. Thus

$$\begin{aligned}
|\operatorname{Orb}_{even}(C_n, \operatorname{Neck}(n, j))^f| =&\frac{1}{2} \left(\binom{\frac{n}{2}}{\frac{j}{2}} + \binom{\frac{n}{2}}{\frac{j}{2}} \right) - |\operatorname{Orb}(C_{\frac{n}{2}}, \operatorname{Neck}(\frac{n}{2}, \frac{j}{2}))^f| \\
=&\begin{cases} \binom{\frac{n}{2}}{\frac{j}{2}} - \binom{\frac{n-2}{4}}{\frac{j-2}{4}}, & j \equiv 2 \mod 4, \\ \binom{\frac{n}{2}}{\frac{j}{2}} - \binom{\frac{n-2}{4}}{\frac{j}{4}}, & j \equiv 0 \mod 4. \end{cases}
\end{aligned}$$

□

PROPOSITION 4.18. *For $\nu_2(n) = 1$ and $\nu_2(j) \geq 1$, we have $\Delta(n, j) = 0$.*

PROOF. By the Corollary 2.3 of Kummer's theorem, for $j \equiv 2 \mod 4$, we have

$$\nu_2 \binom{\frac{n-2}{4}}{\frac{j-2}{4}} = \nu_2 \binom{\frac{n}{2}-1}{\frac{j}{2}-1} = \nu_2 \binom{\frac{n}{2}}{\frac{j}{2}},$$

and for $j \equiv 0 \mod 4$,

$$\nu_2 \binom{\frac{n-2}{4}}{\frac{j}{4}} = \nu_2 \binom{\frac{n}{2}-1}{\frac{j}{2}} = \nu_2 \binom{\frac{n}{2}}{\frac{j}{2}}.$$

By Proposition 4.17,

$$|\operatorname{Orb}_{even}(C_n, \operatorname{Neck}(n, j))^f| \equiv 0 \mod 2.$$

Finally, by Lemma 4.1,

$$\Delta(n, j) = (u - 1) \cdot |\operatorname{Orb}_{even}(C_n, \operatorname{Neck}(n, j))^f| = 0.$$

□

4.7. Proof of Theorem 1.1 when $\nu_2(n) \geq 2$ and $\nu_2(j) \geq 1$. Since both n and j are even, for any $[l] \in \operatorname{Orb}(C_n, \operatorname{Neck}(n, j))_2^f$ there are two beads on any type 2 symmetry axis, and they have to be of the same color. Denote $[l']$ as the subset formed by removing the two beads on a chosen type 2 symmetry axis for $l \in [l]$. If the two beads are blue, then

$$[l'] \in \operatorname{Orb}(C_{n-2}, \operatorname{Neck}(n-2, j-2))_1^f.$$

Otherwise, the two beads are red and

$$[l'] \in \operatorname{Orb}(C_{n-2}, \operatorname{Neck}(n-2, j))_1^f.$$

Now, if we start with $[l'] \in \operatorname{Orb}(C_{n-2}, \operatorname{Neck}(n-2, j-2))_1^f$, then we can recover an element in $\operatorname{Orb}(C_n, \operatorname{Neck}(n, j))_2^f$ by picking a symmetry axis for $l' \in [l']$

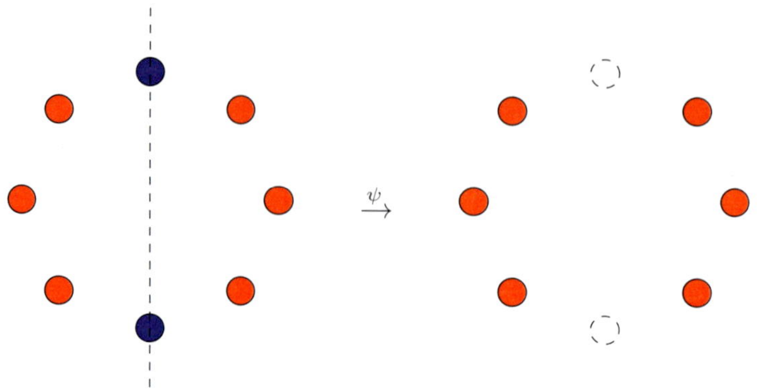

FIGURE 7. By removing the two beads on the symmetry axis, ψ maps an element in $\mathrm{Orb}(C_8, \mathrm{Neck}(8,2))_2^f$ to $\mathrm{Orb}(C_6, \mathrm{Neck}(6,0))_1^f$.

and adding back two blue beads. We can also do a similar procedure for $[l'] \in \mathrm{Orb}(C_{n-2}, \mathrm{Neck}(n-2,j))_1^f$ by adding two red beads. However, the associated map $\psi : \mathrm{Orb}(C_{n-2}, \mathrm{Neck}(n-2,j-2))_1^f \amalg \mathrm{Orb}(C_{n-2}, \mathrm{Neck}(n-2,j))_1^f \to \mathrm{Orb}(C_n, \mathrm{Neck}(n,j))_2^f$ is surjective but not injective[2]. In other words, if we count $\mathrm{Orb}(C_n, \mathrm{Neck}(n,j))_2^f$ via counting the domain, then there will be over-counting. Figure 7 and Figure 8 provide an example of the over-counting phenomenon. To understand the fibers of ψ, i.e., the over-counting, we first notice the following.

LEMMA 4.19. *Every element* $[l]$ *in*

$$\mathrm{Orb}(C_{n-2}, \mathrm{Neck}(n-2,j-2))_1^f \amalg \mathrm{Orb}(C_{n-2}, \mathrm{Neck}(n-2,j))_1^f$$

has exactly one type 1 symmetry axis if $\pi([l])$ *is even, and two symmetry axes of different type if* $\pi([l])$ *is odd.*

PROOF. Take $[l] \in \mathrm{Orb}(C_{n-2}, \mathrm{Neck}(n-2,j-2))_1^f \amalg \mathrm{Orb}(C_{n-2}, \mathrm{Neck}(n-2,j))_1^f]$. By Lemma 4.7, if $\frac{n-2}{\pi([l])}$ is odd, then $[l]$ has exactly one symmetry axis, which has to pass between beads as $\pi([l])$ is even. On the other hand, if $\frac{n-2}{\pi([l])}$ is even, since $n-2 \equiv 2 \mod 4$, $\pi([l])$ is odd, thus the two axes are of a different type, as their distance is a half-integer. □

The only elements $[l] \in \mathrm{Orb}(C_n, \mathrm{Neck}(n,j))_2^f$ that have $|\psi^{-1}([l])| = 2$ are the ones with two different axes both passing through two beads. Viewing on the domain, these are the elements in $\mathrm{Orb}(C_{n-2}, \mathrm{Neck}(n-2,j-2))_1^f$ and

$$\mathrm{Orb}(C_{n-2}, \mathrm{Neck}(n-2,j))_1^f$$

that have two axes of different types. By Lemma 4.19, these are exactly the elements that have an odd period.

The overcounting is not simply

$$|\,\mathrm{Orb}_{odd}(C_{n-2}, \mathrm{Neck}(n-2,j-2))_1^f| + |\,\mathrm{Orb}_{odd}(C_{n-2}, \mathrm{Neck}(n-2,j))_1^f|$$

as there would be an over-counting of the over-counting. This happens in the following situation. Denote the two symmetry axes for $[l]$ in the domain with

[2]For an example of how does the map ψ works, see Figure 7.

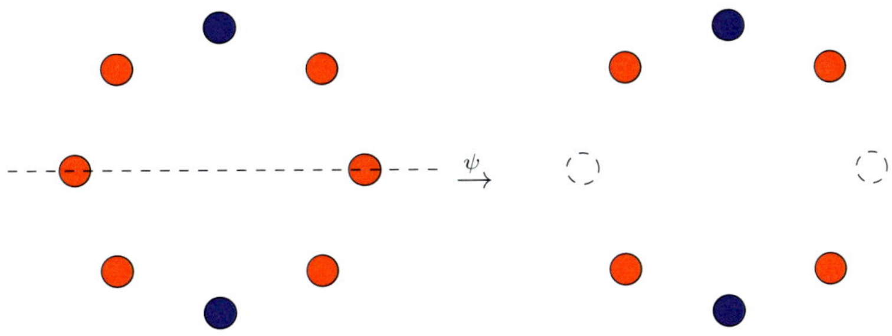

FIGURE 8. The element in Figure 7 has another symmetry axis, which maps to $\mathrm{Orb}(C_6, \mathrm{Neck}(6,2))_1^f$ under ψ.

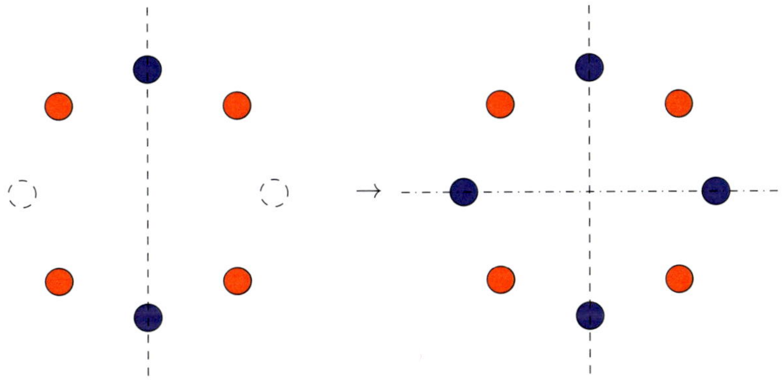

FIGURE 9. An example of the over-counting of over-counting.

$n-2$ beads as $[(l,\sigma_1)]$, $[(l,\sigma_2)]$, which is of type 1 and type 2, respectively. After adding the two beads along $[(l,\sigma_1)]$, the two symmetry axes $[(l,\sigma_1)]$, $[(l,\sigma_2)]$ will both become type 2 symmetry axis in $\psi([l]) \in \mathrm{Orb}(C_n, \mathrm{Neck}(n,j))_2^f$. Therefore, the over-counting of over-counting will happen if $[(l,\sigma_1)]$, $[(l,\sigma_2)]$ become the same axis in $\psi([l])$. A little thought combined with Lemma 4.7 shows this will happen if and only if $\psi([l]) \in \mathrm{Orb}_{odd}(C_n, \mathrm{Neck}(n,j))^f$. For example, if one wants to construct an element in $\mathrm{Orb}(C_8, \mathrm{Neck}(8,4))_2^f$ from the right-hand side of Figure 8, then one needs to add two blue beads to the vacant spots. This becomes a new symmetry axis. However, as shown in Figure 9, this axis is the same as the dashed one from the left-hand side.

PROPOSITION 4.20. *For $\nu_2(n) \geq 2$, and $\nu_2(j) \geq 1$, we have*

$$|\mathrm{Orb}(C_n, \mathrm{Neck}(n,j))_2^f| = \frac{1}{2}\binom{\frac{n}{2}}{\frac{j}{2}} + \frac{1}{2}|\mathrm{Orb}_{odd}(C_n, \mathrm{Neck}(n,j))^f|.$$

PROOF. From the earlier argument in this section, we have

$$\begin{aligned}|\operatorname{Orb}(C_n,\operatorname{Neck}(n,j))_2^f| =& |\operatorname{Orb}(C_{n-2},\operatorname{Neck}(n-2,j-2))_1^f| \\ &+ |\operatorname{Orb}(C_{n-2},\operatorname{Neck}(n-2,j))_1^f| \\ &- \frac{1}{2}\Big(|\operatorname{Orb}_{odd}(C_{n-2},\operatorname{Neck}(n-2,j-2))_1^f| \\ &+ |\operatorname{Orb}_{odd}(C_{n-2},\operatorname{Neck}(n-2,j))_1^f| \\ &- |\operatorname{Orb}_{odd}(C_n,\operatorname{Neck}(n,j))^f|\Big).\end{aligned}$$

Using the facts $|\operatorname{Orb}(C_{n-2},\operatorname{Neck}(n-2,j-2))_1^f| = |\operatorname{Orb}_{even}(C_{n-2},\operatorname{Neck}(n-2,j-2))_1^f| + |\operatorname{Orb}_{odd}(C_{n-2},\operatorname{Neck}(n-2,j-2))_1^f|$, $|\operatorname{Orb}(C_{n-2},\operatorname{Neck}(n-2,j))_1^f| = |\operatorname{Orb}_{even}(C_{n-2},\operatorname{Neck}(n-2,j))_1^f| + |\operatorname{Orb}_{odd}(C_{n-2},\operatorname{Neck}(n-2,j))_1^f|$, and applying Proposition 4.14, we get

$$\begin{aligned}|\operatorname{Orb}(C_n,\operatorname{Neck}(n,j))_2^f| =& \frac{1}{2}\Bigg(\binom{\frac{n}{2}-1}{\frac{j}{2}-1} + \binom{\frac{n}{2}-1}{\frac{j}{2}} \\ &+ |\operatorname{Orb}_{odd}(C_{\frac{n}{2}-1},\operatorname{Neck}(\frac{n}{2}-1,\frac{j}{2}-1))_1^f| \\ &+ |\operatorname{Orb}_{odd}(C_{\frac{n}{2}-1},\operatorname{Neck}(\frac{n}{2}-1,\frac{j}{2}))_1^f| \\ &- |\operatorname{Orb}_{odd}(C_{n-2},\operatorname{Neck}(n-2,j-2))_1^f| \\ &- |\operatorname{Orb}_{odd}(C_{n-2},\operatorname{Neck}(n-2,j))_1^f| \\ &+ |\operatorname{Orb}_{odd}(C_n,\operatorname{Neck}(n,j))^f|\Big) \\ =& \frac{1}{2}\binom{\frac{n}{2}}{\frac{j}{2}} + \frac{1}{2}|\operatorname{Orb}_{odd}(C_n,\operatorname{Neck}(n,j))^f|\end{aligned}$$

where in the second identity, we have used the Pascal identity, and the same argument as in the first sentence of the proof of Lemma 4.15. □

PROPOSITION 4.21. *For $\nu_2(n) \geq 2$ and $\nu_2(j) \geq 1$, we have $\Delta(n,j) = 0$.*

PROOF. Combining Lemma 4.9, Proposition 4.14, and Proposition 4.20, we have

$$|\operatorname{Orb}_{even}(C_n,\operatorname{Neck}(n,j))^f| = \binom{\frac{n}{2}}{\frac{j}{2}} - |\operatorname{Orb}_{even}(C_n,\operatorname{Neck}(n,j))^f|.$$

Now, apply Lemma 4.15, we have

(1) If $\nu_2(n) > \nu_2(j)$,

$$|\operatorname{Orb}_{even}(C_n,\operatorname{Neck}(n,j))^f| = \binom{\frac{n}{2}}{\frac{j}{2}}.$$

(2) If $\nu_2(n) = \nu_2(j)$,

$$|\operatorname{Orb}_{even}(C_n,\operatorname{Neck}(n,j))^f| = \binom{\frac{n}{2}}{\frac{j}{2}} - \binom{\frac{n}{2^{\nu_2(n)+1}} - \frac{1}{2}}{\frac{j}{2^{\nu_2(n)+1}} - \frac{1}{2}}.$$

(3) If $\nu_2(n) < \nu_2(j)$,

$$|\operatorname{Orb}_{even}(C_n, \operatorname{Neck}(n,j))^f|) = \binom{\frac{n}{2}}{\frac{j}{2}} - \binom{\frac{n}{2^{\nu_2(n)+1}} - \frac{1}{2}}{\frac{j}{2^{\nu_2(n)+1}}}.$$

In case (1), by Lucas's theorem, we have

$$|\operatorname{Orb}_{even}(C_n, \operatorname{Neck}(n,j))^f| \equiv 0 \mod 2.$$

In case (2), by Corollary 2.3 of Kummer's theorem,

$$\nu_2\binom{\frac{n}{2^{\nu_2(n)+1}} - \frac{1}{2}}{\frac{j}{2^{\nu_2(n)+1}} - \frac{1}{2}} = \nu_2\binom{\frac{n}{2^{\nu_2(n)}}}{\frac{j}{2^{\nu_2(n)}}} = \nu_2\binom{\frac{n}{2}}{\frac{j}{2}}.$$

Therefore, $|\operatorname{Orb}_{even}(C_n, \operatorname{Neck}(n,j))^f| \equiv 0 \mod 2$. In case (3), since $\frac{j}{2^{\nu_2(n)}}$ is even, and $\frac{n}{2^{\nu_2(n)}}$ is odd, then by Corollary 2.3 and 2.4 of Kummer's theorem

$$\nu_2\binom{\frac{n}{2^{\nu_2(n)}} - 1}{\frac{j}{2^{\nu_2(n)}}} = \nu_2\binom{\frac{n}{2^{\nu_2(n)}}}{\frac{j}{2^{\nu_2(n)}}} = \nu_2\binom{\frac{n}{2}}{\frac{j}{2}}.$$

Therefore, we have

$$|\operatorname{Orb}_{even}(C_n, \operatorname{Neck}(n,j))^f| \equiv 0 \mod 2,$$

and by Lemma 4.1,

$$\Delta(n,j) = (u-1) \cdot |\operatorname{Orb}_{even}(C_n, \operatorname{Neck}(n,j))^f| = 0.$$

□

4.8. Conclusion of the proof of Theorem 1.1. Combining Proposition 4.2, Proposition 4.4, Proposition 4.18, and Proposition 4.21, then invoking Lemma 4.1, we have

$$\binom{L/k}{j} = \binom{n}{j} - (1-u) \cdot \delta(n,j),$$

where $\delta(n,j) = \begin{cases} 1, & \frac{j-1}{2} \prec \frac{n-2}{2}, \\ 0, & \text{else} \end{cases}$ and $\prec$ is as in Definition 4.3 in the beginning of Section 4.3. Finally, we can conclude the proof of Theorem 1.1 by Lucus's theorem.

5. Twisted case - Proof of Theorem 1.2

Let $n = 2j$. The strategy for the proof of Theorem 1.2 is to rewrite

$$\operatorname{Orb}_{even}(C_n, \operatorname{Neck}(n,j)^\tau)$$

in terms of the orbits for the untwisted action that we have studied in the previous sections.

5.1. Necklace interpretation of twisted orbits. We first define a $D_{2j} \times C_2$ action on $\operatorname{Neck}(2j,j)$. The D_{2j} action is defined to be the same as in the untwisted case defined in Section 4.1. The action of $C_2 = \langle e : e^2 = 1\rangle$ commutes with the D_{2j} action, whose generator e is called the *color exchange*. As the name suggests, e exchanges the colors of all the beads. More precisely, it turns the red beads of a necklace to blue, and the blue beads to red. Define the *twisted* action of C_{2j} on $\operatorname{Neck}(2j,j)$ by the map $C_{2j} \to D_{2j} \times C_2$ defined by

$$r \mapsto (r, e)$$

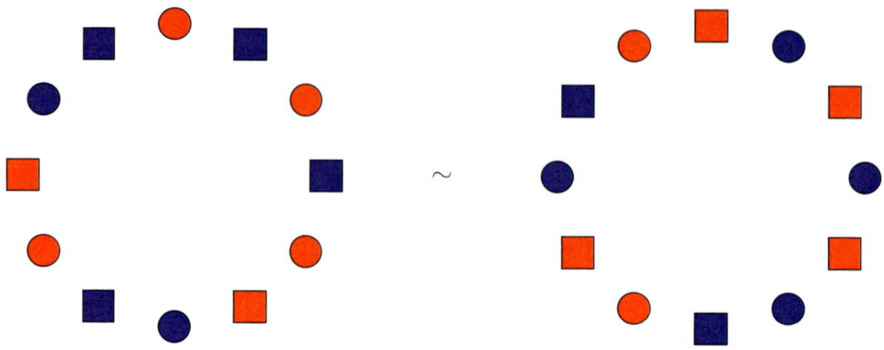

FIGURE 10. An example of elements in $\mathrm{Orb}(C_{12}, \mathrm{Neck}(12,6)^\tau)$.

and the $D_{2j} \times C_2$ action just defined. To distinguish between the twisted and untwisted actions, when we view $\mathrm{Neck}(2j, j)$ with its twisted C_{2j}-action, we will write $\mathrm{Neck}(2j, j)^\tau$. Otherwise, $\mathrm{Neck}(2j, j)$ has the C_{2j}-action from the morphism $C_{2j} \to D_{2j}$, given $r \mapsto r$ as above. We have

$$\binom{L[Q]/k}{j} = \binom{2j}{j} + (u-1)|\,\mathrm{Orb}_{even}(C_{2j}, \mathrm{Neck}(2j,j)^\tau)|,$$

and we will denote

$$\Delta'(2j,j) := \binom{L[Q]/k}{j} - \binom{2j}{j} = (u-1)|\,\mathrm{Orb}_{even}(C_{2j}, \mathrm{Neck}(2j,j)^\tau)|. \tag{5.1}$$

5.2. Reduction to the untwisted case.

LEMMA 5.1. *The elements of* $\mathrm{Orb}(C_{2j}, \mathrm{Neck}(2j,j)^\tau)$ *can be interpreted as triples*

$$\mathrm{Orb}(C_{2j}, \mathrm{Neck}(2j,j)^\tau) \cong \{([l],[l_1],[l_2]) : [l] \in \mathrm{Orb}(C_n, \mathrm{Neck}(n,j)), \phi([l]) = ([l_1],[l_2])\}/\sim,$$

where in $([l],[l_1],[l_2])$, $[l_1]$ *and* $[l_2]$ *are ordered. The decomposition map* ϕ *is defined in* (4.4). *The equivalence relation is defined as* $([l],[l_1],[l_2]) \sim (e\cdot[l], e\cdot[l_2], e\cdot[l_1])$.

PROOF. An element in $\mathrm{Orb}(C_{2j}, \mathrm{Neck}(2j,j)^\tau)$ is uniquely determined by an element $[l]$ in $\mathrm{Orb}(C_n, \mathrm{Neck}(n,j))$ together with a chosen bead in $[l]$. On the other hand, two different chosen beads in $[l]$ will give rise to the same

$$\mathrm{Orb}(C_{2j}, \mathrm{Neck}(2j,j)^\tau)$$

orbits if and only if they both lie in the $[l_1]$ or $[l_2]$, where $\phi([l]) = ([l_1],[l_2])$. Therefore, an element in $\mathrm{Orb}(C_{2j}, \mathrm{Neck}(2j,j)^\tau)$ can be denoted as $([l],[l_1],[l_2])$, where the second entry $[l_1]$ means that the chosen bead lies in $[l_1]$. However, both $([l],[l_1],[l_2])$ and $(e\cdot[l], e\cdot[l_2], e\cdot[l_1])$ give rise to the same element in $\mathrm{Orb}(C_{2j}, \mathrm{Neck}(2j,j)^\tau)$. By quotienting out this equivalence relation, we arrive at the desired isomorphism. □

One way to visualize the triple is shown in Figure 10. Denote the element as $([l],[l_1],[l_2]) \in \mathrm{Orb}(C_{12}, \mathrm{Neck}(12,6)^\tau)$. The circle-shaped beads form $[l_1] \in \mathrm{Orb}(C_6, \mathrm{Neck}(6,2))$, whereas the square-shaped beads form $[l_2] \in \mathrm{Orb}(C_6, \mathrm{Neck}(6,4))$.

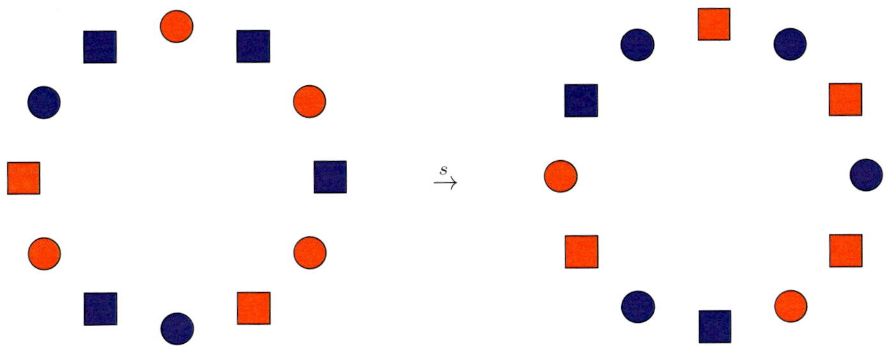

FIGURE 11. An example of swapping.

From now on, we will use the triple $([l],[l_1],[l_2])$ to denote elements in $\operatorname{Orb}(C_{2j},\operatorname{Neck}(2j,j)^\tau)$. Define the twisted period as $\pi'([l],[l_1],[l_2]) := |([l],[l_1],[l_2])|$, i.e., the size of the twisted C_{2j}-orbit. Next, we have a C_2 action on

$$\operatorname{Orb}(C_{2j},\operatorname{Neck}(2j,j)^\tau),$$

which acts by exchanging l_1, l_2. Combinatorically, this action interchanges the two potentially different twisted orbits that could come from the same underlining untwisted orbit $[l]$. We will call this action swapping, and an example is shown in Figure 11. For a set $\mathcal{N}$, we will use $\mathcal{N}^s$ to denote the subset fixed by swapping.

LEMMA 5.2. $\Delta'(2j,j) = (u-1)|\operatorname{Orb}(C_{2j},\operatorname{Neck}(2j,j)^\tau)^s|$, *where*

$$\operatorname{Orb}(C_{2j},\operatorname{Neck}(2j,j)^\tau)^s$$

is the subset of $\operatorname{Orb}_{even}(C_{2j},\operatorname{Neck}(2j,j)^\tau)$ *that is fixed by swapping.*

PROOF. Since the swapping action does not change the twisted period, by (5.1) and $2(u-1)=0$, we got the desired formula. □

LEMMA 5.3. *In* $\operatorname{Orb}(C_{2j},\operatorname{Neck}(2j,j)^\tau)$, $([l],[l_1],[l_2]) = ([l],[l_2],[l_1])$ *if and only if either* $[l_1]=[l_2]$ *or* $[l]=e\cdot[l]$.

PROOF. By the definition of the twisted orbit, there are two possibilities for $([l],[l_1],[l_2]) = ([l],[l_2],[l_1])$. The first is $[l_1]=[l_2]$, and the other is $[l]=e\cdot[l]$, $[l_1] = e\cdot[l_1]$, $[l_2]=e\cdot[l_2]$. □

COROLLARY 5.4. $\Delta'(2j,j)=0$, *for* $\nu_2(j)=0$.

PROOF. Since j is odd, $[l_1]\neq[l_2]$ and $[l_1]\neq e\cdot[l_2]$ for all elements $([l],[l_1],[l_2]) \in \operatorname{Orb}(C_{2j},\operatorname{Neck}(2j,j)^\tau)$. Thus, the C_2-action of swapping acts freely, so

$$|\operatorname{Orb}(C_{2j},\operatorname{Neck}(2j,j)^\tau)^s| \equiv 0 \mod 2.$$

□

From now on, we will assume j is even. Applying the above results to

$$|\operatorname{Orb}(C_{2j},\operatorname{Neck}(2j,j)^\tau)^s|,$$

we have

$$\begin{aligned}|\operatorname{Orb}(C_{2j},\operatorname{Neck}(2j,j)^\tau)^s| =& |([l],[l_1],[l_1]) \in \operatorname{Orb}_{even}(C_{2j},\operatorname{Neck}(2j,j)^\tau), [l]\neq e\cdot[l]| \\ &+ |([l],[l_1],[l_2]) \in \operatorname{Orb}_{even}(C_{2j},\operatorname{Neck}(2j,j)^\tau), [l]=e\cdot[l]|.\end{aligned}$$

For $([l],[l_1],[l_1]) \in \mathrm{Orb}_{even}(C_{2j}, \mathrm{Neck}(2j,j)^\tau)$, $[l] \neq e \cdot [l] \Leftrightarrow [l_1] \neq e \cdot [l_1]$, and notice that if $[l_1] \neq e \cdot [l_1]$, then $2 | \pi'([l],[l_1],[l_1])$. Therefore,

$$\begin{aligned} |\mathrm{Orb}(C_{2j}, \mathrm{Neck}(2j,j)^\tau)^s| = & \sum_{[l] \in \mathrm{Orb}(C_j, \mathrm{Neck}(j, \frac{j}{2})), [l] \neq e \cdot [l]} \frac{|\phi^{-1}([l],[l])|}{2} \\ & + |([l],[l_1],[l_2]) \in \mathrm{Orb}_{even}(C_{2j}, \mathrm{Neck}(2j,j)^\tau), [l] = e \cdot [l]| \end{aligned}$$

To further simplify the second term, we notice the following facts.

LEMMA 5.5. *For* $([l],[l_1],[l_2]) \in \mathrm{Orb}(C_{2j}, \mathrm{Neck}(2j,j)^\tau)$, *we have* $\pi([l]) \equiv 0 \mod 2$.

PROOF. Since $\nu_2(2j) > \nu_2(j)$, assume $2 \nmid \pi([l])$, then the number of blue beads will not be an integer. □

LEMMA 5.6. *If* $[l] = e \cdot [l]$, *then for all* $l \in \mathrm{Neck}(2j,j)$, *we have*

$$e \cdot l = r^{\frac{\pi([l])}{2}} \cdot l.$$

PROOF. Apparently $e = r^m$ for some $m \in \mathbb{Z}/\pi([l])\mathbb{Z}$. Since e changes l, we can take $0 < m < \pi([l])$. On the other hand, since $e^2 = 1$, one has $\pi([l]) | 2m$, thus $m = \frac{\pi([l])}{2}$. □

PROPOSITION 5.7. $\pi'([l],[l_1],[l_2]) = \pi([l])$ *unless* $l = e \cdot [l]$, *and* $\nu_2(\pi([l])) = 1$, *in which case* $\pi'([l],[l_1],[l_2]) = \frac{\pi([l])}{2} \equiv 1 \mod 2$.

PROOF. Since $\pi([l])$ is even, $r^{\pi([l])} = e^{\pi([l])}$, and we always have $\pi'([l],[l_1],[l_2]) \leq \pi([l])$. By the definition of the twisted action, if $\pi'([l],[l_1],[l_2])$ is even, we have $\pi'([l],[l_1],[l_2]) = \pi([l])$. Therefore, the only possibility for $\pi'([l],[l_1],[l_2]) < \pi([l])$ is when $\pi'([l],[l_1],[l_2])$ is odd. In this case, we have $2\pi'([l],[l_1],[l_2]) = \pi([l])$, and thus $\nu_2([l]) = 1$. □

Therefore

$$\begin{aligned} |\mathrm{Orb}(C_{2j}, \mathrm{Neck}(2j,j)^\tau)^s| = & \sum_{[l] \in \mathrm{Orb}(C_j, \mathrm{Neck}(j, \frac{j}{2})), [l] \neq e \cdot [l]} \frac{|\phi^{-1}([l],[l])|}{2} \\ & + |([l],[l_1],[l_2]) \in \mathrm{Orb}(C_{2j}, \mathrm{Neck}(2j,j)^\tau), \\ & \quad [l] = e \cdot [l], \nu_2(\pi([l])) > 1| \\ = & \sum_{[l] \in \mathrm{Orb}(C_j, \mathrm{Neck}(j, \frac{j}{2})), [l] \neq e \cdot [l]} \frac{|\phi^{-1}([l],[l])|}{2} \\ & + |l \in \mathrm{Orb}(C_{2j}, \mathrm{Neck}(2j,j)), [l] = e \cdot [l], \nu_2(\pi([l])) > 1| \end{aligned}$$

where the last formula is an enumeration purely in terms of untwisted orbits.

5.3. Relations to partitions. Consider an element $[l] \in \mathrm{Orb}(C_n, \mathrm{Neck}(n,j))$. One can uniquely write $[l]$ as the cyclic orbit of a partition of n as follows. Call a maximal consecutive segment of beads of l of the same color a *cluster*. By replacing each cluster of l by its number of beads, one obtains the corresponding partition on n. For example, fix $l \in [l]$, if the beads in positions 1 and 5 are red but everything in between is blue, we will replace the 3 blue beads with the number 3. If the cluster is red, we will mark it by adding an underline to distinguish it from the blue ones.

We will use the notation $(\cdots)$ to denote the cyclic equivalent class of partitions. So $[l]$ can be uniquely written as the cyclic orbit of a partition of n as

$$[l] = (\underline{r_1}b_1\underline{r_2}b_2\cdots\underline{r_m}b_m), \quad r_i, b_i \in \mathbb{Z}_{>0},$$

with $\sum_i(r_i+b_i)=n, \sum_i b_i = j$. Under this notation, the action of e is simply

$$e\cdot(\underline{r_1}b_1\underline{r_2}b_2\cdots\underline{r_m}b_m) = (r_1\underline{b_1}r_2\underline{b_2}\cdots r_m\underline{b_m}).$$

Apparently, the length of the partition has to be even. For example, the three elements in Figure 5 can be written as $(\underline{6}4), (\underline{2}1\underline{1}1\underline{2}1\underline{1}1), (\underline{1}1\underline{4}1\underline{1}2)$, respectively.

Denote the set of even length partitions of n with an underlined marking for every other entry such that the sum of unmarked entries is j by $\mathrm{Part}(n,j,\underline{n-j})$. We further denote the set of its cyclic equivalent classes by $\mathrm{Orb}(C,\mathrm{Part}(n,j,\underline{n-j}))$. For an element $[p]\in\mathrm{Orb}(C,\mathrm{Part}(n,j,\underline{n-j}))$, we denote its length by $|p| := \mathrm{length}(p)$, and define its period as $\varpi([p]) = |C_{|p|}\cdot p|$[3].

Denote the set of partitions of j by $\mathrm{Part}(j)$, and similarly denote the set of cyclic equivalent classes and orbits as in the marked case, then we have

LEMMA 5.8. *There is a bijection*

$$\{[l]\in\mathrm{Orb}(C_{2j},\mathrm{Neck}(2j,j)), l = e\cdot[l]\} \cong \{[p]\in\mathrm{Orb}(C,\mathrm{Part}(j)), \nu_2(\varpi([p]))=0\}.$$

PROOF. Let $[l]\in\mathrm{Orb}(C_{2j},\mathrm{Neck}(2j,j))$, and $[p] = (\underline{r_1}b_1\underline{r_2}b_2\cdots\underline{r_m}b_m)$ be its corresponding element in $\mathrm{Orb}(C,\mathrm{Part}(2j,j,\underline{j}))$. If $[l]=e\cdot[l]$, then

$$(\underline{r_1}b_1\underline{r_2}b_2\cdots\underline{r_m}b_m) = (r_1\underline{b_1}r_2\underline{b_2}\cdots r_m\underline{b_m}),$$

and apparently we need $(r_1\cdots r_m) = (b_1\cdots b_m)$. Moreover, by passing to the smaller partition that is formed by taking $\varpi([p])$ consecutive parts, without loss of generality, we can assume the partition is non-periodic, i.e., $\varpi([p]) = 2m$. Then the only element in $\mathrm{Orb}(C,\mathrm{Part}(2j,j,\underline{j}))$ of the form $(\underline{r_1}, r_{i_1}, \underline{r_2}, r_{i_2}, \cdots, \underline{r_m}, r_{i_m})$ that satisfies $[p]=e\cdot[p]$ is

$$(\underline{r_1}r_{\frac{m+3}{2}}\underline{r_2}r_{\frac{m+5}{2}}\cdots r_m\underline{r_{\frac{m+1}{2}}}r_1\underline{r_{\frac{m+3}{2}}}\cdots\underline{r_m}r_{\frac{m+1}{2}}),$$

and m has to be odd. $\square$

Recall that the total number of partitions of $j\in\mathbb{Z}_{>0}$ is 2^{j-1}. So

$$2^{j-1} = \sum_d d|\{[p]\in\mathrm{Orb}(C,\mathrm{Part}(j)), \varpi([p]) = d\}|, \tag{5.2}$$

LEMMA 5.9. *For $j>1$, we have* $|\{l\in\mathrm{Orb}(C_{2j},\mathrm{Neck}(2j,j)), [l] = e\cdot[l], \nu_2(\pi([l]))>1\}| \equiv |\{l\in\mathrm{Orb}(C_{2j},\mathrm{Neck}(2j,j)), [l]=e\cdot[l], \nu_2(\pi([l]))=1\}| \mod 2$.

PROOF. By (5.2), we have

$$\begin{aligned}
&|\{[p]\in\mathrm{Orb}(C,\mathrm{Part}(j)), \nu_2(\varpi([p]))=0\}|\\
&\quad = \sum_{\nu_2(d)=0} d|\{[p]\in\mathrm{Orb}(C,\mathrm{Part}(j)), \varpi([p])=d\}|\\
&\equiv \sum_d d|\{[p]\in\mathrm{Orb}(C,\mathrm{Part}(j)), \varpi([p])=d\}| \mod 2\\
&=2^{j-1}\equiv 0 \mod 2.
\end{aligned}$$

[3]Note there is no direct relation between ϖ and π'.

Then by Lemma 5.8, we have

$$|\{l \in \operatorname{Orb}(C_{2j}, \operatorname{Neck}(2j, j)), [l] = e \cdot [l]\}| \equiv 0 \mod 2.$$

□

Therefore, we have

$$(5.3)\quad \begin{aligned} |\operatorname{Orb}(C_{2j}, \operatorname{Neck}(2j, j)^{\tau})^{s}| \equiv & \sum_{[l] \in \operatorname{Orb}(C_j, \operatorname{Neck}(j, \frac{j}{2})), [l] \neq e \cdot [l]} \frac{|\phi^{-1}([l], [l])|}{2} \\ & + |\{l \in \operatorname{Orb}(C_{2j}, \operatorname{Neck}(2j, j)), [l] = e \cdot [l], \nu_2(\pi([l])) = 1\}| \mod 2. \end{aligned}$$

5.4. Conclusion of the proof of Theorem 1.2. First, we have:

LEMMA 5.10. *For* $[l] \in \operatorname{Orb}(C_j, \operatorname{Neck}(j, \frac{j}{2}))$*, we have* $|\phi^{-1}([l], [l])| = \frac{\pi([l])}{2}$.

PROOF. Similar to Lemma 4.12. □

Then

$$\sum_{[l] \in \operatorname{Orb}(C_j, \operatorname{Neck}(j, \frac{j}{2})), [l] \neq e \cdot [l]} \frac{|\phi^{-1}([l], [l])|}{2} = \sum_{[l] \in \operatorname{Orb}(C_j, \operatorname{Neck}(j, \frac{j}{2})), [l] \neq e \cdot [l]} \frac{|\pi([l])|}{2}$$
$$= \sum_{[l] \in \operatorname{Orb}(C_j, \operatorname{Neck}(j, \frac{j}{2}))} \frac{|\pi([l])|}{2} - \sum_{[l] \in \operatorname{Orb}(C_j, \operatorname{Neck}(j, \frac{j}{2})), [l] = e \cdot [l]} \frac{|\pi([l])|}{2}$$

Since $\pi([l]) \equiv 0 \mod 2$, we have

$$\begin{aligned} \sum_{[l] \in \operatorname{Orb}(C_j, \operatorname{Neck}(j, \frac{j}{2})), [l] = e \cdot [l]} \frac{|\pi([l])|}{2} \equiv & |[l] \in \operatorname{Orb}(C_j, \operatorname{Neck}(j, \frac{j}{2})), \\ & [l] = e \cdot [l], \nu_2(\pi(l)) = 1| \mod 2. \end{aligned}$$

As

$$\begin{aligned} & \{[l] \in \operatorname{Orb}(C_j, \operatorname{Neck}(j, \frac{j}{2})), [l] = e \cdot [l], \nu_2(\pi(l)) = 1\} \\ & = \{[l] \in \operatorname{Orb}(C_j, \operatorname{Neck}(2j, j)), [l] = e \cdot [l], \nu_2(\pi(l)) = 1\}, \end{aligned}$$

we can conclude

PROPOSITION 5.11. $\Delta'(2j, j) = \frac{1}{2}\binom{2j}{j} \cdot (u - 1)$*, for* $j \equiv 0 \mod 2$.

PROOF. By (5.3), and the calculation in this subsection, we have

$$|\operatorname{Orb}(C_{2j}, \operatorname{Neck}(2j, j)^{\tau})^{s}| \equiv \frac{1}{2}\binom{j}{\frac{j}{2}} \mod 2,$$

Then by Lemma 5.2 and Corollary 2.3 of Kummer's theorem, we have

$$\Delta'(2j, j) = \frac{1}{2}\binom{2j}{j} \cdot (u - 1)$$

□

The proof of Theorem 1.2 is concluded by combining Proposition 5.11 and Corollary 5.4.

5.5. Explicit value of twisted binomial coefficients. By Corollary 2.5 of Kummer's theorem, we have

$$\binom{L[Q]/k}{j} = \binom{2j}{j} + (u-1)\delta(j),$$

where $\delta(j) = \begin{cases} 1 & j = 2^m, m \in \mathbb{Z}_{>0} \\ 0 & \text{else} \end{cases}.$

See Example 1.3 for computations with small j.

Acknowledgments

We heartily thank De'Asia Brodie and Zoe Valentine for collaboration on the cases $j = 1, 2, 3$ of Theorem 1.1, and Erwan Brugallé and Rena Chu for useful discussions. KW received support from NSF DMS-2103838 and NSF DMS-2405191. CC is partially supported by DMS-2304981 and DMS-2405191.

References

[1] E. Brugallé and K. Wickelgren, *A quadratic Abramovich-Bertram formula*, 2025. arXiv:2506.17854v1.

[2] P. E. Conner and R. Perlis, *A survey of trace forms of algebraic number fields*, Series in Pure Mathematics, vol. 2, World Scientific Publishing Co., Singapore, 1984, DOI 10.1142/0066. MR761569

[3] E. Cotterill, I. Darago, and C. Han, *Arithmetic inflection formulae for linear series on hyperelliptic curves*, Math. Nachr. **296** (2023), no. 8, 3272–3300, DOI 10.1002/mana.202100229. MR4626882

[4] D. S. Dummit and R. M. Foote, *Abstract algebra*, 3rd ed., John Wiley & Sons, Inc., Hoboken, NJ, 2004. MR2286236

[5] J.-P. Serre, *Cohomological invariants, Witt invariants, and trace forms*, Cohomological invariants in Galois cohomology, Univ. Lecture Ser., vol. 28, Amer. Math. Soc., Providence, RI, 2003, pp. 1–100. Notes by Skip Garibaldi. MR1999384

[6] G. H. Hardy and E. M. Wright, *An introduction to the theory of numbers*, 5th ed., The Clarendon Press, Oxford University Press, New York, 1979. MR568909

[7] J. L. Kass, M. Levine, J. P. Solomon, and K. Wickelgren, *A quadratically enriched count of rational curves*, 2023. *Preprint* ArXiv 2307.01936.

[8] J. L. Kass and K. Wickelgren, *An arithmetic count of the lines on a smooth cubic surface*, Compos. Math. **157** (2021), no. 4, 677–709, DOI 10.1112/s0010437x20007691. MR4247570

[9] E. E. Kummer, *Über die Ergänzungssätze zu den allgemeinen Reciprocitätsgesetzen* (German), J. Reine Angew. Math. **44** (1852), 93–146, DOI 10.1515/crll.1852.44.93. MR1578793

[10] T. Y. Lam, *Introduction to quadratic forms over fields*, Graduate Studies in Mathematics, vol. 67, American Mathematical Society, Providence, RI, 2005, DOI 10.1090/gsm/067. MR2104929

[11] M. Levine, *Aspects of enumerative geometry with quadratic forms*, Doc. Math. **25** (2020), 2179–2239. MR4198841

[12] E. Lucas, *Theorie des Fonctions Numeriques Simplement Periodiques* (French), Amer. J. Math. **1** (1878), no. 4, 289–321, DOI 10.2307/2369373. MR1505176

[13] S. McKean, *An arithmetic enrichment of Bézout's Theorem*, Math. Ann. **379** (2021), no. 1-2, 633–660, DOI 10.1007/s00208-020-02120-3. MR4211099

[14] A. J. Puentes, H. Markwig, S. Pauli, and F. Röhrle, *Arithmetic counts of tropical plane curves and their properties*, Adv. Geom. **24** (2024), no. 4, 553–576, DOI 10.1515/advgeom-2024-0028. MR4813080

[15] M. Wendt, *Oriented Schubert calculus in Chow-Witt rings of Grassmannians*, Motivic homotopy theory and refined enumerative geometry, Contemp. Math., vol. 745, Amer. Math. Soc., [Providence], RI, [2020] ©2020, pp. 217–267, DOI 10.1090/conm/745/15027. MR4071217

Duke University, Durham, North Carolina
Email address: chongyao.chen@duke.edu

Duke University, Durham, North Carolina
Email address: kirsten.wickelgren@duke.edu

Contemporary Mathematics
Volume **842**, 2026
https://doi.org/10.1090/conm/842/16852

Knots, primes and class field theory

Alain Connes and Caterina Consani

ABSTRACT. In this paper, we present a geometric generalization of class field theory, demonstrating how adelic constructions, central to the spectral realization of zeros of L-functions and the geometric framework for explicit formulas in number theory, naturally extend the classical theory. This generalization transitions from the idele class group, which acts as the adelic analog of Galois groups, to a geometric framework associated with schemes and the ring of integers of global fields. This perspective provides a conceptual explanation for the role of the adele class space in the spectral realization of L-function zeros and identifies the idele class group as a generic point in this context. The sector $X_{\mathbb{Q}}$ of the adele class space corresponding to the Riemann zeta function gives the class field counterpart of the scaling topos. The main result is the construction of a functor mapping finite abelian extensions of $\mathbb{Q}$ to finite covers of $X_{\mathbb{Q}}$, with the monodromy of periodic orbits of length $\log p$ under the scaling action corresponding to the Galois action of the Frobenius at the prime p.

Introduction

This paper proposes an extension of class field theory, moving beyond the traditional idele class group description of the abelianized Galois group to a geometric framework of adelic nature. This new perspective establishes a counterpart to Grothendieck's extension of Galois theory to schemes, focusing on those associated with the ring of integers. This shift not only enriches the conceptual foundation of class field theory but also elucidates the role of the adele class space in the spectral realization of L-function zeros and the explicit formulas of number theory.

The geometrization of Galois theory, as pioneered by A. Grothendieck in the 1960s, revealed that the étale site of an algebraic scheme, combined with its profinite fundamental group, provides an algebro-geometric analogue to classical Galois theory. In contrast, class field theory centers on the idele class group of a global field and its relation, via Artin's reciprocity map, to the Galois group of the maximal abelian extension. This number-theoretic perspective already suggests the pivotal role of adelic spaces in capturing the geometric essence of abelian field extensions.

According to C. Chevalley[1]

2020 *Mathematics Subject Classification.* Primary 11R37, 11M06; Secondary 11M55, 14A15, 14F20, 57K10.

Key words and phrases. Knots and primes, étale fundamental group, scheme, class field theory, adele class space, scaling topos, Baum-Connes map, noncommutative geometry.

The second author was supported in part by Simons Collaboration Grant #691493.

[1]C. Chevalley "La Théorie du Corps de Classes" Annals of Math. Vol 41, No. 2 (1940)

> L'objet de la théorie du corps de classes est de montrer comment les extensions abéliennes d'un corps de nombres algébriques K peuvent être déterminées par des éléments tirés de la connaissance de K lui-même; ou, si l'on veut présenter les choses en termes dialectiques, comment un corps possède en soi les éléments de son propre dépassement[2].

It is therefore natural to anticipate that Grothendieck's extension of Galois theory to schemes will, in the context of global fields K, acquire a geometric counterpart of adelic nature—specifically, one involving the ring $\mathbb{A}_K$ of adeles of K. From the very inception of Galois theory, Galois's concept of a *primitive* equation inherently involves not only an abstract group but also, in a fundamental way, the manner in which this group acts on the set of roots of the equation.

The idele class group of a global field K is defined as the quotient group $C_K := K^\times \backslash \mathrm{GL}_1(\mathbb{A}_K)$. Since $\mathrm{GL}_1(\mathbb{A}_K)$ acts on $\mathbb{A}_K$ by multiplication, the most natural space on which C_K acts is the adele class space:

$$Y_K := K^\times \backslash \mathbb{A}_K, \quad g(y) := gy, \quad \forall g \in C_K,\, y \in Y_K.$$

The key notions and properties of this multiplication action of C_K on Y_K (see [**C**]) are as follows:

Isotropy Subgroup: The isotropy subgroup associated with the adeles that vanish exactly at a place v of K is $K_v^\times \subset K^\times \backslash \mathrm{GL}_1(\mathbb{A}_K)$. This subgroup establishes the canonical connection between local and global class field theory.

Explicit Formulas: The action of the isotropy subgroup $K_v^\times$ on the transverse space K_v of the periodic orbit determines the contribution of the place v to the Weil explicit formulas. This contribution is specifically captured as the trace of the Schwartz kernel $k(x,y) = \delta(y - \lambda x)$:

$$\int k(x,x)\,dx = \int \delta(x - \lambda x)\,dx = \frac{1}{|1-\lambda|}.$$

The adele class space Y_K possesses a subtle and delicate structure primarily due to the ergodicity of the action of $K^\times$ on $\mathbb{A}_K$ under its Haar measure. Even when Y_K is viewed naively as a topological space, this complexity remains evident and Y_K appears to be highly singular.

Two theoretical frameworks offer valuable methods for analyzing such spaces, inspired by insights from the geometric case of foliations: **noncommutative geometry** and **topos theory**. Each has distinct merits. Noncommutative geometry leverages the powerful tools of functional analysis and Hilbert space operators, while topos theory operates in close synergy with algebraic geometry and sheaf theory.

The adele class space Y_K was initially studied using noncommutative geometry, particularly through the crossed product noncommutative algebra $C_0(\mathbb{A}_K) \rtimes K^\times$ [**C, CM**].

A notable finding [**CC4**] shows that for $K = \mathbb{Q}$, the quotient $X_\mathbb{Q}$ of the space $Y_\mathbb{Q} := \mathbb{Q}^\times \backslash \mathbb{A}_\mathbb{Q}$ by the maximal compact subgroup $\widehat{\mathbb{Z}}^*$ of $C_\mathbb{Q}$ corresponds to the set of points of a naturally associated topos: the *scaling topos*. This topos is defined

[2]The object of class field theory is to show how the abelian extensions of an algebraic number field K can be determined by elements derived from the knowledge of K itself; or, if one wishes to present it in dialectical terms, how a field possesses within itself the elements of its own surpassing.

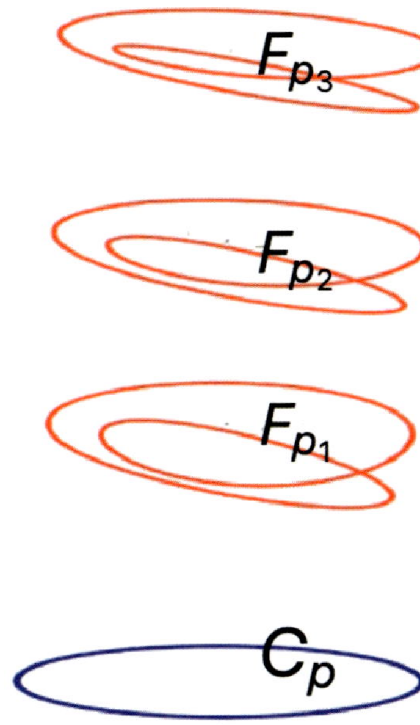

FIGURE 1. A typical example of the inverse image of the periodic orbit C_p is a union of components $F_{\mathfrak{p}_i}$ where the prime p is unramified and factors as a product of prime ideals $(p) = \mathfrak{p}_1\mathfrak{p}_2\mathfrak{p}_3$ in L.

as the semidirect product of the half-line $[0, \infty)$ with the multiplicative action of positive integers. The adelic space $X_{\mathbb{Q}}$, along with its cover

$$Y_{\mathbb{Q}} := \mathbb{Q}^\times \backslash \mathbb{A}_{\mathbb{Q}} \xrightarrow{\pi} X_{\mathbb{Q}} = \mathbb{Q}^\times \backslash \mathbb{A}_{\mathbb{Q}} / \widehat{\mathbb{Z}}^*$$

offers a geometric framework for the conjectural connection between primes and knots. This is realized through the periodic orbits C_p (with length $\log p$ under the idelic scaling action) corresponding to rational primes, and their preimages $\pi^{-1}(C_p)$ (see Figure 1).

The main result of this paper is the construction of a functor that associates finite covers of the space $X_{\mathbb{Q}}$ to finite abelian extensions of $\mathbb{Q}$. For each such extension $\mathbb{Q} \subseteq L$, there is a corresponding open subgroup $W \subset \widehat{\mathbb{Z}}^*$, defined as the kernel of the natural Galois map $\widehat{\mathbb{Z}}^* \to \mathrm{Gal}(L/\mathbb{Q})$, arising from the field inclusions $\mathbb{Q} \subseteq L \subset \mathbb{Q}^{\mathrm{ab}}$. The associated finite cover of $X_{\mathbb{Q}}$ is defined as the double quotient

$$X_{\mathbb{Q}}^L := \mathbb{Q}^\times \backslash \mathbb{A}_{\mathbb{Q}} / W.$$

In section 2 we show that the geometric properties of the finite cover $\pi^L : X_{\mathbb{Q}}^L \to X_{\mathbb{Q}}$ reflect the ramification structure of the extension $\mathbb{Q} \subseteq L$ and that the monodromy of the periodic orbits C_p is determined by the Galois action of the arithmetic Frobenius at the prime p over which the cover is unramified. This justifies the assertion that the maximal abelian extension $\mathbb{Q}^{\mathrm{ab}}$ of the rationals finds its geometric counterpart in the adele class space $Y_{\mathbb{Q}}$, interpreted as the maximal abelian cover of the space $X_{\mathbb{Q}}$.

THEOREM 0.1. The association $L \to \left(\pi^L : X_{\mathbb{Q}}^L \to X_{\mathbb{Q}}\right)$ defines a contravariant functor from finite abelian extensions of $\mathbb{Q}$ to finite covers of $X_{\mathbb{Q}}$. Moreover one has:

(i) The finite set R of places at which the cover ramifies is the union of the archimedean place with the set of primes at which L ramifies.
(ii) Let $p \notin R$ then the monodromy of C_p in $X_{\mathbb{Q}}^L$ is the element of $\mathrm{Gal}(L/\mathbb{Q})$ given by the Frobenius Frob_p.
(iii) The connected components of the inverse image of C_p are circles labeled by the places of L over the prime p.

Furthermore, the "classically visible" geometry of $X_{\mathbb{Q}}$ is captured by the union $\mathcal{X}_{\mathbb{Q}}$ which includes all periodic orbits C_p (as well as a single point corresponding to the archimedean place) and the generic orbit $\eta \simeq \mathbb{R}_+^*$, under the scaling action of $\mathbb{R}_+^*$. The topology of $\mathcal{X}_{\mathbb{Q}}$ corresponds to that of a covering of $\overline{\mathrm{Spec}\,\mathbb{Z}}$, where each place v is covered as follows:

- For a rational prime p, the place v is covered by the periodic orbit C_p, which forms a circle of length $\log p$.
- For the archimedean place $v = \infty$, it is covered by a single point.
- The generic point of $\mathrm{Spec}\,\mathbb{Z}$ is covered by the generic orbit $\eta \subset X_{\mathbb{Q}}$.

The density of the generic point in $\mathrm{Spec}\,\mathbb{Z}$ is mirrored by the density of the generic orbit η in $X_{\mathbb{Q}}$, which in turn reflects the density of ideles within adeles. More precisely the topology of the space of orbits of the scaling action on $\mathcal{X}_{\mathbb{Q}}$ coincides with the topology of $\overline{\mathrm{Spec}\,\mathbb{Z}}$.

The covering $\mathcal{X}_{\mathbb{Q}}$ of $\overline{\mathrm{Spec}\,\mathbb{Z}}$, together with the natural scaling action of $\mathbb{R}_+^*$, can be interpreted as the class field theory counterpart of the points of a genuine curve $\mathscr{C}$ over the analogue of the algebraic closure of a finite field, equipped with the action of the Frobenius automorphisms. This class field theory counterpart to classical algebraic geometry over finite fields provides valuable insights into the essence of the geometry of $\mathscr{C}$ and the nature of the coefficients involved in defining its points. The Riemann-Roch theorem for the ring $\mathbb{Z}$ proven in [**CC6, CC7**], along with the interpretation of $\mathbb{Z}$ as a ring of polynomials over the combinatorial realization $\mathbb{S}$ of the sphere spectrum, suggests heuristically that $\mathscr{C}$ resembles a one-dimensional projective space $\mathbf{P}^1$.

Interpreting $\mathcal{X}_{\mathbb{Q}}$ as the class field theory counterpart of $\mathscr{C}$ suggests intriguing insights. For instance, the unique periodic orbit consisting of a single point is associated with the archimedean place thus the point ∞ appears to be the only one defined over $\mathbb{S}$. Table1 presents an excerpt from this proposed interpretation.

The next diagram highlights the parallel between the perspective in classical algebraic geometry and that in adelic geometry related to the maximal abelian extension of global fields:

$$\begin{array}{ccccc} \widehat{\mathbf{P}}^1_{\mathrm{ab}} & & \mathbb{Q}^{ab} & & \mathcal{Y}_{\mathbb{Q}} \\ {\scriptstyle \text{pro-ab}}\downarrow & & {\scriptstyle \text{Kronecker-Weber}}\Big| {\scriptstyle \widehat{\mathbb{Z}}^*} & & {\scriptstyle \widehat{\mathbb{Z}}^*}\downarrow \\ \mathbf{P}^1 \otimes_{\mathbb{F}_q} \overline{\mathbb{F}}_q & & \mathbb{Q} & & \mathcal{X}_{\mathbb{Q}} \end{array}$$

We now exploit the Mumford-Mazur analogy between knots and primes to investigate the mutual linking of the periodic orbits C_p. The linking number (C_1, C_2) of two disjoint oriented knots in S^3 is determined using the natural morphism

Function field $K = \mathbb{F}_q(t)$	Global field $\mathbb{Q}$	Class field of $\mathbb{Q}$
$\overline{\operatorname{Spec} \mathbb{F}_q[t]}$	$\overline{\operatorname{Spec} \mathbb{Z}}$	$X_{\mathbb{Q}}$
Points of $\mathbf{P}^1 \otimes_{\mathbb{F}_q} \bar{\mathbb{F}}_q$	The curve $\mathscr{C}$	$\mathcal{X}_{\mathbb{Q}}$
Frob_p on the points	Action of Weil group	Scaling action of $\mathbb{R}_+^*$
Frob_p orbits	Places of $\mathbb{Q}$	Scaling orbits C_p
Finite abelian cover of $\mathbf{P}^1$	Finite abelian extension of $\mathbb{Q}$	Finite abelian cover $\mathcal{X}_{\mathbb{Q}}^L$ of $\mathcal{X}_{\mathbb{Q}}$
Pro-Abelian cover $\widehat{\mathbf{P}}^1_{\mathrm{ab}}$	Maximal abelian extension $\mathbb{Q}^{ab}$	$\left(\mathcal{Y}_{\mathbb{Q}} = \varprojlim_L \mathcal{X}_{\mathbb{Q}}^L\right) \to \mathcal{X}_{\mathbb{Q}}$

TABLE 1

Knot C	**Prime** p
Inclusion $C \subset S^3$	$r^* : \operatorname{Spec} \mathbb{F}_p \hookrightarrow \operatorname{Spec} \mathbb{Z}$
Knot complement $X = S^3 - C$	$\operatorname{Spec} \mathbb{Z} \backslash \{p\}$
$\pi_1 \left(S^3 - C\right)^{ab} = \mathbb{Z}$	$\pi_1^{et}(\operatorname{Spec} \mathbb{Z}[\frac{1}{p}])^{ab} = \mathbb{Z}_p^*$
$\pi_1 (C_1) \to \pi_1 \left(S^3 - C_2\right)^{ab}$	$\pi_1^{et} (\operatorname{Spec} \mathbb{F}_p) \to \pi_1^{et}(\operatorname{Spec} \mathbb{Z}[\frac{1}{q}])^{ab}$
Linking Number (C_1, C_2)	$p \in \mathbb{Z}_q^*$

TABLE 2

$\pi_1(C_1) \to \pi_1(S^3 - C_2)^{\mathrm{ab}}$ and the isomorphism $\pi_1(S^3 - C_2)^{\mathrm{ab}} \cong \mathbb{Z}$, which is derived from Alexander-Pontryagin duality combined with the Hurewicz isomorphism. The Mumford-Mazur analogy between knots and rational primes is summarized in Table 2.

For a given rational prime p, the linking with all other (rational) primes q is understood at once as the natural map $r^* : \pi_1^{et}(\operatorname{Spec}\mathbb{F}_p) \to \pi_1^{et}(\operatorname{Spec}\mathbb{Z}_{(p)})^{ab}$. The removal of all primes except p from $\operatorname{Spec}\mathbb{Z}$ is effected algebraically by replacing $\operatorname{Spec}\mathbb{Z}$ with the spectrum of the local ring $\mathbb{Z}_{(p)}$. In our set-up, primes are represented by the periodic orbits C_p, and their linking is revealed explicitly through the following theorem

THEOREM 0.2. *Let p be a rational prime. Let $\operatorname{Frob}_p \in \pi_1^{et}(\operatorname{Spec}(\mathbb{F}_p))$ be the canonical generator.*

(i) *The inverse image $\pi^{-1}(C_p) \subset Y_{\mathbb{Q}}$ of the periodic orbit C_p is canonically isomorphic to the mapping torus of the multiplication by $r^*\{\operatorname{Frob}_p\}$ in the abelianized étale fundamental group $\pi_1^{et}(\operatorname{Spec}\ \mathbb{Z}_{(p)})^{ab}$.*

(ii) *The canonical isomorphism in (i) is equivariant for the action of the idele class group.*

(iii) *The monodromy of the periodic C_p in $\pi^{-1}(C_p) \subset Y_{\mathbb{Q}}$ is equal to the natural map $r^* : \pi_1^{et}(\operatorname{Spec}\mathbb{F}_p) \to \pi_1^{et}(\operatorname{Spec}\mathbb{Z}_{(p)})^{ab}$, and determines the linking of the prime p with all other primes.*

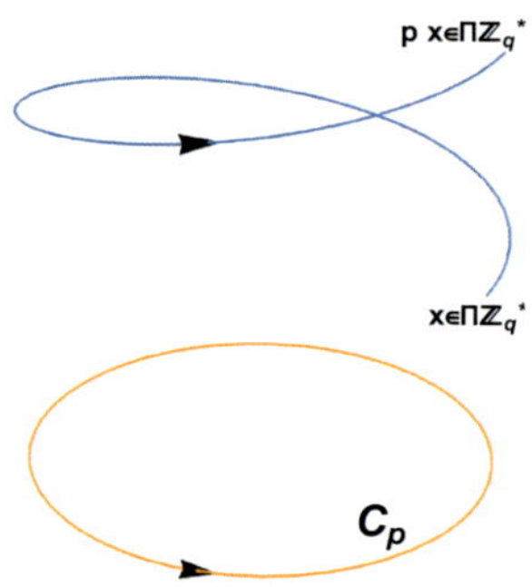

FIGURE 2. The monodromy of the periodic C_p in $\pi^{-1}(C_p) \subset Y_{\mathbb{Q}}$

In section 3 we provide a proof to the fact that the abelianized étale fundamental group $\pi_1^{\rm et}(\operatorname{Spec}\mathbb{Z}_{(p)})^{\rm ab}$ is isomorphic to the product $\prod_{q\neq p}\mathbb{Z}_q^*$. The image $r^*(\operatorname{Frob}_p) \in \pi_1^{\rm et}(\operatorname{Spec}\mathbb{Z}_{(p)})^{\rm ab}$ corresponds to the element p diagonally embedded in the product $\prod_{q\neq p}\mathbb{Z}_q^*$. The injectivity of the map $\mathbb{Z} \ni n \mapsto p^n \in \mathbb{Z}_q^*$ for any $q \neq p$ highlights the non-trivial linking of the periodic orbits C_p and C_q.

The fundamental properties underpinning the analogy between $\overline{\operatorname{Spec}\mathbb{Z}}$ and the 3-sphere S^3 are summarized in the following table:

Sphere S^3	$\overline{\operatorname{Spec}\mathbb{Z}}$
$\pi_1(S^3) = \{1\}$	$\pi_1^{et}(\overline{\operatorname{Spec}\mathbb{Z}}) = \{1\}$
$H^3(S^3, \mathbb{Z}) = \mathbb{Z}$	$H^3(\overline{\operatorname{Spec}\mathbb{Z}}, \mathbf{G}_m) = \mathbb{Q}/\mathbb{Z}$

Let $j : \eta \to \operatorname{Spec}\mathbb{Z}$ be the generic point. A crucial element in the proof of the Artin-Verdier Poincaré duality is the exact sequence of sheaves on Spec $\mathbb{Z}$

$$0 \to \mathbf{G}_m \to j_*\mathbf{G}_{m,\eta} \to \coprod_p \mathbb{Z}|p \to 0$$

Here, $\mathbf{G}_m$ denotes the representable sheaf giving the multiplicative group, and the last term on the right is the sheaf of Cartier divisors. One then utilizes the long exact sequence of cohomology groups

$$0 \to \mathrm{H}^2(\operatorname{Spec}\mathbb{Z}, \mathbf{G}_m) \to \mathrm{H}^2(\eta, \mathbf{G}_{m,\eta}) \xrightarrow{r_1} \bigoplus_p \mathbb{Q}/\mathbb{Z} \to$$

$$\to \mathrm{H}^3(\operatorname{Spec}\mathbb{Z}, \mathbf{G}_m) \to \mathrm{H}^3(\eta, \mathbf{G}_{m,\eta}) \to 0.$$

In this sequence, $\mathrm{H}^3(\operatorname{Spec}\mathbb{Z}, \mathbf{G}_m)$ appears as the cokernel of the map r_1, which, as established by the Brauer-Hasse-Noether theorem, is a single copy of $\mathbb{Q}/\mathbb{Z}$. Due to the presence of the real archimedean place with Brauer group $\mathbb{Z}/2\mathbb{Z}$, there is a subtlety when dealing with higher cohomology groups.

Two aspects of this development are particularly relevant to our context, and we will now address them. The first is the pivotal role of the inclusion $j : \eta \to \operatorname{Spec}\mathbb{Z}$ of the generic point, which suggests a hidden role of the relative cohomology for the pair $(\operatorname{Spec}\mathbb{Z}, \eta)$. In the adelic framework, a similar structure emerges. For instance, the spectral realization of zeros of L-functions can be interpreted as the action of the idele class group on $H^1(Y_\mathbb{Q}, \eta)$, where the generic orbit η (i.e., the idele classes) takes on the role of the generic point [**CM**], Theorem 4.116. A second aspect of the proof of Artin-Verdier duality that is relevant in our context is its reliance solely on abelian extensions of $\mathbb{Q}$. Indeed, O. Gabber demonstrated[3] that the duality continues to hold when the étale site is replaced by the site restricted to abelian extensions.

The considerations above suggest a promising direction of development using topos theory and the analogue of the "abelian site" of Spec $\mathbb{Z}$ over $X_\mathbb{Q}$. Furthermore, the intricate topos-theoretic structure of the scaling topos, together with its structure sheaf, offers the potential to intrinsically understand the topos counterpart of the finite abelian covers of $X_\mathbb{Q}$.

The scaling topos serves as the topos viewpoint of the adelic space $X_\mathbb{Q}$ and, as shown in [**CC4**], its interpretation via an extension of scalars from the arithmetic site (see [**CC3**]) equips the scaling topos with a natural structure sheaf. The sections of this sheaf are piecewise affine, continuous, convex functions. A remarkable consequence of this additional structure is observed in the periodic orbits C_p: each C_p, when equipped with the restriction of the scaling topos's structure sheaf, becomes an analogue of an elliptic curve within the framework of characteristic one.

Noncommutative geometry provides powerful techniques for understanding the complex geometry of the adelic spaces $Y_\mathbb{Q}$ and $X_\mathbb{Q}$. This approach is particularly appealing because it equips us with the necessary tools for the spectral realization of L-functions. A clearer understanding of the adelic spaces $Y_\mathbb{Q}$ and $X_\mathbb{Q}$ can be gained by considering their "semilocal" versions, which are obtained by restricting the places of $\mathbb{Q}$ to a finite subset that includes the archimedean place. Following the presentation of the semilocal setup in Section 4.1, we demonstrate in Section 4.2 that the semilocal algebras constitute a sheaf of algebras over Spec $\mathbb{Z}$. The semilocal

[3]private communication

algebras $\mathcal{S}(\mathbb{A}_S) \rtimes \mathbb{Z}_S^\times$ are constructed as the cross products of the Bruhat-Schwartz algebras $\mathcal{S}(\mathbb{A}_S)$ of functions on semilocal adeles, by the multiplicative groups given by $\mathbb{Z}_S^\times = \mathbf{G}_m(S^c)$, sections of the sheaf $\mathbf{G}_m$ on the open set complement of S in Spec $\mathbb{Z}$. We first show (Proposition 4.2) that the semilocal Bruhat-Schwartz algebra $\mathcal{S}(\mathbb{A}_S)$ form a sheaf $\mathcal{O}$ of commutative algebras over Spec $\mathbb{Z}$. We then obtain the following result which establishes the compatibility of the noncommutative geometric constructions with the algebraic geometry of Spec $\mathbb{Z}$.

THEOREM 0.3. (1) The algebraic cross product $\mathcal{O} \rtimes \mathbf{G}_m$ defines a sheaf of algebras on Spec $\mathbb{Z}$ such that for every finite set of places $S \ni \infty$

$$(\mathcal{O} \rtimes \mathbf{G}_m)\,(S^c) = \mathcal{S}(\mathbb{A}_S) \rtimes \mathbb{Z}_S^\times$$

(2) The stalk of $\mathcal{O} \rtimes \mathbf{G}_m$ at the generic point η is the global cross product $\mathcal{S}(\mathbb{A}_\mathbb{Q}) \rtimes \mathbb{Q}^\times$.

(3) The global sections of $\mathcal{O} \rtimes \mathbf{G}_m$ form the cross product $\mathcal{S}(\mathbb{R}) \rtimes \{\pm 1\}$.

In noncommutative geometry, a fundamental method for handling complex quotients is the interplay between the homotopy-theoretic treatment of such spaces and the cross-product C^*-algebras that encode their topology. The central tool in this context is the assembly map underlying the Baum-Connes conjecture. This conjecture, proven for cross-products by abelian groups [**HK**], is sufficient for computing the K-theory of the C^*-algebras relevant in our context, and was used in [**RM**] to compute the K-theory of the cross product C^*-algebra associated to the adele class space of global fields.

In section 4, we investigate the K-theory of the "visible part" (a precise definition is given there) of $X_\mathbb{Q}$, localized to a finite set of places. Specifically, for concreteness we treat the case $S = \{p, q, \infty\}$, and in that case we compute the K-theory of the C^*-algebra A corresponding to the union, within the semi-local space $X_{\mathbb{Q},S}$, of the generic orbit and the three periodic orbits C_p, C_q, and C_∞. The resulting isomorphism $K_0(A) \simeq \mathbb{Z}^3$ captures the presence of the three periodic orbits, while $K_1(A) \simeq \mathbb{Z}^2$ reflects the one-dimensional nature of the periodic orbits C_p and C_q.

Finally, section 5 provides a general description of the homotopy quotient $\mathcal{Y}_S$, which offers a homotopy-theoretic approach to the semilocal adele class space Y_S. The space $\mathcal{Y}_S$ is locally compact, with dimension equal to the cardinality $\#S$. It is equipped with an action of $\mathbb{R}^n$, where $n = \#S - 1$, by translations. This action defines a codimension 1 lamination, with the space of leaves being Y_S.

1. Notations

Let K be a global field. We denote by $K^\times = K \setminus \{0\}$ the multiplicative group of non-zero elements of K.

The adele ring of a global field K (in particular when $K = \mathbb{Q}$) is the restricted product $\mathbb{A}_K := \prod(K_v, \mathcal{O}_v) \subset \prod K_v$ consisting of the tuples (a_v) where a_v lies in the subring $\mathcal{O}_v \subset K_v$ for all but finitely many places v. The index v ranges in the set Σ_K of all places of the global field K, K_v is the local field completion of K at the place v, and $\mathcal{O}_v$ is the corresponding valuation ring when the place v is non-archimedean.

We denote the adele class space of $\mathbb{Q}$ by $Y_\mathbb{Q} = \mathbb{Q}^\times \backslash \mathbb{A}_\mathbb{Q}$. The action of $\mathbb{Q}^\times$ on the rational adeles is defined through its image under the natural diagonal embedding $\mathbb{Q} \hookrightarrow \mathbb{A}_\mathbb{Q}$, which defines the principal adeles. Throughout this paper we denote by

$X_{\mathbb{Q}}$ the natural quotient $\mathbb{Q}^\times\backslash\mathbb{A}_{\mathbb{Q}}/\widehat{\mathbb{Z}}^*$, where $\widehat{\mathbb{Z}}^* = \prod_{v<\infty}\mathbb{Z}_v^*$ is the subgroup of $\mathbb{A}_{\mathbb{Q}}$ of all p-adic units.

We recall from [**CC4**] that the space of points (up-to isomorphism) of the sector $X_{\mathbb{Q}} = \mathbb{Q}^\times\backslash\mathbb{A}_{\mathbb{Q}}/\widehat{\mathbb{Z}}^*$ of the adele class space $Y_{\mathbb{Q}}$ is in canonical bijection with the space of points of the scaling topos $\mathscr{S}$. This is the Grothendieck topos $[0,\infty)\rtimes\mathbb{N}^\times$ of $\mathbb{N}^\times$-equivariant sheaves of sets on the real interval $[0,\infty)$. The scaling topos is canonically isomorphic to the category $\mathfrak{Sh}(\mathscr{C},J)$ of sheaves of sets on the site $(\mathscr{C},J)$. Here, $\mathscr{C}$ denotes the small category of bounded open (possibly empty) intervals $\Omega\subset[0,\infty)$ including those of the form $[0,a)$ for $a>0$. The multiplicative monoid $\mathbb{N}^\times$ acts on the objects of $\mathscr{C}$ and defines the morphisms between two objects as $\mathrm{Hom}_{\mathscr{C}}(\Omega,\Omega') = \{n\in\mathbb{N}^\times | n\Omega\subset\Omega'\}$. The category $\mathscr{C}$ is endowed with the Grothendieck topology J generated by ordinary coverings of the topological space $[0,\infty)$: we refer to [**CC4**] for more details.

The points of the adelic space $X_{\mathbb{Q}}$ (and also of the scaling topos) are in canonical bijection with abstract rank one groups and rank one subgroups of $\mathbb{R}$.

Let p be a rational prime. The collection of points of $X_{\mathbb{Q}}$ corresponding to subgroups $H\subset\mathbb{R}$ which are abstractly isomorphic to the subgroup $H_p\subset\mathbb{Q}$ of fractions with denominators a power of p determine a subspace $C_p\subset X_{\mathbb{Q}}$ called the (p-)periodic orbit. The map $\mathbb{R}_+^*\to C_p$, $\lambda\mapsto\lambda H_p$ induces an isomorphism $\mathbb{R}_+^*/p^{\mathbb{Z}}\stackrel{\sim}{\to} C_p$ showing that C_p is a circle of length $\log p$. We refer to [**CC4**] for more details.

Finally, we briefly recall the notion of Bruhat-Schwartz function [**Br**]. On a real vector space the Bruhat-Schwartz functions are the smooth functions all of whose derivatives are rapidly decreasing. On a non-archimedean local field K, a Bruhat-Schwartz function is a locally constant function of compact support. The space of Bruhat-Schwartz functions on the adeles $\mathbb{A}_K$ of a global field K is defined to be the restricted tensor product $\bigotimes'_v\mathcal{S}(K_v) = \varinjlim_E\left(\bigotimes_{v\in E}\mathcal{S}(K_v)\right)$ of Bruhat-Schwartz spaces $\mathcal{S}(K_v)$ of local fields, where E is a finite set of places of K. Given a discrete group Γ acting by homeomorphisms on a locally compact space A the cross product $S(A)\ltimes\Gamma$ of a function space $S(A)$ by Γ is an algebra of linear combinations of monomials fU_γ for $f\in S(A)$, $\gamma\in\Gamma$, with multiplication rule $f_1U_{\gamma_1}f_2U_{\gamma_2} = f_1g\,U_{\gamma_1\gamma_2}$, where $g(x) = f_2(\gamma_1^{-1}x)$.

2. Abelian extensions of $\mathbb{Q}$ and finite covers of the adelic space $X_{\mathbb{Q}}$

In this section, we construct a natural functor which associates to a finite abelian extension of the field of rational numbers a covering of the space $X_{\mathbb{Q}}$ with finite fibers. We first define a covering of $X_{\mathbb{Q}}$ as follows

DEFINITION 2.1. Let G be a finite abelian group and $\chi:\widehat{\mathbb{Z}}^*\to G$ be a surjective continuous morphism. We let

$$X_{\mathbb{Q}}^\chi := Y_{\mathbb{Q}}\times_{\widehat{\mathbb{Z}}^*}G,\quad \pi_\chi: X_{\mathbb{Q}}^\chi\to X_{\mathbb{Q}} \tag{2.1}$$

be the quotient of the product $Y_{\mathbb{Q}}\times G$ by the action of $\widehat{\mathbb{Z}}^*$ and π_χ be the first projection on $Y_{\mathbb{Q}}/\widehat{\mathbb{Z}}^* = X_{\mathbb{Q}}$.

For $x\in X_{\mathbb{Q}} = \mathbb{Q}^\times\backslash\mathbb{A}_{\mathbb{Q}}/\widehat{\mathbb{Z}}^*$ in the class of the adele $a=(a_v)\in\mathbb{A}_{\mathbb{Q}}$, the set of places

$$Z(x) = \{v\in\Sigma_{\mathbb{Q}} \mid a_v = 0\}, \tag{2.2}$$

is independent of the choice of a since the elements of $\mathbb{Q}^\times \times \widehat{\mathbb{Z}}^*$ are ideles. The group G acts on $X_{\mathbb{Q}}^\chi$ since the action of $k \in G$ on $Y_{\mathbb{Q}} \times G$ by multiplication $(z,g) \mapsto (z,gk)$ is compatible with the quotient by $\widehat{\mathbb{Z}}^*$. Indeed for $(z',g') \sim (z,g)$ one has $w \in \widehat{\mathbb{Z}}^*$ with $(z',g') = (zw, \chi(w)g)$ and

$$(z', g'k) = (zw, (\chi(w)g)k) = (zw, \chi(w)(gk)) \sim (z, gk)$$

The nature of the action of G is specified by the next lemma.

LEMMA 2.2. Let G, χ and π_χ be as in Definition 2.1 and $x \in X_{\mathbb{Q}}$.

(i) The group G acts transitively on the fiber $F_x = \pi_\chi^{-1}(x)$.

(ii) Assume $\infty \notin Z(x)$, then the stabilizer of any $y \in F_x$ is

$$\chi(H) \subset G, \quad H = \prod_{v \in Z(x)} \mathbb{Z}_v^* \subset \widehat{\mathbb{Z}}^*. \tag{2.3}$$

PROOF. (i) Let $x \in X_{\mathbb{Q}} = Y_{\mathbb{Q}}/\widehat{\mathbb{Z}}^*$ be the class of $y \in Y_{\mathbb{Q}}$. Let $(z,g) \in Y_{\mathbb{Q}} \times G$ such that $\pi_\chi(z,g) = x$. There exists $w \in \widehat{\mathbb{Z}}^*$ such that $wz = y$ and one has, modulo the action of $\widehat{\mathbb{Z}}^*$, $(z,g) \sim (y, \chi(w)g)$. Thus every element of $F_x = \pi_\chi^{-1}(x)$ is equivalent to a pair (y,k) for some $k \in G$ and the action of G on such pairs is transitive.

(ii) Let $(z,g) \in Y_{\mathbb{Q}} \times G$ represent $y \in \pi_\chi^{-1}(x)$. Let $k \in G$ such that $yk = y$. One then has $(z,g) \sim (z,gk)$ modulo the action of $\widehat{\mathbb{Z}}^*$. Thus there exists $w \in \widehat{\mathbb{Z}}^*$ with

$$zw = z, \quad \chi(w)g = gk$$

Let $a \in \mathbb{A}_{\mathbb{Q}}$ in the class of $z \in \mathbb{Q}^\times \backslash \mathbb{A}_{\mathbb{Q}} = Y_{\mathbb{Q}}$. The equality $zw = z$ means that there exists $q \in \mathbb{Q}^\times$ such that $aw = qa$. Since the archimedean component of a is non-zero this implies that $q = 1$, and that for any finite prime p one has $a_p w_p = a_p$. Thus it holds if and only if $w \in H = \prod_{v \in Z(x)} \mathbb{Z}_v^* \subset \widehat{\mathbb{Z}}^*$ and since $k = \chi(w)$ one obtains (ii). ∎

2.1. The covering of the periodic orbits $\pi : \pi^{-1}(C_p) \to C_p$. This part is devoted to demonstrating the non-triviality of the covering restricted on the periodic orbits C_p. To understand the role of the quotient by $\mathbb{Q}^\times$ in the above constructions, we first examine the situation prior to applying this quotient.

LEMMA 2.3. The quotient map $p : \mathbb{A}_{\mathbb{Q}} \to \mathbb{A}_{\mathbb{Q}}/\widehat{\mathbb{Z}}^*$ admits a continuous section

$$\rho : \mathbb{A}_{\mathbb{Q}}/\widehat{\mathbb{Z}}^* \to \mathbb{A}_{\mathbb{Q}}, \quad p \circ \rho(x) = x, \ \forall x \in \mathbb{A}_{\mathbb{Q}}/\widehat{\mathbb{Z}}^*.$$

PROOF. Let $a = (a_v) \in \mathbb{A}_{\mathbb{Q}}$. For each finite place p, there exists an element $w_p \in \mathbb{Z}_p^*$ such that $a_p = w_p \rho_p$ where $\rho_p = 0$ if $a_p = 0$ and is a power of p, $\rho_p = p^n$, when $a_p \neq 0$. Both w_p and the power n are uniquely determined if $a_p \neq 0$. One lets $\rho_\infty = a_\infty$ and one has the equality

$$a = w\rho, \quad w_v = w_p, \ \forall v = p < \infty, \ w_\infty = 1, \ \rho_v = \rho_p, \ \forall v = p < \infty, \rho_\infty = a_\infty$$

The element $\rho(a) \in \mathbb{A}_{\mathbb{Q}}$ is uniquely defined, only depends on the class of a in $\mathbb{A}_{\mathbb{Q}}/\widehat{\mathbb{Z}}^*$ and belongs to this class. The map ρ is continuous since each component $\rho_v(a)$ is a continuous function of a_v. Thus ρ gives the required section. ∎

Next, we consider the division by $\mathbb{Q}^\times$. The cross-section ρ constructed above is not equivariant under the action of $\mathbb{Q}^\times$. This action, which is the diagonal action on adeles, modifies each component $\rho(x)_p$ of $\rho(x)$ by its product $q\rho(x)_p$. Specifically, for a prime q, this results in replacing $\rho(x)_p$ with $q\rho(x)_p$, which is not a power of p when $p \neq q$. To demonstrate that the fibration $\pi : \mathbb{Q}^\times \backslash \mathbb{A}_\mathbb{Q} = Y_\mathbb{Q} \to X_\mathbb{Q}$ is not trivial we examine its restriction to a periodic orbit C_p.

PROPOSITION 2.4. *Let p be a prime and $H = \prod_{v \neq p} \mathbb{Z}_v^*$ with $v < \infty$.*

(i) *The map $\eta : H \times \mathbb{R}_+^* \to \mathbb{A}_\mathbb{Q}$ defined componentwise by*

$$\eta((h,\lambda))_v = h_v, \forall v < \infty,\ v \neq p,\quad \eta((h,\lambda))_p = 0,\quad \eta((h,\lambda))_\infty = \lambda$$

induces a canonical isomorphism $\tilde{\eta}$ of the mapping torus of the homeomorphism of the compact group H given by multiplication by $p \in H$ and the inverse image $\pi^{-1}(C_p)$.

(ii) *The canonical isomorphism $\tilde{\eta}$ as in (i), is equivariant for the action of the idele class group.*

(iii) *The covering $\pi : \pi^{-1}(C_p) \to C_p$ has a non-trivial monodromy given by multiplication by p in H.*

PROOF. (i) We first show that every element of $\pi^{-1}(C_p)$ is in the range of $\tilde{\eta}$. The elements of C_p are the double classes of adeles of the form

$$(2.4) \qquad a(\lambda),\ a(\lambda)_\infty = \lambda > 0,\ a(\lambda)_p = 0,\quad a(\lambda)_v = 1, \forall v < \infty,\ v \neq p$$

The fiber $\pi^{-1}(C_p) \subset \mathbb{Q}^\times \backslash \mathbb{A}_\mathbb{Q}$ is formed of adeles $\alpha = (\alpha_v)$ such that there exist q in $\mathbb{Q}^\times$, $u \in \widehat{\mathbb{Z}}^*$ and $\lambda > 0$ such that $\alpha = qa(\lambda)u$, i.e.

$$(2.5) \qquad \alpha_v = qu_v, \forall v < \infty,\ v \neq p,\quad \alpha_p = 0,\ \alpha_\infty = q\lambda.$$

Let $q = q' \times p^k$ where q' is prime to p. Then modulo $\mathbb{Q}^\times$, α is equivalent to $\alpha' = q'^{-1}\alpha$ and one has

$$(2.6) \qquad \alpha'_v \in \mathbb{Z}_v^*,\ \forall v < \infty,\ v \neq p,\quad \alpha'_p = 0,\ \alpha'_\infty = p^k\lambda$$

This shows that α belongs to the range of $\tilde{\eta}$. Moreover two elements $\alpha, \beta \in H \times \mathbb{R}_+^*$ fulfill $\tilde{\eta}(\alpha) = \tilde{\eta}(\beta)$ if and only if they are in the same class modulo $\mathbb{Q}^\times$, i.e. there exists $q \in \mathbb{Q}^\times$ with $\eta(\beta) = q\eta(\alpha)$. This implies, using (2.4), that $q > 0$ and that $|q|_v = 1$ for $v < \infty$, $v \neq p$ so that $q = p^n$ for some $n \in \mathbb{Z}$. Thus $\tilde{\eta}$

$$(2.7) \qquad \tilde{\eta} : p^\mathbb{Z} \backslash \left(H \times \mathbb{R}_+^*\right) \to \pi^{-1}(C_p)$$

is an isomorphism. In general, one constructs the mapping torus of an homeomorphism $\phi : Y \to Y$ as the flow given by translations of the $\mathbb{R}$ variable in the quotient $(Y \times \mathbb{R})/\mathbb{Z}$ where the generator of $\mathbb{Z}$ acts by the map

$$\psi(y,t) := (\phi(y), t+1).$$

Taking $Y = H$, $\phi(y) = py$ and using the logarithm $\log : \mathbb{R}_+^* \to \mathbb{R}$, the scaling action of $\mathbb{R}_+^*$ on $p^\mathbb{Z} \backslash \left(H \times \mathbb{R}_+^*\right)$ becomes identical to the flow associated to the homeomorphism of the compact group $H = \prod_{v \neq p} \mathbb{Z}_v^*$ given by multiplication by $p \in H$. By using the logarithm with base p one obtains the translation $t \mapsto t+1$.

(ii) The idele class group $C_\mathbb{Q}$ is lifted in ideles as the subgroup $\widehat{\mathbb{Z}}^* \times \mathbb{R}_+^*$. It acts on $H \times \mathbb{R}_+^*$ by multiplication componentwise and similarly on adeles. This shows that $\tilde{\eta}$ is equivariant for the action of $C_\mathbb{Q}$ on both sides.

(iii) The map $\pi : \pi^{-1}(C_p) \to C_p$ is a fibration, whose monodromy can be identified with an element of the automorphism group $\mathrm{Aut}(F)$, describing how the fiber is transformed after traversing a loop in the base space. From (i), i.e. the identification of $\pi^{-1}(C_p)$ with the mapping torus, it follows that this monodromy is given by the multiplication by p in H. Furthermore for each prime $q \neq p$ the map $\mathbb{Z} \to \mathbb{Z}_q^*$ given by $n \mapsto p^n$ is injective since $\mathbb{Q}$ injects in $\mathbb{Q}_q$. A fortiori the map $n \mapsto p^n \in H$ is injective. Thus this monodromy is non-trivial and the map $\pi : \pi^{-1}(C_p) \to C_p$ does not admit a continuous cross section.

$\square$

2.2. Ramification of the cover $\pi_\chi : X_{\mathbb{Q}}^\chi \to X_{\mathbb{Q}}$. We now return to the finite cover π_χ from Definition 2.1 to examine its "ramification". Let $x \in X_{\mathbb{Q}}$ with $\infty \notin Z(x)$. Lemma 2.2 establishes that the group G acts transitively on the fiber $F_x = \pi_\chi^{-1}(x)$, with isotropy described by (2.3). Consequently, one has

$$G \textit{ acts freely transitively on } F_x \iff \chi\Big(\prod_{v \in Z(x)} \mathbb{Z}_v^*\Big) = 1. \tag{2.8}$$

In cases where condition (2.8) is satisfied, the fiber F_x with its G-action behaves like a principal bundle. Otherwise, "ramification" occurs. Let us introduce its definition

DEFINITION 2.5. Let S be a finite set of places. By definition, the morphism $X_{\mathbb{Q}}^\chi \xrightarrow{\pi_\chi} X_{\mathbb{Q}}$ is said to be *unramified outside* S if and only if the following condition holds for every $x \in X_{\mathbb{Q}}$:

$$Z(x) \cap S = \emptyset \implies G \textit{ acts freely transitively on } F_x$$

Since the morphism $\chi : \widehat{\mathbb{Z}}^* \to G$ is continuous and G is finite, the kernel $\ker \chi$ is an open subgroup of $\widehat{\mathbb{Z}}^*$ and hence contains a neighborhhod of 1. The topology of $\widehat{\mathbb{Z}}^* = \prod \mathbb{Z}_p^*$ is the product topology of the infinite product, and a basis of neighborhoods of 1 is formed of product open sets where C is a finite set of primes and the W_p's, with $p \in C$, are open in $\mathbb{Z}_p^*$

$$W = \prod_{p \in C} W_p \times \prod_{q \notin C} \mathbb{Z}_q^*.$$

Thus there exists a finite set C of primes such that $\chi(\prod_{q \notin C} \mathbb{Z}_q^*) = \{1\}$. Let

$$C(\chi) := \{p \in \Sigma_{\mathbb{Q}} \setminus \{\infty\} \mid \chi(\mathbb{Z}_p^*) \neq \{1\}\} \tag{2.9}$$

then the set $C(\chi)$ is finite and $\chi(\prod_{q \notin C(\chi)} \mathbb{Z}_q^*) = \{1\}$ by continuity. In other words, the finite set $C(\chi)$ corresponds to the set of divisors of the *conductor* of χ. The conductor is defined, using the description of $\widehat{\mathbb{Z}}^*$ as the projective limit of the finite groups $(\mathbb{Z}/m\mathbb{Z})^*$, as the gcd of all integers m for which χ factors through the canonical projection $\widehat{\mathbb{Z}}^* \to (\mathbb{Z}/m\mathbb{Z})^*$. The following proposition determines the finite sets of places $S \ni \infty$ outside which π_χ is unramified.

PROPOSITION 2.6. Let $X_{\mathbb{Q}}^\chi \xrightarrow{\pi_\chi} X_{\mathbb{Q}}$ be as in Proposition 2.1. Then for a finite set of places $S \ni \infty$ one has

$$X_{\mathbb{Q}}^\chi \xrightarrow{\pi_\chi} X_{\mathbb{Q}} \textit{ is unramified outside } S \iff C(\chi) \subset S.$$

PROOF. Assume that $C(\chi) \subset S$ then $\chi(\prod_{v \notin S} \mathbb{Z}_v^*) = \{1\}$. It follows that if $Z(x) \cap S = \emptyset$ one has $\chi(\prod_{v \in Z(x)} \mathbb{Z}_v^*) = \{1\}$ and (2.8) shows that G acts freely transitively on F_x. Conversely, if $\exists p \in C(\chi)$ with $p \notin S$, then for x in the periodic orbit C_p, $Z(x) = \{p\}$ fulfills $Z(x) \cap S = \emptyset$ but then, since $\chi(\mathbb{Z}_p^*) \neq \{1\}$, (2.8) shows that G does not act freely on F_x. □

The following corollary highlights the *unavoidable* ramified role of the archimedean place.

COROLLARY 2.7. There exists a smallest finite set S of places outside which π_χ is unramified. This set is given by $C(\chi) \cup \infty$, where $C(\chi)$ is defined in (2.9).

PROOF. By Proposition 2.6, it is enough to show that a finite set S of places outside which π_χ is unramified necessarily contains ∞. Assume $\infty \notin S$ and let $a \in \mathbb{A}_\mathbb{Q}$ be the adele defined componentwise as

$$a_v = 1, \forall v \in S \cup C(\chi), \quad a_v = 0, \ \forall v \notin S \cup C(\chi).$$

Let $x \in X_\mathbb{Q}$ be the class of a modulo $\mathbb{Q}^\times \times \widehat{\mathbb{Z}}^*$. Let us show that the fiber $\pi_\chi^{-1}(x)$ is reduced to a single element. Let $T = S \cup C(\chi)$ and $(z, g) \in \mathbb{A}_\mathbb{Q} \times G$ be such that the class of z modulo $\mathbb{Q}^\times \times \widehat{\mathbb{Z}}^*$ is x. One has $z_v = 0$, $\forall v \notin T$ and in particular $z_\infty = 0$, moreover there exists $(q, w) \in \mathbb{Q}^\times \times \widehat{\mathbb{Z}}^*$ such that $z_v = q w_v$ for all $v \in T$. Then, for any integer n with $(n, p) = 1$ for all $p \in T$ one has

$$z_v = \frac{q}{n}(n_v w_v), \ \forall v \in T.$$

Thus, modulo $\mathbb{Q}^\times \times \widehat{\mathbb{Z}}^*$ one obtains,

$$(z, g) \sim ((a_v n_v w_v), g) \sim (a, \chi((n_v w_v) g).$$

Since one can choose the integer n such that its residue modulo the conductor of χ is arbitrary, this shows that all elements of the fiber $\pi_\chi^{-1}(x)$ are equivalent. □

REMARK 2.8. The subsets of $\mathbb{A}_\mathbb{Q}$ of the form, for $S \subset \Sigma_\mathbb{Q}$ finite,

$$\Omega_S := \{a \in \mathbb{A}_\mathbb{Q} \mid a_v \neq 0, \ \forall v \in S\}$$

are both open (for the topology of $\mathbb{A}_\mathbb{Q}$) and invariant under the multiplicative action of ideles, hence they define open sets in $X_\mathbb{Q}$. The ordered collection of these open sets is similar to the ordered collection of open sets in the Zariski topology of $\overline{\operatorname{Spec} \mathbb{Z}}$.

2.3. Monodromy of π_χ over C_p. Next, we transpose Proposition 2.4 to the finite covers $X_\mathbb{Q}^\chi \xrightarrow{\pi_\chi} X_\mathbb{Q}$ and prove their non-triviality.

THEOREM 2.9. Let $X_\mathbb{Q}^\chi \xrightarrow{\pi_\chi} X_\mathbb{Q}$ be as in Definition 2.1 and $p \notin C(\chi)$ a non-archimedean prime.

(i) The restriction of π_χ to $\pi_\chi^{-1}(C_p)$ is a G-principal bundle over the periodic orbit C_p.

(ii) The isomorphism $\tilde{\eta}$ as in (2.7) induces a canonical isomorphism of $\pi_\chi^{-1}(C_p)$ with the mapping torus of the multiplication by $\chi(p) \in G$, where p is viewed as an element of $\prod_{\ell \in C(\chi)} \mathbb{Z}_\ell^*$.

(iii) The monodromy of C_p in $\pi_\chi^{-1}(C_p)$ is given by the action of the element $\chi(p)$.

PROOF. By construction $X^\chi_{\mathbb{Q}} := Y_{\mathbb{Q}} \times_{\hat{\mathbb{Z}}^*} G$ and the projection π_χ fulfills $\pi_\chi(z,g) = \pi(z)$ where $\pi : Y_{\mathbb{Q}} \longrightarrow X_{\mathbb{Q}}$ is the projection. We thus get

$$\pi_\chi^{-1}(C_p) = \pi^{-1}(C_p) \times_{\hat{\mathbb{Z}}^*} G \tag{2.10}$$

By (2.7), one gets the isomorphism

$$\tilde{\eta} : p^{\mathbb{Z}} \backslash \left(H \times \mathbb{R}_+^*\right) \longrightarrow \pi^{-1}(C_p)$$

where $H = \prod_{v \neq p} \mathbb{Z}_v^*$. Using the map $\chi : H \to G$ and (2.10) one obtains, applying $- \times_{\hat{\mathbb{Z}}^*} G$, an isomorphism

$$\tilde{\eta}_\chi : p^{\mathbb{Z}} \backslash \left(G \times \mathbb{R}_+^*\right) \to \pi_\chi^{-1}(C_p)$$

which identifies $\pi_\chi^{-1}(C_p)$ with the mapping torus of the multiplication by $\chi(p) \in G$. □

Figure 1 shows the general aspect of this mapping torus, depending on the subgroup of G generated by $\chi(p) \in G$. Finally the next corollary shows the non-triviality of the cover π_χ.

COROLLARY 2.10. *Let π_χ be as above. There exists infinitely many primes p for which the restriction of the covering π_χ to $\pi_\chi^{-1}(C_p)$ is non-trivial.*

PROOF. By Dirichlet's density theorem there are infinitely many primes p such that $\chi(p) \neq 1$. □

2.4. The class field functor. We first review briefly the classical results of class field theory for the field $\mathbb{Q}$. The Kronecker-Weber theorem states that every finite abelian extension L of $\mathbb{Q}$ in $\mathbb{C}$ is contained in $\mathbb{Q}(\zeta_n)$ for some n, where ζ_n is a primitive n-th root of 1. The smallest n such that $L \subseteq \mathbb{Q}(\zeta_n)$ is called the conductor m of $L/\mathbb{Q}$. One has a surjective homomorphism

$$(\mathbb{Z}/m\mathbb{Z})^* \cong \mathrm{Gal}(\mathbb{Q}(\zeta_m)/\mathbb{Q}) \xrightarrow{\theta} \mathrm{Gal}(L/\mathbb{Q}).$$

The action of $r \in (\mathbb{Z}/m\mathbb{Z})^*$ is by $\theta(r)(\zeta_m) = \zeta_m^r$. The Artin map associates to each prime p not dividing m an element $\mathrm{Frob}_p \in \mathrm{Gal}(L/\mathbb{Q})$ defined as follows. Let $\wp_L$ be a prime ideal of the ring of integers $\mathcal{O}_L$ in L lying above a rational prime p. By definition the decomposition group of $\wp_L$ is the subgroup of the Galois group $\mathrm{Gal}(L/\mathbb{Q})$ consisting of automorphisms that leave $\wp_L$ invariant:

$$D_p = \{\sigma \in \mathrm{Gal}(L/\mathbb{Q}) \mid \sigma(\wp_L) = \wp_L\}.$$

The decomposition group D_p acts on the residue field $\mathcal{O}_L/\wp_L$, and the kernel of this action is the inertia group. Since L is an abelian extension, the decomposition and inertia groups do not depend on the choice of $\wp_L$ above p, and since p is unramified in L the inertia group is trivial. This provides a canonical isomorphism

$$D_p \cong \mathrm{Gal}(\mathcal{O}_L/\wp_L/\mathbb{F}_p)$$

and thus a canonically defined element $\mathrm{Frob}_p \in D_p \subset \mathrm{Gal}(L/\mathbb{Q})$ corresponding to the Frobenius generator of $\mathrm{Gal}(\mathcal{O}_L/\wp_L/\mathbb{F}_p)$. The key fact then is

FACT 2.11. *The Frobenius Frob_p is equal to the image of p under the map $\theta : (\mathbb{Z}/m\mathbb{Z})^* \to \mathrm{Gal}(L/\mathbb{Q})$.*

We let $\mathbb{Q}^{\mathrm{ab}}$ be the union of all the cyclotomic fields $\mathbb{Q}(\zeta_n) \subset \mathbb{C}$. It is the maximal abelian extension $\mathbb{Q}^{\mathrm{ab}}$ of $\mathbb{Q}$. The map $e : \mathbb{Q}/\mathbb{Z} \to \mathbb{C}$ given by $e(s) := \exp(2\pi i s)$ is a group isomorphism of $\mathbb{Q}/\mathbb{Z}$ with the group of roots of unity in $\mathbb{C}$ and we obtain a canonical isomorphism θ of $\widehat{\mathbb{Z}}^* = \varprojlim(\mathbb{Z}/m\mathbb{Z})^*$ with the Galois group $\mathrm{Gal}(\mathbb{Q}^{\mathrm{ab}}/\mathbb{Q})$. Given a finite abelian extension L of $\mathbb{Q}$ there exists an isomorphism $\sigma : L \to \mathbb{Q}^{\mathrm{ab}}$ with a subfield of $\mathbb{Q}^{\mathrm{ab}}$. The morphism σ is not unique but its range $\sigma(L)$ is unique thus we identify L with this range. We let $W(L) \subset \widehat{\mathbb{Z}}^*$ be the subgroup

$$W(L) = \{g \in \widehat{\mathbb{Z}}^* \mid g(x) = x, \ \forall x \in L\} \tag{2.11}$$

We can now formulate Theorem 0.1 in a precise manner.

THEOREM 2.12. For L a finite abelian extension of $\mathbb{Q}$, let $\chi(L) : \widehat{\mathbb{Z}}^* \to G := \mathrm{Gal}(L/\mathbb{Q})$ be the canonical continuous surjective group morphism.

(i) The association $L \mapsto \gamma(L) := \left(\pi_\chi(L) : X_{\mathbb{Q}}^{\chi} \to X_{\mathbb{Q}}\right)$ (see Definition 2.1) defines a contravariant functor

$$\mathrm{ab}(\mathbb{Q}) \xrightarrow{\gamma} \mathrm{Cov}_{\mathrm{ab}}(X_{\mathbb{Q}})$$

from finite abelian extensions of $\mathbb{Q}$ to finite abelian covers of the space $X_{\mathbb{Q}}$.

(ii) The finite set R of places of $\mathbb{Q}$ at which the cover $\gamma(L)$ ramifies is the union of the archimedean place with the set of primes at which L ramifies.

(iii) Let $p \notin R$ then the monodromy of C_p in $\gamma(L)$ is the element of G given by the arithmetic Frobenius Frob_p.

(iv) The connected components of the inverse image of C_p in $\gamma(L)$ are covering circles labeled by the places of L over the prime p.

PROOF. (i) To define the functor γ on morphisms, we utilize a specific, uniquely determined decomposition of each morphism as a composition of the canonical inclusion and a Galois automorphism. Let L_j, $j = 1, 2$, be finite abelian extensions of $\mathbb{Q}$, and $\chi_j : \widehat{\mathbb{Z}}^* \to G_j$ the canonical continuous surjective group morphisms. Let $r : G_2 \to G_1$ be the restriction morphism of Galois groups. Next, we plan to combine the group morphism r with the action of G_1 on $\gamma(L_1)$. This leads to consider maps $\rho : G_2 \to G_1$ which fulfill the condition

$$\chi_1(w)\rho(g) = \rho(\chi_2(w)g), \ \forall w \in \widehat{\mathbb{Z}}^*, g \in G_2 \tag{2.12}$$

To any such map one associates the map $\tilde{\rho} : X_{\mathbb{Q}}^{\chi_2} \to X_{\mathbb{Q}}^{\chi_1}$ defined by

$$\tilde{\rho}(z, g) = (z, \rho(g)), \ \forall z \in X_{\mathbb{Q}}^{ab}, g \in G_2.$$

We now define the canonical decomposition of morphisms $\sigma : L_1 \to L_2$. One can write uniquely $\sigma = \iota \circ k$ where ι is the canonical inclusion $\iota : L_1 \subset L_2$ and $k \in G_1$. We then define

$$\rho_\sigma(g) := r(g)k, \ \forall g \in G_2.$$

One has $r(\chi_2(w)) = \chi_1(w)$ for all $w \in \widehat{\mathbb{Z}}^*$, and this shows that ρ_σ fulfills (2.12).

To prove (i), it remains to show that given $\sigma : L_1 \to L_2$ and $\sigma' : L_2 \to L_3$, one has

$$\rho_{\sigma' \circ \sigma} = \rho_\sigma \circ \rho_{\sigma'}$$

Let $\sigma'' = \sigma' \circ \sigma : L_1 \to L_3$, $\iota'' : L_1 \to L_3$ the canonical inclusion and $r'' : G_3 \to G_1$ the canonical group morphism (Galois restriction). One has $\iota'' = \iota' \circ \iota$, $r'' = r \circ r'$.

Let now $k \in G_1$ with $\sigma = \iota \circ k$, and $k' \in G_2$ with $\sigma' = \iota' \circ k'$. Then one obtains

$$\sigma'' = \sigma' \circ \sigma = \iota' \circ k' \circ \iota \circ k = \iota' \circ \iota \circ r(k') \circ k = \iota'' \circ k'', \ k'' = r(k')k$$

where the Galois restriction $r : G_2 \to G_1$ fulfills $h \circ \iota = \iota \circ r(h)$ for any $h \in G_2$. By definition one has, for any $g \in G_3$,

$$\rho_{\sigma' \circ \sigma}(g) = r''(g)k'' = r(r'(g))r(k')k = r(r'(g)k')k = \rho_\sigma(r'(g)k') = \rho_\sigma \circ \rho_{\sigma'}(g)$$

and this ends the proof of the functoriality.

(ii) The following statement is a 'standard' fact of the relation occurring between local and global class field theory. Let p be a finite prime, then p ramifies in L if and only if $\chi_L(\mathbb{Z}_p^*) \neq \{1\}$ where $\mathbb{Z}_p^* \subset \widehat{\mathbb{Z}}^*$ is the local ramification group. Thus the finite set of p at which L ramifies coincides with $C(\chi)$ as defined in (2.9). Then (ii) follows from Corollary 2.7.

(iii) Follows from Proposition 2.9 (iii), and the basic fact 2.11 showing that the arithmetic Frobenius $\mathrm{Frob}_p \in G$ coincides with $\chi(p) \in G$.

(iv) By (iii) and (ii) of Proposition 2.9, the inverse image of C_p is isomorphic to the mapping torus of the multiplication by $\mathrm{Frob}_p \in G$. Thus (iv) follows from the fact that p being unramified in the extension, $\mathrm{Frob}_p \in G$ generates the decomposition group D_p.

□

REMARK 2.13. One can also investigate the inverse image $\pi_\chi^{-1}(C_\infty)$ where $C_\infty = \{c\}$ and c is the element of $X_\mathbb{Q}$ represented by an adele a such that $a_\infty = 0$ and $a_v = 1$ for any $v < \infty$. The elements of $\pi_\chi^{-1}(C_\infty)$ are classes modulo $\widehat{\mathbb{Z}}^*$ of pairs $(z, g) \in Y_\mathbb{Q} \times G$ such that $\pi(z) = c$. This equality shows that the class of z in $Y_\mathbb{Q} = \mathbb{Q}^\times \backslash \mathbb{A}_\mathbb{Q}$ contains an element $\tilde{z} \in \mathbb{A}_\mathbb{Q}$ with

$$\tilde{z}_\infty = 0, \quad \tilde{z}_v \in \mathbb{Z}_v^*, \ \forall v < \infty$$

This lift $\tilde{z} \in \mathbb{A}_\mathbb{Q}$ is uniquely determined up to sign since -1 is the only rational number q with $|q|_v = 1 \ \forall v < \infty$. Thus the fiber F_c is the quotient $G/\chi(-1)$ of the group G by the element $\chi(-1)$. This element of the Galois group is the complex conjugation, and thus the quotient $G/\chi(-1)$ is identical with the set of archimedean places together with the transitive Galois action of G.

3. Schemes and adelic spaces: a generalized class field connection

In this section, we establish a conceptual bridge between the adelic spaces arising from our extension of class field theory and the geometric spaces produced by Grothendieck's extension of Galois theory. This link generalizes the classical correspondence between Galois groups and adelic groups, offering a unified perspective on field extensions.

We denote by $\mathbb{Z}_{(p)}$ the local ring describing the stalk of the structure sheaf of Spec $\mathbb{Z}$ at the point p of the spectrum, and by

$$r : \mathbb{Z}_{(p)} \to \mathbb{F}_p$$

the natural residue ring homomorphism. In this section we determine the abelianized fundamental group $\pi_1^{et}(\mathrm{Spec}\ \mathbb{Z}_{(p)})^{ab}$, identify this group with the symmetry group of the fiber $\pi^{-1}(C_p) \subset Y_\mathbb{Q}$ over the periodic orbit C_p and show that the monodromy of C_p in $\pi^{-1}(C_p) \subset Y_\mathbb{Q}$ is the image of the canonical (Frobenius) generator under the map

$$r^* : \pi_1^{et}(\mathrm{Spec}\, \mathbb{F}_p) \to \pi_1^{et}(\mathrm{Spec}\, \mathbb{Z}_{(p)})^{ab}$$

One knows that the étale fundamental group $\pi_1^{et}(\text{Spec } \mathbb{F}_p)$ is isomorphic to the Galois group $\text{Gal}(\bar{\mathbb{F}}_p/\mathbb{F}_p)$ of the algebraic closure $\bar{\mathbb{F}}_p$ of $\mathbb{F}_p$. The latter is an abelian profinite group isomorphic to $\widehat{\mathbb{Z}}$: the isomorphism maps the Frobenius automorphism acting on $\bar{\mathbb{F}}_p$ as $x \mapsto x^p$, to $1 \in \widehat{\mathbb{Z}}$.

Let $A = \mathbb{Z}_{(p)}$, and let A^{sh} denote the strict henselization of A. This is the integral closure of A in the p-typical Witt ring $W(\overline{\mathbb{F}}_p)$ which is a complete discrete valuation ring. Its field of fractions is the completion of $\mathbb{Q}_p^{\text{ur}} := \mathbb{Q}_p(\{\zeta_\ell\}_{(\ell,p)=1})$.

Let K denote the field of fractions of A^{sh}. The following field inclusions hold:

$$\mathbb{Q}(\{\zeta_\ell\}_{(\ell,p)=1}) \subset K \subset \mathbb{Q}_p^{\text{un}}.$$

The etale fundamental group $\pi_1^{\text{et}}(\text{Spec } \mathbb{Z}_{(p)})$ is known to fit into the exact sequence (see [**Stacks**], Lemma 58.11.4):

$$\pi_1^{\text{et}}(\text{Spec } K) \longrightarrow \pi_1^{\text{et}}(\text{Spec}(\mathbb{Q})) \longrightarrow \pi_1^{\text{et}}(\text{Spec}(\mathbb{Z}_{(p)})) \longrightarrow 1.$$

Upon abelianization, this sequence becomes:

$$\pi_1^{\text{et}}(\text{Spec } K)^{\text{ab}} \xrightarrow{\alpha} \pi_1^{\text{et}}(\text{Spec}(\mathbb{Q}))^{\text{ab}} \longrightarrow \pi_1^{\text{et}}(\text{Spec}(\mathbb{Z}_{(p)}))^{\text{ab}} \longrightarrow 1. \tag{3.1}$$

From this, the following result is derived:

PROPOSITION 3.1. Let p be a rational prime. Let $\{\text{Frob}_p\} \in \pi_1^{et}(\text{Spec}\,(\mathbb{F}_p))$ be the canonical generator.

(i) The abelianized étale fundamental group $\pi_1^{et}(\text{Spec } \mathbb{Z}_{(p)})^{ab}$ is isomorphic to the product $\prod_{v \neq p} \mathbb{Z}_v^*$.

(ii) The image $r^*(\text{Frob}_p) \in \pi_1^{et}(\text{Spec } \mathbb{Z}_{(p)})^{ab}$ is given by the element p diagonally embedded in the product $\prod_{v \neq p} \mathbb{Z}_v^*$.

PROOF. By the Kronecker-Weber theorem, the maximal abelian extension of $\mathbb{Q}$ is the cyclotomic field $\mathbb{Q}^{\text{ab}}$. The group of roots of unity $\mathbb{Q}/\mathbb{Z}$ is Pontrjagin dual to the compact additive group $\widehat{\mathbb{Z}}$. The Galois group $\text{Gal}(\mathbb{Q}^{\text{ab}}/\mathbb{Q}) = \widehat{\mathbb{Z}}^*$ acts on roots of unity by the dual of the action of $\widehat{\mathbb{Z}}^*$ on $\widehat{\mathbb{Z}}$ by multiplication.

Let K^{ab} denote the maximal abelian extension of K. By construction one has

$$\pi_1^{\text{et}}(\text{Spec}(K))^{\text{ab}} = \text{Gal}(K^{\text{ab}}/K).$$

Let us simplify the notation by setting $G = \text{Gal}(K^{ab}/K)$. To determine its image in $\text{Gal}(\mathbb{Q}^{ab}/\mathbb{Q})$, it suffices to analyze the action of G on roots of unity. Since $\mathbb{Q}(\{\zeta_\ell\}_{(\ell,p)=1}) \subset K$, G fixes all roots of unity of orders coprime to p. Furthermore, as $K \subset \mathbb{Q}_p^{\text{un}}$ and adjoining a primitive p^n-th root of unity to $\mathbb{Q}_p^{\text{un}}$ produces an abelian Galois extension of degree $(p-1)p^{n-1}$ with Galois group $(\mathbb{Z}/p^n\mathbb{Z})^*$, the same property holds for K. This relationship can be visualized using the following diagram of towers of abelian field extensions, where vertical bars represent inclusions.

$$\begin{array}{ccc}
K(\{\zeta_{p^\infty}\}) & \subset & \mathbb{Q}_p^{un}(\{\zeta_{p^\infty}\}) \\
{\scriptstyle \mathbb{Z}_p^*}\Big| & & \Big|{\scriptstyle \mathbb{Z}_p^*} \\
K & \subset & \mathbb{Q}_p^{un} \\
\Big| & & \Big| \\
\mathbb{Q}(\{\zeta_\ell\}_{(\ell,p)=1}) & & \mathbb{Q}_p
\end{array}$$

In particular, from the inclusion $K(\{\zeta_{p^\infty}\}) \subset K^{ab}$, one obtains that G surjects onto the Galois group $\mathbb{Z}_p^*$ of the extension of K obtained by adjoining all roots of unity of p-power. Furthermore, by construction $\mathbb{Q}^{ab} \subset K(\{\zeta_{p^\infty}\})$ and this ensures that the image of G under the map α in (3.1) is precisely the $\mathbb{Z}_p^*$ factor within $\prod_v \mathbb{Z}_v^* = \mathrm{Gal}(\mathbb{Q}^{ab}/\mathbb{Q})$. The isomorphism $\pi_1^{et}(\mathrm{Spec}\ \mathbb{Z}_{(p)})^{ab} \simeq \prod_{v\neq p} \mathbb{Z}_v^*$ then follows from the exactness of (3.1). (ii) The action of Frob_p on roots of unity is given by raising to the power p. □

With the notations as in section 2.1, we can now match the (number-theoretic) results related to the abelianized étale fundamental group of the scheme Spec $\mathbb{Z}_{(p)}$ with the geometry inherent to the adelic cover $Y_{\mathbb{Q}} \xrightarrow{\pi} X_{\mathbb{Q}}$.

THEOREM 3.2. Let p be a rational prime. Let $\{\mathrm{Frob}_p\} \in \pi_1^{et}(\mathrm{Spec}\,(\mathbb{F}_p))$ be the canonical generator.

(i) The inverse image $\pi^{-1}(C_p) \subset Y_{\mathbb{Q}}$ of the periodic orbit C_p is canonically isomorphic to the mapping torus of the multiplication by $r^*\,\{\mathrm{Frob}_p\}$ in the abelianized étale fundamental group $\pi_1^{et}(\mathrm{Spec}\ \mathbb{Z}_{(p)})^{ab}$.

(ii) The canonical isomorphism in (i) is equivariant for the action of the idele class group.

(iii) The monodromy of the periodic orbit C_p in $\pi^{-1}(C_p) \subset Y_{\mathbb{Q}}$ is equal to the natural map $r^* : \pi_1^{et}\,(\mathrm{Spec}\,\mathbb{F}_p) \to \pi_1^{et}(\mathrm{Spec}\,\mathbb{Z}_{(p)})^{ab}$, and determines the linking of the prime p with all other primes.

PROOF. The proof of this theorem follows from Proposition 2.4 and the isomorphism given by Proposition 3.1 (i) $\pi_1^{et}(\mathrm{Spec}\ \mathbb{Z}_{(p)})^{ab} \simeq \prod_{v\neq p} \mathbb{Z}_v^*$ together with the statement (ii). □

Theorem 3.2 elucidates the geometric aspects of the well-known analogy between primes and knots through the role of the adelic space $X_{\mathbb{Q}}$, its periodic orbits and their liftings to the adele class space of the rationals $Y_{\mathbb{Q}}$ viewed as maximal abelian cover of $X_{\mathbb{Q}}$. The following table summarizes what we have discussed so far and in the introduction[4].

4. The semilocal description of $X_{\mathbb{Q}}$ in NCG

The space $X_{\mathbb{Q}}$, when viewed through the lens of functional analysis, is best understood by focusing on a finite subset of places at a time. Specifically, this involves considering a finite subset $S \subset \Sigma_{\mathbb{Q}}$ of places of $\mathbb{Q}$, which includes the archimedean place ∞. This approach stems from the fact that the function space associated with $X_{\mathbb{Q}}$ is expressed as a cross product and comprises functions defined on the adeles and more specifically belonging to the Bruhat–Schwartz space characterized by their non-trivial dependence on only finitely many places. The Bruhat–Schwartz functions f are finite linear combinations of the products $f = \otimes_v f_v$ over places v of $\mathbb{Q}$, where each f_v is a Bruhat–Schwartz function on the local field $\mathbb{Q}_v$ and $f_v = \mathbf{1}_{\mathcal{O}_v}$ is the characteristic function on the ring of integers $\mathcal{O}_v$ for all but finitely many v. Except for a finite set S of places, the functions reduce to the characteristic

[4]The symbol $\bar{\mathbb{S}}$ stands for the algebraic closure of $\mathbb{S}$, intuitively understood as the spherical $\mathbb{S}$-algebra of the multiplicative monoid formed by complex numbers of modulus ≤ 1 and argument a rational multiple of π.

Etale	**Adelic**
$\pi_1^{et}(\mathrm{Spec}\,\mathbb{Q})^{ab} = \mathrm{Gal}(\mathbb{Q}^{ab}/\mathbb{Q})$	$(\mathrm{GL}_1(\mathbb{A}_\mathbb{Q})/\mathbb{Q}^\times)\,/\,(\mathrm{GL}_1(\mathbb{A}_\mathbb{Q})/\mathbb{Q}^\times)_0$
Local Weil group	$\mathbb{Q}_v^*$ = isotropy in $Y_\mathbb{Q} = \mathbb{Q}^\times\backslash\mathbb{A}_\mathbb{Q}$
$r^* : \mathrm{Spec}\,\mathbb{F}_p \hookrightarrow \mathrm{Spec}\,\mathbb{Z}$	Periodic orbit C_p in $X_\mathbb{Q}$
$\pi_1^{et}\,(\mathrm{Spec}\,\mathbb{F}_p)$	$\pi_1(C_p)$ topological
$\pi_1^{et}(\mathrm{Spec}\,\mathbb{Z}_{(p)})^{ab} \simeq \prod_{q\neq p}\mathbb{Z}_q^*$	$\mathrm{Aut}(\pi^{-1}(C_p)) = \prod_{q\neq p}\mathbb{Z}_q^*$
$\pi_1^{et}\,(\mathrm{Spec}\,\mathbb{F}_p) \to \pi_1^{et}(\mathrm{Spec}\,\mathbb{Z}_{(p)})^{ab}$	Monodromy of C_p in $\pi^{-1}(C_p) \subset Y_\mathbb{Q}$
$\overline{\mathrm{Spec}\,\mathbb{Z}}$ extended to $\overline{\mathbb{S}}$	$\mathcal{X}_\mathbb{Q}$ = first 2 layers of stratification of $X_\mathbb{Q}$
Generic point $\eta \in \mathrm{Spec}\ \mathbb{Z}$	Generic orbit $\eta \subset X_\mathbb{Q}$

functions of the maximal compact subring of the restricted infinite product of local fields $\prod'_{v\notin S}\mathbb{Q}_v$. Consequently, the functional analysis focuses on the locally compact rings

$$\mathbb{A}_S = \prod_{v\in S}\mathbb{Q}_v.$$

Moreover, the group involved in the cross product, which globally corresponds to $\mathbb{Q}^\times$, naturally restricts to a subgroup $\Gamma := \mathbb{Z}_S^\times \subset \mathbb{Q}^\times$ that depends on the finite set S. After presenting the semilocal setup in section 4.1, we show in section 4.2 that the semilocal algebras form a sheaf of algebras over Spec $\mathbb{Z}$. We then describe in section 4.3 the natural stratification of the semilocal spaces $Y_{\mathbb{Q},S}$ and $X_{\mathbb{Q},S}$. In section 4.4 we compute the K-theory of the C^*-algebra associated to the first two stratas.

4.1. The semilocal setup. The ring $\mathbb{A}_S$ includes $\mathbb{Q}$ as a subring via the diagonal embedding. Let $\mathbb{Z}_S$ denote the ring $\mathbb{Z}$ localized at $S\backslash\{\infty\}$. It is the subring of $\mathbb{Q}$ consisting of rational numbers whose denominators involve only primes $p \in S$. Explicitly,

$$\mathbb{Z}_S = \{q \in \mathbb{Q} \mid |q|_v \leq 1\,,\ \forall v \notin S\}.$$

The group Γ_S, of the invertible elements of $\mathbb{Z}_S$, can be expressed as

$$\Gamma_S = \{\pm p_1^{n_1}\cdots p_k^{n_k} \mid p_j \in S\setminus\{\infty\},\ n_j \in \mathbb{Z}\}. \tag{4.1}$$

This multiplicative group is generated by ±1 and the primes in S. Structurally, it is the direct product of a copy of $\mathbb{Z}$ for each prime in S and the group $\mathbb{Z}/2\mathbb{Z}$ corresponding to the sign ±1. Conceptually, Γ_S is the group $\mathbf{G}_m(S^c)$ of sections of the sheaf $\mathbf{G}_m$ on the open set $S^c \subset$ Spec $\mathbb{Z}$, complement of S.

The semilocal version $Y_{\mathbb{Q},S}$ of the adèle class space is defined as the quotient

$$Y_{\mathbb{Q},S} := \Gamma_S\backslash\mathbb{A}_S,$$

while the semilocal version of $X_{\mathbb{Q}}$ is given by

$$X_{\mathbb{Q},S} := \Gamma_S \backslash \mathbb{A}_S / \widehat{\mathbb{Z}}^*(S), \quad \widehat{\mathbb{Z}}^*(S) = \prod_{v \in S \backslash \{\infty\}} \mathbb{Z}_v^*.$$

All the components of the global setting, including $\mathbb{Q}$, $\mathbb{Q}^\times$, the adeles $\mathbb{A}_{\mathbb{Q}}$, the compact ring $\widehat{\mathbb{Z}}$, its unit group $\widehat{\mathbb{Z}}^*$, the idele class group $C_{\mathbb{Q}} := \mathrm{GL}_1(\mathbb{A}_{\mathbb{Q}})/\mathbb{Q}^\times$, and the double quotient $X_{\mathbb{Q}} = \mathbb{Q}^\times \backslash \mathbb{A}_{\mathbb{Q}} / \widehat{\mathbb{Z}}^*$, admit a natural semilocal counterpart. These semilocal analogs are summarized in the following table:

Global	**Semilocal**
$\mathbb{Q}$	$\mathbb{Z}_S = \{q \in \mathbb{Q} \,:\, \lvert q\rvert_v \leq 1\,,\ \forall v \notin S\}$
$\mathbb{Q}^\times$	$\mathbb{Z}_S^\times = \{\pm p_1^{n_1} \cdots p_k^{n_k} \,:\, p_j \in S \setminus \{\infty\}\,,\ n_j \in \mathbb{Z}\}$
$\mathbb{A}_{\mathbb{Q}}$	$\mathbb{A}_S = \prod_{v \in S} \mathbb{Q}_v$
$\widehat{\mathbb{Z}}$	$\widehat{\mathbb{Z}}(S) = \prod_{v \in S \backslash \{\infty\}} \mathbb{Z}_v$
$\widehat{\mathbb{Z}}^*$	$\widehat{\mathbb{Z}}^*(S) = \prod_{v \in S \backslash \{\infty\}} \mathbb{Z}_v^*$
$C_{\mathbb{Q}}$	$C_{\mathbb{Q},S} := \mathrm{GL}_1(\mathbb{A}_S)/\mathbb{Z}_S^\times$
$X_{\mathbb{Q}}$	$X_{\mathbb{Q},S} := \mathbb{Z}_S^\times \backslash \mathbb{A}_S / \widehat{\mathbb{Z}}^*(S)$

4.2. Sheaf on Spec $\mathbb{Z}$ **of semilocal algebras.** We shall first characterize Bruhat-Schwartz functions of the form $f = 1_{\mathbb{Z}_v} \otimes g$ by the following properties which are "local" at v.

LEMMA 4.1. *Let F be a finite set of places $F \subset \Sigma_{\mathbb{Q}}$, with $\infty \in F$. Let $v \in F$ be non-archimedean. Then a Bruhat-Schwartz function $f \in \mathcal{S}(\mathbb{A}_F)$ is of the form $f = 1_{\mathbb{Z}_v} \otimes g$ for some $g \in \mathcal{S}(\mathbb{A}_{F \backslash \{v\}})$ if and only if it fulfills the following two properties*

$$a \in \mathbb{A}_F,\ \lvert a_v\rvert > 1 \ \Rightarrow\ f(a) = 0 \tag{4.2}$$

$$f(a + \alpha) = f(a),\ \forall \alpha \in \mathbb{Q}_v,\ \lvert\alpha\rvert \leq 1. \tag{4.3}$$

PROOF. Let $f = 1_{\mathbb{Z}_v} \otimes g$, then (4.2) holds, the second property (4.3) holds since $\mathbb{Z}_v$ is a group under addition and its characteristic function is invariant under the group translations. Conversely now, assume that a Bruhat-Schwartz function f fulfills the above two properties. Let us consider the function $g(y) := f(0, y)$ for $y \in \mathbb{A}_{F \backslash \{v\}}$, it is a Bruhat-Schwartz function. We show that $f = 1_{\mathbb{Z}_v} \otimes g$. Let $(x, y) \in \mathbb{Q}_v \times \mathbb{A}_{F \backslash \{v\}}$. If $\lvert x\rvert > 1$, then both f and $1_{\mathbb{Z}_v} \otimes g$ vanish (using for f the first property (4.2)). If $\lvert x\rvert \leq 1$, then by the second property (4.3) one has

$$f(x, y) = f(0, y) = g(y) = (1_{\mathbb{Z}_v} \otimes g)(x, y).$$

The desired factorization follows. $\square$

The Zariski topology of Spec $\mathbb{Z}$ allows one to better understand the role of the semilocal Bruhat-Schwartz algebras and the transitions maps.

PROPOSITION 4.2. For a finite set of places $S \subset \Sigma_{\mathbb{Q}}$, with $\infty \in S$, let $\mathcal{S}(\mathbb{A}_S)$ denote the Bruhat-Schwartz space of functions on the semilocal adeles $\mathbb{A}_S$.

(1) For any pair of finite sets $\infty \in S \subset S' \subset \Sigma_{\mathbb{Q}}$, let $R_{S',S}$ be the maximal compact subring of $\prod_{v\in S'\setminus S} \mathbb{Q}_v$. Then the natural maps

$$\gamma(S,S') : \mathcal{S}(\mathbb{A}_S) \to \mathcal{S}(\mathbb{A}_{S'}) \quad f \mapsto 1_{R_{S',S}} \otimes f$$

define a sheaf $\mathcal{O}$ of algebras on Spec $\mathbb{Z}$.

(2) The stalk of the sheaf $\mathcal{O}$ at the generic point $\eta \in$ Spec $\mathbb{Z}$ is the Bruhat-Schwartz algebra of functions on $\mathbb{A}_{\mathbb{Q}}$.

(3) The global sections $\Gamma(\text{Spec } \mathbb{Z}, \mathcal{O})$ form the Schwartz space

$$\mathcal{S}(\mathbb{R}) = \mathcal{S}(\mathbb{A}_{\{\infty\}}).$$

PROOF. (1) A presheaf of sets on Spec $\mathbb{Z}$ is a contravariant functor on the small category of Zariski open subsets of Spec $\mathbb{Z}$. This is the category whose objects are the empty set and the complements F^c of finite subsets $F \subset$ Spec $\mathbb{Z}$. Equivalently, a presheaf of sets on Spec $\mathbb{Z}$ can be seen as a covariant functor from the category of finite sets of places $S \ni \infty$. By construction, the maps $\gamma(S,S')$ define morphisms of algebras which respect the composition of inclusions. Thus $\mathcal{O}$ is a presheaf of algebras. In order to test the sheaf condition we first consider the stalk at the generic point, i.e.

$$\mathcal{O}_\eta := \varinjlim \mathcal{S}(\mathbb{A}_S).$$

By construction this is the global Bruhat-Schwartz space $\mathcal{S}(\mathbb{A}_{\mathbb{Q}})$ of functions on adeles. Thus, in particular, (2) holds. The fact that the restriction maps $\gamma(S,S')$ are injective allows us to view $\mathcal{O}$ as a sub-presheaf of the constant sheaf associated to $\mathcal{S}(\mathbb{A}_{\mathbb{Q}})$. Here we use the connectedness of Spec $\mathbb{Z}$ to state that the sections of the constant sheaf on any non-empty open subset coincide with $\mathcal{S}(\mathbb{A}_{\mathbb{Q}})$. To conclude that $\mathcal{O} \subset \mathcal{S}(\mathbb{A}_{\mathbb{Q}})$ is a sheaf one applies the following fact

PROPOSITION ([**MM**], II, Proposition 1). If F is a sheaf on a topological space X, then a subfunctor $S \subset F$ is a subsheaf if and only if, for every open set U and every element $f \in F(U)$, and every open covering $U = \bigcup U_i$, one has $f \in S(U)$ if and only if $f|_{U_i} \in S(U_i)$ for all i.

Thus, to prove (1) it suffices to show that for any given finite sets of places $S_j \ni \infty$, with $S = \cap S_j$, and for any $f \in \mathcal{S}(\mathbb{A}_{\mathbb{Q}})$, f belongs to $\mathcal{O}(S^c)$ when f belongs to each $\mathcal{O}(S_j^c)$.

Let $f \in \mathcal{S}(\mathbb{A}_{\mathbb{Q}})$, there exists a finite set $F \supset S$ of places such that $f \in \mathcal{S}(\mathbb{A}_F)$. The maximal compact subring $R_{F\setminus S}$ of $\prod_{v\in F\setminus S} \mathbb{Q}_v$ is the product of the maximal compact subrings $R_v = \mathbb{Z}_v \subset \mathbb{Q}_v$, thus the characteristic function $1_{R_{F\setminus S}}$ is the tensor product $\otimes 1_{R_v}$. Therefore one is reduced to show that for each $v \in F \setminus S$ the Bruhat-Schwartz function f on $\mathbb{Q}_v \times \mathbb{A}_{F\setminus\{v\}}$ is of the form $1_{R_v} \otimes g$, for some Bruhat-Schwartz function g on $\mathbb{A}_{F\setminus\{v\}}$. Since $S = \cap S_j$ one has $v \notin S_j$ for some j and thus since $f \in \mathcal{O}(S_j^c)$, Lemma 4.1 gives the desired factorization.

(2) The global sections $\Gamma(\text{Spec } \mathbb{Z}, \mathcal{O})$ correspond to the set $S = \{\infty\}$.

□

The identification $\Gamma_S = \mathbf{G}_m(S^c)$ supplies the conceptual meaning of the groups Γ_S as sections of the sheaf in groups $\mathbf{G}_m$ on Spec $\mathbb{Z}$.

One can then easily transpose Proposition 4.2 to the algebraic cross products as follows

THEOREM 4.3. For a finite set of places $\Sigma_\mathbb{Q} \supset S \ni \infty$, let $\mathcal{S}(\mathbb{A}_S) \rtimes \mathbb{Z}_S^\times$ be the algebraic cross product of the Bruhat-Schwartz algebra $\mathcal{S}(\mathbb{A}_S)$ by the action of $\mathbb{Z}_S^\times$.

(1) The algebraic cross product $\mathcal{O} \rtimes \mathbf{G}_m$ defines a sheaf of algebras on Spec $\mathbb{Z}$ such that

$$(\mathcal{O} \rtimes \mathbf{G}_m)(S^c) = \mathcal{S}(\mathbb{A}_S) \rtimes \mathbb{Z}_S^\times.$$

(2) The stalk of the sheaf $\mathcal{O} \rtimes \mathbf{G}_m$ at the generic point $\eta \in$ Spec $\mathbb{Z}$ is the global cross product $\mathcal{S}(\mathbb{A}_\mathbb{Q}) \rtimes \mathbb{Q}^\times$.

(3) The global sections of $\mathcal{O} \rtimes \mathbf{G}_m$ form the cross product $\mathcal{S}(\mathbb{R}) \rtimes \{\pm 1\}$.

PROOF. By construction the algebraic cross products $\mathcal{S}(\mathbb{A}_S) \rtimes \mathbb{Z}_S^\times$ define a presheaf $\mathcal{O} \rtimes \mathbf{G}_m$ of algebras which is a sub-presheaf of the constant sheaf given by the global cross product $\mathcal{S}(\mathbb{A}_\mathbb{Q}) \rtimes \mathbb{Q}^\times$. To prove that $\mathcal{O} \rtimes \mathbf{G}_m$ is a sheaf we proceed as in the proof of Proposition 4.2. Let $h \in \mathcal{S}(\mathbb{A}_\mathbb{Q}) \rtimes \mathbb{Q}^\times$, then h can be uniquely written as a finite sum

$$h = \sum f_j U_{q_j}, \quad q_j \in \mathbb{Q}^\times, \quad f_j \in \mathcal{S}(\mathbb{A}_\mathbb{Q}), \quad f_j \neq 0.$$

If $h \in (\mathcal{O} \rtimes \mathbf{G}_m)(S_k^c)$ for finite sets of places $S_k \ni \infty$ with $\cap S_k = S$, one has $q_j \in \cap \mathbf{G}_m(S_k^c)$ and hence $q_j \in \mathbf{G}_m(S^c)$ for all j. Moreover, for all j one has $f_j \in \mathcal{O}(S_k^c)$ for all k and Proposition 4.2 shows that $f_j \in \mathcal{O}(S^c)$. This shows that $\mathcal{O} \rtimes \mathbf{G}_m$ is a sheaf. The proof of the other statements is straightforward. □

4.3. The stratification of the space $Y_{\mathbb{Q},S}$. For $x \in X_{\mathbb{Q},S}$, represented by the adele $a = (a_v) \in \mathbb{A}_S$, the set of places

$$Z(x) = \{v \in S \mid a_v = 0\}$$

is well-defined and does not depend on the choice of the representative a, as noted after (2.2). The same property holds for $x \in Y_{\mathbb{Q},S}$. For each $v \in S$, the subset defined by the condition $v \in Z(x)$ is closed in the quotient topology.

Define $\nu(x) := \#Z(x)$. For any integer n, the condition $\nu(x) \geq n$ defines a closed subset of both $Y_{\mathbb{Q},S}$ and $X_{\mathbb{Q},S}$ and open subsets

$$Y_{\mathbb{Q},S}^{(n)} := \{x \in Y_{\mathbb{Q},S} \mid \#Z(x) < n\}, \quad X_{\mathbb{Q},S}^{(n)} := \{x \in X_{\mathbb{Q},S} \mid \#Z(x) < n\}.$$

We now proceed to describe the natural stratification of the space $Y_{\mathbb{Q},S}$. For simplicity we denote the group Γ_S of (4.1), by Γ.

PROPOSITION 4.4. Let $\pi_S : Y_{\mathbb{Q},S} \to X_{\mathbb{Q},S}$ denote the quotient map. Then

(i) The orbits of the action of the group $C_{\mathbb{Q},S}$ on $Y_{\mathbb{Q},S}$ are indexed by the subsets of S as follows:

$$\Omega_Z := \Gamma \backslash \{(a_v) \in \mathbb{A}_S \mid a_v = 0, \ \forall v \in Z, \ a_v \neq 0 \qquad \forall v \notin Z\}, \ \forall Z \subset S.$$

(ii) The orbit $\Omega_{\{p\}}$ associated to the subset $\{p\} \subset S$ is the inverse image $\pi_S^{-1}(C_p)$ of the periodic orbit C_p of length $\log p$.

(iii) The orbit $\pi_S^{-1}(C_p)$ is equivariantly isomorphic to the mapping torus of the multiplication by p in the compact group $\prod_{q \in S, q \neq p} \mathbb{Z}_q^*$.

PROOF. (i) This follows from the transitivity of the action of $\mathrm{GL}_1(\mathbb{Q}_v)$ on $\mathbb{Q}_v \setminus \{0\}$ for any place v.

(ii) The periodic orbit C_p, with length $\log p$ under the action of the scaling group $\mathbb{R}_+^*$ on $X_{\mathbb{Q},S}$, is given by $\pi_S(\Omega_{\{p\}}) \subset X_{\mathbb{Q},S}$ and is identical to the quotient $p^{\mathbb{Z}}\backslash\mathbb{R}_+^*$.

(iii) The set $\Omega_{\{p\}}$ satisfies $\Omega_{\{p\}} = \pi_S^{-1}(C_p)$. Furthermore, the group Γ decomposes as $\Gamma = p^{\mathbb{Z}} \times \Gamma'$, where

$$\Gamma' = \{\pm p_1^{n_1} \cdots p_k^{n_k} \mid p_j \in S \setminus \{p, \infty\},\ n_j \in \mathbb{Z}\}.$$

The action of Γ' on $\{0\}_p \times \left(\prod_{q\in S, q\neq p} \mathbb{Q}_q \setminus \{0\}\right) \times \mathbb{R}^*$ has a fundamental domain $\{0\}_p \times \prod_{q\in S,q\neq p} \mathbb{Z}_q^* \times \mathbb{R}_+^*$, which remains invariant under the action of $p^{\mathbb{Z}}$.

The quotient

$$p^{\mathbb{Z}}\backslash\left(\prod_{q\in S,q\neq p} \mathbb{Z}_q^* \times \mathbb{R}_+^*\right)$$

is the mapping torus corresponding to multiplication by p on the compact group $\prod_{q\in S,q\neq p} \mathbb{Z}_q^*$.

□

This description reveals that C_p lies at the boundary of the free orbit $\pi_S(\Omega_\emptyset) \subset X_{\mathbb{Q},S}$. The condition $\nu(x) < 2$ defines an open subset Ω of $Y_{\mathbb{Q},S}$, whose structure, along with its counterpart in $X_{\mathbb{Q},S}$, is readily accessible. The latter is the union of the generic orbit with the periodic orbits C_v for $v \in S$.

4.4. *K*-theory of the C^*-algebra of the two first stratas. We provide a detailed proof of the following statement, which was stated without proof in [**CC8**]: "The simplest meaningful computation of the K-theory of the C^*-algebras involved pertains to the cross product A associated with the union in $X_{\mathbb{Q},S}$ of the generic orbit and the three periodic orbits C_p, C_q, C_∞. The result is $K_0(A) \simeq \mathbb{Z}^3$, reflecting the presence of the three periodic orbits, while $K_1(A) \simeq \mathbb{Z}^2$, capturing the one-dimensionality of the periodic orbits C_p and C_q."

We consider the C^*-algebra associated with the open subset formed by the union of the generic orbit and the periodic orbits. As a topological space, this subset is locally compact. We describe it explicitly in the semilocal setting of two primes.

Let $S = \{p, q, \infty\}$, and consider the open subspace

$$\Omega \subset \mathbb{A}_S = \mathbb{Q}_p \times \mathbb{Q}_q \times \mathbb{R}$$

defined by adeles having at most one zero component. Its complement in $\mathbb{A}_S$ consists of the union of three closed subsets corresponding to adeles with at least two zero components.

Next, we divide Ω by the action of the compact group $G = \mathbb{Z}_p^* \times \mathbb{Z}_q^*$. This yields a locally compact space

$$Z := \Omega/G,$$

which is the union of the following four subspaces:

(1) $Z_\emptyset = \left(\mathbb{Q}_p^* \times \mathbb{Q}_q^* \times \mathbb{R}^*\right)/G \simeq p^{\mathbb{Z}} \times q^{\mathbb{Z}} \times \{\pm 1\} \times \mathbb{R}_+^*$.
(2) $Z_p = \left(\{0\} \times \mathbb{Q}_q^* \times \mathbb{R}^*\right)/G \simeq \{0\} \times q^{\mathbb{Z}} \times \{\pm 1\} \times \mathbb{R}_+^*$.
(3) $Z_q = \left(\mathbb{Q}_p^* \times \{0\} \times \mathbb{R}^*\right)/G \simeq p^{\mathbb{Z}} \times \{0\} \times \{\pm 1\} \times \mathbb{R}_+^*$.

(4) $Z_\infty = \left(\mathbb{Q}_p^* \times \mathbb{Q}_q^* \times \{0\}\right)/G \simeq p^{\mathbb{Z}} \times q^{\mathbb{Z}} \times \{0\}$.

Next, we describe the cross product C^*-algebra $A = C_0(Z) \ltimes \Gamma$, $\Gamma = \Gamma_S$ as in (4.1).

LEMMA 4.5. *The cross product C^*-algebras are, up to Morita equivalence, with $\mathcal{K}$ the compact operators,*

(1) $A_\emptyset = C_0(Z_\emptyset) \ltimes \Gamma = \mathcal{K} \otimes C_0(\mathbb{R}_+^*)$.
(2) $A_p = C_0(Z_p) \ltimes \Gamma = \mathcal{K} \otimes C(C_p)$
(3) $A_q = C_0(Z_q) \ltimes \Gamma = \mathcal{K} \otimes C(C_q)$
(4) $A_\infty = C_0(Z_\infty) \ltimes \Gamma = \mathcal{K} \otimes C^*(\mathbb{Z}/2\mathbb{Z})$.

One has an exact sequence of C^*-algebras of the form

$$0 \to A_\emptyset \overset{\iota}{\to} A \overset{\rho}{\to} (A_p \oplus A_q \oplus A_\infty) \to 0$$

None of the involved C^*-algebras is unital but the C^*-algebra $B = A_p \oplus A_q \oplus A_\infty$ on the right in the above sequence is Morita equivalent to the unital C^*-algebra $\mathbf{C} := C(C_p) \oplus C(C_q) \oplus C^*(\mathbb{Z}/2\mathbb{Z})$. We thus get the exact hexagon of K-theory groups

$$\begin{array}{ccccc} & K_0(A_\emptyset) & \xrightarrow{\ \iota_*\ } & K_0(A) & \\ {\scriptstyle\delta_1}\nearrow & & & & \searrow{\scriptstyle\rho_*} \\ K_1(B) & & & & K_0(B) \\ {\scriptstyle\rho_*}\nwarrow & & & & \swarrow{\scriptstyle\delta_0} \\ & K_1(A) & \xleftarrow[\ \iota_*\ ]{} & K_1(A_\emptyset) & \end{array} \tag{4.4}$$

One has

(1) $K_0(A_\emptyset) = K_0(C_0(\mathbb{R}_+^*)) = \{0\}$. $K_1(A_\emptyset) = K_1(C_0(\mathbb{R}_+^*)) = \mathbb{Z}$.
(2) $K_0(A_p) = K_0(C(C_p)) = \mathbb{Z}$, $K_1(A_p) = K_1(C(C_p)) = \mathbb{Z}$.
(3) $K_0(A_q) = K_0(C(C_q)) = \mathbb{Z}$, $K_1(A_q) = K_1(C(C_q)) = \mathbb{Z}$.
(4) $K_0(A_\infty) = K_0(C^*(\mathbb{Z}/2\mathbb{Z})) = \mathbb{Z}^2$, $K_1(A_\infty) = K_1(C^*(\mathbb{Z}/2\mathbb{Z})) = \{0\}$.

From this, we deduce that $K_0(B) = \mathbb{Z}^4$ and $K_1(B) = \mathbb{Z}^2$. Starting with $K_0(A_\emptyset) = \{0\}$, it follows that the map

$$\rho_* : K_0(A) \to K_0(B)$$

is injective. To determine the range of this map, which is a subgroup $R \subset \mathbb{Z}^4$ and hence a free abelian group, we need to analyze the connecting map

$$\delta_0 : K_0(B) \to K_1(A_\emptyset) = \mathbb{Z}.$$

LEMMA 4.6. *Let $C := C_0(\mathbb{R}) \ltimes \mathbb{Z}/2\mathbb{Z}$ denote the cross product by the symmetry $x \mapsto -x$, and let $r : C \to C^*(\mathbb{Z}/2\mathbb{Z})$ be the morphism associated with evaluation at $0 \in \mathbb{R}$. Then, with $J := \ker(r)$, the connecting map*

$$\delta_0 : K_0(C^*(\mathbb{Z}/2\mathbb{Z})) \to K_1(J) = \mathbb{Z}$$

is surjective.

PROOF. An element of $C = C_0(\mathbb{R}) \ltimes \mathbb{Z}/2\mathbb{Z}$ can be written as $f + gU$, where $f, g \in C_0(\mathbb{R})$, $U^2 = 1$, and $UfU^*(x) = f(-x)$ for all $x \in \mathbb{R}$. To this element, we associate the matrix-valued function:

$$x \mapsto \begin{pmatrix} f(x) & g(x) \\ g(-x) & f(-x) \end{pmatrix}.$$

The multiplication rule for these matrices is given by:

$$\begin{pmatrix} f(x) & g(x) \\ g(-x) & f(-x) \end{pmatrix} \begin{pmatrix} h(x) & k(x) \\ k(-x) & h(-x) \end{pmatrix} =$$
$$= \begin{pmatrix} f(x)h(x) + g(x)k(-x) & f(x)k(x) + g(x)h(-x) \\ f(-x)k(-x) + g(-x)h(x) & f(-x)h(-x) + g(-x)k(x) \end{pmatrix}.$$

This multiplication corresponds to the relation:

$$(f + gU)(h + kU) = (fh + gk^s) + (fk + gh^s)U,$$

where $k^s(x) = k(-x)$ and $h^s(x) = h(-x)$. Thus, this construction provides an isomorphism of C with the sub-C^*-algebra of $C_0(\mathbb{R}_+, M_2(\mathbb{C}))$ consisting of matrices whose value at $x = 0$ lies in $C^*(\mathbb{Z}/2\mathbb{Z}) \subset M_2(\mathbb{C})$, i.e., matrices of the form:

$$a + bU = \begin{pmatrix} a & b \\ b & a \end{pmatrix}.$$

The following equivalence shows the diagonalization of these matrices:

$$\begin{pmatrix} 1 & 1 \\ -1 & 1 \end{pmatrix} \begin{pmatrix} a & b \\ b & a \end{pmatrix} \begin{pmatrix} \frac{1}{2} & -\frac{1}{2} \\ \frac{1}{2} & \frac{1}{2} \end{pmatrix} = \begin{pmatrix} a+b & 0 \\ 0 & a-b \end{pmatrix}.$$

Using a unitary conjugation, we can reduce to the sub-C^*-algebra of $C_0(\mathbb{R}_+, M_2(\mathbb{C}))$ formed of matrices whose value at $x = 0$ is diagonal, i.e., belonging to $D \subset M_2(\mathbb{C})$.

To complete the analysis, we examine the connecting map

$$\delta_0 : K_0(D) \to K_1(C_0((0,\infty), M_2(\mathbb{C}))).$$

As shown in [**Bl**] (page 77), this map is defined by associating to an idempotent $e \in D$ the loop:

$$(0,\infty) \ni x \mapsto \exp(2\pi i f(x)) \in M_2(\mathbb{C}),$$

where $f \in C_0(\mathbb{R}_+, M_2(\mathbb{C}))$ satisfies $f(0) = e$.

Choosing e as the element of D with diagonal values $(1, 0)$, and extending it as a diagonal matrix with elements $(t(x), 0)$, where t is a continuous function varying from $t(0) = 1$ to $t(\infty) = 0$, produces a generator of $K_1(C_0((0,\infty), M_2(\mathbb{C})))$.

This computation mirrors the case of $K_1(C_0([0,\infty))) = \{0\}$. Indeed, the associated six-term exact sequence displays as:

$$\begin{array}{ccccc} & K_0(C_0((0,\infty))) = \{0\} & \longrightarrow & K_0(C_0([0,\infty))) & \\ {\scriptstyle \delta_1}\nearrow & & & & \searrow \\ K_1(\mathbb{C}) = \{0\} & & & & K_0(\mathbb{C}) = \mathbb{Z} \\ \nwarrow & & & & \swarrow {\scriptstyle \delta_0} \\ & K_1(C_0([0,\infty))) = \{0\} & \longleftarrow & K_1(C_0((0,\infty))) = \mathbb{Z} & \end{array}$$

thus it is evident that δ_0 is surjective. □

We can now complete the terms in the exact sequence (4.4) as follows:

$$\begin{array}{ccccc}
 & & K_0(A_\emptyset)=\{0\} \xrightarrow{\ \iota_*\ } K_0(A) & & \\
 & \nearrow^{\delta_1} & & \searrow^{\rho_*} & \\
K_1(B)=\mathbb{Z}^2 & & & & K_0(B)=\mathbb{Z}^4 \\
 & \nwarrow^{\rho_*} & & \swarrow^{\delta_0} & \\
 & & K_1(A) \xleftarrow[\ \iota_*\ ]{} K_1(A_\emptyset)=\mathbb{Z} & &
\end{array}$$

Next, we will show that the map $\delta_0 : K_0(B) \to K_1(A_\emptyset)$ is surjective. Let $A_1 \subset A$ be the subalgebra (in fact, ideal) of A, which is the cross product $A_1 = C_0(Z_1) \ltimes \Gamma$, where $Z_1 = Z_\emptyset \cup Z_\infty$. We have the identification:

$$A_1 = \mathcal{K} \otimes (C_0(\mathbb{R}) \ltimes \mathbb{Z}/2\mathbb{Z}),$$

and the following commutative diagram:

$$\begin{array}{ccccc}
A_\emptyset & \xrightarrow{\ \iota\ } & A_1 & \xrightarrow{\ \rho_1\ } & A_\infty \\
{\scriptstyle \mathrm{Id}}\downarrow & & \downarrow{\scriptstyle \subset} & & \downarrow{\scriptstyle \subset} \\
A_\emptyset & \xrightarrow[\ \iota\ ]{} & A & \xrightarrow[\ \rho\ ]{} & B
\end{array}$$

where the map $\rho_1 : A_1 \to A_\infty$ is given by $\mathrm{Id}_{\mathcal{K}} \otimes r$, with r being the restriction map $r : C \to C^*(\mathbb{Z}/2\mathbb{Z})$ as in Lemma 4.6. By the lemma, we obtain a projection $e \in A_\infty$, whose image under the connecting map δ_0 to $K_1(A_\emptyset) = \mathbb{Z}$ is the generator of $\mathbb{Z}$. Specifically, $\delta_0([e]) = [\exp(2\pi i h)] \in K_1(A_\emptyset)$, where $\rho_1(h) = e$. We can lift e to A by choosing the same $h \in A$, so we conclude that the connecting map $\delta_0 : K_0(B) \to K_1(A_\emptyset)$ is surjective.

5. The classifying space $\Gamma\backslash(\mathbb{A}_S \times \underline{E\Gamma})$ and its codimension 1 foliation

Let $S \ni \infty$ be a finite set of places of $\mathbb{Q}$, $\Gamma = \Gamma_S$ as in (4.1). The C^*-algebra $\mathcal{A} = C_0(\mathbb{A}_S) \rtimes \Gamma$, associated with the noncommutative space $Y_{\mathbb{Q},S} := \Gamma\backslash\mathbb{A}_S$, is not of type I when S contains at least two distinct primes (see [**CM**], Lemma 2.28, Chapter II). To understand the K-theory of $\mathcal{A}$, we consider the assembly map [**BCH**], which computes the K-theory of the C^*-algebra $\mathcal{A}$ in terms of the universal proper action $\underline{E\Gamma}$ of Γ. Since Γ is commutative, all definitions of the cross product C^*-algebra coincide with the reduced cross product, so we omit the subscript r in the notation C^*_r. Moreover, the assembly map is an isomorphism. The group Γ is $\mathbb{Z}^n \times \{\pm 1\}$, where n is the number of distinct primes in S, and the universal space for proper actions, $\underline{E\Gamma}$, is $\mathbb{R}^n$, where $\mathbb{Z}^n$ acts by translations and $\{\pm 1\}$ acts trivially. This leads to the space

$$\mathcal{Y}_S = \Gamma\backslash(\mathbb{A}_S \times \underline{E\Gamma}),$$

which is the quotient of $\mathbb{A}_S \times \mathbb{R}^n$ by the diagonal action of Γ, given by

$$\mathcal{Y}_S = \Gamma\backslash\left(\prod_{v\in S}\mathbb{Q}_v \times \mathbb{R}^n\right).$$

This space resolves the noncommutative nature of the quotient $\Gamma\backslash\mathbb{A}_S$.

PROPOSITION 5.1. (i) The space $\mathcal{Y}_S$ is locally compact and has dimension $\#S$.

(ii) The action of $\mathbb{R}^n$ by translations on the space $\mathcal{Y}_S$ defines a codimension 1 lamination of $\mathcal{Y}_S$, with the space of leaves being Y_S.

PROOF. (i) The properness of the action of Γ on $\underline{E\Gamma}$ implies that the action of Γ on $\mathbb{A}_S \times \underline{E\Gamma}$ is also proper. Thus, $\mathcal{Y}_S$ is Hausdorff, and since $\mathbb{A}_S \times \underline{E\Gamma}$ is locally compact, it follows that $\mathcal{Y}_S$ is locally compact. The dimension of $\prod_{v\in S} \mathbb{Q}_v \times \mathbb{R}^n$ is $n+1 = \#S$, and this dimension is preserved under the quotient.

(ii) The orbits of the $\mathbb{R}^n$-action by translation on $\prod_{v\in S} \mathbb{Q}_v \times \mathbb{R}^n$ define a lamination that is invariant under the action of Γ, and hence descends to the quotient $\mathcal{Y}_S$. The space of leaves of this lamination is $\prod_{v\in S} \mathbb{Q}_v$, and its quotient by Γ is Y_S.

□

References

[BCH] P. Baum, A. Connes, and N. Higson, *Classifying space for proper actions and K-theory of group C^*-algebras*, C^*-algebras: 1943–1993 (San Antonio, TX, 1993), Contemp. Math., vol. 167, Amer. Math. Soc., Providence, RI, 1994, pp. 240–291, DOI 10.1090/conm/167/1292018. MR1292018

[Bl] B. Blackadar, *K-theory for operator algebras*, Mathematical Sciences Research Institute Publications, vol. 5, Springer-Verlag, New York, 1986, DOI 10.1007/978-1-4613-9572-0. MR859867

[Br] F. Bruhat, *Distributions sur un groupe localement compact et applications à l'étude des représentations des groupes $\wp$-adiques* (French), Bull. Soc. Math. France **89** (1961), 43–75. MR140941

[C] A. Connes, *Trace formula in noncommutative geometry and the zeros of the Riemann zeta function*, Selecta Math. (N.S.) **5** (1999), no. 1, 29–106, DOI 10.1007/s000290050042. MR1694895

[CC1] A. Connes and C. Consani, *Schemes over $\mathbb{F}_1$ and zeta functions*, Compos. Math. **146** (2010), no. 6, 1383–1415, DOI 10.1112/S0010437X09004692. MR2735370

[CC2] A. Connes and C. Consani, *From monoids to hyperstructures: in search of an absolute arithmetic*, Casimir force, Casimir operators and the Riemann hypothesis, Walter de Gruyter, Berlin, 2010, pp. 147–198. MR2777715

[CC3] A. Connes and C. Consani, *Geometry of the arithmetic site*, Adv. Math. **291** (2016), 274–329, DOI 10.1016/j.aim.2015.11.045. MR3459019

[CC4] A. Connes and C. Consani, *Geometry of the scaling site*, Selecta Math. (N.S.) **23** (2017), no. 3, 1803–1850, DOI 10.1007/s00029-017-0313-y. MR3663595

[CC5] A. Connes and C. Consani, *The Riemann-Roch strategy: complex lift of the scaling site*, Advances in noncommutative geometry—on the occasion of Alain Connes' 70th birthday, Springer, Cham, [2019] ©2019, pp. 53–125, DOI 10.1007/978-3-030-29597-4_2. MR4300552

[CC6] A. Connes and C. Consani, *Riemann-Roch for the ring $\mathbb{Z}$* (English, with English and French summaries), C. R. Math. Acad. Sci. Paris **362** (2024), 229–235, DOI 10.5802/crmath.543. MR4745331

[CC7] A. Connes and C. Consani, *Riemann-Roch for $\overline{\operatorname{Spec}\mathbb{Z}}$*, Bull. Sci. Math. **187** (2023), Paper No. 103293, 29, DOI 10.1016/j.bulsci.2023.103293. MR4609943

[CC8] A. Connes and C. Consani, *Knots, Primes and the Adele Class Space*, arXiv:2401.08401

[CM] A. Connes and M. Marcolli, *Noncommutative geometry, quantum fields and motives*, American Mathematical Society Colloquium Publications, vol. 55, American Mathematical Society, Providence, RI; Hindustan Book Agency, New Delhi, 2008, DOI 10.1090/coll/055. MR2371808

[HK] N. Higson and G. Kasparov, *Operator K-theory for groups which act properly and isometrically on Hilbert space*, Electron. Res. Announc. Amer. Math. Soc. **3** (1997), 131–142, DOI 10.1090/S1079-6762-97-00038-3. MR1487204

[L] H. Lenstra, *Galois theory for schemes.* First edition: 1985 (Mathematisch Instituut, Universiteit van Amsterdam) Second edition: 1997 (Department of Mathematics, University of California at Berkeley) Electronic third edition: 2008

[M] Y. Manin *Introduction into theory of schemes.* Translated from the Russian and edited by Dimitry Leites. Abdus Salam School of Mathematical Sciences Lahore, Pakistan. (2009).

[Ma0] B. Mazur, *Remarks on the Alexander Polynomial,* (1963), https://bpb-us-e1.wpmucdn.com/sites.harvard.edu/dist/a/189/files/2023/01/Remarks-on-the-Alexander-Polynomial.pdf

[Ma1] B. Mazur, *Primes, Knots and Po*, Lecture notes for the conference "Geometry, Topology and Group Theory" in honor of the 80th birthday of Valentin Poenaru, July 2012. https://www-fourier.ujf-grenoble.fr/~funar/CONFERENCES/article_barry.pdf

[Ma2] B. Mazur, *Bridges between Geometry and Number Theory*, talk at the conference "Unifying themes in geometry" como 2021 https://www.youtube.com/watch?v=U3EzqIYgqEw

[Ma3] B. Mazur, *Thoughts about primes and knots*, (2021), https://www.youtube.com/watch?v=KTVEFwRbuzU

[MM] S. MacLane, I. Moerdijk, *S*heaves in geometry and logics, A first introduction to topos theory. Corrected reprint of the 1992 edition. Universitext. Springer-Verlag, New York, 1994. xii+629 pp.

[RM] R. Meyer, *The cyclic homology and K-theory of certain adelic crossed products*, arXiv:math/0312305.

[Mo1] M. Morishita, *Analogies between knots and primes, 3-manifolds and number rings [translation of MR2208305]*, Sugaku Expositions **23** (2010), no. 1, 1–30. Sugaku expositions. MR2605747

[Mo2] M. Morishita, *Knots and primes*, Universitext, Springer, London, 2012. An introduction to arithmetic topology, DOI 10.1007/978-1-4471-2158-9. MR2905431

[Stacks] *The Stacks project* 2024, https://stacks.math.columbia.edu

Collège de France, 3 Rue d'Ulm, 75005 Paris, France; and I.H.E.S., France
Email address: alain@connes.org

Department of Mathematics, Johns Hopkins University, Baltimore, Maryland 21218
Email address: cconsan1@jhu.edu

Contemporary Mathematics
Volume **842**, 2026
https://doi.org/10.1090/conm/842/16853

The Grassmannian complex and Goncharov's motivic complex in weight 4

Herbert Gangl

To Spencer Bloch's 80th birthday, in admiration

ABSTRACT. For a field F and a given integer $n > 1$, Goncharov (1995) has given a motivic complex $\Gamma_F(n)$, generalising the Bloch-Suslin complex (Bloch, 2000; Suslin, 1990) for $n = 2$, which he expects to rationally compute the weight n motivic cohomology of Spec F, hence its algebraic K-groups in Adams weight n, and he was also led to—conjecturally quasiisomorphic—'thickened' complexes thereof.

These complexes involve tensor products of higher polylogarithm groups, the latter having been linked to the geometry of certain configurations in Goncharov's proof of Zagier's Polylogarithm Conjecture for weight 3, and an analogous picture has long been envisioned by Goncharov (1995) for higher weight.

We provide a partial morphism in weight 4 by giving three out of four maps for configurations in general position. We moreover give partial results for the leftmost square.

1. Introduction

Let F be a field (typically $F = \mathbf{C}$, a number field F or a rational function field $F(t_1, \ldots, t_r)$ over the latter). We will consider configurations of $2n$ points in $\mathbb{P}^{n-1}(F)$ in general position (i.e. any $k \leqslant n$ points are linearly independent) as studied by Suslin and Goncharov. Specifically, for $n = 4$ we focus on the configurations of up to eight points in $\mathbb{P}^3(F)$. We define $\tilde{C}_N(4)$ as the free abelian group generated by configurations of N points in $\mathbb{P}^3(F)$ in general position, and let $C_N(4) = \tilde{C}_N(4)/\operatorname{PGL}_4(F)$ be the group of coinvariants under the action of $G = \operatorname{PGL}_4(F)$, which acts by projective transformations. There is a natural differential map $\partial_N : C_N(4) \to C_{N-1}(4)$ defined by alternatingly omitting one of the points.

Our goal is to relate the resulting complex to Goncharov's so-called motivic complex $\Gamma_F(4)$ (e.g. [**13**], [**15**]) which is defined with the help of certain single-valued variants $\mathcal{L}_n(z)$ of the classical polylogarithms $\operatorname{Li}_n(z) = \sum_{n>0} z^k/k^n$. The idea is to use higher polylogarithm groups[1] $B_n(F) = \mathbf{Z}[F]/\langle$all functional equations of $\mathcal{L}_n\rangle$, together with their tensor (or wedge) products, relating them by certain coboundary maps. The latter are ultimately inspired by a cobracket in an associated 'motivic Lie

[1]Goncharov [**13**] calls them "higher Bloch groups", which alas clashes with other authors (e.g. Suslin who coined the notion "Bloch group", and Zagier [**19**]) for a much smaller subgroup thereof. Some authors call $B_2(F)$ the "pre-Bloch group". We follow Cathelineau's convention [**2**].

coalgebra' originally envisaged by Beilinson, which is known to exist for a number field F [**7**]—to form a complex that computes certain (graded pieces of) algebraic K-groups and hence the associated motivic cohomology groups.

There are some intricate combinatorial problems that make it rather hard to produce a morphism that relates the two complexes, where the most right hand map is rather straightforward. We define two of the other three maps involved in such a way that the corresponding squares are indeed commutative.

A correct definition of the fourth map making the remaining square on the left commutative should give a "quadruple ratio" akin to Goncharov's triple ratio in the weight 3 case and would be key in a particularly satisfactory proof of said conjecture. We are able to give an—alas only partial—solution of this problem in that we can make the diagram commutative for certain degenerate configurations but not for generic ones. In collaboration with Charlton and Radchenko [**4**] we were subsequently able to give an explicit (albeit complicated) quadruple ratio, invoking several hundred orbits under the action of a rather large group.

Moreover, we show in §3 that the coboundary map in the associated Chevalley-Eilenberg Lie coalgebra complex annihilates (a dual of) $f_7(4)\circ d_8(\mathbf{v})$ for each generator $\mathbf{v}$ in $C_8(4)$, by showing its vanishing under a specific eight-fold antisymmetrisation on its symbol. We also check that this combination satisfies a Chen type integrability condition and hence is expressible as the coboundary of a combination of weight 4 multiple polylogarithms, albeit not explicitly.

According to Dan [**5**], one can reduce any weight 4 multiple polylogarithm to an explicit combination of the function $\mathrm{Li}_{3,1}(x,y) = \sum_{0<k<\ell} x^k y^\ell / k^3\ell$ (or rather its cousin $I_{3,1}(z_1,z_2) = \mathrm{Li}_{3,1}(z_2/z_1, 1/z_2)$ arising from its integral representation) and Li_4. Furthermore, he reduces any combination of $I_{3,1}$-terms with vanishing $\bigwedge^2 B_2(F)$-component to a sum of combinations of the form $\sum_{i=1}^5 I_{3,1}\big(V_i(x,y),z\big)$ where the $V_i(x,y)$ denote the terms in the five term relation for the dilogarithm. Finally, a long-standing conjecture of Goncharov held that one can write any of the latter five term combinations explicitly in terms of Li_4 alone, and this was solved by this author in [**11**], Thm 17.

Overall the above provides the existence of a map $C_8(4)$ to $B_4(F)$ but does not give an explicit form for it.

A similar morphism of complexes, again without an explicit map on the very left, has been given by Goncharov and Rudenko in their inspiring paper [**17**], §7+9, where they use Goncharov's motivic correlators and a surprising connection to cluster algebras allowing them to give much more structure to the main result of [**11**] which, in conjunction with many other insights, permits them to give a proof of Zagier's Polylogarithm Conjecture in weight 4. See also [**10**].

2. Towards a morphism of complexes

NOTATION. For a field F, we will be looking at many cross ratios (of four points on the projective line over F), triple ratios (of six points in a projective plane), projected cross ratios and projected triple ratios as well as 4×4-determinants where the columns arise from vectors in affine 4-space F^4.

We adopt Goncharov's notation from [**15**]. Configurations in $C_N(4)$ are ordered sets of N vectors (by abuse of notation we will also sometimes call then points) $v_i \in F^4$ in general position (i.e. each four of them give a non-zero determinant), viewed up to the diagonal action of the general linear group $GL_4(F)$.

In particular, an expression consisting of four indices $(i_1\, i_2\, i_3\, i_4)$ is shorthand for $\Delta(v_{i_1}, v_{i_2}, v_{i_3}, v_{i_4}) = |v_{i_1}\, v_{i_2}\, v_{i_3}\, v_{i_4}|$, i.e. to the determinant of the 4×4-matrix whose columns are given by the points v_{i_k}.

All the expressions in the following are homogeneous in each vector (i.e. they will occur in any given expression as often in the numerator as in the denominator), so we can view the vectors also as points in $\mathbf{P}^3(F)$.

Furthermore, an ordered set of four points x_i in $\mathbf{P}^1(F)$ in general position has a well-known invariant, their cross-ratio r_2, where our convention is

$$\mathrm{r}_2(x_1, x_2, x_3, x_4) = \frac{(x_1 - x_3)(x_2 - x_4)}{(x_1 - x_4)(x_2 - x_3)},$$

which we also abbreviate further by $(1\,2\,3\,4)$ where the x_i are understood.

For an ordered set of six points P_j $(j = 1, \dots, 6)$ in $\mathbf{P}^2(F)$ in general position Goncharov, using a clever simplification by Zagier, has invented the triple ratio, which is the antisymmetrisation under the symmetric group $\mathcal{S}_6$ of the following expression r_3', again denoting Δ by $|\cdot|$ for short,

$$\mathrm{r}_3'(P_1, P_2, P_3, P_4, P_5, P_6) = \frac{|P_1P_2P_4| \cdot |P_2P_3P_5| \cdot |P_3P_1P_6|}{|P_1P_2P_5| \cdot |P_2P_3P_6| \cdot |P_3P_1P_4|},$$

which we also abbreviate by $(P_1P_2P_3P_4P_5P_6)$, or even, provided the context is clear, simply by the index vector $(1\,2\,3\,4\,5\,6)$.

The only other abbreviations we are using are

(1) the projected cross ratio of six points in F^4, denoted by sequences of six indices separated by a bar where we project four points (written to the right of the bar) from the span of two points (written to the left of the bar) onto any generic plane in F^4, so with the above shorthand we have (one recognises the formula for the cross ratio after simply fixing the two vectors indicated by i_1 and i_2)

$$(i_1\, i_2 \mid i_3\, i_4\, i_5\, i_6) = \frac{|i_1 i_2 i_3 i_5|\, |i_1 i_2 i_4 i_6|}{|i_1 i_2 i_3 i_6|\, |i_1 i_2 i_4 i_5|};$$

(2) the projected triple ratio term of seven points in F^4, denoted by sequences of seven indices separated by a bar where we project six points (written to the right of the bar) from a seventh point (written to the left of the bar) onto any generic hyperplane in F^4, so with the above shorthand we have (one recognises the formula for the triple ratio term after simply fixing the vector indicated by i_1)

$$(i_1 \mid i_2\; i_3\, i_4\, i_5\, i_6\, i_7) = \frac{|i_1 i_2 i_3 i_5|\, |i_1 i_3 i_4 i_6|\, |i_1 i_4 i_2 i_7|}{|i_1 i_2 i_3 i_6|\, |i_1 i_3 i_4 i_7|\, |i_1 i_4 i_2 i_5|}.$$

Finally, a subscript $(\dots)_n$ for $n = 2, 3$ is shorthand for $\{(\dots)\}_n$ indicating a generator in $B_n(F) = \mathbf{Z}[F]/R_n(F)$, where $R_n(F)$ is a formal version of the group of functional equations of $\mathcal{L}_n$ (see e.g. [**13**], p.214, 217).

By $\mathrm{Alt}_n\big(f(v_1, \dots, v_n)\big) = \sum_{\sigma \in \mathcal{S}_n} (-1)^{|\sigma|} f(v_{\sigma(1)}, \dots, v_{\sigma(n)})$, for a function f on n vectors in F^4, we understand the alternation under the symmetric group $\mathcal{S}_n$. Note that we adopt the convention that we do not divide by $n!$.

2.1. Defining maps from configurations to Goncharov's polylogarithmic motivic complex. With the notations as in the introduction we can draw the diagram as follows, where the lower left hand group $\mathcal{G}_4(F)$ is defined as a quotient of $\mathbf{Z}[F] \oplus \bigwedge^2 \mathbf{Z}[F]$ by a certain group of relations that plays no role in the following.

$$\begin{array}{ccccccc} C_8(4) & \xrightarrow{d_8} & C_7(4) & \xrightarrow{d_7} & C_6(4) & \xrightarrow{d_6} & C_5(4) \\ \downarrow{\scriptstyle f_8(4)} & & \downarrow{\scriptstyle f_7(4)} & & \downarrow{\scriptstyle f_6(4)} & & \downarrow{\scriptstyle f_5(4)} \\ \mathcal{G}_4(F) & \xrightarrow{\partial_8} & \begin{array}{c} B_3(F)\otimes F^\times \\ \oplus \bigwedge^2 B_2(F) \end{array} & \xrightarrow{\partial_7} & B_2(F)\otimes \bigwedge^2 F^\times & \xrightarrow{\partial_6} & \bigwedge^4 F^\times . \end{array} \tag{2.1}$$

Here the map ∂_6 is defined on generators as $\{a\}_2 \otimes b \wedge c \mapsto a \wedge (1-a) \wedge b \wedge c$, ∂_7 is given on respective generators by $\{a\}_3 \otimes b \mapsto \{a\}_2 \otimes a \wedge b$ as well as by $\{a\}_2 \wedge \{b\}_2 \mapsto \{a\}_2 \wedge (b \wedge (1-b)) - \{b\}_2 \wedge (a \wedge (1-a))$, and ∂_8 sends a generator $\{a\}$ to $\{a\}_3 \otimes a$ and $\{a,b\}$ to $\{a\}_2 \wedge \{b\}_2$.

Ignoring torsion. Note that some of the maps involved have denominators (with only small primes involved), so we should strictly speaking tensor the groups by $\mathbf{Z}[1/N]$ for a certain small N (in our formulas below the largest denominator is 7 and $N = 7!$ should certainly suffice). For ease of notation we take this as understood.

The vertical maps in the diagram are defined as follows:

2.1.1. *The map $f_5(4)$.* The right hand map is simply (note the sign) given by

$$f_5(4) : (1, \ldots, 5) \mapsto -\operatorname{Alt}_5\Big((1234) \wedge (1235) \wedge (1245) \wedge (1345)\Big).$$

2.1.2. *The map $f_6(4)$.* The map $f_6(4)$ can be given as a linear combination of 3 orbits, with coefficients 2, 2 and 1, respectively.

More precisely, we have

$$f_6(4) : (1, \ldots, 6) \longmapsto \operatorname{Alt}_6 \begin{pmatrix} 2\ (12|3456)_2 \otimes (1234) \wedge (1356) \\ +1\ (12|3456)_2 \otimes (1345) \wedge (2345) \\ +2\ (12|3456)_2 \otimes (1234) \wedge (1345) \end{pmatrix}.$$

Note that this combination is not unique—see Remark 2.2 below.

2.1.3. *The map $f_7(4)$.* Again there are ambiguities in choosing orbits under the alternation, the possibly most convenient one being the following. We define

$$f_{22}\big((1,2,3,4,5,6,7)\big) = \operatorname{Alt}_7\left((12|3456)_2 \wedge (34|1257)_2\right)$$

and then put

$$f_7(4) : (1, \ldots, 7) \longmapsto \frac{3}{7} f_{22} + 2 f_{31},$$

where

$$f_{31}\big((1,2,3,4,5,6,7)\big) = \operatorname{Alt}_7\big((1|234567)_3 \otimes (2356)\big).$$

2.1.4. *The map $f_8(4)$.* The definition for this map obtained so far for certain *degenerate* configurations are somewhat complicated: For a specific configuration with three parameters the computer found a matching Li_4–expression with 1260 terms (and integer coefficients of modulus $\leqslant 318$).

2.1.5. *A compatibility condition for $f_8(4)$.* A first condition to consider at this stage is that $f_7(4) \circ d_8$ for each generator $\mathbf{v} = (v_1, \ldots, v_8)$ vanishes under the coboundary map, by checking that its associated symbol vanishes under a specific eight-fold antisymmetry, where the symbol $\mathcal{S}$ replaces $\{a\}_2$ by $a \wedge (1-a)$ and $\{a\}_3$ by $\big((a \wedge (1-a)\big) \otimes a$. It turns out that it is equivalent but somewhat less cumbersome to pass to the dual picture where d_8 is replaced by the other Grassmannian differential

$d_8' : C_8(4) \to C_7(3)$, $(v_1, \dots, v_8) \mapsto \sum_{j=1}^{8} (-1)^j (v_j | v_1, \dots, \widehat{v_j}, \dots, v_8)$ (i.e. which amounts to alternatingly sum the projections of the v_i ($i \neq j$) from one of the eight points v_j of a configuration in $C_8(4)$, giving a configuration of 7 points in $\mathbf{P}^2(F)$ each).

A second property to check is that the combination $f_7(3) \circ d_8'(\mathbf{v})$ is actually integrable in the sense of Chen. In §3 below we indicate a—rather computational—proof that $f_7(3) \circ d_8'(\mathbf{v})$ satisfies this integrability. An independent computer check of this result has also been obtained by Radchenko (private communication). This integrability ensures that a combination in $\mathbf{Z}[F] \oplus \bigwedge^2 \mathbf{Z}[F]$ (corresponding to a multiple polylogarithm combination in weight 4) maps to $f_7(3) \circ d_8'(\mathbf{v})$, and then the main result from [**11**] guarantees, in view of Dan's Théorème 4 [**5**], that a combination in the smaller group $\mathbf{Z}[F]$ (corresponding to a *classical* polylogarithm combination in weight 4) suffices, and hence that we can find a map $f_8(4)$ with the prescribed properties, albeit the procedure does not give an explicit candidate.

2.2. Relating the maps $f_m(4)$.

2.2.1. *Relating* $f_5(4)$ *and* $f_6(4)$. Considering the composite maps $C_6(4) \to B_2(F) \otimes \bigwedge^2 F^\times$, we find the following statement showing that the right hand square of (2.1) commutes.

THEOREM 2.1. $\partial_6 \circ f_6(4) = f_5(4) \circ d_6$.

PROOF. We will introduce a convenient notation by associating to a term $(abcd)$ its "complement/dual" inside (123456), so e.g. $(1346) \leftrightsquigarrow (25)$, $(5436) \leftrightsquigarrow (12)$ etc. Note that we are allowed to reorder the indices (e.g. we can replace the 4×4-determinant $(bacd)$ by $(abcd)$ up to invoking a possible sign which we can safely ignore, as it only affects 2-torsion).

1. *The first term:* $\mathrm{Alt}_6\big(2\ (12|3456)_2 \otimes (1234) \wedge (1356)\big)$.

For the first of the three terms under the Alt_6-operator, we consider the boundary of $(12|3456)_2$, i.e. $(12|3456) \wedge (12|3546) \in F^\times \wedge F^\times$, which can be written as

$$\big((1235) - (1236) + (1246) - (1245)\big) \wedge \big((1234) - (1236) + (1256) - (1254)\big).$$

After wedging this expression with $(1234) \wedge (1356)$ on the right and passing to the "complement notation" as indicated above, we get the following contributions (for convenience we also replace wedges by bars), all of which are understood to be under the Alt_6 operator

$$\begin{array}{llll} +46|56|56|24 & +\ 46|34|56|24 & -\ 46|45|56|24 & -\ 46|36|56|24 \\ -45|56|56|24 & -\ 45|34|56|24 & +\ 45|45|56|24 & +\ 45|36|56|24 \\ +35|56|56|24 & +\ 35|34|56|24 & -\ 35|45|56|24 & -\ 35|36|56|24 \\ -36|56|56|24 & -\ 36|34|56|24 & +\ 36|45|56|24 & +\ 36|36|56|24. \end{array}$$

Now the terms in the first column of this tableau can be ignored as we have a factor $56 \wedge 56$ in each. The first and last of the red terms vanish for similar reasons while the remaining two red terms cancel due to the antisymmetry property of the wedge $(45|36|\cdots+36|45|\cdots=0)$. So we are left with eight terms

$$\begin{aligned}&+\ 46|34|56|24\ -\ 46|45|56|24\ -\ 46|36|56|24\\ &-\ 45|34|56|24\\ &+\ 35|34|56|24\ -\ 35|45|56|24\ -\ 35|36|56|24\\ &-\ 36|34|56|24,\end{aligned}$$

which we bring into a slightly more convenient form by permuting the four factors, possibly invoking a sign which we mark below in red

$$\begin{aligned}&+\ 24|34|46|65\ -\ 24|45|46|65\ +\ 24|46|36|65\\ &-\ 24|34|45|56\\ &+\ 24|43|35|56\ -\ 24|45|53|56\ +\ 24|35|36|56\\ &-\ 24|34|36|56.\end{aligned}$$

Under the alternation Alt_6 some of the terms cancel while others are identified. More precisely, leftover terms 3, 4 and 6 are identified with leftover term 1 using the permutations $(4\,6)(2\,5)$, $(5\,6)$, and $(4\,6\,2\,5)$, respectively (all occurring with the same sign), and leftover terms 8 is identified via $(5\,6)$ with term 5. Term 2 is fixed under the odd permutation (13), while term 7 is fixed under the odd permutation (24); hence the Alt_6–orbits of the latter two vanish.

Overall, the boundary of the first term (i.e. of $\mathrm{Alt}_6\big((12|3456)_2\otimes(1234)\wedge(1235)\big)$) becomes

$$4\big(24|34|46|65\big)+2\big(24|43|35|56\big).$$

2. *The second term:* $\mathrm{Alt}_6\big(+1\ (12|3456)_2\otimes(1345)\wedge(2345)\big)$.
For the second term, we find similarly

$$\begin{aligned}&+46|56|56|26\ +\ 46|34|56|26\ -\ 46|45|56|26\ -\ 46|36|56|26\\ &-45|56|56|26\ -\ 45|34|56|26\ +\ 45|45|56|26\ +\ 45|36|56|26\\ &+35|56|56|26\ +\ 35|34|56|26\ -\ 35|45|56|26\ -\ 35|36|56|26\\ &-36|56|56|26\ -\ 36|34|56|26\ +\ 36|45|56|26\ +\ 36|36|56|26,\end{aligned}$$

giving after suitable permutation of factors (again, eight of the terms cancel for reasons of antisymmetry)

$$\begin{aligned}&-\ 34|46|65|26\ +\ 45|46|65|26\ -\ 26|36|46|56\\ &+\ 34|45|56|62\\ &-\ 43|35|56|62\ -\ 35|45|56|26\ -\ 35|36|56|26\\ &+\ 34|36|56|26.\end{aligned}$$

In the latter expression, leftover terms 4 and 5 are identified via $(3\,4)$, while terms 1 and 6 cancel, and terms 2 and 7 are each fixed under an odd permutation, hence their alternation vanishes. So we obtain for the boundary of the second term, i.e. of $\mathrm{Alt}_6\big((12|3456)_2\otimes(1234)\wedge(3456)\big)$, the contribution

$$\big(34|36|56|26\big)+2\big(34|45|56|62\big)-\big(26|36|46|56\big),$$

or equivalently, applying the odd permutation $(3\,6\,4\,5)$ to the first term and $(6\,5\,3\,2)$ to the second one, we can write the result as

$$\big(65|64|34|24\big)+2\big(24|43|35|56\big)-\big(26|36|46|56\big).$$

3. *The third term:* $\mathrm{Alt}_6\left(+2\ (12|3456)_2 \otimes (1234) \wedge (1345)\right)$. Finally, for the third term we obtain

$$\begin{aligned}
&+46|56|26|16 \ + \ 46|34|26|16 \ - \ 46|45|26|16 \ - \ 46|36|26|16\\
&-45|56|26|16 \ - \ 45|34|26|16 \ + \ 45|45|26|16 \ + \ 45|36|26|16\\
&+35|56|26|16 \ + \ 35|34|26|16 \ - \ 35|45|26|16 \ - \ 35|36|26|16\\
&-36|56|26|16 \ - \ 36|34|26|16 \ + \ 36|45|26|16 \ + \ 36|36|26|16,
\end{aligned}$$

and in slightly adapted form

$$\begin{aligned}
&-16|26|46|56 \ - \ 34|46|26|16 + \ 45|46|26|16 \ - \ 16|26|36|46\\
&-45|56|26|16 \ - \ 45|34|26|16\\
&+35|56|26|16 + 35|34|26|16 - 35|45|26|16 \ - \ 35|36|26|16\\
&-16|26|56|36 \ + \ 34|36|26|16.
\end{aligned}$$

This time the terms in the first column are non-zero, and we have three different types of terms, all of which arise with compatible sign. Leftover terms 6, 8 and 9 are fixed under an odd permutation each, hence do not contribute. Terms 1, 4 and 11 combine to multiplicity 3, while the remaining six terms combine to multiplicity 6, so we obtain for the boundary of the third term the contribution

$$-6(34|46|26|16) - 3(16|26|46|56).$$

4. *Combining the three contributions.* Now it remains to compare the terms under the alternation. Putting

$$[a, b, c] = \mathrm{Alt}_6\left(a(24|34|46|65) + b(24|43|35|56) + c(26|36|46|56)\right),$$

we find $[4, 2, 0]$, $[-1, -2, -1]$ and $[-6, 0, -3]$ for the terms in 1., 2. and 3., respectively.

Noting that $2[4, 2, 0] + 2[-1, -2, 1] + [-6, 0, -3] = [0, 0, -1]$, and taking the corresponding complement notation (e.g. 26 is replaced by 1345 etc.) we obtain for the LHS $-\mathrm{Alt}_6((1345) \wedge (1245) \wedge (1235) \wedge (1234)$, showing that the diagram commutes. □

There is an equivalent way to write the map $f_6(4)$ using the following claim.

REMARK 2.2. The following combination in $B_2(F) \otimes \bigwedge^2 F^\times$ has vanishing boundary, i.e.

$$\partial_6\,\mathrm{Alt}_6\left((12|3456)_2 \otimes \left(3((1345) \wedge (1234)) + (1345) \wedge (2345) + (1345) \wedge (3456)\right)\right) = 0.$$

PROOF. From a similar analysis as used in the proof of Theorem 2.1, using the same shorthands, we find that alternating the term $(12|3456)_2 \otimes (1345) \wedge (3456)$ can be expressed as $[3, -6, 0]$. Now use from the proof of that theorem that the alternations of the terms $3(12|3456)_2 \otimes (1345) \wedge (1234)$ and $(12|3456)_2 \otimes (1345) \wedge (2345)$ can be written in the same shorthand as $3\,[1, 2, 1]$ and $[-6, 0, -3]$, respectively. □

2.2.2. *Relating $f_6(4)$ and $f_7(4)$ (the middle square of (2.1)).* We have the following compatibility of maps $C_7(4) \to B_3(F) \otimes F^\times$, showing that the centre square of (2.1) commutes.

THEOREM 2.3. $f_6(4) \circ d = \partial \circ f_7(4)$.

PROOF. *LHS.* We first compute the left hand side for an arbitrary configuration of seven vectors, using the definition of $f_6(4)$, to get

$$\begin{aligned} f_6(4)\circ d\big((123467)\big) = {} & \ 2\ \mathrm{Alt}_7(12|3456)_2\otimes(1234)\ \wedge\ (1345)\\ & + 2\ \mathrm{Alt}_7(12|3456)_2\otimes(1234)\ \wedge\ (1356)\\ & + \ \mathrm{Alt}_7(12|3456)_2\otimes(1345)\ \wedge\ (2345). \end{aligned}$$

We rewrite below the three contributions under the Alt–sign using the permutations $(2\,5\,4)$, $(2\,3)$ and $(3\,1\,4\,2\,5)$, respectively, to obtain

$$\begin{aligned} f_6(4)\circ d\big((123467)\big) = {} & \ 2\ \mathrm{Alt}_7(15|3246)_2\otimes(1532)\ \wedge\ (1324)\\ & + 2\ \mathrm{Alt}_7(13|2456)_2\otimes(1234)\ \wedge\ (1256)\\ & + \ \mathrm{Alt}_7(45|1236)_2\otimes(4123)\ \wedge\ (5123). \end{aligned}$$

Note that the terms in the respective wedge products come in two types: they either overlap in two entries (the second sum) or in three entries (first and last sum). Hence it is useful to collect the ones of the same type; in fact we also apply for later convenience the permutation $(1\,3)(6\,7)$ to the first sum. Moreover, we note the stabilisers that will fix the expressions or turn them into their negatives: we can write the last sum as

$$\mathrm{Alt}_7(45|1237)_2\otimes(1235)\ \wedge\ (1234)$$

(note that we swapped the terms in the wedge product but also applied the odd permutation $(6\,7)$, hence keep the sign). This expression is symmetric in the indices $4,5$ and antisymmetric in the three indices $1,2,3$ as well as in the indices $6,7$; so we can think of regrouping terms in the sum in this manner and combine first and third sum to

$$\mathrm{Sym}_{(4\,5)}\,\mathrm{Alt}_{\langle(1\,2),(1\,3),(6\,7)\rangle}\Big(\big((2(35|1247)+(45|1237)\big)\otimes 1235\wedge 1234\Big)\ +\ \text{symm.},$$

where "+ symm." denotes the sum over the different Alt_7–translates of that expression. Similarly, the second sum can be written as

$$\mathrm{Alt}_{(3\,5)(4\,6)}\,\mathrm{Alt}_{\langle(1\,2),(3\,4),(5\,6)\rangle}\big((13|2456)\otimes(1256)\wedge(1234)\big)\ +\ \text{symm.},$$

where this time the antisymmetrization is over a group of order 4, generated by the two involutions (given in the indices of Alt) whose cycle forms are $(3\,5)(4\,6)$ and $(1\,2)(3\,4)(5\,6)$, respectively. It will suffice to match these regrouped terms in order to prove commutativity of the square.

In summary, the left hand side can be written as

$$\begin{aligned} &\Big(\mathrm{Sym}_{(4\,5)}\,\mathrm{Alt}_{\langle(1\,2),(1\,3),(6\,7)\rangle}\Big(\big((2(35|1247)+(45|1237)\big)\otimes(1235)\wedge(1234)\Big)\ +\ \text{symm.}\Big)\\ &\qquad +\Big(\ \mathrm{Alt}_{(3\,5)(4\,6)}\,\mathrm{Alt}_{\langle(1\,2),(3\,4),(5\,6)\rangle}\Big((13|2456)\otimes(1256)\wedge(1234)\Big)\ +\ \text{symm.}\Big). \end{aligned}$$

RHS. We will now analyse the two expressions arising from taking the boundary of the two terms in $f_7(4)$.

Factoring the first term $\partial f_{31}\big((1234567)\big)$ we get $(1|234567)=\frac{|1235|\cdot|1346|\cdot|1427|}{|1236|\cdot|1347|\cdot|1425|}$, where the factors are 4×4–determinants and so we find

LEMMA 2.4.

$$\partial\,\mathrm{Alt}_7\Big((1|234567)_3\otimes(2356)\Big) = \mathrm{Alt}_7(1|234567)_2\otimes\Big((1235)\wedge(2356)-(1236)\wedge(2356) + (1346)\wedge(2356)-(1347)\wedge(2356) + (1427)\wedge(2356)-(1425)\wedge(2356)\Big).$$

(Proof is straightforward.)

Grouping again by types of "overlaps" of indices of the two tensor factors on the right, we find that the first two terms $(1235)\wedge(2356)$ and $(1236)\wedge(2356)$ have three overlapping indices ($\{2,3,5\}$ and $\{2,3,6\}$, respectively) and are combined into the first line below, terms #3 and #6 have overlap 2 and are combined into the second line below, and finally terms #4 and #5 have overlap 1 and are combined into the third line below. In summary, we can write the first term on the RHS ($=\partial\circ\mathrm{Alt}_7\left((1|234567)_3\otimes(2356)\right)$) as

$$\begin{aligned}&2\,\mathrm{Sym}_{(4\,5)}\,\mathrm{Alt}_{\langle(1\,2),(1\,3),(6\,7)\rangle}\Big((5|236147)_2+(5|236417)_2\Big)\otimes(1235)\wedge(1234)+\text{symm.}\\ &-2\,\mathrm{Alt}_{(3\,5)(4\,6)}\,\mathrm{Alt}_{(1\,2)(3\,4)(5\,6)}\Big((3|614527)_2+(5|236147)_2\Big)\otimes(1256)\wedge(1234)+\text{symm.}\\ &-4\,\mathrm{Alt}_7\Big((1|234567)_2\otimes(1347)\wedge(2356)\Big).\end{aligned}$$

It turns out that the last line (i.e. the contribution with overlap 1) vanishes.

PROPOSITION 2.5. *We have*

$$\mathrm{Alt}_7\Big((1|234567)_2\otimes(1347)\wedge(2356)\Big)=0.$$

PROOF. We expand the expression in a similar way to what we did for the above terms into

$$\mathrm{Alt}_{(1\,2)(4\,5)(6\,7)}\,\mathrm{Alt}_{\langle(1\,4),(4\,7),(2\,5),(5\,6)\rangle}\Big((1|234567)_2\otimes(1347)\wedge(2356)\Big)\;+\;\text{symm.}$$

and are then reduced to show that the individual sums already vanish. For this we use (2.2) below. ☐

Similarly, the second term $\partial\circ\frac{3}{7}\,\mathrm{Alt}_7\Big((12|3456)_2\otimes(34|1257)_2\Big)$ of the RHS is by definition

$$\frac{3}{7}\,\mathrm{Alt}_7\Big((12|3456)_2\otimes(34|1257)\wedge(34|1527)\;-\;(34|1257)_2\otimes(12|3456)\wedge(12|3546)\Big)$$

and we can rewrite it as follows.

PROPOSITION 2.6. *The expression* $-\partial\circ\frac{3}{7}\,\mathrm{Alt}_7\Big((12|3456)_2\otimes(34|1257)_2\Big)$ *equals*

$$2\,\mathrm{Sym}_{(4\,5)}\,\mathrm{Alt}_{\langle(1\,2),(1\,3),(6\,7)\rangle}\Big(\big((45|1237)_2+2(46|1237)_2\big)\otimes(1254)\wedge(1234)\Big)+\text{symm.}$$

PROOF. The differential ∂_7 of $\{x\}_2\wedge\{y\}_2\in\bigwedge^2B_2(F)$ is defined as follows: $\{x\}_2\otimes y\wedge(1-y)\;-\;\{y\}_2\otimes x\wedge(1-x)\in B_2(F)\otimes\bigwedge^2F^\times$, so up to the involution $(1\,3)(2\,4)(6\,7)$ we obtain twice the same terms, so using that $1-\mathrm{r}_2(x_1,x_2,x_3,x_4)=\mathrm{r}_2(x_1,x_3,x_2,x_4)$ we obtain

$$-\partial\circ\frac{3}{7}\,\mathrm{Alt}_7\Big((12|3456)_2\otimes(34|1257)_2\Big)=2\,(34|1257)_2\otimes(12|3456)\wedge(12|3546)$$

and the latter can be written as

$$2\ \mathrm{Alt}_7(34|1257)_2 \otimes \begin{pmatrix} (1235)\wedge(1234) \\ +(1235)\wedge(1256) \\ -(1235)\wedge(1236) \\ -(1235)\wedge(1254) \\ +(1246)\wedge(1234) \\ +(1246)\wedge(1256) \\ -(1246)\wedge(1236) \\ -(1246)\wedge(1254) \\ -(1236)\wedge(1234) \\ -(1236)\wedge(1256) \\ -(1245)\wedge(1234) \\ -(1245)\wedge(1256) \end{pmatrix} = 2\ \mathrm{Alt}_7 \begin{pmatrix} (34|1257)_2 \\ +(56|1237)_2 \\ +(36|1257)_2 \\ +(54|1237)_2 \\ +(34|1267)_2 \\ +(64|1257)_2 \\ +(54|1267)_2 \\ +(63|1257)_2 \\ +(34|1267)_2 \\ +(46|1257)_2 \\ +(53|1247)_2 \\ +(64|1237)_2 \end{pmatrix} \otimes (1235)\wedge(1234)$$

where we apply the following permutations, respectively: #2: $(3\,5)(4\,6)$; #3: $(4\,6)$; #4: $(3\,5)$; #5: $(3\,4)(5\,6)$; #6: $(3\,6)$; #7: $(6\,3\,5)$; #8: $(6\,4\,3)$; #9: $(5\,6)$; #10: $(6\,3)$; #11: $(3\,4)$; #12: $(6\,5\,3)$.

Note that terms #1 and #11 give the same orbit as do terms #2 and #12, terms #3 and #8, terms #5 and #9 as well as terms #6 and #10, so we are left with only seven different terms as follows

$$2\ \mathrm{Alt}_7 \begin{pmatrix} 2(34|1257)_2 \\ +2(56|1237)_2 \\ +2(36|1257)_2 \\ +(54|1237)_2 \\ +2(34|1267)_2 \\ +2(64|1257)_2 \\ +(54|1267)_2 \end{pmatrix} \otimes (1235)\wedge(1234).$$

Now we need to use a couple of functional equations to write each of these summands (or rather their $\mathrm{Sym}_{(4\,5)}\,\mathrm{Alt}_{\langle(1\,2),(2\,3),(6\,7)\rangle}$–orbits) in terms of a generating orbit set. It turns out that we need three such orbits, we choose as representatives $(45|1237)_2$, $(46|1237)_2$ and $(46|1257)_2$. As a shorthand, we will then express each of the six summands as a linear combination of the given three generators and identify each with the corresponding coefficient vector. E.g. $(34|1257)_2 \leftrightsquigarrow [\frac{1}{3}, \frac{1}{3}, -1]$ denotes

$$\mathrm{SymAlt}(34|1257)_2 = \mathrm{SymAlt}\big(\frac{1}{3}(45|1237)_2 + \frac{1}{3}(46|1237)_2 - (46|1257)_2\big),$$

where SymAlt is shorthand for $\mathrm{Sym}_{(4\,5)}\,\mathrm{Alt}_{\langle(1\,2),(2\,3),(6\,7)\rangle}$. Then we find the respective correspondences

$$\begin{aligned}
2(34|1257)_2 &\rightsquigarrow [\tfrac{2}{3}, \tfrac{2}{3}, -2]\\
+2(56|1237)_2 &\rightsquigarrow [0, 2, 0]\\
+2(36|1257)_2 &\rightsquigarrow [0, \tfrac{2}{3}, 0]\\
+(54|1237)_2 &\rightsquigarrow [0, \tfrac{4}{3}, 0]\\
+2(34|1267)_2 &\rightsquigarrow [1, 0, 0]\\
+2(64|1257)_2 &\rightsquigarrow [\tfrac{2}{3}, 0, 0]\\
+(54|1267)_2 &\rightsquigarrow [0, 0, 2]
\end{aligned}$$

which arise from the following functional equations (valid after applying SymAlt)

$$\begin{aligned}
3(45|1267)_2 &= 2(45|1237)_2,\\
3(34|1267)_2 &= 2(46|1237)_2,\\
3(36|1257)_2 &= (46|1237)_2,\\
3(34|1257)_2 &= (45|1237)_2 + (46|1237)_2 - 3(46|1257)_2,\\
3(35|1267)_2 &= 2(46|1237)_2.
\end{aligned}$$

Adding up the right hand vectors gives $-\frac{7}{3}[1, 2, 0]$ which then immediately translates into the claim of the proposition. □

In order to finish the proof of the theorem, we need to compare the three different types of contributions according to the number of overlapping indices (either one, two or three) in the respective rightmost two wedge factors.

(1) For #overlaps= 1 there is no contribution from the RHS, so the contribution from the LHS has to vanish. Indeed we have

LEMMA 2.7.

$$\mathrm{Alt}_{(1\,2)(4\,5)(6\,7)}\,\mathrm{Alt}_{\langle(1\,4),(4\,7),(2\,5)(5\,6)\rangle}(1|234567)_2 = 0. \tag{2.2}$$

(Proof see below, FE A.3.)

(2) For #overlaps= 2 there are contributions from the LHS and from the first term of the RHS. Demanding that their difference vanishes amounts to the following statement.

LEMMA 2.8.

$$\mathrm{Alt}_{(3\,5)(4\,6)}\,\mathrm{Alt}_{\langle(1\,2),(3\,4),(5\,6)\rangle}\big((13|2456)_2 + (3|614527)_2 + (5|236147)_2\big) = 0.$$

(Proof see below, FE A.4.)

(3) For #overlaps= 3 there are contributions from the LHS and both terms on the RHS. Demanding that their difference vanishes amounts to the following statement.

LEMMA 2.9.

$$\begin{aligned}
\mathrm{Sym}_{(4\,5)}\,\mathrm{Alt}_{\langle(1\,2),(2\,3),(6\,7)\rangle}\big((35|1247)_2 - (46|1237)_2&\\
- (5|236147)_2 - (5|236417)_2\big) &= 0.
\end{aligned}$$

(Proof see below, FE A.5.)

Invoking these three relations we obtain that each of the contributions on the LHS equals the corresponding contribution of the RHS, which proves the theorem. □

REMARK 2.10. Given that there is some ambiguity involved in the choice of $f_7(4)$, it is conceivable that there is a more suitable candidate for it. In particular, in Goncharov's analysis of the morphism (2.1) using Aomoto polylogarithms there are other orbits involved in his definition of $f_7(4)$. Moreover, it would be interesting to compare our morphism of complexes to a very related one in [**17**], §7, which invokes insight from cluster algebras.

3. Vanishing of the $\bigwedge^2 B_2$–component and an integrability condition

3.1. The $\bigwedge^2 B_2$–component. For the final (leftmost) square of the main diagram, we give the following partial results. For a generator $\mathbf{v} = (v_1, \ldots, v_8) \in C_8(4)$ the image under $f_7(4) \circ d_8$ is expected to be representable as the image under ∂_8 of a formal "depth 1" combination $\sum_i n_i[x_i]$, and hence the a priori larger complex $\mathcal{G}_4(F) \xrightarrow{\partial_8} B_3(F) \otimes F^\times \oplus \bigwedge^2 B_2(F) \xrightarrow{\partial_7} \ldots$ should be quasiisomorphic to $B_4(F) \xrightarrow{\partial_8} B_3(F) \otimes F^\times \xrightarrow{\partial_7}$ (same). Goncharov ([**13**] and [**15**]) shows that one can replace any individual $\{a\}_2 \wedge \{b\}_2$ by a combination of terms in $B_3(F) \otimes F^\times$ with the same ∂_7-image, hence we can indeed land in $B_3(F) \otimes F^\times$ instead. But he also shows ([**12**], Thm. 4.7) that there is no natural map, in terms of formulas, from $\bigwedge^2 B_2(F)$ to $B_3(F) \otimes F^\times$.

In order to check that the outcome of $\xi = f_7(4) \circ d_8(\mathbf{v})$ indeed lies in $\partial_8(B_4(F))$, we first check that it has no genuine contribution to depth 2 by showing that the coboundary δ of ξ in the associated motivic Lie coalgebra is trivial, which in our case boils down to a check—already profitably used in spectacular fashion by Goncharov, Spradlin, Vergu and Volovich in [**16**] to drastically reduce a very complicated expression found by Del Duca, Duhr and Smirnov [**6**] for a central integral in $N = 4$ supersymmetric Yang Mills theory—that ξ vanishes under a specific eight-fold symmetry, due to Goncharov, which is outlined below. A second property needed is that the resulting expression in $B_3(F) \otimes F^\times$ is actually integrable in the sense of Chen. This guarantees that it arises as the ∂_8-image of an element of $\mathcal{G}_4(F)$. Finally, a theorem of (now Rumanian President) Dan ([**5**], Théorème 4), together with the main result of [**11**], implies that an element in $\partial_8 \mathcal{G}_4(F)$ with trivial coboundary as above arises from a depth 1 term in weight 4, i.e. it already lies in the smaller group $\partial_8 B_4(F)$, albeit it does not give an explicit expression.

Preparations. When dealing with $C_8(4)$, and in particular when dealing with determinants of 4×4-matrices, duality simply replaces any 4-element subset of $\{1, 2, \ldots, 8\}$ of indices by the complementary set of indices, while preserving the multiplication and division of determinants. We still ignore torsion (in particular 2-torsion) in the following, so we can neglect signs of these determinants and hence write sequences of four indices representing such determinants in the natural ascending order.

We also note that we drop the checks that our expressions are indeed invariant when multiplying any individual vector with a non-zero scalar, and instead refer to Goncharov who has elegantly provided very similar checks in [**13**] and [**15**].

Finally, we will refer to the eightfold antisymmetrisation which mimics the coboundary map in the Chevalley-Eilenberg complex in weight 4 as "applying δ".

3.1.1. *The part from f_{31}.* We now consider the term corresponding to the second contribution f_{31} of the dual ξ' of ξ. In particular we will deal with 3×3–determinants in this dual situation which makes the task slightly less cumbersome. This amounts to analysing the symbol attached to the alternating sum Alt_7 of the following expression where each number $1, \ldots, 7$ stands for an associated point in $\mathbf{P}^2(F)$

$$(123456)_3 \otimes |147|,$$

and where $(\ldots)_3$ is a shorthand for the triple ratio (consisting of 6! terms) of the six points, viewed as an element of the polylogarithmic group $B_3(F)$. For a triple ratio term a, the numerator of $(1-a)$ factors into a 3×3-determinant and a 6×6-determinant, indicating that we have to consider two types of factors here.

In order to check the coboundary vanishing of ξ', we map all the expressions into $(F^\times)^{\otimes 4}$ and introduce the following 'symbol' $\mathcal{S}$: $\mathcal{S}$ maps $\{a\}_3 \otimes b$ to the antisymmetrisation of $(1-a) \otimes a \otimes a \otimes b$ with respect to the first two slots, i.e. it associates $(1-a) \otimes a \otimes a \otimes b \;-\; a \otimes (1-a) \otimes a \otimes b)$, and it maps $\{a\}_2 \wedge \{b\}_2$ to $\big(a \wedge (1-a)\big) \wedge (b \wedge (1-b))$, obvious shorthand for a sum of eight elementary tensors.

Type 1: In the decomposition of $1-x$, where x is one of Goncharov's triple ratios, there are 'new factors' arising (certain 6×6-determinants). For each such 'new factor' the contribution to $\mathcal{S} \circ f_7(3) \circ d_8'(\mathbf{v})$ is zero. In order to see this, note that anti-symmetrising with respect to the group of order eight generated by the transpositions $(1\,4)$, $(2\,5)$ and $(3\,6)$ fixes the new factor as well as the right hand tensor factor $|147|$, and the corresponding eight terms add up to zero in the same way as for Goncharov-Zagier's (cf. the comment in [**12**] before Theorem 3.10) 840-term relation for the trilogarithm.

Type 2: All the other factors are 3×3–determinants. We are left with considering the Alt_7–alternation of

$$\left(\frac{|123|}{|125|\,|236|\,|314|} \wedge \frac{|124|\,|235|\,|316|}{|125|\,|236|\,|314|}\right) \otimes \frac{|124|\,|235|\,|316|}{|125|\,|236|\,|314|} \otimes |147|.$$

Ignoring the terms with Type-1-factors, each tensor factor of any of the remaining terms, after expansion, becomes simply a single 3×3–determinant.

After expanding in this way we are dealing with $36(=(3+3)\times(3+3))$ terms containing the leftover numerator $|123|$ of the leftmost tensor factor, together with $54(=3 \times 3 \times (3+3))$ terms containing instead one of the denominator factors of that same tensor factor. (Note that, due to the antisymmetry in the first two slots, we can ignore terms involving only denominator factors from the first two slots.)

Among the Alt_7–orbits of the resulting 90 expressions there are only 34 terms which are actually non-zero. We will consider those in more detail. Two of the orbits occur with multiplicity 5, four others occur with multiplicity 2, the remaining 16 ones only occur once.

For completeness' sake we reproduce representatives for all the leftover orbits, together with their multiplicity (we drop the determinant bars and replace tensor or wedge signs by a comma for ease of notation), while introducing reference labels in square brackets

$$\begin{array}{rr}
5(123,124,135,246)[*], & +5(123,124,135,356)[\mathrm{X}], \\
+2(123,124,125,146)[2\mathrm{A}], & -2(123,124,125,156)[-2\mathrm{B}], \\
+2(123,124,135,146)[2\mathrm{E}], & +2(123,124,135,156)[*], \\
+(123,124,135,126)[\mathrm{A}], & +(123,124,135,136)[\mathrm{B}], \\
+(123,124,145,126)[\mathrm{A}], & +(123,124,145,146)[\mathrm{B}], \\
+(123,124,145,236)[*], & +(123,124,145,456)[\mathrm{X}], \\
+(123,145,124,126)[\mathrm{A}], & +(123,145,124,146)[-\mathrm{A}], \\
+(123,145,124,236)[\mathrm{F}], & +(123,145,124,456)[-\mathrm{F}], \\
-(123,145,246,137)[\mathrm{G}], & -(123,145,246,157)[-\mathrm{G}], \\
-(123,145,246,237)[\mathrm{H}], & -(123,145,246,457)[-\mathrm{H}], \\
-(123,145,246,267)[\mathrm{I}], & -(123,145,246,467)[-\mathrm{I}].
\end{array}$$

Each of these elementary tensors $a \otimes b \otimes c \otimes d$ represents an eightfold combination arising from alternating the first two slots, alternating the last two slots, and furthermore alternating under the double swap (slot 1$\leftrightarrow$ slot 3, slot 2$\leftrightarrow$ slot 4).

Three orbits among the above 34 ones, marked with a [$*$], vanish under δ: apply the cycle $(1\,2)(3\,4)(5\,6)$ to both the multiplicity 5 orbits $(123, 124, 135, 246)$ and the multiplicity 1 orbit $(123, 145, 145, 236)$, and apply the cycle $(2\,5)(4\,6)$ to the multiplicity 2 orbit $(123, 124, 135, 156)$.

Under δ some of the remaining $26(= 34-5-2-1)$ orbits agree, possibly up to sign only, and we are left with 8 orbit types only, denoted by roman letters (A, B, E, F, G, H, I, X) as follows, where we indicate the contributions from multiplicities by a superscript. The three orbits marked F, G and H in the three last lines above cancel pairwise (use the permutation $(2\,4)(3\,5)$ in each case).

Type A: $(+^2)$, $(+)$, $(+)$, $(+)$, $(-)$, overall multiplicity $4(= 2+1+1+1-1)$;
Type B: $(-^2)$, $(+)$, $(+)$, overall multiplicity 0;
Type E: $(+^2)$, overall multiplicity 2;
Type F, G, H, I: $(+)$, $(-)$, overall 0;
Type X: $(+^5)$, $(+)$, overall multiplicity 6.

Finally, since Type A and Type E consist of expressions whose four factors contain a common index, they vanish once we compose with the boundary map d' as the latter provides each factor with a second common index and the transposition swapping these two indices is an odd permutation fixing the expression.

Upshot: The only type that gives a non-zero contribution to $\delta \circ \mathcal{S} \circ f_{31} \circ d'_8(\mathbf{v})$ is type X, and it occurs with coefficient 6.

3.1.2. *The part from f_{22}.* We consider the term corresponding to the dual of the second contribution f_{22}. This amounts to analysing the Alt_7–alternation of

$$\left(\frac{|123|\,|145|}{|125|\,|143|} \wedge \frac{|124|\,|135|}{|125|\,|134|}\right) \wedge \left(\frac{|215|\,|267|}{|217|\,|265|} \wedge \frac{|216|\,|257|}{|217|\,|256|}\right).$$

which gives us $12^2 = 144$ terms (again, we do not need to consider contributions if the two leftmost factors—or the two rightmost factors—both arise from the denominator).

Under Alt_7 there are 89 non-zero orbits leftover, which are grouped into orbits of multiplicities 9 (2 such), 7 (5 such), 5 (2 such), 2 (7 such) and 1 (12 such).

$$\begin{array}{cc}
9(123,124,456,145),[-\mathrm{X}] & 9(123,124,356,135),[-\mathrm{X}] \\
7(123,124,156,157),[\mathrm{D}] & -7(123,124,156,125),[\mathrm{B}] \\
7(123,124,145,146),[\mathrm{B}] & 7(123,124,135,136),[\mathrm{B}] \\
7(123,124,125,156),[\mathrm{B}] & -5(123,124,145,456),[-\mathrm{X}] \\
-5(123,124,135,356),[-\mathrm{X}] & -2(123,124,156,257)[*], \\
2(123,124,156,145)[*], & 2(123,124,156,135)[*], \\
-2(123,124,145,246)[*], & 2(123,124,145,156)[*], \\
-2(123,124,135,236)[*], & 2(123,124,135,156)[*], \\
123,124,456,157,[\mathrm{I}] & 123,124,356,157,[\mathrm{I}] \\
123,124,345,146,[\mathrm{C}] & 123,124,345,136,[\mathrm{C}] \\
-(123,124,156,457),[-\mathrm{I}] & -(123,124,156,357),[-\mathrm{I}] \\
-(123,124,145,346),[-\mathrm{C}] & -(123,124,145,126),[-\mathrm{A}] \\
-(123,124,135,346),[-\mathrm{C}] & -(123,124,135,126),[-\mathrm{A}] \\
123,124,125,146,[\mathrm{A}] & 123,124,125,136,[\mathrm{A}].
\end{array}$$

After applying δ, precisely the seven multiplicity 2 orbits (marked by $[*]$) vanish.

Furthermore, under δ some of the remaining $75(=89-2\cdot 7)$ orbits agree, possibly up to sign only, and we are left with 6 orbit types only (four of which agree with orbit types for f_{31}).

Type A: $(+)$, $(+)$, $(-)$, $(-)$, overall multiplicity 0;
Type B: $(+^7)$, $(+^7)$, $(+^7)$ $(+^7)$, overall multiplicity 28;
Type C, I: $(+)$, $(+)$, $(-)$, $(-)$, overall multiplicity 0;
Type D: $(+^7)$, overall 7;
Type X: $(-^9)$, $(-^9)$, $(-^5)$, $(-^5)$, overall multiplicity -28.

Finally, since Type B and Type D consist of expressions whose four factors contain a common index, they vanish under d' as above.

Upshot: The only type that will contribute is type X, and it occurs with coefficient -28, so combining with the result above it is now clear how to cancel the type X contributions. "Dualising the indices" as indicated above we get the following theorem. Note that the linear combination given is 12 times the map $f_7(4)$ above.

THEOREM 3.1. *The following linear combination vanishes under $\delta\circ\mathcal{S}$:*

$$\mathrm{Alt}_8\left(28\Big((1|234567)_3\otimes(1258)\Big)\;+\;6\Big((81|2345)_2\wedge(82|1567)_2\Big)\right).$$

3.2. Integrability. We now consider the above orbits after applying d_8', which essentially amounts to adding an eighth common index to the determinant indices of each of the four factors. Note that all the resulting orbits, with the possible exception of type X orbits, either cancel each other or they vanish under the combined antisymmetrisation Alt_8 and under swapping the first two or the last two tensor factors. Using the criterion for integrability as adapted from Chen, as e.g. given in [**9**] (3.17), one can check that also the type X orbit is indeed integrable to a weight 4 multiple polylogarithm.

For the f_{31}–part, we need to apply the map sending $\{a\}_3\otimes b\in B_3(F)\otimes F^\times$ to $\big((1-a)\otimes a\,-\,a\otimes(1-a)\big)\otimes\big(d\log a\wedge d\log b\big)$ and check that the image of the type X orbit vanishes.

For the f_{22}–part, we have to apply δ to $\{a\}_2 \wedge \{b\}_2$ and further map it (up to overall sign) to the tensor product of $a \wedge b$ with the sum of four symmetric tensors, with rational functions as coefficients, given by

$$\frac{1}{ab}\big((1-a)\odot(1-b)\big)+\frac{1}{a(1-b)}\big((1-a)\odot b\big)+\frac{1}{(1-a)b}\big(a\odot(1-b)\big)+\frac{1}{(1-a)(1-b)}(a\odot b).$$

Here we have used the notation $x \odot y = x\otimes y + y\otimes x$ for the symmetric tensor. The package [**8**] can be used to check that the corresponding antisymmetrised expressions vanish: for the former one only needs to check the integrability with respect to the last two tensor factors as the other possibilities vanish, anyway, while for the latter only the two middle factors have to be tested. The cancellations are rather non-trivial and rely on (differences of) projected Plücker relations.

Dualising, we finally get the result that $f_7(4) \circ d$ maps any configuration in $C_8(4)$ to an integrable expression.

Comments: The results from the above paragraph imply that we can attach to each configuration in $C_8(4)$ a weight 4 multiple polylogarithm. As already indicated in the introduction, Dan [**5**] showed that any multiple polylogarithmic expression in weight 4 can be explicitly reduced[2] to one in I_{31} and Li_4. He also showed that the exactness of the complex $0 \to B_4(F)_{\mathbf{Q}} \to \mathcal{H}_4(F)_{\mathbf{Q}} \to^{\delta_4} \bigwedge^2 B_2(F)_{\mathbf{Q}} \to 0$, where $\mathcal{H}_4(F)_{\mathbf{Q}}$ denotes the vector space of formal multiple polylogarithms in weight 4 and δ_4 an associated coboundary map, would follow provided one can show a conjecture of Goncharov stating that $\sum_{i=1}^{5} I_{31}(V(x,y),z)$, where the $V_i(x,y)$ denote the five term relation for the dilogarithm, can be expressed in terms of Li_4 only. A solution to this conjecture was given in [**11**], Thm 17. Therefore we conclude that we can attach to each configuration in $C_8(4)$ a linear combination of Li_4 terms, i.e. a map $f_8(4)$ completing the left hand square. While an explicit version for it is still elusive, with Charlton and Radchenko we give a complicated map in [**4**], a "quadruple ratio" (consisting of 368 orbits under the symmetric group $\mathcal{S}_8$), which constitutes a close cousin. A similar (non-explicit) result has been given by Goncharov and Rudenko in [**17**] in their proof of Zagier's Conjecture in weight 4. Moreover, with Radchenko we have obtained partial results pertaining to certain *degenerate* configurations which allow us to define $f_8(4)$ explicitly in those situations. In particular, for a specific configuration in $C_8(4)$ with three parameters a, b, c, consisting of the four standard basis vectors $\mathbf{e}_i$ $(i = 1, \ldots, 4)$ together with $(1,-1,0,0)$, $(0,1,-1,0)$, $(0,0,1,-1)$ and $(-b,c,-1,a)$ the computer found a matching Li_4–expression with 1260 terms (and integer coefficients of modulus $\leqslant 318$).

Acknowledgments

We want to thank Sasha Goncharov for providing an early draft of [**14**], in which he already gave partial answers, thereby initiating the investigation, and Danylo Radchenko for providing his Mesa routine and for many helpful discussions. In order to derive and check the results given in this paper we used Goncharov's symbol for iterated integrals as it was implemented in Mathematica by Duhr [**8**] for our joint paper [**9**]. The author is very grateful for the hospitality in particular of the Max-Planck-Institute and also the Hausdorff-Institute for Mathematics in Bonn where the maps and proofs were obtained during our stay in 2013 and subsequently

[2] A couple of misprints in his formula were originally corrected in collaboration with Duhr, and treated more thoroughly by Charlton in his thesis [**3**], Thm. 5.2.5 and Rem. 5.2.6, and in [**4**].

refined in 2015 and 2018. He also would like to thank the anonymous referee for flagging a number of issues.

Appendix A. Functional equations

In this section we collect and prove a couple of functional equations for the dilogarithm needed in the proof that the centre square of (2.1) commutes.

LEMMA A.1. *There are the following symmetries among the terms* $(\dots)_2$*:*

(1) $(12|3456)_2$ *is symmetric with respect to the permutation* $(1\,2)$ *and anti-symmetric with respect to permutations of 3,4,5,6, so e.g.* $(12|3456)_2 = -(12|4356)_2 = (12|4536)_2$ *etc.*

(2) $(1|234567)_2 = (1|342675)_2 = (1|423756)_2 = -(1|324657)_2 = -(1|432765)_2 = -(1|243576)_2$.

(3) *The (projected) triple ratio* $(1|234567)$ *is a product of two (doubly) projected cross ratios in the following three ways:*

$$(1|234567) = \frac{(12|3457)}{(13|2467)} = \frac{(13|4265)}{(14|3275)} = \frac{(14|2376)}{(12|4356)}.$$

(4) *We have a five term relation involving two triple ratios and two projected cross ratios:*

$$(1|234567)_2 - (1|237564)_2 = -\Big(\frac{(12|3574)}{(13|2674)}\Big)_2 + (13|2764)_2 - (12|3754)_2.$$

[Proof: Write $x = (13|2764)$ *and* $y = (12|3754)$*, then the five terms are expressed as follows:* $(1|234567) = \frac{(12|3457)}{(13|2467)} = \frac{1-x^{-1}}{1-y^{-1}}$, $(1|237564) = \frac{(12|3754)}{(13|2764)} = \frac{y}{x}$ *and* $\frac{(12|3574)}{(13|2674)} = \frac{1-y}{1-x}$*, so the above reduces to the five term relation in the form*

$$\Big(\frac{1-x^{-1}}{1-y^{-1}}\Big)_2 - \Big(\frac{y}{x}\Big)_2 = -\Big(\frac{1-y}{1-x}\Big)_2 + (x)_2 - (y)_2.]$$

Furthermore, we note that the terms of the form $(a\,b\,c\,d)$ *occurring in a tensor or wedge factor are a shorthand for a* 4×4*-determinant and hence are, up to 2-torsion, invariant under permutation.*

FE A.2. *Variants of the five term relation.*

$$2\,\mathrm{Alt}_{\langle(6\,7),(1\,2\,3)\rangle}(56|1237)_2 = 3\,\mathrm{Alt}_{\langle(6\,7),(1\,2\,3)\rangle}(52|3716)_2. \tag{A.1}$$

$$\mathrm{Alt}_{(6\,7)}(5|236147)_2 = -(53|2746)_2 + (52|3716)_2. \tag{A.2}$$

FE A.3.

$$\mathrm{Alt}_{(1\,2)(4\,5)(6\,7)}\,\mathrm{Alt}_{\langle(1\,4),(4\,7),(2\,5)(5\,6)\rangle}(1|234567)_2 = 0. \tag{A.3}$$

(We skip the proof which is similar to but easier than the ones for FE A.4 and FE A.5 below.)

FE A.4.

$$\mathrm{Alt}_{(3\,5)(4\,6)}\,\mathrm{Alt}_{\langle(1\,2),(3\,4),(5\,6)\rangle}\big((13|2456)_2 + (3|614527)_2 + (5|236147)_2\big) = 0.$$

PROOF. The second term $(3|614527)_2$ can be written via Lemma A.1 (2), as $(3|146275)_2 = \left(\frac{31|4625}{34|1675}\right)_2$ and invoking Lemma A.1 (4) we find

$$\mathrm{Alt}_{(5\,6)}(3|146275)_2 = -\left(\frac{31|4256}{34|1756}\right)_2 + (34|1576)_2 - (31|4526)_2.$$

Similarly, the third term $(5|236147)_2$ can be rewritten under the alternation sign as $-(3|254167)_2$ (use the permutation $(3\,5)(4\,6)$) and hence via Lemma A.1 (2) as $-(3|425716)_2$ and we get an analogous five term relation

$$-\,\mathrm{Alt}_{(5\,6)}(3|425716)_2 = \left(\frac{34|2765}{32|4165}\right)_2 - (32|4615)_2 + (34|2675)_2.$$

Now the first terms on the right in the above two five term equations turn out to be negatives of each other, hence they cancel in the sum. (One can also check that the Alt-orbit of that first term on the right vanishes as it is invariant under the odd permutation $(1\,2)$: write it out as a product of determinants to get $\frac{|3126|\cdot|3475|}{|3125|\cdot|3476|}$.)

Therefore the original sum can be replaced by the sum

$$\mathrm{Alt}_{(3\,5)(4\,6)}\,\mathrm{Alt}_{\langle(1\,2),(3\,4),(5\,6)\rangle}$$
$$\left((13|2456)_2 + \frac{1}{2}\big((34|1576)_2 - (31|4526)_2\big) + \frac{1}{2}\big(-(32|4615)_2 + (34|2675)_2\big)\right).$$

But $\mathrm{Alt}_{(1\,2)}(34|1576)_2 = -\,\mathrm{Alt}_{(1\,2)}(34|2576)_2 = \mathrm{Alt}_{(1\,2)}(34|2675)_2$; for the second equality we use the symmetries in Lemma A.1 (1) and similarly $\mathrm{Alt}_{(1\,2)}(31|4526)_2 = \mathrm{Alt}_{(1\,2)}(32|4615)_2$, so we can combine the five summands to three terms, all with coefficient ±1, in fact to

$$\big((13|2456)_2 + (34|1576)_2 - (31|4526)_2\big)$$

but the first and last of these cancel in view of Lemma A.1 (1) while the middle one is invariant under the odd permutation $(3\,4)$ and hence its Alt-orbit vanishes. In summary, the original sum indeed vanishes as claimed. □

FE A.5.

$$\mathrm{Sym}_{(4\,5)}\,\mathrm{Alt}_{\langle(1\,2),(2\,3),(6\,7)\rangle}\big((35|1247)_2 - (46|1237)_2 - (5|236147)_2 - (5|236417)_2\big) = 0.$$

PROOF. Adding (A.2) to its variant where 1 and 4 are swapped we get

$$-(5|236147)_2 - (5|236417)_2 = -\frac{1}{2}(53|2746)_2 + \frac{1}{2}(52|3716)_2 - \frac{1}{2}(53|2716)_2 + \frac{1}{2}(52|3746)_2$$

and so, alternating with respect to $(2\,3)$ we find

$$\mathrm{Alt}_{(2\,3)}\big(-(5|236147)_2 - (5|236417)_2\big) = \mathrm{Alt}_{(2\,3)}\big(-(53|2746)_2 - (53|2716)_2\big).$$

Furthermore, antisymmetrising with respect to $(4\,5)$ and then invoking (A.1) we find

$$\mathrm{Alt}_{(4\,5)}\big(-(46|1237)_2\big) = \mathrm{Alt}_{(4\,5)}\big((56|1237)_2\big) = \frac{3}{2}\,\mathrm{Alt}_{(4\,5)}(53|2716)_2,$$

so we can now write the combination in question as

$$\begin{aligned}
&\mathrm{Sym}_{(4\,5)}\,\mathrm{Alt}_{\langle(1\,2),(2\,3),(6\,7)\rangle}\big((35|1247)_2-(46|1237)_2\\
&\qquad\qquad\qquad\qquad -(5|236147)_2-(5|236417)_2\big)\\
&=\mathrm{Sym}_{(4\,5)}\,\mathrm{Alt}_{\langle(1\,2),(2\,3),(6\,7)\rangle}\big((35|1247)_2+\frac{3}{2}(53|2716)_2\\
&\qquad\qquad\qquad\qquad -(53|2746)_2-(53|2716)_2\big).
\end{aligned}$$

We combine the second and fourth term to $\frac{1}{2}(53|2716)_2$ and use that the operator $\mathrm{Alt}_{\langle(1\,2),(2\,3),(6\,7)\rangle}$ antisymmetrises over $(1\,2)$ and also over $(6\,7)$, so under the alternation sign we can replace $(35|1247)_2$ by $\frac{1}{2}(35|1247)_2-\frac{1}{2}(35|1246)_2$ and $-(53|2746)_2$ by $-\frac{1}{2}(53|2746)_2+\frac{1}{2}(53|1746)_2$. Hence we obtain that the expression under the alternation sign reduces to a standard five term relation (with fixed projection points 3 and 5) as follows, with the obvious new ad hoc notation $(53|27164)$,

$$0=\text{`}\partial\big((53|27164)\big)\text{'}=(53|7164)_2-(53|2164)_2+(53|2764)_2-(53|2714)_2+(53|2716)_2,$$

thereby proving the statement. □

References

[1] S. J. Bloch, *Higher regulators, algebraic K-theory, and zeta functions of elliptic curves*, CRM Monograph Series, vol. 11, American Mathematical Society, Providence, RI, 2000, DOI 10.1090/crmm/011. MR1760901

[2] J.-L. Cathelineau, *Homologie du groupe linéaire et polylogarithmes (d'après A. B. Goncharov et d'autres)* (French, with French summary), Astérisque **216** (1993), Exp. No. 772, 5, 311–341. Séminaire Bourbaki, Vol. 1992/93. MR1246402

[3] S. Charlton, *Identities arising from coproducts on multiple zeta values and multiple polylogarithms*, PhD Thesis Durham (2016).

[4] S. Charlton, H. Gangl, and D. Radchenko, *Explicit formulas for grassmannian polylogarithms*, arXiv:1909.13869 (2019), `https://arxiv.org/abs/1909.13869`, available at `1909.13869`.

[5] N. Dan, *Sur la conjecture de Zagier pour $n=4$*, arXiv:0809.3984 (2008), `http://arxiv.org/abs/0809.3984v1`.

[6] V. Del Duca, C. Duhr, and V. A. Smirnov, *An analytic result for the two-loop hexagon Wilson loop in $\mathcal{N}=4$ SYM*, J. High Energy Phys. **3** (2010), 099, 17, DOI 10.1007/JHEP03(2010)099. MR2653448

[7] P. Deligne and A. B. Goncharov, *Groupes fondamentaux motiviques de Tate mixte* (French, with English and French summaries), Ann. Sci. École Norm. Sup. (4) **38** (2005), no. 1, 1–56, DOI 10.1016/j.ansens.2004.11.001. MR2136480

[8] C. Duhr and F. Dulat, *PolyLogTools—polylogs for the masses*, J. High Energy Phys. **8** (2019), 135, 56, DOI 10.1007/jhep08(2019)135. MR4013407

[9] C. Duhr, H. Gangl, and J. R. Rhodes, *From polygons and symbols to polylogarithmic functions*, J. High Energy Phys. **10** (2012), 075, front matter + 77, DOI 10.1007/JHEP10(2012)075. MR3033860

[10] C. Dupont, *Progrès récents sur la conjecture de Zagier et le programme de Goncharov [d'après Goncharov, Rudenko, Gangl,...]* (French), Astérisque **430** (2021), Exp. No. 1176, 295–343. Séminaire Bourbaki. Vol. 2019/2021. Exposés 1166–1180. MR4439195

[11] H. Gangl, *Multiple polylogarithms in weight 4*, arXiv:1609.05557 (2016), available at `1609.05557`.

[12] A. B. Goncharov, *Polylogarithms and motivic Galois groups*, Motives (Seattle, WA, 1991), Proc. Sympos. Pure Math., vol. 55, Amer. Math. Soc., Providence, RI, 1994, pp. 43–96, DOI 10.1090/pspum/055.2/1265551. MR1265551

[13] A. B. Goncharov, *Geometry of configurations, polylogarithms, and motivic cohomology*, Adv. Math. **114** (1995), no. 2, 197–318, DOI 10.1006/aima.1995.1045. MR1348706

[14] A. B. Goncharov, *Special values of Hasse-Weil L-functions and generalized Eisenstein-Kronecker series*, 1996. unpublished notes.

[15] A. B. Goncharov, *Geometry of the trilogarithm and the motivic Lie algebra of a field*, Regulators in analysis, geometry and number theory, Progr. Math., vol. 171, Birkhäuser Boston, Boston, MA, 2000, pp. 127–165. MR1724890

[16] A. B. Goncharov, M. Spradlin, C. Vergu, and A. Volovich, *Classical polylogarithms for amplitudes and Wilson loops*, Phys. Rev. Lett. **105** (2010), no. 15, 151605, 4, DOI 10.1103/PhysRevLett.105.151605. MR2734443

[17] A. B. Goncharov and D. Rudenko, *Motivic correlators, cluster varieties and Zagier's conjecture on $\zeta_F(4)$*, arXiv:1803.08585 (2018), `http://arxiv.org/abs/1803.08585`, available at `1803.08585`.

[18] A. A. Suslin, *K_3 of a field, and the Bloch group* (Russian), Trudy Mat. Inst. Steklov. **183** (1990), 180–199, 229. Translated in Proc. Steklov Inst. Math. **1991**, no. 4, 217–239; Galois theory, rings, algebraic groups and their applications (Russian). MR1092031

[19] D. Zagier, *Polylogarithms, Dedekind zeta functions and the algebraic K-theory of fields*, Arithmetic algebraic geometry (Texel, 1989), Progr. Math., vol. 89, Birkhäuser Boston, Boston, MA, 1991, pp. 391–430, DOI 10.1007/978-1-4612-0457-2_19. MR1085270

Mathematical Sciences & Computer Science Building, Upper Mountjoy, Durham University, DH1 3LE

Email address: `herbert.gangl@durham.ac.uk`

Contemporary Mathematics
Volume **842**, 2026
https://doi.org/10.1090/conm/842/16854

A finite regulator map for the Bloch group

Herbert Gangl and Maxim Kontsevich

Abstract. For a number field F and an N-th root of unity q, put $k_N = F(q)$ and $K_N = \sqrt[N]{F}$. We give two homomorphisms mapping an element $\sum_i [x_i]$ in the Bloch group of F to the quotient group $K_N^\times / k_N^\times (K_N^\times)^N$. The first one is $\prod_i f(x_i; q)$ with $f(a;q) = \prod_{j=1}^N (1-aq^j)^j$, whereas the second assigns $\mathrm{Trace}(M_N)^N / \det(M_N)$, where the $N \times N$-matrix M_N arises as a product of Morita equivalences relating central simple algebras associated to the terms x_i.

1. Motivation[1]

For odd $N \in \mathbb{N}$, choose an N-th root $y = 2^{1/N}$ of 2 and a primitive N-th root of 1, denoted $\zeta = \zeta_N$. Experiments show that we have the following conjectural identity

$$\prod_{j=1}^{N-1} (\zeta^j y - 1)^j \stackrel{?}{=} \Big(\frac{1}{N} \sum_{k=1}^{N} \zeta^{k^2} \sum_{i=1}^{N} \prod_{j=i}^{N-1} (\zeta^j y - 1) \Big)^N. \tag{1.1}$$

This identity is related to the well-known evaluation of the dilogarithm function $\mathrm{Li}_2(z) = \sum_{m>0} z^m/m^2$ at the point $\frac{1}{2}$, given by

$$\mathrm{Li}_2\Big(\frac{1}{2}\Big) = \frac{\pi^2}{12} - \frac{1}{2}\log^2(2).$$

There are only six further non-trivial evaluations of the dilogarithm known (cf. [**6**])[2], half of which involving the golden ratio $\rho = \frac{\sqrt{5}-1}{2}$, stating that

$$\mathrm{Li}_2(\rho) = \frac{\pi^2}{10} - \log^2(\rho),$$

and the latter is related to an identity for ρ similar to (1.1).

The purpose of this note is to explain how such canonical choices for these N-th roots arise naturally. This leads to a suggestion of a map which might be called "étale dilogarithm".

2020 *Mathematics Subject Classification.* Primary 11R70, 12G05, 57K16.

[1] *Disclaimer:* This paper was essentially written in the early 2000s. It ought to be read in conjunction with the problem stated by the second named author in 2006 [**3**], and some of its contents have been superseded by related work of Calegari, Garoufalidis and Zagier [**2**]. Speculations on the asymptotics, being outdated, have been removed. For a recent far more general prediction see [**4**], §8.

[2] The coefficient $\frac{1}{2}$ of $\log(\frac{1+\sqrt{5}}{2})$ in the evaluation for $\mathrm{Li}_2(\frac{-1-\sqrt{5}}{2})$ in loc.cit. should read -1.

2. Outline of results

Let F be a number field. It was envisaged by Bloch [**1**] and proved by Suslin [**5**] that the "indecomposable" algebraic K-group $\mathrm{K}_3^{\mathrm{ind}}(F)$, defined as the quotient of Quillen's K_3-group $\mathrm{K}_3(F)$ by Milnor's one $\mathrm{K}_3^{\mathrm{M}}(F)$, has a rather explicit avatar, at least after tensoring with $\mathbb{Q}$, as the Bloch group

$$B(F) = \ker \partial_2 / \langle \text{specialisations of the 5-term relation} \rangle$$

with $\partial_2 : \mathbb{Z}[F] \to \bigwedge^2 F^\times$, where $\bigwedge^2 A = A \otimes A / \langle a \otimes b + b \otimes a \rangle$, defined on generators by $[x] \mapsto (1-x) \wedge x$, and (one of the variants of) the 5-term relation is given by

$$[x] + [y] + \Big[\frac{1-y}{x}\Big] + \Big[\frac{x+y-1}{xy}\Big] + \Big[\frac{1-x}{y}\Big]. \tag{2.1}$$

More precisely, a theorem of Suslin shows that $B(F)$ differs from $\mathrm{K}_3^{\mathrm{ind}}(F)$ by a torsion group which for a number field F consists essentially of its roots of unity, a finite subgroup of $F^\times$, so one does not lose much by restricting oneself to $B(F)$.

We describe, for any $N \in \mathbb{N}$, a map

$$L_2^{\text{ét}} = L_2^{\text{ét},N} : B(F)/\text{torsion} \to K_N^\times / k_N^\times (K_N^\times)^N,$$

where $k_N = F(\sqrt[N]{1})$ and $K_N = \sqrt[N]{F}$ (typically a huge Kummer extension).

Using this fact, we attach to each element $\xi = \sum_i [x_i] \in \mathbb{Z}[F]$ an expression $L_2^{\text{ét}}(\xi)$ in K_N in such a way that, given $\xi \in \ker \partial_2$, together with a compatible choice of N-th roots of the x_i and $1 - x_i$, we expect that $L_2^{\text{ét}}(\xi)$ is essentially an N-th power in K_N. We want to think of the expression $L_2^{\text{ét}}(\xi)^{1/N}$ as an "étale regulator value" in motivic weight 2, hence related to the dilogarithm.

2.1. Our results. The map factors through a quotient $K_N^\times / k_N^\times (K_N^\times)^N$, because the five term relation for the dilogarithm is mapped to $(K_N^\times)^N$ (Theorem 3.1 below).

In fact, we get two maps which should be closely related.

1. The more straightforward map simply takes $\xi = \sum_{i \in I} [x_i]$ and attaches, for given N and a "compatible choice" of N-th roots $x_i^{1/N}$, together with $q = 1^{1/N}$ a primitive N-th root of 1, the expression $\Phi(\xi, q)$ where we define
$$\Phi(\xi, q) = \prod_{i \in I} \prod_{j=1}^{N} (1 - x_i^{1/N} q^j)^j.$$
2. The second map involves Azumaya algebras. In fact, as we are working over a field, these can be realised as central simple algebras. For any $\xi = \sum_i [x_i]$, we construct a sequence of algebras and Morita equivalences between them, where each of the latter corresponds to precisely one of the terms $[x_i]$.
3. If $\xi \in \ker \partial_2$ then the assignment of algebras is subject to the requirement that the final algebra in this sequence of isomorphic ones *agrees* with the one we started out with. The composition of the Morita equivalences involved provides a matrix M_N (of size $N \times N$) and we then put $\xi_N = \mathrm{Trace}(M_N)^N / \det(M_N)$. We show that, for the five term relation, the latter quotient ξ_N is an N-th power (cf. the proof of Theorem 3.1).
4. We illustrate the construction for a non-trivial element in the Bloch group for the field $\mathbb{Q}(\sqrt{-7})$.

We note that Calegari, Garoufalidis and Zagier [**2**] have independently studied a very similar map to $\Phi(\xi, q)$, and they proved a beautiful application to "Nahm's Conjecture" relating torsion elements in $B(F)$ to modular q-hypergeometric series.

3. A category $\mathcal{C}$ of algebras reflecting generators and relations in K_2 of a field

3.1. The category $\mathcal{C}$. We will work with the following tensor (i.e. symmetric monoidal) category $\mathcal{C}$.

Its **objects** $Ob(\mathcal{C})$ are given as follows:
for each pair $(x, y) \in \mathbb{C}^\times \times \mathbb{C}^\times$ there is an object $\mathcal{E}_{x,y} \in Ob(\mathcal{C})$;
for each sequence $\big((x_1, y_1), \dots, (x_m, y_m)\big)$ $(m \geq 0)$ there is an object

$$\mathcal{E}_{x_1,y_1} \otimes \cdots \otimes \mathcal{E}_{x_m,y_m}.$$

Its **Hom sets** are freely generated by the following three types mimicking the generators and relations in the Milnor K-group K_2 of a suitable field.

(1) For any $m \geq 0$ there is a certain group Γ_m, given as an extension of the symplectic group $Sp(2m, \mathbb{Z})$ by $\mathbb{Z}^{2m}$, i.e.

$$1 \longrightarrow \mathbb{Z}^{2m} \longrightarrow \Gamma_m \overset{pr_m}{\longrightarrow} Sp(2m, \mathbb{Z}) \longrightarrow 1,$$

with the following property: $\forall \gamma \in \Gamma_m$, $\forall x_1, y_1, \dots, x_m, y_m \in \mathbb{C}^\times$, there is an isomorphism

$$i_{\gamma;x_1,y_1,\dots,x_m,y_m} \; : \; \bigotimes_{i=1}^{m} \mathcal{E}_{x_i,y_i} \overset{\sim}{\longrightarrow} \bigotimes_{i=1}^{m} \mathcal{E}_{x'_i,y'_i},$$

where $pr_m(\gamma) \in Sp(2m, \mathbb{Z})$ maps $(x_1, y_1, \dots, x_m, y_m) \in \big(\mathbb{C}^\times\big)^{2m} = \mathbb{Z}^{2m} \otimes \mathbb{C}^\times$ to $(x'_1, y'_1, \dots, x'_m, y'_m) \in \big(\mathbb{C}^\times\big)^{2m}$.

The extension group Γ_m is realised as the group of automorphisms of the algebra

$$\mathbb{C}(q)\langle \alpha_1^{\pm}, \dots, \alpha_{2m}^{\pm}\rangle \big/ I$$

over the function field $\mathbb{C}(q)$, where the ideal I is generated by the symplectic relations $\alpha_{2i-1}\alpha_{2i} = q\alpha_{2i}\alpha_{2i-1}$ $(i = 1, \dots, m)$, together with the commutation relations $\alpha_j\alpha_k = \alpha_k\alpha_j$ for all the other pairs (j, k). More precisely, we associate to $(a_i)_i \in \mathbb{Z}^{2m}$ and $(c_{ij})_{i,j} \in Sp(2m, \mathbb{Z})$ the automorphism defined by mapping (we use the notation $\rightarrow$ to indicate the ordered product of the non-commuting variables α_j)

$$\alpha_i \mapsto q^{a_i} \prod_{j=1,\dots,2m}^{\rightarrow} \alpha_j^{c_{ij}}, \quad (i = 1, \dots, 2m).$$

(2) We impose an "antisymmetry" isomorphism, as well as its dual relation

$$\forall x, y \in \mathbb{C}^\times \qquad \tau_{x,y} : \mathcal{E}_{x,y} \otimes \mathcal{E}_{y,x} \longrightarrow \mathbb{1}$$
$$\tau^\vee_{x,y} : \mathbb{1} \longrightarrow \mathcal{E}_{x,y} \otimes \mathcal{E}_{y,x}.$$

(3) Finally, we impose specific isomorphisms reflecting "Steinberg relations"

$$\forall x, y \in \mathbb{C}^\times, \quad x \neq 1, \qquad \sigma_{x,y} \; : \; \mathcal{E}_{x,y} \overset{\sim}{\longrightarrow} \mathcal{E}_{x,(1-x)y}.$$

3.2. Mapping $\mathcal{C}$ to the category of (finitely generated) associative algebras. The above tensor category $\mathcal{C}$ can now be "realised", for any integer $N \geq 1$ and any choice of primitive N-th root of unity $q \in \mathbb{C}^\times$, in the category of finitely generated associative $\mathbb{C}$–algebras $\mathcal{A}$. We recall that the objects of $\mathcal{A}$ are given by

$$Ob(\mathcal{A}) : \text{a finitely generated associative } \mathbb{C}\text{–algebra } A$$
$$(\text{e.g. } A \cong Mat(k \times k, \mathbb{C}), \text{ for some } k \geq 1),$$

and its Hom sets are

$$Hom_{\mathcal{A}}(A, B) = \{ \text{ isomorphism classes of pairs } (M, m),$$
$$\text{where } M \in A\text{-}Mod\text{-}B \text{ is an } (A, B)\text{-bimodule, together with some}$$
$$\text{distinguished element } m \in M \text{ realising Morita equivalence}\}.$$

So if $A \xrightarrow{\sim} Mat(k_1 \times k_1, \mathbb{C})$ and $B \xrightarrow{\sim} Mat(k_2 \times k_2, \mathbb{C})$, then $M \cong \mathbb{C}^{k_1} \otimes \mathbb{C}^{k_2}$ (the isomorphism is defined up to a scalar in $\mathbb{C}^\times$), and m is a $k_1 \times k_2$-matrix (again up to multiplying by a constant in $\mathbb{C}^\times$).

Also, the tensor $\otimes$ in $\mathcal{A}$ is the usual tensor product.

Now $\mathcal{A}$ contains a full subcategory $\mathcal{S}$ (Algebras, isomorphisms) whose objects are isomorphic to *matrix algebras* and whose [iso]morphisms are given as follows: in the case where $k_1 = k_2$ and m is invertible, a homomorphism class $[(M, m)] \in Hom_{\mathcal{A}}(A, B)$ gives rise to an isomorphism $A \to B$ where $a \in A \mapsto b \in B$ if $am = mb \in M$.

We map our category $\mathcal{C}$ above to $\mathcal{S} \subset \mathcal{A}$, for each N and q, via the following $\otimes$-functor:

$$F_q : \mathcal{C} \longrightarrow \mathcal{S} \subset \mathcal{A},$$

defined on objects by

$$\mathcal{E}_{x,y} \mapsto A_{x,y;q} := \mathbb{C}(q)\langle \xi, \eta \rangle / \langle \xi^N = x, \ \eta^N = y, \ \eta\xi = q\xi\eta \rangle,$$

and on the three types of morphisms, respectively, as follows.

(1) For each $\gamma \in \Gamma_m$, putting $(x'_1, y'_1, \dots, x'_m, y'_m) = pr_m(\gamma)(x_1, y_1, \dots, x_m, y_m)$, we have an isomorphism

$$\gamma_{m;x_1,y_1,\dots,x_m,y_m} : A_{x_1,y_1;q} \otimes \cdots \otimes A_{x_m,y_m;q} \cong A_{x'_1,y'_1;q} \otimes \cdots \otimes A_{x'_m,y'_m;q}.$$

(2) For any $x, y \in \mathbb{C}^\times$ we have a canonical isomorphism

$$A_{x,y;q} \xrightarrow{\sim} A^{\mathrm{op}}_{y,x;q}.$$

For any algebra A there are bimodules

$$M \in A \otimes A^{\mathrm{op}} - mod - \mathbb{C}, \quad m \in M, \qquad M' \in \mathbb{C} - mod - A^{\mathrm{op}} \otimes A, \quad m' \in M',$$

such that

$$M = M' \cong A, \qquad m = m' = \mathbb{1}.$$

(3) For any $x, y \in \mathbb{C}^\times$ we have an isomorphism of algebras

$$A_{x,y;q} \xrightarrow{\sim} A_{x,(1-x)y;q},$$
$$\xi, \eta \longrightarrow \xi, (1 - \xi)\eta.$$

3.3. Two equivalence relations on the morphisms in $\mathcal{C}$.

3.3.1. *The first equivalence relation* $\sim$. Using the $\otimes$-functor F_q above, we impose the following equivalence relation on morphisms in $\mathcal{C}$: if there are two isomorphisms

$$\varphi,\ \psi : \otimes_i \mathcal{E}_{x_i,y_i} \longmapsto \otimes_i \mathcal{E}_{x'_i,y'_i},$$

then we define

$$\varphi \sim \psi \quad \text{if} \quad F_q(\varphi) = F_q(\psi) \quad \forall N, q.$$

On the one hand, we get a (surjective) map to the second K-group of $\mathbb{C}$:

$$\begin{aligned} Ob(\mathcal{C}/\sim) &\longrightarrow \mathrm{K}_2(\mathbb{C}) \\ \otimes_i \mathcal{E}_{x_i,y_i} &\longmapsto \sum_i x_i \wedge y_i \in \bigwedge^2 \mathbb{C}^\times / J, \end{aligned}$$

where J denotes the subgroup generated by the $z \wedge (1-z)$ for $z \in \mathbb{C}^\times \setminus \{1\}$. Here we use the fact that we can also write a presentation of $\mathrm{K}_2(\mathbb{C})$ via wedge products rather than tensor products if we put $\bigwedge^2 \mathbb{C}^\times = \mathbb{C}^\times \otimes \mathbb{C}^\times / \langle x \otimes y + y \otimes x \rangle$.

Notice that all morphisms in $\mathcal{C}/\sim$ fiber over $\mathrm{K}_2(\mathbb{C})$.

On the other hand, if we more carefully keep track of the contributions to the wedge product, then for any pair of equivalent (sequences of) morphisms (φ, ψ) we attach an element in the Bloch group of $\mathbb{C}$. Indeed, we can view the composition $\psi^{-1} \circ \varphi$ as an endomorphism χ of $\otimes_i \mathcal{E}_{x_i,y_i}$, and F_q maps it to J, hence we can find $z_j \in \mathbb{C}$ such that χ can be realised as a composition of isomorphisms of type (3), replacing $\mathcal{E}_{z_j,w_j}$ by $\mathcal{E}_{z_j,(1-z_j)w_j}$, and those of type (2) swapping the arguments. Their difference has to vanish under ∂_2, i.e.

$$\sum_j [z_j] \in \ker\left(\partial_2 : \mathbb{Z}[\mathbb{C}] \longrightarrow \bigwedge^2 \mathbb{C}^\times\right).$$

3.3.2. *The second equivalence relation* $\approx$. But we also impose a stronger equivalence relation

$$\varphi \approx \psi \ \text{ provided } \ \varphi \sim \psi \ \textit{and} \text{ the associated element in the Bloch group is zero.}$$

There is a natural (projection) map

$$\mathcal{C}/\approx \longrightarrow \mathcal{C}/\sim,$$

and the Hom set $Hom_{\mathcal{C}/\approx}$ for the stronger relation fibres over the Hom set $Hom_{\mathcal{C}/\sim}$ of the weaker one, i.e. we can think of the former as a principal homogeneous space over the latter with fibre being the Bloch group of $\mathbb{C}$.

3.4. Explicit matrices for the Azumaya algebras. Now given x_i and y_i $(i = 1, \ldots, m)$, we choose, for fixed $N > 0$ and q an N-th root of unity, N-th roots $\sqrt[N]{x_i}$, $\sqrt[N]{y_i}$, compatibly with multiplicative relations, i.e. $\prod_i x_i^{a_i} y_i^{b_i} = 1$ entails $\prod_i \sqrt[N]{x_i}^{a_i} \sqrt[N]{y_i}^{b_i} = 1$. Then we identify

$$A_{x_i,y_i;q} \xrightarrow{\sim} \text{Matrix algebra } Mat(N \times N, \mathbb{C}) \text{ with}$$

$$\xi_i = x_i^{1/N} \begin{pmatrix} 1 & & & \\ & q & & \\ & & q^2 & \\ & & & \ddots \end{pmatrix}, \qquad \eta_i = y_i^{1/N} \begin{pmatrix} & & & 1 \\ 1 & & & \\ & 1 & & \\ & & \ddots & \end{pmatrix}.$$

In this case, we can associate to the isomorphism class $[(Mat(N\times N,\mathbb{C}),m)]$ certain explicit matrices, e.g. for a morphism of type (3) above we have

$$M_N=(1-x_i^{1/N})\begin{pmatrix}1 & & & \\ & (1-qx_i^{1/N}) & & \\ & & (1-qx_i^{1/N})(1-q^2x_i^{1/N}) & \\ & & & \ddots\end{pmatrix}(q^{ij})_{i,j},$$

where the matrix $(q^{ij})_{i,j}$ is a Vandermonde matrix.

CONJECTURE. *If two morphisms in $\mathcal{C}$ are equivalent under $\approx$ then the products of the explicit matrices above are equal (up to an N-th power).*

3.5. Trace and determinant. Finally we can attach, to any $\varphi\in End_{\mathcal{C}}(\otimes\,\mathcal{E}_{x_i,y_i})$ and an N-th root of unity q, a matrix M_N attached in a similar way to the above, i.e., multiplying the associated matrices for the individual moves of type (3) in a realization of an endomorphism of some $\otimes_i\mathcal{E}_{x_i,y_i}$, and hence we associate numerical invariants, the trace $\mathrm{Tr}(M_N)\in K_N$ as well as the determinant $\mathrm{Det}(M_N)$.

3.6. Well-definedness of the map. We first establish that the relations in the Bloch group map to zero in the quotient. More precisely, with the above notation we can show the following.

THEOREM 3.1. *Let F be a number field, $N\geq 1$ and let q be any choice of a primitive N-th root of unity. For the five term relation (2.1), written as $\sum_i[t_i]$, the trace $\mathrm{Tr}(M_N)$ of the associated matrix M_N is a number $Z_N\in F\big(q,\{t_i^{1/N},(1-t)_i^{1/N}\}_i\big)$, and $\mathrm{Det}(M_N)=(NZ_N)^N$.*

PROOF. For each $x,y\in F^\times$ we consider the cyclic algebra $A_{x,y;q}$ generated over F by ξ, η, subject to the relations $\xi^N=x$, $\eta^N=y$ and $\eta\xi=q\xi\eta$ (q commuting with both ξ and η).

If furthermore $y\neq 1$ then there is an algebra isomorphism

$$A_{x,y;q}\overset{\simeq}{\longrightarrow}A_{y,\frac{1-y}{x};q},\tag{3.1}$$

given by

$$(\xi,\eta)\longmapsto(\eta,\xi^{-1}(1-\eta)\cdot q^{-1})=:(\tilde\xi,\tilde\eta),\tag{3.2}$$

whose inverse is $(\tilde\xi,\tilde\eta)\mapsto\big((1-\tilde\xi)\tilde\eta^{-1}\cdot q^{-1},\tilde\xi\big)$.

In keeping with the above conventions, we put $\tilde x=y$ and $\tilde y=\frac{1-y}{x}$ and then identify $A_{x,y;q}$ with $A_{\tilde x,\tilde y;q}$ by checking that the latter has generators $\tilde\xi$, $\tilde\eta$ subject to $\tilde\xi^N=\tilde x$, $\tilde\eta^N=\tilde y$ and $\tilde\eta\tilde\xi=q\tilde\xi\tilde\eta$. Indeed, $\tilde\xi^N=\eta^N=y=\tilde x$ and

$$\begin{aligned}\tilde\eta^N&=\underbrace{\xi^{-1}(1-\eta)\cdots\xi^{-1}(1-\eta)}_{N\text{ factors}}\cdot q^{-N}=\xi^{-N}(1-q^{N-1}\eta)\cdots(1-q^{-1}\eta)(1-\eta)\\&=\xi^{-N}(1-\eta^N)=x^{-1}(1-y)=\tilde y.\end{aligned}$$

Here we have used that $\xi^{-1}f(\eta)=f(q\eta)\xi^{-1}$ and $f(\eta)\xi^{-1}=\xi^{-1}f(q^{-1}\eta)$ for any polynomial $f(\eta)$ in η. Finally, since $\xi^{-1}\eta=q\eta\xi^{-1}$, we have

$$\tilde\eta\tilde\xi=\xi^{-1}(1-\eta)\eta\cdot q^{-1}=\xi^{-1}\eta(1-\eta)\cdot q^{-1}=q\,\eta\,\xi^{-1}(1-\eta)\cdot q^{-1}=q\tilde\xi\tilde\eta.$$

The automorphism of $\mathbb{Q}(x,y)$ given by $(x,y)\mapsto(y,\frac{1-y}{x})$ has order 5 and one obtains an automorphism of $A_{x,y;q}$ by composing the respective five isomorphisms of the type in (3.1). By choosing the N-th roots of $(t_i)_i = \left(x, y, \frac{1-y}{x}, \frac{x+y-1}{xy}, \frac{1-x}{y}\right)$ compatibly, we find that this composition equals the identity.

LEMMA 3.2. *Choose* $\tau_i = \sqrt[N]{t_i}$ *such that* $\tau_{i-1}\tau_{i+1} = (1-\tau_i)q^{-1}$ ($i \mod 5$). *Then the product of the corresponding five algebra isomorphisms* $\varphi_i : A_{t_{i-1},t_i;q} \xrightarrow{\sim} A_{t_i,t_{i+1};q}$ *equals the identity.*

PROOF. Writing the terms τ_i in sequence, there are four neighbouring terms that are immediate from (3.2) and its inverse, giving

$$(\tau_{i-2}, \tau_{i-1}, \tau_i, \tau_{i+1}) = \left((1-\xi)\eta^{-1}q^{-1}, \, \xi, \, \eta, \, \xi^{-1}(1-\eta)q^{-1}\right).$$

It remains to check that $\tau_{i-3} = \tau_{i+2}$. Indeed,

$$\begin{aligned}\tau_{i-3} &= \left(1-(1-\xi)\eta^{-1}q^{-1}\right)\xi^{-1}q^{-1} = \left(1-q^{-1}\eta^{-1}+q^{-1}\xi\eta^{-1}\right)\xi^{-1}\eta^{-1} \\ &= q^{-1}(\xi^{-1} - q^{-1}\eta^{-1}\xi^{-1} + q^{-1}\xi\,\underbrace{\eta^{-1}\xi^{-1}}_{=q\xi^{-1}\eta^{-1}}),\end{aligned}$$

which indeed agrees with

$$\begin{aligned}\tau_{i+2} &= \eta^{-1}\left(1-\xi^{-1}(1-\eta)q^{-1}\right)q^{-1} = \eta^{-1}\left(1-q^{-1}\xi^{-1}+q^{-1}\xi^{-1}\eta\right)q^{-1} \\ &= q^{-1}(\eta^{-1} - q^{-1}\eta^{-1}\xi^{-1} + \underbrace{q^{-1}\eta^{-1}\xi^{-1}}_{=\xi^{-1}\eta^{-1}}\eta).\end{aligned}$$

□

Since we want to attach a numerical invariant to a given 5-term relation, we consider matrix representations for the associated algebras. A typical one is suggested by the commutation relation $\eta\xi = q\xi\eta$. Define

$$\rho_{x,y;q} : A_{x,y;q} \to Mat(N \times N, F') = End(V_{x,y;q})$$

for some extension field F' of F and some N-dimensional F'-vector space $V_{x,y;q}$, by choosing a basis $(e_j)_j \simeq \mathbb{Z}/N\mathbb{Z}$ and, after a further choice of N-th roots $x^{1/N}$, $y^{1/N}$, we can take the endomorphism

$$\xi : e_j \mapsto q^j x^{1/N} e_j, \qquad \eta : e_j \mapsto y^{1/N} e_{j+1}$$

(one checks immediately the commutation relation above, as well as $\xi^N = x \cdot Id$, $\eta^N = y \cdot Id$).

We can think of the five rational functions t_i arising from the 5-cycle above as coordinates on the moduli space $\mathcal{M}_{0,5}$ of 5 points on $\mathbb{P}^1$ up to the standard action of $PGL_2(K)$. (Note that the constraints on the t_i are given by $1 - t_i = t_{i-1}t_{i+1}$ (with $i \mod 5$).)

More generally, in the same way we can impose coordinates on the variety $\sqrt[N]{\mathcal{M}_{0,5}}$ with coordinate functionals s_i ($i \bmod 5$) such that $s_i \in F^\times \setminus \mu_N$ (μ_N denoting the N-th roots of unity) and subject to the constraints $s_{i-1}^N s_{i+1}^N = 1 - s_i^N$. (One should think that s_i^N corresponds to t_i above.) Thus we can view $\sqrt[N]{\mathcal{M}_{0,5}}$ as an N^5-fold covering of $\mathcal{M}_{0,5}$, with Galois group $\left(\mathbb{Z}/N\mathbb{Z}\right)^5$ acting by $s_i \mapsto q^{a_i}s_i$, $a_i \in \mathbb{Z}/N\mathbb{Z}$ ($i = 1, \ldots, 5$). For each choice of such $(s_i)_i$ we obtain a representation $\rho_{s_i,s_{i+1}}$ in the above way, where the basis vectors $e_j^{(i,i+1)} \in V_{i,i+1}$ are acted upon as follows ($j \mod N$)

$$\tau_i : e_j^{(i,i+1)} \longmapsto q^j s_i e_j^{(i,i+1)}, \qquad \tau_{i+1} : e_j^{(i,i+1)} \longmapsto s_{i+1} e_{j+1}^{(i,i+1)}.$$

Now identify explicitly each isomorphism $\phi_i : V_{i-1,i} \overset{\sim}{\longrightarrow} V_{i,i+1}$ since then the composition $\prod_{i \bmod 5} \phi_i$ automatically acts as a scalar (matrix) $Z = Z(s_1, \ldots, s_5)$.

Denote the basis vectors $e_j^{(i-1,i)}$ on the left by ℓ_j and the $e_j^{(i,i+1)}$ on the right by r_j. With the notations above, we find on the left

$$\xi : \ell_j \longmapsto q^j x^{1/N} \ell_j, \qquad \eta : \ell_j \longmapsto y^{1/N} \ell_{j+1},$$

while on the right we get (cf. above)

$$\tilde{\xi} = \eta : r_j \longmapsto q^j y^{1/N} r_j, \qquad \tilde{\eta} = q^{-1}\xi^{-1}(1-\eta) : r_j \longmapsto \Big(\frac{1-y}{x}\Big)^{1/N} r_{j+1}.$$

It is clear that r_j is an eigenvector of η with eigenvalue $q^j y^{1/N}$. But the same is true for $\sum_i q^{-ij}\ell_i$. (For the proof, shift the summation index and use $q^{-(i-1)j} = q^j \cdot q^{-ij}$.) Therefore we make the Ansatz

$$r_j = c_j \cdot \sum_i q^{-ij} \ell_i =: c_j \cdot F_j$$

for some constant $c_j \in \sqrt[N]{F}$; here F_j should be thought of as a "Fourier transform". Then we use the transformation property under $q^{-1}\xi^{-1}(1-\eta) : c_j F_j \longmapsto c_{j+1}F_{j+1}$ which we know produces a factor $\big(\frac{1-y}{x}\big)^{1/N}$. This implies, after normalizing $c_0 = 1$, the constants successively as $c_{-1} = q^{-1}(1-y)^{-1/N} \cdot (1 - y^{1/N})$, $c_{-2} = q^{-2}(1-y)^{-2/N} \cdot (1-y^{1/N})(1-q^{-1}y^{1/N})$, etc., and hence we can rewrite the product as

$$\prod_{j\in\mathbb{Z}/N\mathbb{Z}} c_j = \Big(q(1-y)^{1/N}\Big)^{-\binom{N}{2}} \cdot (1-y^{1/N})^{N-1}(1-q^{-1}y^{1/N})^{N-2} \cdots (1-q^{-(N-2)}y^{1/N}))^1.$$

On the one hand, the product over the $N-1$ righthand factors is reexpressed, after rearranging the factors backwards, as

$$\begin{aligned}
&(1-y^{1/N})^{N-1}(1-q^{-1}y^{1/N})^{N-2} \cdots (1-q^{-(N-3)}y^{1/N}))^2(1-q^{-(N-2)}y^{1/N}))^1\\
&= (1-q^2y^{1/N}))^1(1-q^3y^{1/N}))^2 \cdots (1-q^{N-1}y^{1/N}))^{N-2}(1-q^Ny^{1/N}))^{N-1}\\
&= \frac{(1-q^1y^{1/N}))(1-q^2y^{1/N}))^2(1-q^3y^{1/N}))^3 \cdots (1-q^{N-1}y^{1/N}))^{N-1}}{(1-q^1y^{1/N}))^1(1-q^2y^{1/N}))^1(1-q^3y^{1/N}))^1 \cdots (1-q^{N-1}y^{1/N}))^1} \cdot \frac{(1-q^Ny^{1/N}))^N}{(1-q^Ny^{1/N}))^1},
\end{aligned}$$

where the denominator now simplifies to $\prod_j (1-q^jy^{1/N}) = 1-y$ and we can rewrite the left numerator as $f(y^{1/N};q)$, where

$$f(t;q) := \prod_{j=1}^{N-1} (1-q^jt)^j.$$

(Equivalently, we can let the product run also over $j=0$, at least for $t \neq 1$, and for $t=1$ it is then still consistent with the usual convention $0^0 = 1$.)

We further note that

$$q^{\frac{N(N-1)}{2}} = (-1)^{N+1}, \tag{3.3}$$

and that the two exponents $\binom{N}{2}$ and N of the factor $(1-y)$ add up to $\binom{N+1}{2}$.

In summary, we obtain

$$\prod_{j\in\mathbb{Z}/N\mathbb{Z}} c_j = (-1)^{N+1} \cdot \frac{(1-y^{1/N})^N}{\Big((1-y)^{1/N}\Big)^{\frac{N(N+1)}{2}}} \cdot f(y^{1/N};q).$$

On the other hand, we can compute the determinant of the product of the Jacobians of all the transformations from "left" (ℓ_j) to "right" (r_j):

$$\frac{\bigwedge_{j=0}^{N-1} \mathrm{d}\, r_j}{\bigwedge_{j=0}^{N-1} \mathrm{d}\, \ell_j} = \Big(\prod_j c_j\Big) \cdot \frac{\bigwedge_{j=0}^{N-1} \mathrm{d}\, F_j}{\bigwedge_{j=0}^{N-1} \mathrm{d}\, \ell_j},$$

the second factor on the right being a Vandermonde determinant, more precisely

$$\prod_{0 \le i < j \le N-1} (q^j - q^i) = (q^1-1)^{N-1}(q^2-1)^{N-2} \cdots (q^{N-1}-1)^1 \cdot q^{(N-1)(N-2)/2}.$$

Here the rightmost factor in turn can be simplified, in view of (3.3), to

$$q^{(N-1)(N-2)/2} = q^{N(N-1)/2} q^{-(N-1)} = (-1)^{N+1} q,$$

and the product of the remaining $N-1$ factors equals

$$\frac{(1-q^1)^N(1-q^2)^N \cdots (1-q^{N-1})^N}{(1-q^1)(1-q^2)^2 \cdots (1-q^{N-1})^{N-1}} (-1)^{\frac{N(N-1)}{2}} = \frac{N^N (-1)^{\frac{N(N-1)}{2}}}{f(1;q)}.$$

Therefore we get overall

$$\prod_{0 \le i < j \le N-1} (q^j - q^i) = \frac{N^N (-1)^{\frac{N(N-1)}{2}}}{f(1;q)} \cdot (-1)^{N+1} q.$$

Together with the above result on $\prod c_j$, we find that, at least up to $f(1;q)$ and a root of unity, the product $\prod_{i=1}^5 f(s_i;q)$ is an N-th power in the underlying field, more precisely, it is (essentially) equal to the N-th power of

$$\frac{\prod_i s_i^{N+1}}{\prod_i (1-s_i)} Z_N^{-1},$$

where

$$N \cdot Z_N = \sum_{j_1,\ldots,j_5 \in \mathbb{Z}/N\mathbb{Z}} q^{\sum_{i\in\mathbb{Z}/5\mathbb{Z}} j_i j_{i+1}} \cdot \prod_{i \in \mathbb{Z}/5\mathbb{Z}} (q s_{i-1} s_{i+1}))^{-j_i} (s_i; q^{-1})_{j_i},$$

with the usual q-Pochhammer notation $(s;q)_a = (1-s)(1-qs)\cdots(1-q^{a-1}s)$ $(a \ge 1)$. □

In view of the theorem, the map $L_2^{\text{ét},N}$ factors through $\mathbb{Z}[F']/\langle \text{5-term relations} \rangle$, where $F' = F(\{t_i^{1/N}\}_i)$, and may be thought of as a "finite/étale regulator map" for $\ker \partial_2$ and hence for $\mathrm{K}_3(F')$.

REMARK. It may be interesting to give details on $f(1;q)$, which plays the role of $\mathrm{Li}_2(1)$. Experimentally, one has (for $N > 3$ a prime, while for $N = 3$ a slightly different formula holds: the right hand side needs to be multiplied by q)

$$\prod_{j=1}^{N-1} (1-q^j)^j = \left(\frac{2}{N}\right) N^{\frac{N-1}{2}} \sum_{k=1}^{N-1} \left(\frac{k}{N}\right) q^k$$

with the usual notation for the Kronecker symbol. (Can we think of it as some kind of "finite/étale period"?)

REMARK. There is a natural way to produce the map suggested above, by "differentiating the distribution relation" in the following sense: let $\partial_2 : \mathbb{Z}[F] \to \bigwedge^2 F^\times$ be the usual map given on generators by $[t] \mapsto t \wedge (1-t)$, then

$$\partial_2\Big(\frac{1}{N}[t^N] - \sum_{j=0}^{N-1}[\zeta^i t]\Big) = \prod_{j=1}^{N-1}\big(\zeta \otimes (1-\zeta^j t)^j\big) = \zeta \otimes \prod_{j=1}^{N-1}(1-\zeta^j t)^j.$$

4. Non-vanishing elements in $B(F)$

4.1. An example with $F = \mathbb{Q}(\sqrt{-7})$. We want to associate the above trace to non-trivial elements in the Bloch group. Let α and β be algebraic numbers satisfying

$$1-\alpha = -\frac{1}{\beta}, \qquad 1-\beta = \alpha^2\beta.$$

This can be realized over the field $F = \mathbb{Q}(\sqrt{-7})$, where $\alpha = \frac{1-\sqrt{-7}}{2}$ (cf. [**6**], §4, or [**7**]).

Then $\rho = 2[\alpha] + [1-\beta]$ lies in the kernel of the above map ∂_2.

The Beilinson regulator of ρ in weight 2 (which we identify here with the Bloch-Wigner dilogarithm for simplicity) is a special L-value, more precisely it equals $\frac{3\sqrt{7}^3}{4\pi^2}\zeta_F(2)$.

Remark. How to characterize Bloch elements which are linear combinations of specialisations of the 5-term relation, as opposed to non-trivial elements in the Bloch group? If we write

$$\Phi(a,q) = \prod_{j=1}^{N}(1-aq^j)^j,$$

then we look at the quotient

$$\frac{\Phi(x_i^{1/N}q,q)}{\Phi(x_i^{1/N},q)} = \frac{(1-x_i^{1/N})^N}{(1-x_i)} = \frac{(1-x_i^{1/N})^N}{\prod_j x_j^{a_{ij}}}.$$

Now a functional equation/five term relation is characterised by having the product (over i) of the corresponding quotients above being an N-th power.

Hence in a sense an element in $B(F)$ gives a canonical extension field: Any $N > 0$ and $\sum_i [x_i] \in \mathbb{Z}[F]$ give rise to $F(\sqrt[N]{1}, \{\sqrt[N]{x_i}, \sqrt[N]{1-x_i}\}_i)$, with compatibly chosen roots for the x_i and $1-x_i$.

Example (cont'd): We consider the above example in detail. The algebras arising from this process (on the left), together with an ordered basis (on the right), are

$$\begin{aligned}
\mathcal{E}_{\alpha,-1} &\leftrightarrow \quad \xi_1,\ \eta_1,\\
\mathcal{E}_{\alpha,-(1-\alpha)} &\leftrightarrow \quad \xi_2,\ \eta_2,\\
\mathcal{E}_{\alpha,1/\beta} &\leftrightarrow \text{the same}\\
\mathcal{E}_{1/\alpha,\beta} &\leftrightarrow \quad \xi_3,\ \eta_3,\\
\mathcal{E}_{(1-\beta)/\alpha,\beta} &\leftrightarrow \quad \xi_4,\ \eta_4,\\
\mathcal{E}_{\alpha\beta,\beta} &\leftrightarrow \text{the same}\\
\mathcal{E}_{\alpha,\beta} &\leftrightarrow \quad \xi_5,\ \eta_5,\\
\mathcal{E}_{\alpha,\beta(1-\alpha)} &\leftrightarrow \quad \xi_6,\ \eta_6,\\
\mathcal{E}_{\alpha,-1} &\leftrightarrow \text{the same.}
\end{aligned}$$

The following transitions highlight the steps where we encounter a new Steinberg symbol $z \wedge (1-z)$ when passing from one algebra to the next,

$$\begin{aligned}
&(\xi_1,\ \eta_1) \overset{\alpha\wedge(1-\alpha)}{\longleftrightarrow} (\xi_2,\ \eta_2),\\
&(\xi_3,\ \eta_3) \overset{(1-\beta)\wedge\beta}{\longleftrightarrow} (\xi_4,\ \eta_4),\\
&(\xi_5,\ \eta_5) \overset{\alpha\wedge(1-\alpha)}{\longleftrightarrow} (\xi_6,\ \eta_6).
\end{aligned} \tag{4.1}$$

We choose the following N-th roots (for consistency of signs, take N odd): $q = \sqrt[N]{1}$, $q' = \sqrt[N]{-1} = -1$, $\sqrt[N]{\alpha}$ and $\sqrt[N]{\beta}$. Then we need to make sure that we choose the roots of $1-\alpha$ and of $1-\beta$ as follows:

$$\sqrt[N]{1-\alpha} = -\frac{1}{\sqrt[N]{\beta}},$$

$$\sqrt[N]{1-\beta} = (\sqrt[N]{\alpha})^2\, \sqrt[N]{\beta}.$$

For the generators ξ_i, η_i, we find the following (starting with $i=1$), recalling that we have $\xi_i\eta_i = q\eta_i\xi_i$):

$$\xi_1 = \sqrt[N]{\alpha}\begin{pmatrix} 1 & & & \\ & q & & \\ & & q^2 & \\ & & & \ddots \end{pmatrix}, \qquad \eta_1 = q'\begin{pmatrix} 0 & & & & 1\\ 1 & 0 & & & \\ & 1 & 0 & & \\ & & \ddots & \ddots & \\ & & & 1 & 0 \end{pmatrix}.$$

For $i > 1$, we get successively the following conditions:

$$\xi_2 = T_{12}\,\xi_1\,T_{12}^{-1}, \qquad \eta_2 = T_{12}\,\eta_1(1-\xi_1)\,T_{12}^{-1}.$$

$$\xi_3 = T_{23}\,\xi_2^{-1}\,T_{23}^{-1}, \qquad \eta_3 = T_{23}\,\eta_2^{-1}\,T_{23}^{-1}.$$

$$\xi_4 = T_{34}\,\xi_3(1-\eta_3)\,T_{34}^{-1}, \qquad \eta_4 = T_{34}\,\eta_3\,T_{34}^{-1}.$$

$$\xi_5 = T_{45}\,\xi_4\eta_4^{-1}\,T_{45}^{-1}, \qquad \eta_5 = T_{45}\,\eta_4\,T_{45}^{-1}.$$

$$\xi_6 = T_{56}\,\xi_5\,T_{56}^{-1}, \qquad \eta_6 = T_{56}\,\eta_5(1-\xi_5)\,T_{56}^{-1}.$$

We guess the form of T_{23} as

$$\begin{pmatrix} 1 & & & \\ & & & 1 \\ & & \iddots & \\ & 1 & & \\ 1 & & & \end{pmatrix}.$$

Looking at the eigenvectors and their eigenvalues for T_{12}, say, we find

$$T_{12}^{-1} = \mathrm{diag}\begin{pmatrix} 1 \\ -(1-\sqrt[N]{\alpha})\sqrt[N]{\beta} \\ +(1-\sqrt[N]{\alpha})(1-q\sqrt[N]{\alpha})\sqrt[N]{\beta}^2 \\ \vdots \end{pmatrix} = \mathrm{diag}\Big((-1)^i\sqrt[N]{\beta}^i \prod_{0\le j<i}(1-q^j\sqrt[N]{\alpha})\Big).$$

Moreover, we get the same matrix for the last step, i.e., $T_{56} = T_{12}$. The transformation T_{45}^{-1} is given by cyclic shifts of the first row vector $(q^{\binom{i}{2}})_{i=1,\dots,n}$, i.e. as $(q^{\binom{i-j}{2}})_{i,j}$. Finally, the arguably most complicated one is T_{34}^{-1}, where in the (i,j)-entry $(0\le i,j<N)$ we get

$$\frac{q^{\binom{i-j+1}{2}}\sqrt[N]{\beta}^{i-j}}{\prod_{k=1}^{i-j}(q^k-\sqrt[N]{\alpha}^2\sqrt[N]{\beta})}.$$

Now we take the product of all the five transformation matrices involved, i.e.,

$$M_N := T_{56}T_{45}T_{34}T_{23}T_{12},$$

and consider its trace and determinant. The resulting $\xi_N = \mathrm{Trace}(M_N)^N/\det(M_N)$ ought to be related, in a sense, to the corresponding “étale regulator” for the Bloch element $2[\alpha]+[1-\beta]$ where the three contributions stem from (4.1).

A numerical example. For $N=5$ in the above example, we can recognise the numerical value $2\xi_5 \doteq -62.8929250152668144\ldots + 100.462207922407031\ldots\cdot i$ as an algebraic integer of degree 8 and of norm $(2^4\cdot 331)^5$.

Furthermore, numerical checks appear to relate the limit $\lim_{N\to\infty}\log|\xi_N|^{1/N^2}$ with the Beilinson-Borel regulator in weight 2 for the field $\mathbb{Q}(\sqrt{-7})$. For a recent account of a much more general prediction see [**4**], §8.

Acknowledgments

The first named author is grateful to the IHES in Bures-sur-Yvette and the MPI for Mathematics in Bonn for their hospitality and ideal working conditions, with thanks to Qizhen Sang for spotting inconsistencies in a prior draft.

References

[1] S. J. Bloch, *Higher regulators, algebraic K-theory, and zeta functions of elliptic curves*, CRM Monograph Series, vol. 11, American Mathematical Society, Providence, RI, 2000, DOI 10.1090/crmm/011. MR1760901

[2] F. Calegari, S. Garoufalidis, and D. Zagier, *Bloch groups, algebraic K-theory, units, and Nahm's conjecture* (English, with English and French summaries), Ann. Sci. Éc. Norm. Supér. (4) **56** (2023), no. 2, 383–426, DOI 10.24033/asens.2537. MR4598726

[3] Kontsevich, M., *Sums of products of q-factorials over rational polyhedra as fake quantum invariants of 3-manifolds*, CTQM Problem List (Dec 2006), `https://www.ctqm.au.dk/PL/`

[4] Kontsevich, M., Soibelman, Y., *Holomorphic Floer theory I: exponential integrals in finite and infinite dimensions*, `arXiv:2402.07343`.

[5] A. A. Suslin, K_3 *of a field, and the Bloch group* (Russian), Trudy Mat. Inst. Steklov. **183** (1990), 180–199, 229. Translated in Proc. Steklov Inst. Math. **1991**, no. 4, 217–239; Galois theory, rings, algebraic groups and their applications (Russian). MR1092031

[6] D. Zagier, *The dilogarithm function*, Frontiers in number theory, physics, and geometry. II, Springer, Berlin, 2007, pp. 3–65, DOI 10.1007/978-3-540-30308-4_1. MR2290758

[7] D. Zagier and H. Gangl, *Classical and elliptic polylogarithms and special values of L-series*, The arithmetic and geometry of algebraic cycles (Banff, AB, 1998), NATO Sci. Ser. C Math. Phys. Sci., vol. 548, Kluwer Acad. Publ., Dordrecht, 2000, pp. 561–615. MR1744961

Mathematical Sciences & Computer Science Building, Upper Mountjoy, Durham University, DH1 3LE

Email address: herbert.gangl@durham.ac.uk

Institut des Hautes Etudes Scientifiques, 35 Route de Chartres, 91440 Bures-sur-Yvette

Email address: maxim@ihes.fr

Contemporary Mathematics
Volume **842**, 2026
https://doi.org/10.1090/conm/842/16855

Calculating Archimedean height pairing via generalized cross-ratio

Alexander B. Goncharov

ABSTRACT. I answer a question which Spencer Bloch asked during the *Regulators V* conference in Pisa.

During the last several years, Spencer Bloch emphasized that Archimedean hight pairings between two homologically trivial cycles of complimentary arithmetic dimensions, see (1) for the precise condition, or the closely related classes in the biextension, should lead to interesting analogs of the classical cross ratio. In this note I calculate such a generalized cross-ratio for planes in P^n.

Let $\mathrm{A}_1, \mathrm{A}_2$ (respectively $\mathrm{B}_1, \mathrm{B}_2$) be $a-$dimensional (respectively $b-$dimensional) planes in $\mathbb{C}\,\mathrm{P}^n$, where

$$a + b + 1 = n. \tag{1}$$

We assume that $\mathrm{A}_i \cap \mathrm{B}_j = \emptyset$. Let us define a number[1]

$$r(\mathrm{A}_1, \mathrm{A}_2; \mathrm{B}_1, \mathrm{B}_2) \in \mathbb{C}^\times.$$

Let V_{n+1} be a complex vector space, and $\mathbb{C}\,\mathrm{P}^n = \mathrm{P}(V_{n+1} - \{0\})$. Denote by $\widehat{\mathrm{A}}_i$ and $\widehat{\mathrm{B}}_j$ the subspaces in V_{n+1} of dimensions $a+1$ and $b+1$ projecting onto the planes $\mathrm{A}_i, \mathrm{B}_j$, $i, j = 1, 2$. Pick non-zero volume elements $\alpha_i \in \det(\widehat{\mathrm{A}}_i)$ and $\beta_j \in \det(\widehat{\mathrm{B}}_j)$. Since $\mathrm{A}_i \oplus \mathrm{B}_j = V_{n+1}$, we can define

$$(\mathrm{A}_i, \mathrm{B}_j) := \alpha_i \wedge \beta_j \in \det(V_{n+1}).$$

DEFINITION 0.1. *We set*

$$r(\mathrm{A}_1, \mathrm{A}_2; \mathrm{B}_1, \mathrm{B}_2) := \frac{(\mathrm{A}_1, \mathrm{B}_1)(\mathrm{A}_2, \mathrm{B}_2)}{(\mathrm{A}_1, \mathrm{B}_2)(\mathrm{A}_2, \mathrm{B}_1)} \in \mathbb{C}^\times.$$

Evidently this number does not depend on the choices of the volume forms α_i, β_j. Clearly one has

$$r(\mathrm{A}_1, \mathrm{A}_2; \mathrm{B}_1, \mathrm{B}_2) \cdot r(\mathrm{A}_2, \mathrm{A}_3; \mathrm{B}_1, \mathrm{B}_2) = r(\mathrm{A}_1, \mathrm{A}_3; \mathrm{B}_1, \mathrm{B}_2). \tag{2}$$

$$r(\mathrm{A}_1, \mathrm{A}_2; \mathrm{B}_1, \mathrm{B}_2) = r(\mathrm{B}_1, \mathrm{B}_2; \mathrm{A}_1, \mathrm{A}_2).$$

Let me recall the definition of the Archimedean height pairing for a compact $n-$dimensional Kähler manifold M [**B**], [**Bl**]. Let $\Delta \subset M \times M$ be the diagonal.

This work was supported by the NSF grant DMS-2153059.

[1]The construction below works for any field k.

Denote by $\mathcal{H}_\Delta \in \mathcal{A}^{n,n}(M \times M)$ the harmonic representative of its cohomology class. There exists a Green current $G_{M\times M} \in \mathcal{A}^{n-1,n-1}(M)$ such that

$$(2\pi i)^{-1}\partial\overline{\partial}G_{M\times M} = \delta_\Delta - \mathcal{H}_\Delta.$$

DEFINITION 0.2. *Given homologically trivial cycles $\mathcal{A}, \mathcal{B} \subset M$ with proper intersection, such that* $\dim\mathcal{A} + \dim\mathcal{B} + 1 = \dim M$ *the Archimedean height pairing $\langle\mathcal{A}, \mathcal{B}\rangle$ is defined by*

$$\langle\mathcal{A}, \mathcal{B}\rangle := \int_{\mathcal{A}\times\mathcal{B}} G_{M\times M}. \tag{3}$$

Any other current $G_{M\times M}$ solving the equation differs by an exact current, leaving the integrals the same. Consider the currents

$$G_\mathcal{A} := \int_\mathcal{A} G_{M\times M}, \qquad G_\mathcal{B} := \int_\mathcal{B} G_{M\times M}.$$

They satisfy the differential equations

$$(2\pi i)^{-1}\partial\overline{\partial}G_\mathcal{A} = \delta_\mathcal{A}, \quad (2\pi i)^{-1}\partial\overline{\partial}G_\mathcal{B} = \delta_\mathcal{B}.$$

Therefore we have

$$\langle\mathcal{A}, \mathcal{B}\rangle = \int_\mathcal{B} G_\mathcal{A} = \int_\mathcal{A} G_\mathcal{B}. \tag{4}$$

In particular, we have

$$\langle \mathrm{A}_1 - \mathrm{A}_2, \mathrm{B}_1 - \mathrm{B}_2\rangle := \Big(\int_{\mathrm{A}_1(\mathbb{C})} - \int_{\mathrm{A}_2(\mathbb{C})}\Big) G_{\mathrm{B}_1 - \mathrm{B}_2}.$$

THEOREM 0.3. *One has*

$$\langle \mathrm{A}_1 - \mathrm{A}_2, \mathrm{B}_1 - \mathrm{B}_2\rangle = \log|r(\mathrm{A}_1, \mathrm{A}_2; \mathrm{B}_1, \mathrm{B}_2)|.$$

PROOF. If $n = 1$, the Green function $G_\mathcal{A}$ for the divisor $\mathcal{A} = \{0\} - \{\infty\}$ is $\log|z|$. So its evaluation on the divisor $\mathcal{B} = \{z\} - \{1\}$ is $\log|z|$. Assume that $n > 1$. Then using the symmetry we can assume without loss of generality that the planes A_i are of codimension ≥ 2.

Take generic hyperplanes $\mathrm{H}_1, \mathrm{H}_2$ containing the cycles $\mathrm{A}_1, \mathrm{A}_2$ respectively. Their intersection $\mathrm{H}_1 \cap \mathrm{H}_2$ is of codimension 2. Take a generic subspace $\mathrm{A} \subset \mathrm{H}_1 \cap \mathrm{H}_2$ of the same dimension as A_1. Let

$$\mathrm{B}_i' := \mathrm{H}_1 \cap \mathrm{B}_i; \qquad \mathrm{B}_i'' := \mathrm{H}_2 \cap \mathrm{B}_i, \quad i = 1, 2.$$

Then we have the following quadruples of planes in the hyperplanes H_1 and H_2:

$$(\mathrm{A}, \mathrm{A}_1; \mathrm{B}_1', \mathrm{B}_2') \subset \mathrm{H}_1; \qquad (\mathrm{A}_2, \mathrm{A}; \mathrm{B}_1'', \mathrm{B}_2'') \subset \mathrm{H}_2\,.$$

Let us use the notation $\langle *, *\rangle_\mathrm{H}$ and $r_\mathrm{H}(\dots)$ working with planes in the projective space H.

LEMMA 0.4. *The invariant $r(\mathrm{A}_1, \mathrm{A}; \mathrm{B}_1, \mathrm{B}_2)$ remains the same if we pass to the hyperplane $H_1 \subset \mathbb{C}\mathrm{P}^n$ and replace $\mathrm{B}_1, \mathrm{B}_2$ by their intersections with H_1, and similarly for H_2:*

$$\begin{aligned} r_{\mathbb{C}\mathrm{P}^n}(\mathrm{A}_1, \mathrm{A}; \mathrm{B}_1, \mathrm{B}_2) &= r_{\mathrm{H}_1}(\mathrm{A}_1, \mathrm{A}; \mathrm{B}_1', \mathrm{B}_2'); \\ r_{\mathbb{C}\mathrm{P}^n}(\mathrm{A}, \mathrm{A}_2; \mathrm{B}_1, \mathrm{B}_2) &= r_{\mathrm{H}_2}(\mathrm{A}, \mathrm{A}_2; \mathrm{B}_1'', \mathrm{B}_2''). \end{aligned} \tag{5}$$

PROOF. It is sufficient to show that given a hyperplane H containing subspaces A_1, A_2, and setting $B_i' = H \cap B_i$, we get $r_{\mathbb{C}P^n}(A_1, A_2; B_1, B_2) = r_H(A_1, A_2; B_1', B_2')$. Take volume forms $\omega_1, \ldots, \omega_4$ in H, and a linear functional f in $\mathbb{C}^{n+1}$ annihilating H. We have the following identity, where each fraction is a number in $\mathbb{C}^\times$, given by the ratio of two volume forms:

$$\frac{\omega_1 \wedge f}{\omega_2 \wedge f} \cdot \frac{\omega_3 \wedge f}{\omega_4 \wedge f} = \frac{\omega_1}{\omega_2} \cdot \frac{\omega_3}{\omega_4}.$$

The claim follows immediately from this. □

By the induction on the dimension of the projective space H_i we have

$$\begin{aligned} \langle A_1 - A, B_1' - B_2' \rangle_{H_1} &= \log |r_{H_1}(A_1, A; B_1', B_2')|; \\ \langle A - A_2, B_1'' - B_2'' \rangle_{H_2} &= \log |r_{H_2}(A, A_2; B_1'', B_2'')|. \end{aligned} \tag{6}$$

LEMMA 0.5.

$$\begin{aligned} \langle A_1 - A, B_1' - B_2' \rangle_{H_1} &= \langle A_1 - A, B_1 - B_2 \rangle_{\mathbb{C}P^n}; \\ \langle A - A_2, B_1'' - B_2'' \rangle_{H_2} &= \langle A - A_2, B_1 - B_2 \rangle_{\mathbb{C}P^n}. \end{aligned} \tag{7}$$

PROOF. It is sufficient to check the first identity, and it follows easily from the definitions. An alternative proof follows from Lemma 0.7 below. □

Substituting this to the left hand sides of identities (6), adding the obtained identities and using the multiplicativity (2) and Lemma 0.4, we get:

$$\langle A_1 - A, B_1' - B_2' \rangle_{H_1} + \langle A - A_2, B_1'' - B_2'' \rangle_{H_2} = \langle A_1 - A_2, B_1 - B_2 \rangle_{\mathbb{C}P^n}. \tag{8}$$

Therefore the left hand side in (8) is equal to

$$\begin{aligned} &\langle A_1 - A, B_1' - B_2' \rangle_{H_1} + \langle A - A_2, B_1'' - B_2'' \rangle_{H_2} \\ =&\langle A_1 - A, B_1 - B_2 \rangle_{\mathbb{C}P^n} + \langle A - A_2, B_1 - B_2 \rangle_{\mathbb{C}P^n} \\ =&\langle A_1 - A_2, B_1 - B_2 \rangle_{\mathbb{C}P^n}. \end{aligned} \tag{9}$$

So by the induction the claim boils down to $\mathbb{C}P^1$, where it is well known. □

Let us lift the Archimedean height pairing to an invariant in $\mathbb{C}^\times \otimes \mathbb{Q}$. Since odd dimensional cohomology of $\mathbb{C}P^n$ vanish, we can avoid biextensions and proceed as follows. Assume that the planes A_i, B_j do not intersect. Then the well known arguments provide an exact sequence[2]

$$0 \longrightarrow \mathbb{Q}(1) \longrightarrow H^{2b+1}(\mathbb{C}P^n - A_1 \cup A_2, B_1 \cup B_2; \mathbb{Q}(b)) \longrightarrow \mathbb{Q}(0) \longrightarrow 0.$$

Considered, say, as a mixed Hodge structure over $\mathbb{Q}$, this extension defines a class

$$\mathrm{cl}(A_1 - A_2, B_1 - B_2) \in \mathbb{C}^\times \otimes \mathbb{Q}.$$

THEOREM 0.6. *Under the assumptions above, we have*

$$\mathrm{cl}(A_1 - A_2, B_1 - B_2) = r(A_1, A_2; B_1, B_2).$$

[2]Indeed, there are maps

$$\begin{aligned} &H^{2b+1}(\mathbb{C}P^n - A_1 \cup A_2, B_1 \cup B_2; \mathbb{Q}(b)) \longrightarrow H^{2b+1}(\mathbb{C}P^n - A_1 \cup A_2; \mathbb{Q}(b)) = \mathbb{Q}(0). \\ &H^{2b+1}(\mathbb{C}P^n, B_1 \cup B_2; \mathbb{Q}(b)) = \mathbb{Q}(1) \longrightarrow H^{2b+1}(\mathbb{C}P^n - A_1 \cup A_2, B_1 \cup B_2; \mathbb{Q}(b)). \end{aligned} \tag{10}$$

Here the second equality in the first row follows from the exact sequence

$$H^{2b}(\mathbb{C}P^n, \mathbb{Q}(b)) \longrightarrow H^{2b}(A, \mathbb{Q}(b)) = \mathbb{Q}^{\oplus 2} \longrightarrow H^{2b+1}(\mathbb{C}P^n - A, \mathbb{Q}(b)) \longrightarrow 0.$$

The first equality in the second row is similar.

PROOF. We use the same notation as in the proof of Theorem 0.3. The $n = 1$ is the classical case. Let us assume that $n > 1$. By the induction on the dimension of the projective space H_i we have

$$\begin{aligned} \mathrm{cl}(A_1 - \mathrm{A}, \mathrm{B}_1' - \mathrm{B}_2')_{\mathrm{H}_1} &= r_{\mathrm{H}_1}(\mathrm{A}_1, \mathrm{A}; \mathrm{B}_1', \mathrm{B}_2'); \\ \mathrm{cl}(\mathrm{A} - \mathrm{A}_2, \mathrm{B}_1'' - \mathrm{B}_2'')_{\mathrm{H}_2} &= r_{\mathrm{H}_2}(\mathrm{A}, \mathrm{A}_2; \mathrm{B}_1'', \mathrm{B}_2''). \end{aligned} \tag{11}$$

LEMMA 0.7.

$$\begin{aligned} \mathrm{cl}(\mathrm{A}_1 - \mathrm{A}, \mathrm{B}_1' - \mathrm{B}_2')_{\mathrm{H}_1} &= \mathrm{cl}(\mathrm{A}_1 - \mathrm{A}, \mathrm{B}_1 - \mathrm{B}_2)_{\mathbb{C}\mathrm{P}^n}; \\ \mathrm{cl}(\mathrm{A} - \mathrm{A}_2, \mathrm{B}_1'' - \mathrm{B}_2'')_{\mathrm{H}_2} &= \mathrm{cl}(\mathrm{A} - \mathrm{A}_2, \mathrm{B}_1 - \mathrm{B}_2)_{\mathbb{C}\mathrm{P}^n}. \end{aligned} \tag{12}$$

PROOF. Consider the map of the cohomology induced by the embedding $\mathrm{H}_1 \hookrightarrow \mathbb{C}\mathrm{P}^n$:

(13)

$$H^{2b+1}(\mathbb{C}\mathrm{P}^n - \mathrm{A}_1 \cup \mathrm{A}_2, \mathrm{B}_1 \cup \mathrm{B}_2; \mathbb{Q}(b)) \longrightarrow H^{2b+1}(\mathrm{H} - \mathrm{A}_1 \cup \mathrm{A}, \mathrm{B}_1' \cup \mathrm{B}_2'; \mathbb{Q}(b)).$$

It induces an isomorphisms on the sub-object $\mathbb{Q}(1)$ and the quotient object $\mathbb{Q}(0)$. Therefore it is an isomorphism, and thus preserves the extension class. The second identity is similar. □

Substituting this to the left hand sides of identities (11), adding the obtained identities and using the multiplicativity (2) and Lemma 0.4, we get:

$$\mathrm{cl}(\mathrm{A}_1 - \mathrm{A}, \mathrm{B}_1' - \mathrm{B}_2')_{\mathrm{H}_1} + \mathrm{cl}(\mathrm{A} - \mathrm{A}_2, \mathrm{B}_1'' - \mathrm{B}_2'')_{\mathrm{H}_2} = \mathrm{cl}(\mathrm{A}_1 - \mathrm{A}_2, \mathrm{B}_1 - \mathrm{B}_2)_{\mathbb{C}\mathrm{P}^n}. \tag{14}$$

□

Finally, it is well known that the Archimedean height pairing is related to the extension class by

$$\log|\mathrm{cl}(\mathrm{A}_1 - \mathrm{A}_2, \mathrm{B}_1 - \mathrm{B}_2)| = \langle \mathrm{A}_1 - \mathrm{A}_2, \mathrm{B}_1 - \mathrm{B}_2 \rangle.$$

Acknowledgment

I am grateful to the referee for useful comments. I am grateful to Spencer Bloch for discussions during the conference, and to the organisers for the invitation to Pisa.

References

[B] A. A. Beĭlinson, *Height pairing between algebraic cycles*, K-theory, arithmetic and geometry (Moscow, 1984), Lecture Notes in Math., vol. 1289, Springer, Berlin, 1987, pp. 1–25, DOI 10.1007/BFb0078364. MR923131

[Bl] S. Bloch, *Height pairings for algebraic cycles*, Proceedings of the Luminy conference on algebraic K-theory (Luminy, 1983), J. Pure Appl. Algebra **34** (1984), no. 2-3, 119–145, DOI 10.1016/0022-4049(84)90032-X. MR772054

[Bl2] S. Bloch, *Cycles and biextensions*, Algebraic K-theory and algebraic number theory (Honolulu, HI, 1987), Contemp. Math., vol. 83, Amer. Math. Soc., Providence, RI, 1989, pp. 19–30, DOI 10.1090/conm/083/991974. MR991974

Contemporary Mathematics
Volume **842**, 2026
https://doi.org/10.1090/conm/842/16856

Atypical Hodge loci

Phillip Griffiths

ABSTRACT. In the recent works of a number of people there has emerged a beautiful new perspective on the arithmetic properties of Hodge structures. A central result in that development appears in a paper by Baldi, Klingler, and Ullmo. In this expository work we will explain that result and give a proof. The main conceptual step is to formulate Noether-Lefschetz loci in terms of intersections of period images with Mumford-Tate subdomains of period domains. The main technical step is to use the alignment of the Hodge and root space decompositions of the Lie algebras of the associated groups and from it to use the integrability conditions associated to a Pfaffian PDE system. These integrability conditions "explain" the generally present excess intersection property associated to the integral varieties of a pair of Pfaffian exterior differential systems.

Hodge theory may be said to include three types of applications to algebraic geometry; namely to

- the geometric properties of algebraic varieties;
- the topology of algebraic varieties;
- the arithmetic properties of algebraic varieties.

This paper will discuss a topic in the third item.

Recently there has been a flurry of activity concerning the arithmetic properties, especially functional transcendence, of periods of integrals of algebraic differential forms. A summary of this theory appears in the proceedings of the ICM 2020 paper [**Kli23**]. In this expository paper we will explain and discuss a proof of one of the main theorems in the work [**BKU24**]. This beautiful result plays a central role in the recent developments in functional transcendence theory. In the proof in loc. cit. of this result there are two main Hodge theoretic components, one conceptual and the other Lie theoretic. The conceptual one is to use intersections of period images with Mumford-Tate subdomains instead of the more traditional Noether-Lefschetz loci. The Lie theoretic one is to use the compatible alignments of the Hodge and root space decompositions of the Lie algebras of the automorphism groups of the period domain and Mumford-Tate subdomain to produce the non-transversality of the integrability conditions associated to a pair of Pfaffian exterior differential systems.

2020 *Mathematics Subject Classification.* Primary 14-XX, 22-XX, 42-XX, 53-XX.

This paper is based on [**BKU24**] and related works given in the references in that work, and on discussions with Mark Green, Matt Kerr and Colleen Robles. In particular I would like to thank Colleen Robles for help with the technical arguments used in the proof of the main result.

This paper grew out of a talk given at the Regulators V Conference held June 3–15, 2024, at Pisa. A main purpose of both the talk and this paper is to help draw attention to some of the Hodge theoretic aspects of the developments in arithmetic algebraic geometry that are discussed in [**Kli23**], [**BKU24**] and in a number of subsequent works including [**BKT20**], [**BBKT24**], [**Chi21**], [**HP16**], [**KO21**], [**Urb24**]. Hopefully it will also help serve as an invitation for Hodge theorists and others to this very rich and currently very active development.

1. Introduction

We will be concerned with the question

What can one say about Hodge loci?

Specifically, we consider the situation where

- B is a smooth, connected quasi-projective variety;
- $\mathbb{V} \to B$ is the local system underlying a variation of polarized Hodge structure of weight n;
- $\mathrm{HL}(B)$ is the set of points $b \in B$ where there are more Hodge classes in the tensor algebra $\mathbb{V}_b^{\otimes} := \overset{k,\ell}{\oplus}(\mathbb{V}_b^{\otimes^k} \otimes V_b^{*\otimes \ell})$ than there are at a very general point of B.

A central question is

What can we say about $\mathrm{HL}(B)$*?*

In [**CDK95**] it is proved that $\mathrm{HL}(B)$ is a countable union of algebraic varieties. Very informally stated, denoting by $\mathrm{HL}(B)_{\mathrm{pos}}$ the positive dimensional components of $\mathrm{HL}(B)$ the result of [**BKU24**] is

(1.1) *For $n \geqq 3$ and aside from exceptional cases, every irreducible component of* $\mathrm{HL}(B)_{\mathrm{pos}}$ *has less than the expected codimension.*[1]

That means that if there are Hodge classes that vary in a positive dimensional family, then there are strictly more of these than suggested by a standard dimension count. As explained below the reason for this will be a consequence of the integrability conditions arising from the differential constraint of the variation of Hodge structure. In calculating these a key ingredient is properties relating the Hodge and root space decompositions of a semi-simple Hodge Lie algebra [**Rob14**].

Although we shall not discuss it, in [**BKU24**] it is proved that if (1.2) is satisfied, then $\mathrm{HL}(B)_{\mathrm{pos}}$ is a *finite* union of irreducible algebraic varieties.

The *geometric case* is when the variation of Hodge structure arises from the cohomology along the fibres of a smooth family $\mathcal{X} \to B$ of projective varieties. In this case, assuming the Hodge conjecture the result gives that if there are algebraic cycles $Z_b \subset X_b$ whose cohomology classes are not found on a general X_b and that non-trivially vary with b, then there are strictly more such cycles than one naively expects to find.

An interesting point is that whereas in general integrability conditions *decrease* the expected dimension of the space of solutions to a single system of differential equations, for the pair of EDS's in the case at hand this dimension is actually *increased.* The mechanism behind this will be illustrated in Example 2 below.

[1]For subvarieties A, B of a variety C the expected codimension in C of $A \cap B$ is $\operatorname{codim}_C A + \operatorname{codim}_C B$; i.e., that which would be obtained if the intersection were transverse.

With the notation to be explained below a sufficient condition for (1.1) to hold is

$$\mathfrak{g}^{-k,k} \neq 0 \quad \text{for some} \quad k \geqq 3. \tag{1.2}$$

In [**BKU24**] it is shown that these conditions hold for smooth hypersurfaces $X \subset \mathbb{P}^{n+1}$ where $n \geqq 3$, $d \geqq 5$ and $(d) \neq (4,5)$. We will give the argument for this in Section 5.

In Section 5 we will also recall the definition of the coupling length $\zeta(\mathfrak{A})$ where $\mathfrak{A} \subset \mathfrak{g}^{-1,1}$ is the image of the differential of the period mapping at a general point ([**VZ05**]). There we will show that

$$\zeta(\mathfrak{A}) \geqq 3 \implies (1.2). \tag{1.3}$$

Examples where the coupling length $\zeta(\mathfrak{A}) \geqq 3$ include the moduli of Calabi-Yaus of dimension $\geqq 3$ whose Yukaya coupling is non-zero.

The case

$$\mathfrak{g}^{-k,k} = 0 \quad \text{for} \quad k \geqq 3 \tag{1.4}$$

includes the *classical case* where the period domain is Hermitian symmetric. The case $\mathfrak{g}^{-2,2} \neq 0$ but $\mathfrak{g}^{-k,k} = 0$ for $k \geqq 3$ includes the case of weight $n = 2$ Hodge structures; here the differential constraint is non-trivial but the corresponding integrability conditions do not enter into the dimension count. In this situation classical results in [**Gre88**] [**Voi07a**], and [**Voi07b**] give that in many cases Noether-Lefschetz loci are analytically dense in moduli. The classical case and the case of weight $n = 2$ are discussed in detail in [**BKU24**].

The conditions to have (1.4) may be expressed as a non-linear PDE system for the variation of Hodge structure. At least in principle the E. Cartan theory of exterior differential systems gives an algorithmic procedure for determining the space of formal solutions to this system ([**BCG$^+$91**]). It would seem of interest to carry this out in the present situation.

2. Two examples

We consider the geometric case where $\mathcal{X} \to B$ is a family of surfaces with $X_{b_0} = X$ and $\lambda \in \mathrm{Hg}^1(X)_{\mathrm{prim}}$ is a primitive Hodge class.

(2.1) **Q**: *How many conditions is it for* λ *to vary with* X *as a Hodge class?*

We restrict to a neighborhood U of $b_0 \in B$ so that $\lambda \in H^2(X_b)_{\mathrm{prim}}$ is well defined. Setting

$$\mathrm{NL}_\lambda = \{b \in U : \lambda \in \mathrm{Hg}^1(X_b)\}$$

we are asking *What is the codimension of* NL_λ *in* U*?* For this there is the classical estimate

$$\operatorname{codim}_U \mathrm{NL}_\lambda \leqq h^{2,0}(X). \tag{2.2}$$

This bound is achieved; e.g., for smooth surfaces $X \subset \mathbb{P}^3$ with $d = \deg X \geqq 4$ (cf. [**Gre88**] and [**Gre89**]).[2] In this case one also has

$$d - 3 \leqq \operatorname{codim}_B \mathrm{NL}_\lambda \tag{2.3}$$

with equality holding only for surfaces containing a line. It is also proved in loc. cit. that $\mathrm{HL}(B)$ is analytically dense where B is the moduli space.

[2] A general treatment of Noether-Lefschetz loci given in [**Gre94**] and in [**Voi07a**], [**Voi07b**].

For the second example taken from [**GG06**] and [**GGK12**] we let $\mathcal{X} \to B$ be a family of 4-folds with $\lambda \in \mathrm{Hg}^2(X)_{\mathrm{prim}}$ a primitive Hodge class for $X = X_{b_0}$ and ask the same question. The analogue of (2.2) is

$$\operatorname{codim}_B \mathrm{NL}_\lambda \leqq h^{4,0}(X) + h^{3,1}(X). \tag{2.4}$$

However due to transversality of the period mapping we have that for the first order variation of X in a direction $\theta \in T := T_{b_0}B$ the product $\theta \cdot \lambda \in H^{1,3}(X)$ and therefore for any $\omega \in H^{4,0}(X)$

$$\langle \omega, \theta \cdot \lambda \rangle = 0. \tag{2.5}$$

Thus a refinement of (2.4) is

$$\operatorname{codim}_B \mathrm{NL}_\lambda \leqq h^{3,1}(X). \tag{2.6}$$

We refer to the right-hand side of (2.6) as the *naive expected codimension* of NL_λ in B.

At this juncture a new consideration enters. Setting

$$\begin{aligned} T_\lambda &:= \{\theta \in T : \theta \cdot \lambda = 0 \quad \text{in} \quad H^{1,3}(X)\}, \\ \sigma(\lambda) &= \mathrm{Image}\{T_\lambda \otimes H^{4,0}(X) \to H^{3,1}(X)\} \end{aligned} \tag{2.7}$$

we have

$$\operatorname{codim}_B \mathrm{NL}_\lambda \leqq h^{3,1}(X) - \dim \sigma(\lambda). \tag{2.8}$$

PROOF. For $\theta \in T_\lambda$ and any $\theta' \in T$, $\omega \in H^{4,0}(X)$

$$\begin{aligned} \langle \theta \cdot \omega, \theta' \lambda \rangle &= -\langle \omega, \theta\theta' \lambda \rangle \\ &= -\langle \omega, \theta'\theta \lambda \rangle \\ &= 0 \end{aligned}$$

where the second step follows from the integrability condition $\theta\theta' = \theta'\theta$ arising from transversality. □

Assuming that the map defining $\sigma(\lambda)$ in (2.7) is non-zero we see that due to integrability the actual codimension of NL_λ is strictly less than the naive expected codimension.

One may show that the estimate (2.8) is sharp; e.g., by taking $X \subset \mathbb{P}^5$ a hypersurface of degree 6 containing a 2-plane Λ and for $\lambda \in \mathrm{Hg}^2(X)_{\mathrm{prim}}$ the primitive part of the class of Λ (cf. [**GG06**]).

Finally there is a discussion of Noether-Lefschetz loci in [**GGK12**, §III.C]. A general version of the second example above is given there in III.C.5.

3. The main result

(i) *Hodge structures and Mumford-Tate groups.*[3] A *polarized Hodge structure* of weight n is given by the data $(V, Q, F^\bullet)$ where

- V is a $\mathbb{Q}$-vector space and $Q: V \otimes V \to \mathbb{Q}$ is a non-degenerate bilinear form with $Q(u, v) = (-1)^n Q(v, u)$;

[3] A general reference for Hodge theory is [**CMSP17**]. For Mumford-Tate groups we will generally follow the presentation and notations given in [**GGK12**].

- $F^n \subset F^{n-1} \subset \cdots \subset F^0 \subset V_{\mathbb{C}}$ is a *Hodge filtration* satisfying
$$F^p \oplus \overline{F}^{n-p+1} \xrightarrow{\sim} V_{\mathbb{C}}, \quad 0 \leqq p \leqq n;$$
and
- the two Hodge-Riemann bilinear relations are satisfied (we do not need their explicit form).

Setting
$$V^{p,q} = F^p \cap \overline{F}^q$$
the second condition above is equivalent to the *Hodge decomposition*

$$V_{\mathbb{C}} = \oplus V^{p,q}, \quad \overline{V^{p,q}} = V^{q,p}. \tag{3.1}$$

Using Q we have an identification

$$V \cong V^*. \tag{3.2}$$

We will generally omit reference to Q, its presence being understood.

A Hodge structure of weight n may equally be given by a homomorphism
$$\varphi : S^1 \to \mathrm{GL}(V_{\mathbb{R}})$$
such that its action on $V_{\mathbb{C}}$ is determined by the *Weil operator*
$$\varphi(i)v = i^{p-q}v, \quad v \in V^{p,q}.$$

When the weight $n = 2m$ the *Hodge classes* are
$$\mathrm{Hg}^m(V) = V^{m,m} \cap V,$$
the rational vectors of type (m, m). These are the rational classes that are fixed by the Weil operator.

We denote by
$$V^{\otimes} := \overset{k,\ell}{\oplus} \left((\overset{k}{\otimes} V) \otimes (\overset{\ell}{\otimes} V^*) \right)$$
the tensor algebra of V. The algebra of Hodge tensors involves tensor products of both V and its dual V^*. Using Q we may consider only powers of V. Although somewhat artificial this simplifies the notations and involves no loss of generality. We then denote by
$$\mathrm{Hg}^{\bullet}(V^{\otimes}) := \overset{k}{\oplus} \mathrm{Hg}^{k\,n/2}(\otimes^k V)$$
the sub-algebra of Hodge tensors.

DEFINITION. The *Mumford-Tate group* $\mathrm{MT}(V)$ is the smallest $\mathbb{Q}$-algebraic subgroup of $\mathrm{Aut}(V, Q)$ such that $\varphi(S^1) \subset \mathrm{MT}(V)(\mathbb{R})$. Equivalently (cf. (I.B.1) in [**GGK12**])

$$\mathrm{MT}(V) \subset \mathrm{Aut}(V, Q) \quad \text{is the subgroup that fixes} \quad \mathrm{Hg}^{\bullet}(V^{\otimes}). \tag{3.3}$$

The Mumford-Tate group is a reductive $\mathbb{Q}$-algebraic group denoted here by G. Its Lie algebra
$$\mathfrak{g} \subset \mathrm{End}(V, Q)$$
is a *Hodge Lie algebra*; i.e., it has a Hodge structure of weight zero with Hodge decomposition
$$\mathfrak{g}_{\mathbb{C}} = \oplus \mathfrak{g}^{-k,k}$$
where
$$\mathfrak{g}^{-k} := \mathfrak{g}^{-k,k} = \{A \in \mathfrak{g}_{\mathbb{C}} \quad \text{such that} \quad A : V^{p,q} \to V^{p-k,q+k}\}.$$

We note that after possibly passing a finite cover the real Lie group

$$G(\mathbb{R}) = G_1 \times \cdots \times G_k \times T$$

is a product of simple Lie groups with a compact torus. However we will not have a corresponding product decomposition of G.

Let $\mathfrak{g}_{\mathbb{R}} = \oplus \mathfrak{g}_{i,\mathbb{R}}$ be the decomposition of $\mathfrak{g}_{\mathbb{R}}$ into simple factors. Following [**BKU24**] we have the

DEFINITION 3.4. The *level* $\ell(\mathfrak{g})$ is the smallest k such that all $\mathfrak{g}_i^k \neq 0$.

(ii) *Variation of Hodge structure.* This is given by the data $(\mathbb{V}, \mathcal{F}^\bullet; B)$ where

- $\mathbb{V} \to B$ is a local system over a smooth, connected quasi-projective variety B;
- $\mathcal{F}^\bullet$ is a filtration of $\mathcal{V} = \mathbb{V}_{\mathbb{C}} \otimes_{\mathbb{C}} \mathcal{O}_B$ by holomorphic sub-bundles that induce on each $\mathbb{V}_b$ a Hodge structure, and where for ∇ the Gauss-Manin connection corresponding to $\mathbb{V} \subset \mathcal{V}$ the transversality condition

$$\nabla \mathcal{F}^p \subset \mathcal{F}^{p-1} \otimes \Omega^1_B \tag{3.5}$$

is satisfied.

It is understood that there is a horizontal section Q of $(\mathbb{V}_{\mathbb{Q}} \otimes \mathbb{V}_{\mathbb{Q}})^*$ that polarizes the Hodge structures.

At each point of B there is an algebra of Hodge tensors and Mumford-Tate group. Outside of a countable union of proper subvarieties of B these algebras are locally constant. We denote by $V = \mathbb{V}_{b_0}$ the fibre of $\mathbb{V}$ at such a very general point and by $G \subset \mathrm{Aut}(V, Q)$ the corresponding Mumford-Tate group of the variation of Hodge structure.

The action of $\pi_1(B, b_0)$ on V induces the monodromy group $\Gamma \subset \mathrm{Aut}(V, Q)$. It is known that $\Gamma \subset G$ and in [**GGK12**] there is the general structure theorem III.A.1 (cf. Corollary III.A.2) describing their relation. In this works in order to isolate the central points we will make the assumption

(3.6) *G is a simple $\mathbb{Q}$-algebraic group equal to the $\mathbb{Q}$-Zariski closure $\overline{\Gamma^{\mathbb{Q}}}$ of the monodromy group.*

As explained in loc. cit. the general case may be reduced to the case in which up to a finite cover G and Γ are products of factors G_i and Γ_i where either $\overline{\Gamma_i^{\mathbb{Q}}} = G_i$ or Γ_i is trivial. The induced variations of Hodge structure corresponding to the latter are constant and will not contribute to the end result.

(iii) *Period mappings.* Given (V, Q) the set of polarized Hodge structures with given Hodge numbers $h^{p,q} = \dim V^{p,q}$ is a homogeneous complex manifold called a period domain. The set of those polarized Hodge structures whose Mumford-Tate group is contained in G gives a *Mumford-Tate domain*

$$D = G(\mathbb{R})/G_0$$

where $G(\mathbb{R})$ is the real Lie group associated to G and G_0 is a compact subgroup. Following [**BKU24**] for Γ an arithmetic subgroup of G the quotient $\Gamma \backslash D$ is called a *Hodge variety.*

Associated to a variation of Hodge structure with Mumford-Tate group G there is a period mapping

$$\Phi : B \to \Gamma \backslash D. \tag{3.7}$$

It may be assumed that Φ is proper with image $P \subset \Gamma\backslash D$ a quasi-projective variety [**BBT23**].[4] In order to isolate the essential points in the following discussion we will make the assumption

(3.8) *P is smooth, and we identify $\Phi(B) = P \subset \Gamma\backslash D$.*

All of the results discussed below hold without this assumption, and the proofs may be given using standard technical modifications in the general case where (3.8) may not hold. The one situation that requires extra care is when there is an irreducible subvariety $Z \subset B$ such that $\Phi(Z)$ is contained in the singular locus of $\Phi(B)$ and where the generic Mumford-Tate along Z is smaller than G.

For $b \in B$ and using a lift to D of $\Phi(b)$ the tangent space to D at the point is identified with $\mathfrak{g}_{\mathbb{C}}/F^0\mathfrak{g}_{\mathbb{C}}$. Using (3.5) the differential of the period mapping is

$$\Phi_* : T_bB \to F^{-1}\mathfrak{g}_{\mathbb{C}}/F^0\mathfrak{g}_{\mathbb{C}}. \tag{3.9}$$

We may identify the right-hand side of (3.9) with $\mathfrak{g}^{-1,1}$. For later use we have (cf. [**CMSP17**])

$$\Phi_*(T_bB) := \mathfrak{A} \text{ is an abelian sub-algebra of } \mathfrak{g}^{-1,1}. \tag{3.10}$$

The "abelian" is a consequence of the integrability conditions imposed by the transversality property (3.9).

It is interesting to note that given a Hodge Lie algebra and abelian sub-algebra $\mathfrak{A} \subset \mathfrak{g}^{-1,1}$ there is a local VHS with $\mathfrak{A}$ as its tangent space. Thus there are no "higher" integrability conditions beyond having an abelian sub-algebra of $\mathfrak{g}^{-1,1}$.

(iv) *Hodge loci.* We are interested in proper, irreducible subvarieties $Z \subset B$ along which the corresponding Hodge structures have extra Hodge tensors. Equivalently the Lie algebra of the Mumford-Tate group H at a general point of Z should be strictly contained in that of G:

$$\mathfrak{h} \subsetneqq \mathfrak{g}.$$

A basic observation is

(3.11) *If $D_H \subset D$ is the $H(\mathbb{R})$-orbit of a very general point of Z, then for $\Gamma_H = \Gamma \cap H$ we have*

$$\Phi(Z) \subset \Gamma_H\backslash D_H.$$

DEFINITION ([**BKU24**]). If $H \subset G$ is a Mumford-Tate subgroup, then

$$\Phi^{-1}(\Phi(B) \cap (\Gamma_H\backslash D_H))^0 \tag{3.12}$$

is a *special subvariety* of B.

The exponent ^{0} means to take an irreducible component of the intersection. Recalling the notation $P = \Phi(B)$ we set

$$P_H = (P \cap (\Gamma_H\backslash D_H))^0. \tag{3.13}$$

The subvariety Z may not be maximal with Mumford-Tate group H. In order to consider irreducible subvarieties that have extra Hodge classes rather than a particular Z one should use the intersection (3.12).

The standard codimension of an intersection estimate gives

$$\operatorname{codim}_{\Gamma\backslash D}(\Phi(B) \cap (\Gamma_H\backslash D_H))^0 \leqq \operatorname{codim}_{\Gamma\backslash D}(\Gamma_H\backslash D_H) + \operatorname{codim}_{\Gamma\backslash D}\Phi(B), \tag{3.14}$$

[4] In the final Reprise section of this paper we will comment on the proof of this result; this argument brings new ideas into the study of period mappings.

or in the notation (3.13)

$$\text{(3.15)} \qquad \operatorname{codim}_{\Gamma\backslash D} P_H \leqq \operatorname{codim}_{\Gamma\backslash D}(\Gamma_H\backslash D_H) + \operatorname{codim}_{\Gamma\backslash D} P.$$

DEFINITION. The subvariety $\Phi^{-1}(\Gamma_H\backslash D_H) \subset B$ is *atypical* if we have strict inequality in (3.14).

In other words, atypical means there are strictly more Hodge tensors than suggested by the standard dimension count.

MAIN THEOREM 3.16. *If the variation of Hodge structure has level at least three, then every positive dimensional special subvariety is atypical.*

As will be explained below, the proof will be to show that the condition for $\mathfrak{g}$ to have level at least three implies that the integrability conditions arising from transversality are non-trivial for *every* positive dimensional, special subvariety of B.

4. Proof of the main result

(i) *Sketch of the argument for Theorem* 3.16. The strategy is to assume equality in (3.14), or equivalently in (3.15), and from this infer that

$$\text{(4.1)} \qquad \mathfrak{h}^- = \mathfrak{g}^-.$$

This implies that

$$\text{(4.2)} \qquad D_H = D,$$

i.e., the special subvariety is all of B.

We let $\widetilde{P} \subset D$ be the inverse image in D of $P \subset \Gamma\backslash D$ and $\widetilde{P}_H = (\widetilde{P} \cap D_H)^0$. Here recall that the $(\ )^0$ means taking an irreducible component of $P \cap D_H$. From (3.15) we have

$$\text{(4.3)} \qquad \operatorname{codim}_D(\widetilde{P}_H) = \operatorname{codim}_D \widetilde{P} + \operatorname{codim}_D D_H.$$

Using the assumption (3.8) we may work infinitesimally in the tangent space $T_0 D \cong \mathfrak{g}^-$ with $T_0 D_H \cong \mathfrak{h}^-$. This gives

$$\begin{aligned} \operatorname{codim}_D \widetilde{P}_H &= \dim \mathfrak{g}^- - \dim T_0 \widetilde{P}_H, \\ \operatorname{codim}_D \widetilde{P} &= \dim \mathfrak{g}^- - \dim T_0 \widetilde{P}, \\ \operatorname{codim}_D D_H &= \dim \mathfrak{g}^- - \dim \mathfrak{h}^-. \end{aligned}$$

Then (4.3) yields

$$\dim T_0 \widetilde{P} - \dim T_0 \widetilde{P} \cap \mathfrak{h}^- = \dim \mathfrak{g}^- - \dim \mathfrak{h}^-.$$

Rewrite this as

$$\sum_{p \geqq 2} \dim \mathfrak{h}^{-p} + \operatorname{codim}_{\mathfrak{h}^{-1}} T_0 \widetilde{P}_H = \sum_{p \geqq 2} \dim \mathfrak{g}^{-p} + \operatorname{codim}_{\mathfrak{g}^{-1}} T_0 \widetilde{P}.$$

Since $\dim \mathfrak{h}^{-p} \leqq \dim \mathfrak{g}^{-p}$ and $\operatorname{codim}_{\mathfrak{h}^{-1}} T_0\widetilde{P}_H \leqq \operatorname{codim}_{\mathfrak{g}^{-1}} T_0\widetilde{P}$ this forces

$$\mathfrak{h}^{-p} = \mathfrak{g}^{-p}, \qquad p \geqq 2. \tag{4.4}$$

$$T_0 D_H = T_0 D$$

and which then implies (4.2).[5]

At this point the idea is that from the condition in (4.4) the Lie algebra $\mathcal{L}$ generated by the $\mathfrak{h}^{-k}$, $|k| \geqq 2$ is equal to the Lie algebra generated by the $\mathfrak{g}^{-k}$, $|k| \geqq 2$. Here $\mathfrak{g} = \oplus \mathfrak{g}^{-k,k}$ is the Hodge decomposition of $\mathfrak{g}_{\mathbb{C}}$ and $\mathfrak{g}^{-k} := \mathfrak{g}^{-k,k}$. If we can show that

$$\ell(\mathfrak{g}) \geqq 3 \text{ implies that } \mathcal{L}^{-} = \mathfrak{g}^{-},$$

then we are done. For this it would suffice to show that $\mathfrak{h}^{-1} = \mathfrak{g}^{-1}$. This would be the case if we show that the Hodge and root space decompositions of $\mathfrak{g}, \mathfrak{h}$ align, and then that the assumption $\ell(\mathfrak{g}) \geqq 3$ implies that all the simple roots of $\mathfrak{g}$ are also roots of $\mathfrak{h}$.[6]

(ii) *Proof of Theorem* 3.16 *under the assumption* (4.5) *below.* We will first give the argument under the assumption

$$\mathfrak{g}^{\pm} \quad \text{bracket generates all of} \quad \mathfrak{g}_{\mathbb{C}}. \tag{4.5}$$

This assumption may not be satisfied, but we shall show that in the particular circumstances at hand an adaptation of the argument assuming it gives the result.

Recall that the semi-simple Lie algebra $\mathfrak{g}$ has a weight zero polarized Hodge structure with the Hodge decomposition

$$\mathfrak{g}_{\mathbb{C}} = \oplus\, \mathfrak{g}^{k}, \qquad \mathfrak{g}^{-k} = \mathfrak{g}^{-k,k} = \overline{\mathfrak{g}^{k}}.$$

that was noted above. Let $\mathfrak{g}'_{\mathbb{C}}$ be the complex Lie sub-algebra generated by $\mathfrak{g}^{\pm}$. Then $\mathfrak{g}'_{\mathbb{C}}$ is the complexification of a real sub-algebra $\mathfrak{g}'_{\mathbb{R}}$. Writing

$$\mathfrak{g}_{\mathbb{R}} = \mathfrak{g}'_{\mathbb{R}} \oplus \mathfrak{l}_{\mathbb{R}}$$

and denoting by $^{\perp}$ the orthogonal relative to the Cartan-Killing form on $\mathfrak{g}$ it follows that

$$\mathfrak{l}_R = (\mathfrak{g}'_{\mathbb{R}})^{\perp} \subset \mathfrak{g}^0_{\mathbb{R}}$$

is an ideal in $\mathfrak{g}$. Since $\mathfrak{g}$ is semi-simple and $[\mathfrak{g}^k, \mathfrak{g}^\ell] \subset \mathfrak{g}^{k+\ell}$ it follows that $\mathfrak{l}_R$ is a direct sum of factors $\widetilde{\mathfrak{g}}_i$ of $\mathfrak{g}$ where all $\widetilde{\mathfrak{g}}_i^{\pm} = (0)$. This implies that $\exp \mathfrak{l}_R$ acts trivially on D and so we may assume that $\mathfrak{l}_R$ is trivial.

We then have

- $\mathfrak{g}$ is a Hodge Lie algebra;
- $\mathfrak{g}^+$ is a nilpotent sub-algebra of $\mathfrak{g}$;
- $\mathfrak{g}^+ \oplus \mathfrak{g}^0$ is a parabolic sub-algebra with Levi factor $\mathfrak{g}^0$;
- the center of $\mathfrak{g}^0$ is contained in a Cartan sub-algebra $\mathfrak{t}$, and, fixing a Borel subalgebra, we denote by β_i, $i \in I$, the positive simple $\mathfrak{t}$-roots of $\mathfrak{g}^+$ with corresponding root spaces $\mathfrak{g}_{\beta_i}$.

LEMMA 4.6. $\mathfrak{g}^+$ *is generated by* $\mathfrak{g}^1$ *if and only if* $\mathfrak{g}_{\beta_i} \subset \mathfrak{g}^1$ *for all* $i \in I$.

PROOF. To establish the implication $\Longrightarrow$ we note that since every positive root is $\sum n_i \beta_i$ with $0 \leqq n_i \in \mathbb{Z}$, if some $\mathfrak{g}_{\beta_i} \not\subset \mathfrak{g}^1$, then the algebra generated by $\mathfrak{g}^1$ will not contain $\mathfrak{g}_{\beta_i}$. For the converse see Theorem 3.2.1 (1), and definition 3.1.2, in [**CS09**] □

[5] We note that (4.4) is vacuous in the classical case.

[6] For the Lie theoretic aspects of Hodge theory that will be used here we suggest [**Rob14**].

Now let $\mathcal{L} \subset \mathfrak{g}$ be the sub-algebra of $\mathfrak{g}$ generated by the $\mathfrak{h}^{\pm i}$, $i \geqq 2$. It is a sub-Hodge structure and by assumption (4.5) is satisfied.

In the proof of the following proposition there are three basic facts to keep in mind.

(1) We have

$$[\mathfrak{g}_\alpha, \mathfrak{g}_\beta] = \begin{cases} \mathfrak{g}_{\alpha+\beta}, & \alpha+\beta \in \Delta(\mathfrak{g}), \\ \{0\}, & \alpha+\beta \notin \Delta(\mathfrak{g}). \end{cases}$$

(2) From $[\mathfrak{g}^k, \mathfrak{g}^\ell] \subset \mathfrak{g}^{k+\ell}$ we see that each $\mathfrak{g}^k$ is an $\mathfrak{g}^0$-module. Since $\mathfrak{t} \subset \mathfrak{g}^0$, this implies that each $\mathfrak{g}^k$, is a direct sum of root spaces; let $\Delta(\mathfrak{g}^k)$ denote the corresponding roots. (In the case $k = 0$, we must include the "zero root space" $\mathfrak{t} \subset \mathfrak{g}^0$.)

(3) Let $\Delta_0(\mathfrak{g}) = \{\beta_1, \ldots, \beta_r\}$ denote the simple roots, then $\Delta_0(\mathfrak{g}^k) = \Delta(\mathfrak{g}^k) \cap \Delta_0(\mathfrak{g})$ denotes the simple roots β_i with $\mathfrak{g}_{\beta_i} \subset \mathfrak{g}^k$. The $\mathfrak{g}^0$-module $\mathfrak{g}^1$ decomposes as a direct sum $\oplus V_{\beta_i}$ of irreducible $\mathfrak{g}_0$-submodules. The V_{β_i} are indexed by the simple roots $\beta_i \in \Delta_0(\mathfrak{g}^1)$. The root space $\mathfrak{g}_{\beta_i}$ is the lowest weight line in V_{β_i}. We have $0 \neq \mathfrak{g}_\alpha \subset V_{\beta_i}$, if and only if $\alpha \equiv \beta_i$ modulo $\Delta(\mathfrak{g}^0)$.

PROPOSITION 4.7. *Assume that* $\mathfrak{g}^1$ *generates* $\mathfrak{g}^+$ *under the Lie bracket; the level* $\ell(\mathfrak{g}) \geq 3$*; and that* $\mathfrak{h}^k = \mathfrak{g}^k$ *for all* $k \geq 2$*. Then* $\mathfrak{h}^+ = \mathfrak{g}^+$.

The proposition is a corollary of the following lemma.

LEMMA 4.8. *Assume that* $\mathfrak{g}^1$ *generates* $\mathfrak{g}^+$*, and that the level* $\ell(\mathfrak{g}) \geq 3$*. Let* $\tilde{\mathfrak{g}}$ *be the subalgebra generated by* $\mathfrak{g}^{\pm k}, k \geq 2$*. Then* $\tilde{\mathfrak{g}}$ *contains* $\mathfrak{g}^{\pm 1}$.

PROOF. It suffices to show that $\mathfrak{g}^1 \subset \tilde{\mathfrak{g}}$. Each $\mathfrak{g}^k$ is a $\mathfrak{g}_0$-module. It follows from the Jacobi identity that $\tilde{\mathfrak{g}}$ is a $\mathfrak{g}^0$-module. In particular, $\tilde{\mathfrak{g}} = \oplus \tilde{\mathfrak{g}}^k$, where $\tilde{\mathfrak{g}}^k = \mathfrak{g}^k \cap \tilde{\mathfrak{g}}$. We have $\tilde{\mathfrak{g}}^k = \mathfrak{g}^k$ for all $|k| \geq 2$.

Let $\mathfrak{g}^1 = \tilde{\mathfrak{g}}^1 \oplus V^1$ be the $\mathfrak{g}^0$-module decomposition.

We claim that $[\tilde{\mathfrak{g}}^1, V^1] = 0$. Assume not. Then there exists $\mathfrak{g}_\alpha \subset \tilde{\mathfrak{g}}^1$ and $\mathfrak{g}_{\alpha'} \subset V^1$ so that $0 \neq \mathfrak{g}_{\alpha+\alpha'} = [\mathfrak{g}_\alpha, \mathfrak{g}_{\alpha'}] \subset \mathfrak{g}^2 = \tilde{\mathfrak{g}}^2$. Then $\mathfrak{g}_{\alpha'} = [\mathfrak{g}_{\alpha+\alpha'}, \mathfrak{g}_{-\alpha}] \subset [\tilde{\mathfrak{g}}, \tilde{\mathfrak{g}}] = \tilde{\mathfrak{g}}$, a contradiction.

So $\mathfrak{g}^2 = [\mathfrak{g}^1, \mathfrak{g}^1] = [\tilde{\mathfrak{g}}^1, \tilde{\mathfrak{g}}^1] + [V^1, V^1]$. We claim that the sum is necessarily a direct sum

$$\mathfrak{g}^2 = [\mathfrak{g}^1, \mathfrak{g}^1] = [\tilde{\mathfrak{g}}^1, \tilde{\mathfrak{g}}^1] \oplus [V^1, V^1].$$

To see why note that the root spaces $\mathfrak{g}_\alpha$ of $[\mathfrak{g}^1, \mathfrak{g}^1]$ are all of the form $\alpha \equiv \beta_i + \beta_j$ modulo $\Delta(\mathfrak{g}_0)$ with $\beta_i, \beta_j \in \Delta_0(\tilde{\mathfrak{g}}^1)$. Likewise the root spaces $\mathfrak{g}_{\alpha'}$ of $[V^1, V^1]$ are all of the form $\alpha' \equiv \beta_i' + \beta_j'$ modulo $\Delta(\mathfrak{g}_0)$ with $\beta_i', \beta_j' \in \Delta_0(V^1)$. The claim then follows from the fact that $\Delta_0(\tilde{\mathfrak{g}}^1)$ and $\Delta_0(V^1)$ are disjoint sets.

Next, the Jacobi identity implies

$$\mathfrak{g}^3 = [\mathfrak{g}^1, [\mathfrak{g}^1, \mathfrak{g}^1]] = [\tilde{\mathfrak{g}}^1, [\tilde{\mathfrak{g}}^1, \tilde{\mathfrak{g}}^1]] \oplus [V^1, [V^1, V^1]].$$

Continuing inductively, we see that

$$\mathfrak{g}^+ = \tilde{\mathfrak{g}}^+ \oplus V^+,$$

with $\tilde{\mathfrak{g}}^+$ the algebra generated by $\tilde{\mathfrak{g}}^1$, and V^+ the algebra generated by V^1. We claim that this forces $V^+ = 0$. Assuming the claim for the moment, we have $\tilde{\mathfrak{g}}^1 = \mathfrak{g}^1$. This establishes the lemma.

To see why the claim $V^+ = 0$ holds, note that $\Delta(\mathfrak{g}^+)$ must contain the highest root $\tilde{\alpha}$. The coefficient of $\beta_i \in \Delta_0(\mathfrak{g})$ in $\tilde{\alpha}$ is necessarily nonzero for *every* simple root β_i. But if $\tilde{\alpha} \in \Delta(\tilde{\mathfrak{g}}^+)$, then the coefficient of $\beta_i \in \Delta_0(V^1)$ is necessarily zero. That is, $\tilde{\alpha} \in \Delta(\tilde{\mathfrak{g}}^+)$ implies $\Delta_0(V^1) = \emptyset$ (which is equivalent to $V^1 = 0$). Likewise if $\tilde{\alpha} \in \Delta(V^+)$, then $\Delta_0(\tilde{\mathfrak{g}}^1) = \emptyset$. So one of $\Delta_0(\tilde{\mathfrak{g}}^1)$ or $\Delta_0(V^1)$ must be empty. By hypothesis $[\mathfrak{g}^1, \mathfrak{g}^2] = \mathfrak{g}^3 \neq 0$; this implies $0 \neq [\mathfrak{g}^3, \mathfrak{g}^{-2}] \subset \tilde{\mathfrak{g}}^1$. □

The final step using $\ell(\mathfrak{g}) \geqq 3$ is the

LEMMA 4.9. $\ell(\mathfrak{g}) \geqq 3 \implies [\mathfrak{g}^{-2}, \mathfrak{g}^3] \subset \mathfrak{h} \cap \mathfrak{g}^1$ *is non-zero.*

PROOF. Since $\mathfrak{g}^1$ generates $\mathfrak{g}^+$ and $\mathfrak{g}^1$ is spanned by simple positive root vectors, there exist simple positive roots $\beta_1, \beta_2, \beta_3$ such that $\beta_1 + \beta_2 + \beta_3$ is a root. Then $\mathfrak{g}_{-\beta_1-\beta_2} \in \mathfrak{g}^{-2}$ and

$$\left[\mathfrak{g}_{\beta_1+\beta_2+\beta_3}, \mathfrak{g}_{-\beta_1-\beta_2}\right] = \mathfrak{g}_{\beta_3} \neq 0.$$ □

(iii) *Discussion of assumption of* (4.5). The fact that we may assume (4.5) uses the following result from Proposition 3.10 in [**Rob14**],[7]

THEOREM 4.10. *Let G be a Mumford-Tate group and $D_G = G(\mathbb{R})/H_0$ a Mumford-Tate domain. Any point $o \in D_G$ defines a weight zero Hodge structure on the Lie algebra $\mathfrak{g}$ of G. Let $\widetilde{\mathfrak{g}} \subset \mathfrak{g}$ be the real semi-simple Lie subalgebra generated by $\mathfrak{g}^{-1,1} \oplus \mathfrak{g}^{1,-1}$. Then $\widetilde{\mathfrak{g}} \subset \mathfrak{g}$ is a real sub-Hodge structure. Any connected integral manifold of the horizontal sub-bundle corresponding to $\mathfrak{g}^{-1,1} \subset T_0 D_G$ is contained in $D_{\widetilde{G}} = \widetilde{G}(\mathbb{R})o$ where $\widetilde{G}(\mathbb{R})$ is the connected real Lie group with Lie algebra $\widetilde{\mathfrak{g}}$. The horizontal sub-bundle $T^h D_{\widetilde{G}} \subset TD_{\widetilde{G}}$ is bracket generating; equivalently, $\widetilde{\mathfrak{g}}^{-1,1}$ generates $\widetilde{\mathfrak{g}}^-$ under Lie bracket.*

The general picture is that $\widetilde{\mathfrak{g}}$ generates an integrable sub-bundle of TD_G and the $D_{\widetilde{G}} = \widetilde{G}(\mathbb{R})_o$ above is a leaf of the corresponding foliation.

Using this theorem the assumption (4.5) may be dropped thus completing the proof of Theorem 3.16. □

The result 4.10, specifically the last sentence, is the key to where the integrability conditions imposed by transversality kick in to give atypicality of Hodge loci.

We also note that in [**Rob14**] it is shown that for the grading element $E \in \mathfrak{h}$ defined by condition $[E, X] = pX$ for all $X \in \mathfrak{h}^{p,-p}$ (thus $E \in$ center of $\mathfrak{h}^{0,0}$)

$$\ell(\mathfrak{h}) = \widetilde{\alpha}(E) \tag{4.11}$$

where $\widetilde{\alpha}$ is the highest root.

Using this we can see that it is *not* always the case that $\mathfrak{h}^{-1,1}$ generates $\mathfrak{h}^{-,+}$. To construct examples let

$$\varphi : S^1 \to H(\mathbb{R})$$

be the circle defining the complex structure on D_H. Then for the E defined above, we can choose a Cartan subalgebra $\mathfrak{t} \subset \mathfrak{h}_{\mathbb{C}}$ and a set $\Delta \subset \mathfrak{t}$ of simple roots such that $E \in \mathfrak{t}$ and $0 \leqq \alpha(E) \in \mathbb{Z}$ for all $\alpha \in \Delta$. Then it can be shown (loc. cit.) that

$$\mathfrak{h}^{1,-1} \text{ generates } \mathfrak{h}^{+,-} \iff \alpha(E) \in \{0, 1\}, \qquad \alpha \in \Delta. \tag{4.12}$$

Denote by $E_{\mathfrak{g}}$ the grading element for $\mathfrak{g}$. Then from (4.11) we have

$$\ell(\mathfrak{g}) \leqq 2 \iff \widetilde{\alpha}(E_{\mathfrak{g}}) \leqq 2 \tag{4.13}$$

where $\widetilde{\alpha}$ is the highest root. Thus, *if* we know the Mumford-Tate Lie algebra $\mathfrak{g}$, then this provides a test for when the main theorem applies.

[7] The instances of $\mathbb{Q}$ in loc. cit. are typos; the $\mathbb{Q}$ should be $\mathbb{R}$.

5. Reprise

(i) In the second example above if we assume that the algebra $\mathrm{Hg}^\bullet(X)_{\mathrm{prim}}$ corresponding to $H^4(X)_{\mathrm{prim}}$ is generated by Q and λ, then

$$G_\lambda := \{g \in \mathrm{Aut}(V,Q) : g\lambda = \lambda\}$$

is the Mumford-Tate group. In (3.14) we take $H = G_\lambda$, $\Gamma_\lambda = \Gamma \cap G_\lambda$ and set $D_H = D_\lambda \subset D$. Then it can be shown that

(5.1)

$$\operatorname{codim}_{\Gamma\backslash D}(\Phi(B) \cap (\Gamma_\lambda\backslash D_\lambda)) = \operatorname{codim}_{\Gamma\backslash D}(\Gamma_\lambda\backslash D_\lambda) + \operatorname{codim}_{\Gamma\backslash D}\Phi(B) - \dim\sigma(\lambda).$$

Thus $\dim\sigma(\lambda)$ is the correction term needed to convert the inequality (3.14) into an equality.

(ii) Referring to the first example, for $d = \deg X \geqq 5$

$$d - 3 = \operatorname{codim} \mathrm{NL}_\lambda < h^{2,0}(X) \tag{5.2}$$

holds only for X's containing a line Λ. In this case the strict inequality holds for geometric, not Hodge theoretic, reasons. Namely, if $\omega \in H^0(\Omega^2_X)$ has divisor $(\omega) \supset \Lambda$ containing the line, then for any $\theta \in T$ we have

$$\langle \omega, \theta\cdot\lambda\rangle = 0$$

due to $\theta\cdot\omega\big|_\lambda = 0$ at the form level.

This phenomenon is general; we may say that the correction term needed to have equality in (3.15) is always positive if $\mathfrak{g}^3 \neq (0)$, and in particular cases it may be greater than it is for a general Hodge locus in B due to geometric reasons peculiar to the particular Hodge locus.

These considerations raise the following

Q: Given $(V, Q, F^\bullet)$ with Mumford-Tate group G where $\mathfrak{g}^3 \neq 0$, is there a *uniform bound* depending only $\mathfrak{g}$ for the correction term needed to convert (3.14) into an equality?

(iii) An example where $\mathfrak{g}^3 = (0)$ is given by a variation of polarized Hodge structures of the form $\mathbb{V}_b = \mathbb{V}'_b \otimes \mathbb{V}''_b$ where each of the factors has weight 2. As discussed above the condition $\mathfrak{g}^3 = 0$ is a non-linear PDE and the general Cartan theory of exterior differential systems gives in principle an algorithmic process for determining all local solutions. The example given just above is the only one we know in weight $n = 4$ where $\mathfrak{g}^{-1,1} \neq 0$ and $\mathfrak{g}^3 = 0$.

(iv) *Hypersurface example.* Let $X \subset \mathbb{P}^{n+1}$ be a smooth degree d hypersurface given by an equation

$$F(x) = 0 \tag{5.3}$$

where $x = [x_0, \dots, x_{n+1}]$ and $F(x)$ is homogeneous of degree d. For

$$\begin{cases} S^\bullet = \mathbb{C}[x_0, \cdots, x_{n+1}], \\ J^\bullet_F = \text{Jacobian ideal } \{F_{x_0}, \cdots, F_{x_{n+1}}\}, \\ R^\bullet = S^\bullet / J^\bullet_F \end{cases}$$

it is well known ([**Gre94**]) that there is an isomorphism

$$H^{p,n-p}(X)_{\mathrm{prim}} \cong R^{(n-p)d+n-2}.$$

Moreover the tangent space to the family of X's modulo projective equivalence is

$$T \cong R^d$$

and the maps

$$((\mathrm{V}.4)_p) \qquad T \otimes H^{p,n-p}(X)_{\mathrm{prim}} \to H^{p-1,n-p+1}(X)_{\mathrm{prim}}$$

are given by multiplication of polynomials

$$((\mathrm{V}.5)_p) \qquad R^d \otimes R^{(n-p)d+n-2} \to R^{(n-p+1)d+n-2}.$$

Finally since X is non-singular, it follows from Macaulay's theorem that

(5.6) *the mappings* $(V.5)_p$ *are non-zero whenever both sides are non-zero*

(cf. [**Gre94**]).

If G is the Mumford-Tate group for the period mapping of X's as above, then we have

$$R^d \to \mathfrak{g}^{-1,1} \subset F^{-1}\operatorname{End}(V,Q).$$

The image of this map is an *abelian* sub-algebra $\mathfrak{A} \subset \mathfrak{g}^{-1,1} \subset \mathfrak{g}$. There is an induced map

$$\operatorname{Sym}^k \mathfrak{A} \to \mathfrak{g}^{-k,k}. \tag{5.7}$$

This then gives

$$\operatorname{Sym}^k R^d \to \mathfrak{g}^{-k,k} \subset \oplus \operatorname{Hom}\left(H^{p,n-p}(X)_{\mathrm{prim}}, H^{n-k,n-p+k}(X)_{\mathrm{prim}}\right)$$

which is just the map

$$\operatorname{Sym}^k R^d \otimes R^{(n-p)d+n-2} \to R^{(n-p+k)d+n-2}$$

given by multiplication of polynomials. From (5.6) we may conclude that the map is non-zero whenever both sides are non-zero, which then gives that $\mathfrak{g}^{-3,3} \neq 0$ for $n \geqq 3, d > 3$. □

(v) *The coupling length:* This is defined by $\zeta(\mathfrak{A}) = \max\{m : \operatorname{Sym}^m \mathfrak{A} \to \operatorname{Hom}(\mathbb{V}_b^{n,0}, \mathbb{V}_b^{n-m,m}) \text{ is} \neq 0\}$ at a general point of b. The same argument as in the hypersurfaces example then gives (1.3) above. There are many examples where this holds.

(vi) The main Theorem III.14 from [**BKU24**] is closely related to the Ax-Schanuel conjecture, which has been established in [**BT19**]. To state the result we keep the above notations and denote by $\check{D}$ the compact dual to the Mumford-Tate domain D. Then $\check{D} = G(\mathbb{C})/P$ is a homogeneous rational variety.

THEOREM 5.8 ([**BT19**]). *Let* $W \subset B \times \check{D}$ *be an algebraic subvariety and* U *and irreducible component of* $W \cap B \times_{\Gamma\backslash D} \check{D}$ *such that*

$$\operatorname{codim}_{B\times\check{D}}(U) < \operatorname{codim}_{B\times\check{D}}(W) + \operatorname{codim}_{B\times\check{D}}(B \times_{\Gamma\backslash D} \check{D}).$$

Then the projection of U *to* B *is contained in a proper Mumford-Tate subvariety.*

In the geometric case this says that the analytic locus in B where the periods satisfy a given set of algebraic relations must be of the expected codimension unless there is a reduction in the generic Mumford-Tate group.

(vii) Finally we will comment on the proof in [**BBT23**] that the image P of a period mapping

$$\Phi : B \to \Gamma\backslash D$$

is a quasi-projective algebraic variety.

- In the classical case there is the Satake-Baily-Borel completion $\overline{\Gamma\backslash D}_{\mathrm{SBB}}$ of $\Gamma\backslash D$. It is a projective algebraic variety, and Borel's extension theorem gives that (after possibly passing to a finite branched cover of $\overline{B}$) Φ extends to a morphism
$$\overline{\Phi}:\overline{B}\to\overline{\Gamma\backslash D}_{\mathrm{SBB}}$$
of algebraic varieties. This gives the result.
- In the non-classical case the Hodge variety $\Gamma\backslash D$ is not an algebraic variety; on it the only global meromorphic functions are constant.
- At this juncture the new concept of a *definable o-minimal structure* enters the picture. A non-compact analytic variety can have such a structure only if it is "tame at infinity" (cf. [**BKT20**]).
- For definable o-minimal maps of algebraic varieties to analytic varieties having a definable o-minimal structure there are algebraicity results; among other things the image is algebraic.
- The by now classical analysis of degenerations of Hodge structures ([**Sch73**], [**CKS86**], [**Kas85**]) is then used to show that the images of period mappings have a definable o-minimal structure.
- Finally the line bundle $L := \overset{p}{\otimes}\det F^p$ may be shown to induce an ample line bundle on the image $\Phi(B)\subset\Gamma\backslash D$, which is then a quasi-projective algebraic variety.[8]

The use of definable o-minimal structures has introduced a whole new set of ideas into Hodge theory and into the arithmetic properties of a variation of Hodge structure and period mappings ([**Kli23**]).

References

[BBKT24] B. Bakker, Y. Brunebarbe, B. Klingler, and J. Tsimerman, *Definability of mixed period maps*, J. Eur. Math. Soc. (JEMS) **26** (2024), no. 6, 2191–2209, DOI 10.4171/jems/1319. MR4742809

[BBT23] B. Bakker, Y. Brunebarbe, and J. Tsimerman, *o-minimal GAGA and a conjecture of Griffiths*, Invent. Math. **232** (2023), no. 1, 163–228, DOI 10.1007/s00222-022-01166-1. MR4557401

[BCG$^+$91] R. L. Bryant, S. S. Chern, R. B. Gardner, H. L. Goldschmidt, and P. A. Griffiths, *Exterior differential systems*, Mathematical Sciences Research Institute Publications, vol. 18, Springer-Verlag, New York, 1991, DOI 10.1007/978-1-4613-9714-4. MR1083148

[BFMT] Benjamin Bakker, Stefano Filipazzi, Mirko Mauri, and Jacob Tsimerman, *Baily-Borel compactifications of period images and the B-semiampleness conjecture.*

[BKT20] B. Bakker, B. Klingler, and J. Tsimerman, *Tame topology of arithmetic quotients and algebraicity of Hodge loci*, J. Amer. Math. Soc. **33** (2020), no. 4, 917–939, DOI 10.1090/jams/952. MR4155216

[BKU24] G. Baldi, B. Klingler, and E. Ullmo, *On the distribution of the Hodge locus*, Invent. Math. **235** (2024), no. 2, 441–487, DOI 10.1007/s00222-023-01226-0. MR4689371

[BT19] B. Bakker and J. Tsimerman, *The Ax-Schanuel conjecture for variations of Hodge structures*, Invent. Math. **217** (2019), no. 1, 77–94, DOI 10.1007/s00222-019-00863-8. MR3958791

[CDK95] E. Cattani, P. Deligne, and A. Kaplan, *On the locus of Hodge classes*, J. Amer. Math. Soc. **8** (1995), no. 2, 483–506, DOI 10.2307/2152824. MR1273413

[Chi21] K. C. T. Chiu, *Ax-Schanuel for variations of mixed Hodge structures*, 2021.

[CKS86] E. Cattani, A. Kaplan, and W. Schmid, *Degeneration of Hodge structures*, Ann. of Math. (2) **123** (1986), no. 3, 457–535, DOI 10.2307/1971333. MR840721

[8]It has recently been proved [**BFMT**] that the canonical extension $L_e\to\bar{B}$ of $L\to B$ is semi-ample; thus $\mathrm{Proj}(L_e)$ defines a canonical projective completion of $\Phi(B)$ giving an extension of the classical Baily-Borel-Satake compactification of the image of period mappings.

[CMSP17] J. Carlson, S. Müller-Stach, and C. Peters, *Period mappings and period domains*, Cambridge Studies in Advanced Mathematics, vol. 168, Cambridge University Press, Cambridge, 2017. Second edition of [MR2012297]. MR3727160

[CS09] A. Čap and J. Slovák, *Parabolic geometries. I*, Mathematical Surveys and Monographs, vol. 154, American Mathematical Society, Providence, RI, 2009. Background and general theory, DOI 10.1090/surv/154. MR2532439

[GG06] M. Green and P. Griffiths, *Algebraic cycles and singularities of normal functions. II*, Inspired by S. S. Chern, Nankai Tracts Math., vol. 11, World Sci. Publ., Hackensack, NJ, 2006, pp. 179–268, DOI 10.1142/9789812772688_0009. MR2313336

[GGK12] M. Green, P. Griffiths, and M. Kerr, *Mumford-Tate groups and domains*, Annals of Mathematics Studies, vol. 183, Princeton University Press, Princeton, NJ, 2012. Their geometry and arithmetic. MR2918237

[Gre88] M. L. Green, *A new proof of the explicit Noether-Lefschetz theorem*, J. Differential Geom. **27** (1988), no. 1, 155–159. MR918461

[Gre89] M. L. Green, *Components of maximal dimension in the Noether-Lefschetz locus*, J. Differential Geom. **29** (1989), no. 2, 295–302. MR982176

[Gre94] M. L. Green, *Infinitesimal methods in Hodge theory*, Algebraic cycles and Hodge theory (Torino, 1993), Lecture Notes in Math., vol. 1594, Springer, Berlin, 1994, pp. 1–92, DOI 10.1007/978-3-540-49046-3_1. MR1335239

[HP16] P. Habegger and J. Pila, *O-minimality and certain atypical intersections* (English, with English and French summaries), Ann. Sci. Éc. Norm. Supér. (4) **49** (2016), no. 4, 813–858, DOI 10.24033/asens.2296. MR3552014

[Kas85] M. Kashiwara, *The asymptotic behavior of a variation of polarized Hodge structure*, Publ. Res. Inst. Math. Sci. **21** (1985), no. 4, 853–875, DOI 10.2977/prims/1195178935. MR817170

[Kli23] B. Klingler, *Hodge theory, between algebraicity and transcendence*, ICM—International Congress of Mathematicians. Vol. 3. Sections 1–4, EMS Press, Berlin, [2023] ©2023, pp. 2250–2284. MR4680317

[KO21] B. Klingler and A. Otwinowska, *On the closure of the Hodge locus of positive period dimension*, Invent. Math. **225** (2021), no. 3, 857–883, DOI 10.1007/s00222-021-01042-4. MR4296351

[Rob14] C. Robles, *Schubert varieties as variations of Hodge structure*, Selecta Math. (N.S.) **20** (2014), no. 3, 719–768, DOI 10.1007/s00029-014-0148-8. MR3217458

[Sch73] W. Schmid, *Variation of Hodge structure: the singularities of the period mapping*, Invent. Math. **22** (1973), 211–319, DOI 10.1007/BF01389674. MR382272

[Urb24] D. Urbanik, *On the transcendence of period images*, J. Differential Geom. **127** (2024), no. 2, 615–662, DOI 10.4310/jdg/1717772423. MR4756091

[Voi07a] C. Voisin, *Hodge theory and complex algebraic geometry. I*, Reprint of the 2002 English edition, Cambridge Studies in Advanced Mathematics, vol. 76, Cambridge University Press, Cambridge, 2007. Translated from the French by Leila Schneps. MR2451566

[Voi07b] C. Voisin, *Hodge theory and complex algebraic geometry. II*, English edition, Cambridge Studies in Advanced Mathematics, vol. 77, Cambridge University Press, Cambridge, 2007. Translated from the French by Leila Schneps. MR2449178

[VZ05] E. Viehweg and K. Zuo, *Complex multiplication, Griffiths-Yukawa couplings, and rigidity for families of hypersurfaces*, J. Algebraic Geom. **14** (2005), no. 3, 481–528, DOI 10.1090/S1056-3911-05-00400-5. MR2129008

Institute for Advanced Study, Einstein Drive, Princeton, New Jersey 08540; and Department of Mathematics, University of Miami, Coral Gables, Florida 33124

Email address: pg@ias.edu

Contemporary Mathematics
Volume **842**, 2026
https://doi.org/10.1090/conm/842/16857

Motivic irrationality proofs

Matt Kerr

ABSTRACT. We exhibit geometric conditions on a family of toric hypersurfaces under which the value of a canonical normal function at a point of maximal unipotent monodromy is irrational.

1. Introduction

The limit of a generalized normal function at a point where the underlying variation of Hodge structure degenerates, as recently studied in [**7K**], turns out to have an unexpected arithmetic application. R. Apéry's famous proof (see [**vdP**]) of irrationality of $\zeta(3) := \sum_{k\geq 1} k^{-3}$ relies on the existence of rapidly divergent sequences $a_m \in \mathbb{Z}$, $b_m \in \mathbb{Q}$ (the latter having denominators of bounded growth) with $2a_m\zeta(3)+b_m$ converging rapidly to zero. (Here we have swapped the meanings of a_m and b_m in [**vdP**] and multiplied the latter by -2.) Beukers, Peters and Stienstra [**Be**, **BP**, **Pe**, **PS**] geometrically repackaged much of the proof, noting for instance that the generating function $\sum_{m\geq 0} a_m\lambda^m =: A(\lambda)$ records the period of a holomorphic 2-form on a family of $K3$ surfaces $\{X^\lambda\}_{\lambda\in\mathbb{P}^1}$, hence must satisfy a Picard-Fuchs differential equation $D_{\mathrm{PF}}A(\lambda) = 0$.

Behind the remaining details of the irrationality proof lurks a family of cycles in (algebraic) K_3 of the $K3$. The associated higher normal function $\tilde{V}(\lambda)$ has special value $\tilde{V}(0) = -2\zeta(3)$, and satisfies the inhomogeneous equation $D_{\mathrm{PF}}\tilde{V}(\lambda) = Y(\lambda)$, where Y denotes the Yukawa coupling.[1] Setting $\sum_{m\geq 1} b_m\lambda^m := A(\lambda)\tilde{V}(0) - \tilde{V}(\lambda)$, one deduces from this a recurrence relation on the $\{b_m\}$ which (as presented here) give "half" of the required bound on denominator growth. The other "half" comes from the *Fermi family* of $K3$s studied, but not linked to the Apéry proof, in [**PS**]. Finally, the behavior of the cycles at singular members of the family $\{X^\lambda\}$ shows that $\tilde{V}(\lambda)$ has no monodromy about the conifold singular fiber closest to $\lambda = 0$, implying the "overconvergence" of its power series and thus the rapid convergence of $2a_m\zeta(3) + b_m \to 0$.

In this paper, we reveal a general criterion for the irrationality of special values of certain higher normal functions. (An expository, much less technical account of the proof for $\zeta(3)$ will appear elsewhere.) Given a Laurent polynomial $\phi(x_1, \dots, x_n)$

2020 *Mathematics Subject Classification.* Primary 14C25, 14C30, 19E15, 11J82.

[1] See (3.1) and (3.4) for the definitions of $\tilde{V}$ and Y. The inhomogeneous equation, differently normalized, appears in (3.7). The branch of $\tilde{V}$ that we are evaluating at 0 is described in §3.1 and §4.

with reflexive Newton polytope Δ, the equation $\phi(\underline{x}) = \lambda$ defines a family of Calabi-Yau hypersurfaces $\{X^\lambda\}$ in the toric variety $\mathbb{P}_\Delta$. Associated to this family is a pure irreducible variation of Hodge structure $\mathcal{V}_\phi$ of weight $(n-1)$ over a Zariski open $\mathcal{U} \subset \mathbb{P}^1$, together with a canonical section $\{\tilde{\omega}^\lambda\}$ of the Hodge line bundle $\mathcal{F}^{n-1}\mathcal{V}_\phi$. We call ϕ *tempered* if the coordinate symbol $\{x_1, \ldots, x_n\}$ lifts to a family of motivic cohomology classes $\Xi^\lambda \in H^n_{\mathcal{M}}(X^\lambda, \mathbb{Q}(n))$ on the family, producing an extension

$$0 \to \mathcal{V}_\phi(n) \to \mathcal{E}_\phi \to \mathbb{Q}(0)_{\mathcal{U}} \to 0 \tag{1.1}$$

of admissible variations of mixed Hodge structure over $\mathcal{U}$. (This temperedness typically holds, for example, for Landau-Ginzburg models constructed from Minkowski polynomials [**dS**].) Applying a variant of this hypothesis allows us to construct a canonical truncated higher normal function $\tilde{V}(\lambda)$ on $\mathbb{P}^1 \backslash \phi(\mathbb{R}_-^{\times n})$ by pairing the regulator class of Ξ^λ (i.e., the extension class of (1.1)) with $\tilde{\omega}^\lambda$ (see Theorem 4.2).[2]

To arrange $\tilde{V}(0) \notin \mathbb{Q}$, we must impose several additional conditions on ϕ, roughly as follows (see Theorem 3.2 and Prop. 3.3):

- the local system of periods of $\tilde{\omega}^\lambda$ must be of rank n, admit an isomorphism to its pullback by $\lambda \mapsto C/\lambda$ (for some $C \in \mathbb{Q}^\times$), and have two mild singularities apart from 0 and ∞, one of which is very far from 0;
- $\phi(-\underline{x})$ has positive integer coefficients, and the Picard-Fuchs operator associated to $\tilde{\omega}^\lambda$ (suitably normalized) is integral; and
- a finite $(r:1)$ pullback of the family X^λ can be presented as a family of toric hypersurfaces in $\mathbb{P}_\Diamond$, where $\Diamond$ is a "facile" polytope (Definition 2.2).

The role of the last condition is to produce a basis of periods whose power-series coefficients have the right denominator bounds (see Corollary 2.6). This basis is closely tied to mirror symmetry [**HLY**] and the Frobenius method [**IKSY**]; for $n \geq 4$, Theorem 2.4 uncovers a surprising arithmetic implication of the Hyperplane Conjecture [**HLY**, **LZ**], which may be of independent interest. We also remark that, assuming only temperedness, the higher normal function $\tilde{V}(\lambda)$ can always be written as one of the chain-integral solutions of [**HLYZ**]. Moreover, its asymptotics at $\lambda = \infty$ are related to the Frobenius constants and gamma-classes of [**GGI**, **GI**, **GZ**, **BV**, **Ke2**], while $\tilde{V}(0)$ may be interpreted as an Apéry constant as studied in [**Ga**, **Go**, **GKS**].

In the last section, we exhibit Laurent polynomials which satisfy all these conditions for $n = 1$, 2, and 3, recovering irrationality of $\log(1 + b^{-1})$ $(b \in \mathbb{N})$, $\zeta(2)$, and $\zeta(3)$. We also propose relaxations of some of the conditions, together with specific families of polynomials, for making contact with results on linear forms in more than one odd zeta value—for instance [**Va**], [**Z1**], [**BR**], and especially [**Br**], whose basic cellular integrals on $\mathcal{M}_{0,n+3}$ are the power series coefficients of a $\tilde{V}(\lambda)$ as above (see §5.4 for this connection).

While the results on $\zeta(2)$ and $\zeta(3)$ complete, in a way, the story begun in [**BP**], the reader familiar with those works will notice (perhaps deplore?) the complete lack of reference to modular forms in what follows. The omission is strategic, as a weight-$(n-1)$ VHS $\mathcal{V}_\phi$ with maximal unipotent and conifold monodromies

[2]This $\tilde{V}$ may be viewed as a period of the extension *dual* to (1.1). In Hodge theory, the extension (1.1) is typically called a (higher) normal function; the term "truncation" refers to the pairing just described.

cannot have a modular parametrization[3] for $n \geq 4$. This is, of course, precisely where we hope to stimulate the search for examples with Theorem 3.2, starting with the increasingly sophisticated databases of polytopes, local systems, Calabi-Yau differential operators and their geometric realizations [**AESZ**, **ASZ**, **Fano1**, **Fano2**, **DM**]. (It is also the setting where, in [**Ke2**, **GKS**] and articles to come, we will extend the methods developed below in connection with Apéry and gamma constants of Fano varieties.)

The same reader may be puzzled by our reference to "splitting up the bound" on denominators of b_m, as in the Apéry $\zeta(3)$ story one simply shows that $2(L_m)^3 b_m \in \mathbb{Z}$, where $L_m := \mathrm{lcm}\{1, 2, \ldots, m\}$. What we are able to show under *general* hypotheses, using techniques from Hodge theory and mirror symmetry, is instead that (for some unknown fixed $\varepsilon \in \mathbb{N}$) both $\varepsilon(L_{rm})^n b_m$ and $\varepsilon(m!)^n b_m$ are integers for all m. Though weaker than the above bound in the Apéry $\zeta(3)$ ($r = 2$, $n = 3$) case, this is still enough for an irrationality proof. With that said, the results of this article are intended to be a narrow proof of concept for an approach to irrationality proofs via motivic cohomology and Hodge theory, rather than to be optimal as respects either methodology or hypotheses. In particular, the removal of the restriction to normal functions arising from K_n of Calabi-Yau $(n-1)$-folds will be considered in a subsequent work.[4]

We freely (though infrequently) use the notation and terminology of regulator currents (e.g. T_Ξ, Ω_Ξ, R_Ξ) and toric varieties throughout, as reviewed for instance in §§1-2 of [**DK**]. The heart of this paper is §3 (together with the examples in §5), and no more than a quick perusal of the technical §2 is needed on a first reading. The difference in notation ($\Diamond$ vs. Δ) for polytopes in §2 and §3 is intended to draw a bright line between the family $X^\lambda \subset \mathbb{P}_\Delta$ and its facile r-cover $\mathrm{X}_\xi \subset \mathbb{P}_\Diamond$. Similarly, before bothering with §4 (where the regulator-current calculus has its main appearance) the reader may prefer to look at how it is used in §§5.2–5.3.

Throughout we shall write ∂_t resp. δ_t for $\frac{\partial}{\partial t}$ resp. $t\frac{\partial}{\partial t}$.

Acknowledgments

It is my pleasure to thank S. Bloch, F. Brown, C. Doran, V. Golyshev, B. Lian, V. Maillot, F. Rodriguez-Villegas, D. Zagier and W. Zudilin for helpful conversations and correspondence—some recent, and some further back. I thank the referee as well for their expository suggestions. This work was partially supported by NSF FRG Grant DMS-1361147 and Simons Collaboration Grant 634268.

2. Facile polytopes

Let $\Diamond, \Diamond^\circ$ be a dual pair of reflexive polytopes in $\mathbb{R}^n$, admitting regular projective triangulations $\mathcal{T}, \mathcal{T}^\circ$. Take Σ, Σ° to be the fans on these triangulations, and $\mathbb{P}_{\Diamond^\circ}, \mathbb{P}_\Diamond$ (respectively—note the reversal) the toric Fano n-folds determined by the

[3]in the sense of global monodromy factoring through a representation of SL_2. The presence of maximal unipotent monodromy forces a copy of the nth symmetric power irrep, whence action by a transvection resp. reflection T with $\mathrm{rank}(T - I) = 1$ is only possible for $n = 2$ resp. 3.

[4]Two particularly noteworthy articles which have appeared since this paper was written, and with which we hope to make contact, are [**BZ**] and [**CDT**]. The new strategy for irrationality proofs introduced in [**CDT**], in particular, makes use of "overconvergent" power series like our $\tilde{V}_\phi$. Indeed, one of the functions used in [op. cit.] to prove irrationality of $L(\chi_{-3}, 2)$ is essentially the higher normal function arising from one of Zagier's examples (cf. Remark 5.1).

fans. (See [**CK**, Chap. 3-4] for definitions of reflexive polytopes, $\mathbb{P}_\Diamond$, and the like.) Write[5]

$$\mathcal{A} = \mathcal{A}_\Diamond := \Diamond \cap \mathbb{Z}^n = \{\underline{v}^{(0)} = \underline{0},\, \underline{v}^{(1)}, \ldots, \underline{v}^{(N)}\}$$

for the integer points, and

$$\mathbb{L} := \left\{ \underline{\ell} = \left(\ell_0 = -\Sigma_{i=1}^N \ell_i,\, \ell_1, \ldots, \ell_N\right) \in \mathbb{Z}^{N+1} \,\middle|\, \Sigma_{i=1}^N \ell_i \underline{v}^{(i)} = \underline{0} \right\}$$

for the lattice of integral relations on them. Let $x_1, \ldots, x_n$ denote the coordinates on $\mathbb{G}_m^n$ and also their extensions to rational functions on $\mathbb{P}_\Diamond$ and $\mathbb{P}_{\Diamond^\circ}$. The irreducible components $\{D_i\}_{i=1}^N$ of $\mathbb{P}_{\Diamond^\circ} \backslash \mathbb{G}_m^n = \mathbb{D}_{\Diamond^\circ}$ (with $\deg_{D_i}(x_k) = v_k^{(i)}$) generate $H^2(\mathbb{P}_{\Diamond^\circ}, \mathbb{Z})$, and $K_{\mathbb{P}_{\Diamond^\circ}} = -\sum_{i=1}^N D_i =: D_0$.

Assume that the Mori cone $\mathcal{M} = \mathcal{M}_{\Diamond^\circ} \subset H_2(\mathbb{P}_{\Diamond^\circ}, \mathbb{R})$ (of classes of effective curves) is regular simplicial, so that $\mathbb{M} := \mathcal{M} \cap H_2(\mathbb{P}_{\Diamond^\circ}, \mathbb{Z}) = \mathbb{Z}_{\geq 0}\langle C_1, \ldots, C_{M=N-n}\rangle$. Write $C_j \mapsto \underline{\ell}^{(j)} = C_j \cdot \underline{D}$ for the images under

$$H_2(\mathbb{P}_{\Diamond^\circ}, \mathbb{Z}) \overset{\cong}{\to} \mathbb{L} \overset{\cong}{\to} \operatorname{Hom}\left(H^2(\mathbb{P}_{\Diamond^\circ}, \mathbb{Z}), \mathbb{Z}\right),$$

so that the dual (nef) cone $\overline{\mathcal{K}} \subset H^2(\mathbb{P}_{\Diamond^\circ}, \mathbb{R})$ has $\mathbb{K} := \overline{\mathcal{K}} \cap H^2(\mathbb{P}_{\Diamond^\circ}, \mathbb{Z}) = \mathbb{Z}_{\geq 0}\langle J_1, \ldots, J_M\rangle$ with $J_k \cdot C_j = \delta_{kj}$ and $D_i = \sum_{k=1}^M \ell_i^{(k)} J_k$. (To compute the $\underline{\ell}^{(k)}$, one may use primitive collections as in [**LZ**].)

Let $F_{\underline{\mathsf{a}}}(\underline{x}) := \sum_{i=0}^N \mathsf{a}_i \underline{x}^{\underline{v}^{(i)}}$, and $X_{\underline{\mathsf{a}}}^\times := \{F_{\underline{\mathsf{a}}}(\underline{x}) = \underline{0}\} \subset \mathbb{G}_m^n$ with (CY $(n-1)$-fold) Zariski closure $X_{\underline{\mathsf{a}}} \overset{\imath}{\hookrightarrow} \mathbb{P}_\Diamond$; take $X^\circ \subset \mathbb{P}_{\Diamond^\circ}$ to be any smooth, $\Diamond^\circ$-regular anticanonical hypersurface. We consider the $\mathbb{A}$-*periods* ($\mathbb{A} = \mathbb{Z}, \mathbb{Q}$, or $\mathbb{C}$) of

$$\omega_{\underline{\mathsf{a}}} = \operatorname{Res}_{X_{\underline{\mathsf{a}}}} \Omega_{\underline{\mathsf{a}}} = \tfrac{\mathsf{a}_0}{(2\pi\mathbf{i})^{n-1}} \operatorname{Res}_{X_{\underline{\mathsf{a}}}} \left(\tfrac{dx_1/x_1 \wedge \cdots \wedge dx_n/x_n}{F_{\underline{\mathsf{a}}}(\underline{x})} \right) \in \Omega^{n-1}(X_{\underline{\mathsf{a}}}),$$

which are the multivalued functions $\pi_\gamma(\underline{\mathsf{a}}) = \int_\gamma \omega_{\underline{\mathsf{a}}}$ ($\gamma \in H_{n-1}(X_{\underline{\mathsf{a}}}, \mathbb{A})$) defined on a Zariski open in the affine $(N+1)$-space with coordinates $(\mathsf{a}_0, \ldots, \mathsf{a}_N)$. These periods are known to solve the GKZ system

$$\tau_{\mathrm{GKZ}}^{\Diamond}: \quad \begin{array}{c} \left\{ \left(\prod_{\ell_i > 0} \partial_{\mathsf{a}_i}^{\ell_i} - \prod_{\ell_i < 0} \partial_{\mathsf{a}_i}^{-\ell_i} \right) \frac{1}{\mathsf{a}_0} \right\}_{\underline{\ell} \in \mathbb{L}}, \\ \left\{ \left(\Sigma_{i=0}^N v_j^{(i)} \delta_{\mathsf{a}_i} \right) \frac{1}{\mathsf{a}_0} \right\}_{j=1,\ldots,n}, \quad \left(\Sigma_{i=0}^N \delta_{\mathsf{a}_i} + 1 \right) \frac{1}{\mathsf{a}_0}, \end{array}$$

whose remaining solutions are the other integrals of $\frac{1}{2\pi\mathbf{i}} \Omega_{\underline{\mathsf{a}}}$ over *relative* cycles in $H_n\left(\mathbb{P}_\Diamond \backslash X_{\underline{\mathsf{a}}}, \mathbb{D}_\Diamond \backslash \mathbb{D}_\Diamond \cap X_{\underline{\mathsf{a}}}; \mathbb{C}\right)$ [**HLYZ**]. We are interested in the behavior of the π_γ in the *large complex-structure limit*, i.e., where the $t_k := \underline{\mathsf{a}}^{\underline{\ell}^{(k)}}$ are sufficiently small. Our conclusion will be Theorem 2.4 below.

A formula for the solutions to $\tau_{\mathrm{GKZ}}^{\Diamond}$ in the large complex structure limit was given by [**HLY**]: writing $\tau_i := \frac{\log(t_i)}{2\pi\mathbf{i}}$, $\mathcal{O} := \mathbb{C}[[\underline{t}]][\underline{\tau}]$, they are precisely the functions $\psi(\mathcal{B}_\Diamond) \in \mathcal{O}$ where $\psi \in H^\bullet(\mathbb{P}_{\Diamond^\circ}, \mathbb{C})^\vee$ and

$$\mathcal{B}_\Diamond := \sum_{\underline{\ell} \in \mathbb{M}} \mathcal{B}_{\underline{\ell}}(\underline{D}) \underline{\mathsf{a}}^{\underline{\ell} + \underline{D}} = \sum_{\underline{n} \in \mathbb{Z}_{\geq 0}^M} \mathbb{B}_{\underline{n}}(\underline{J}) \underline{t}^{\underline{n} + \underline{J}} \in H^\bullet(\mathbb{P}_{\Diamond^\circ}, \mathcal{O}),$$

[5] In this section vectors are denoted with an underline, and their components with subscripts (viz., $\underline{v} = (v_1, \ldots, v_n)$).

with $\mathbb{B}_{\underline{n}}(\underline{J}) := \mathcal{B}_{\sum n_k \underline{\ell}^{(k)}}\left(\sum \underline{\ell}^{(k)} J_k\right)$ and

$$(2.1)\quad \mathcal{B}_{\underline{\ell}}(\underline{D}) := \frac{\prod_{i=1:\ell_i<0}^{N} D_i(D_i-1)\cdots(D_i+\ell_i+1)}{\prod_{i=1:\ell_i>0}^{N}(D_i+1)\cdots(D_i+\ell_i)} \times (D_0-1)\cdots(D_0+\ell_0).$$

According to the *Hyperplane Conjecture* [**HLY, LZ**], $\psi(\mathcal{B}_{\Diamond})$ is a $\mathbb{C}$-period (in the above sense) precisely when ψ belongs to

$$\mathrm{im}\left\{\imath_*^{\bullet} : H^{\bullet}(X^{\circ}, \mathbb{C})^{\vee} \to H^{\bullet}(\mathbb{P}_{\Diamond^{\circ}}, \mathbb{C})^{\vee}\right\}.$$

More explicitly, for each $\underline{\kappa} := (\kappa_1, \ldots, \kappa_M) \in \mathbb{Z}_{\geq 0}^M$ with $|\underline{\kappa}| := \sum \kappa_j \leq n$, we compute

$$B^{\underline{\kappa}}(\underline{t}, \underline{\tau}) := \tfrac{1}{(2\pi\mathbf{i})^{|\underline{\kappa}|}}\left(\partial_{J_1}^{\kappa_1} \cdots \partial_{J_M}^{\kappa_M} \mathcal{B}_{\Diamond}\right)\Big|_{\underline{J}=\underline{0}} = \sum_{\underline{\kappa}'+\underline{\kappa}''=\underline{\kappa}} (2\pi\mathbf{i})^{-|\underline{\kappa}''|} \underline{\tau}^{\underline{\kappa}'} \sum_{\underline{n}\in\mathbb{Z}_{\geq 0}^M} b_{\underline{n}}^{\underline{\kappa}''} \underline{t}^{\underline{n}}$$

where $b_{\underline{n}}^{\underline{\kappa}''} = \left(\partial_{J_1}^{\kappa_1''} \cdots \partial_{J_M}^{\kappa_M''} \mathbb{B}_{\underline{n}}\right)(\underline{0})$. Given bases $\{\psi_\ell^r\} \subset H_{2r}(\mathbb{P}_{\Diamond^{\circ}}, \mathbb{Z})$ resp. $\{\hat{\psi}_\ell^r\} \subset \mathrm{im}(\imath_*^{2r})_{\mathbb{Z}}$ for each r, we obtain $\mathbb{C}$-bases for the solutions to $\tau_{\mathrm{GKZ}}^{\Diamond}$ resp. for the $\mathbb{C}$-periods (assuming the Hyperplane Conjecture) which are $\mathbb{Z}$-linear combinations of the $\{B^{\underline{\kappa}}\}$. That is, writing

$$B^{\underline{0}}(\underline{t}) = \sum b_{\underline{n}}^{\underline{0}} \underline{t}^{\underline{n}} =: \sum a_{\underline{n}} \underline{t}^{\underline{n}} =: A(\underline{t})$$

for the period holomorphic in $\underline{t}$, we have

$$(2.2)\quad \overset{(\wedge)}{\psi}{}_\ell^r(\mathcal{B}_{\Diamond}) = A(\underline{t})\, \overset{(\wedge)}{P}{}_\ell^r(\underline{\tau}) + \sum_{|\underline{\kappa}'|<r} (2\pi\mathbf{i})^{|\underline{\kappa}'|-r} \underline{\tau}^{\underline{\kappa}'} \sum_{\underline{n}} \left(\sum_{|\underline{\kappa}''|=r-|\underline{\kappa}'|} \overset{(\wedge)}{c}{}_{\underline{\kappa}',\underline{\kappa}''} b_{\underline{n}}^{\underline{\kappa}''} \right) \underline{t}^{\underline{n}}$$

where $\overset{(\wedge)}{P}{}_\ell^r$ are $\mathbb{Z}$-homogeneous polynomials of degree r and $\overset{(\wedge)}{c}{}_{\underline{\kappa}',\underline{\kappa}''} \in \mathbb{Z}$.

REMARK 2.1.

(i) The full assertion for the $\mathbb{C}$-periods holds without the Hyperplane Conjecture for $r \leq \min\left\{\frac{n-1}{2}, 1\right\}$. Writing $\mathcal{A}' \subset \mathcal{A}$ for the points (if any) interior to facets, the periods cannot depend on the $\{\log(\mathsf{a}_i)\}_{\underline{v}^{(i)}\in\mathcal{A}'}$ because taking $\mathsf{a}_i \to 0$ does not make $X_{\underline{\mathsf{a}}}$ singular. Moreover, X° avoids the corresponding (exceptional) $\{D_i\}$; so there are $M - |\mathcal{A}'|$ independent $\{\hat{\psi}_\ell^1\}$ (with leading terms $A(\underline{t}) \times$ the $M - |\mathcal{A}'|$ independent linear combinations of the $\{\tau_j\}$ with no such $\log(\mathsf{a}_i)$'s). But these must all be periods, since by mirror symmetry (see [**AGM**] and [**Ir**, Thm. 6.9]) there are $h_{\mathrm{van}}^{n-2,1}(X_{\underline{\mathsf{a}}}) = h_{\mathrm{tor}}^{1,1}(X^{\circ}) = M - |\mathcal{A}'|$ independent periods of degree 1 in $\underline{\tau}$.

(ii) Denote the (unipotent) local monodromies in t_i by T_i and their logarithms by N_i. Applying r of these $\{N_i\}$ to a $\mathbb{Z}$-period with "$\log^r(\underline{t})$" leading term must yield a $\mathbb{Z}$-multiple of $A(\underline{t})$. So (a fixed integer multiple of) this period must be a $\mathbb{Z}$-linear combination of the $\{\psi_\ell^r(\mathcal{B}_{\Diamond})\}$ plus a $\mathbb{C}$-linear combination of the $\{\psi_\ell^{r'}(\mathcal{B}_{\Diamond})\}_{r'<r}$. If all of the $\mathbb{C}$-linear combinations that can appear are themselves $\mathbb{C}$-periods, they can be subtracted off. So by (i), one in fact has a basis of $\mathbb{C}$-periods of the form (2.2) for $r \leq \min\{\frac{n+1}{2}, 2\}$.

Call $\underline{\ell}(\underline{n}) := \sum n_k \underline{\ell}^{(k)}$ *effective* (resp. *quasi-effective*) if all $\ell_i(\underline{n})$ $(i > 0)$ are ≥ 0 (resp. at most one < 0). Clearly the

$$a_{\underline{n}} = \begin{cases} \frac{(-\ell_0(\underline{n}))!}{\prod_{i>0} \ell_i(\underline{n})!} & \underline{\ell}(\underline{n}) \text{ effective} \\ 0 & \text{otherwise,} \end{cases}$$

being multinomial coefficients, are all integers. Now using (2.1):

- If $\underline{\ell}(\underline{n})$ is effective, then the $\left(\partial_{D_0}^{r_0} \cdots \partial_{D_N}^{r_N} \mathcal{B}_{\underline{\ell}(\underline{n})}\right)(\underline{0}) / a_{\underline{n}}$ are $\mathbb{Z}$-linear combinations of products of the form $\prod_{i=0}^{N} \prod_{j=1}^{|\ell_i(\underline{n})|} j^{-p_j^{(i)}}$, where $\underline{p}^{(i)} \in \mathbb{Z}_{\geq 0}^{|\ell_i(\underline{n})|}$ and $|\underline{p}^{(i)}| = r_i$. Since $b_{\underline{n}}^{\underline{\kappa}''} / a_{\underline{n}}$ is a $\mathbb{Z}$-linear combination of these with $|\underline{r}| = |\underline{\kappa}''|$, $b_{\underline{n}}^{\underline{\kappa}''}$ can be written as $\frac{P}{Q}$ $(P, Q \in \mathbb{Z})$ with $Q \mid (L_{|\ell_0(\underline{n})|})^{|\underline{\kappa}''|}$, where
$$L_s := \operatorname{lcm}\{1, \ldots, s\}.$$
- If $\underline{\ell}(\underline{n})$ is quasi-effective, with (say) $\ell_1(\underline{n}) < 0$, then $\left(\partial_{D_1} \mathcal{B}_{\underline{\ell}(\underline{n})}\right)(\underline{0}) =$
$$\frac{(-\ell_0(\underline{n}))!\,(-\ell_1(\underline{n}) - 1)!}{\prod_{i>1} \ell_i(\underline{n})!} = \frac{(\Sigma_{i>1}\ell_i(\underline{n}))!}{\Pi_{i>1}\ell_i(\underline{n})!} \Big/ \frac{(\Sigma_{i>1}\ell_i(\underline{n}))!}{(-\ell_0(\underline{n}))!\,(-\ell_1(\underline{n}) - 1)!}$$
is the quotient of a multinomial coefficient by an integer of the form $A!/(B-1)!(A-B)!$, which always divides L_A [**Na**, Thm. 3(i)]. Repeated differentiation as above now shows that $b_{\underline{n}}^{\underline{\kappa}''} = P/Q$ with $Q \mid (L_{|\underline{\ell}(\underline{n})|^+})^{|\underline{\kappa}''|}$, where $|\underline{\ell}|^+ := \sum_{i:\, \ell_i > 0} \ell_i \; (\geq -\ell_0)$.

DEFINITION 2.2. A reflexive polytope $\Diamond \subset \mathbb{R}^n$ is *facile* if:
- $\Diamond, \Diamond^\circ$ admit regular projective triangulations;
- the Mori cone $\mathcal{M}_{\Diamond^\circ}$ is regular simplicial, with generators $\underline{\ell}^{(k)}$;
- $\underline{\ell}(\underline{n}) := \sum n_k \underline{\ell}^{(k)}$ is quasi-effective for each $\underline{n} \in \mathbb{Z}_{\geq 0}^M$; and
- $n \leq 3$, or the Hyperplane Conjecture holds for $\Diamond$.

EXAMPLE 2.3. The convex hulls of $\{(1,0), (0,1), (1,1), (-1,-1)\}$ in $\mathbb{R}^2$ and $\{(\pm 1, 0, 0), (0, \pm 1, 0), (0, 0, \pm 1)\}$ in $\mathbb{R}^3$ are facile. These polytopes underlie the "facile 2-coverings" (terminology defined at the beginning of §3) in §§5.2–5.3, which compensate for the fact that the two Apéry polytopes are not facile.

Now let $W_\bullet$ denote the monodromy weight filtration for the large complex structure limit (i.e. for the sum of the monodromy logarithms), with

$$h_i := \operatorname{rk}\left(\operatorname{Gr}_{2i}^W H^{n-1}(X_{\underline{a}})\right) = \operatorname{rk}\left(\operatorname{Gr}_{-2i}^W H_{n-1}(X_{\underline{a}})\right) = \operatorname{rk}\left(H^{i,n-i-1}(X_{\underline{a}})\right)$$

($\underline{a}$ very general), which is 1 for $i = 0$ or $n - 1$. We have shown:

THEOREM 2.4. *If $\Diamond$ is facile, there exists a basis of $H_{n-1}(X_{\underline{a}}, \mathbb{C})$, of the form $\gamma_\mu^r \in W_{2(r-n+1)} H_{n-1}(X_{\underline{a}}, \mathbb{C})$ $(r = 0, \ldots, n-1;\ \mu = 1, \ldots, h_r)$, with periods*

$$\pi_\mu^r := \int_{\gamma_\mu^r} \omega_{\underline{a}} = \sum_{|\underline{\kappa}| = r} c_{\mu,\underline{\kappa}}^r \underline{T}^{\underline{\kappa}} \sum a_{\underline{n}} \underline{t}^{\underline{n}} + \sum_{|\underline{\kappa}| < r} \frac{1}{(2\pi \mathbf{i})^{r - |\underline{\kappa}|}} \underline{T}^{\underline{\kappa}} \sum_{\underline{n} \in \mathbb{Z}_{\geq 0}^M} \beta_{\mu,\underline{n}}^{r,\underline{\kappa}} \underline{t}^{\underline{n}}, \tag{2.3}$$

where each $a_{\underline{n}} \in \mathbb{Z}$, each $c_{\mu,\underline{\kappa}}^r \in \mathbb{Z}$, each $\beta_{\mu,\underline{n}}^{r,\underline{\kappa}}$ is of the form $\frac{P}{Q}$ $(P, Q \in \mathbb{Z})$ with $Q \mid (L_{|\underline{\ell}(\underline{n})|^+})^{r - |\underline{\kappa}|}$, and (for each r, μ) some $c_{\mu,\underline{\kappa}}^r \neq 0$. Moreover, this basis becomes $\mathbb{Q}$-rational in the associated graded $\oplus \operatorname{Gr}_{2i}^W$.

For the irrationality proofs, we need to apply this to certain 1-parameter families. We revise the notation for Laurent polynomials from $F_{\underline{\mathsf{a}}}$ to Φ to emphasize the specialization.

DEFINITION 2.5. A *facile CY pencil* is a family of anticanonical hypersurfaces $\mathrm{X}_\xi = \overline{\{\Phi(\xi)=0\}} \subset \mathbb{P}_\Diamond$ parametrized by $\xi \in \mathbb{P}^1$, where:
• $\Diamond \subset \mathbb{R}^n$ is a facile polytope;
• $\Phi(\xi, \underline{x}) := \sum_{\underline{m}\in\mathcal{A}_\Diamond} \xi^{o_{\underline{m}}} P_{\underline{m}}(\xi)\underline{x}^{\underline{m}}$, with $o_0 = 0$ and $o_{\underline{m}} > 0$ for $\underline{m} \neq \underline{0}$, $P_{\underline{m}}(\xi) \in \mathbb{Z}[\xi]$ with $\gcd_{\underline{m}\in\mathcal{A}}\{P_{\underline{m}}(\xi)\} = 1$, and (if $n > 1$) $\prod_{i=0}^N P_{\underline{v}^{(i)}}(0)^{\ell_i^{(k)}} = 1$ ($\forall k$), $P_{\underline{0}}(0) = 1$;
• the VHS on $H^{n-1}(\mathrm{X}_\xi)$ is pure, with a factor[6] $H^{n-1}_{\mathrm{tr}}(\mathrm{X}_\xi) =: \mathcal{W}_\Phi$ with Hodge numbers $h^{n-1,0} = h^{n-2,1} = \cdots = h^{0,n-1} = 1$; and
• $\sum_{i=1}^N o_{\underline{v}^{(i)}} \ell_i(\underline{n}) \geq |\underline{\ell}(\underline{n})|^+$ for all $\underline{n} \in \mathbb{Z}^M_{\geq 0}$ (automatic if $\underline{\ell}(\underline{n})$ effective).

In particular, note that $\mathcal{W}_\Phi$ has maximal unipotent monodromy at $\xi = 0$. Replacing the $\{\mathsf{a}_i\}$ in $t_k = \underline{\mathsf{a}}^{\underline{\ell}^{(k)}}$ by the corresponding coefficients of Φ, we get $t_k = \xi^{\sum o_i \ell_i^{(k)}} g_k(\xi)$ (with $g_k(\xi) \in 1+\xi\mathbb{Z}[[\xi]]$) and $2\pi\mathbf{i}\tau_k = \log(t_k) = (\sum o_i\ell_i^{(k)})\log\xi + \sum_{m>0}\frac{h_{km}}{m}\xi^m$ (with $h_{km} \in \mathbb{Z}$). Substituting into (2.3) and normalizing yields at once the

COROLLARY 2.6. *Near $\xi = 0$, any facile CY pencil admits a multivalued basis $\{\gamma_j\}_{j=0}^{n-1}$ of $\mathcal{W}^\vee_{\phi,\mathbb{C}} \cong H^{\mathrm{tr}}_{n-1}(\mathrm{X}_\xi, \mathbb{C})$ ($\mathbb{Q}$-rational in $\mathrm{Gr}^W_\bullet$), holomorphic functions $f^{(j)}(\xi) = \sum_{m\geq 0} \mathfrak{f}_m^{(j)}\xi^m$, and an integer $\varepsilon \in \mathbb{N}$, such that the periods*

$$\Pi_\ell(\xi) := \int_{\gamma_\ell} \varpi_\xi := \int_{\gamma_r} \tfrac{1}{(2\pi\mathbf{i})^{n-1}}\mathrm{Res}_{\mathrm{X}_\xi}\left(\tfrac{dx_1/x_1\wedge\cdots\wedge dx_n/x_n}{\Phi(\xi,\underline{x})}\right)$$

take the form

$$\Pi_\ell(\xi) = \frac{1}{(2\pi\mathbf{i})^\ell}\sum_{j=0}^{\ell}\frac{1}{j!}\log^j(\xi) f^{(\ell-j)}(\xi),$$

with each $\varepsilon L^j_m \mathfrak{f}_m^{(j)} \in \mathbb{Z}$, and $\mathfrak{f}_0^{(0)} = 1$.

Note that in this scenario, γ_j generates $\mathrm{Gr}^W_{2(j-n+1)}$, the monodromy logarithm N sends $\gamma_{n-1} \mapsto \gamma_{n-2} \mapsto \cdots \mapsto \gamma_1 \mapsto \gamma_0$, and γ_0 generates the local invariant cycles in $H^{\mathrm{tr}}_{n-1}(\mathrm{X}_\xi, \mathbb{Z})$.

3. Very special values

Let $\phi \in \mathbb{Z}[x_1^{\pm 1}, \ldots, x_n^{\pm 1}]$ ($n \geq 1$) be an integral Laurent polynomial, with reflexive Newton polytope $\Delta = \Delta_\phi$, and $\mathbb{P}_\Delta$ be the (possibly singular) toric variety associated to a maximal projective triangulation of Δ° [**Ba**]. We begin by defining several notions we shall require for the general irrationality statement.

Denote by $\mathcal{X}_\phi \overset{\rho}{\to} \mathbb{P}^1$ the Zariski closure in $\mathbb{P}_\Delta \times \mathbb{P}^1_t$ of $\mathcal{X}^\times_\phi := \{1 - t\phi(\underline{x}) = 0\} \subset \mathbb{G}_m^n \times (\mathbb{P}^1_t \backslash \{0\})$, and write $t = \lambda^{-1}$, $X_{\phi,t} := X^\lambda_\phi := \rho^{-1}(t)$. (Note that ρ is given by $1/\phi$ resp. ϕ when working in t resp. λ.) We shall call ϕ *involutive* if there exists a birational map $\mathcal{I} : \mathcal{X}_\phi \dashrightarrow \mathcal{X}_\phi$ over $t \mapsto \pm t^{-1}$, defined over $\mathbb{Q}$. Further, ϕ is said to admit a *facile r-cover* ($r \in \mathbb{N}$) if there is a facile CY pencil $\mathrm{X} = \{\Phi(\xi, \underline{y}) = 0\} \subset \mathbb{P}_\Diamond \times \mathbb{P}^1_\xi$ and a dominant (generically $r : 1$) rational map

[6] "tr" stands for "transcendental" (motivated by the $n = 3$ case)

$\mathcal{J} : \mathrm{X} \dashrightarrow \mathcal{X}_\phi$ over $\xi \mapsto \xi^r = t$. We say that ϕ is of *conifold type* if[7] the nonzero critical values of $\phi : (\mathbb{C}^\times)^n \to \mathbb{C}$ underlie (isolated) ordinary double points in the fibers.

While we will not assume that $1 - t\phi$ is Δ-regular for general t, or even that the generic fiber of ρ is smooth, we shall impose the conditions that the variation $\mathcal{H}_\phi$ underlain by $R^{n-1}\rho_*\mathbb{Q}$ is pure (of weight $(n-1)$),[8] and that the minimal sub-VHS $\mathcal{V}_\phi$ containing the class of

$$\omega_{\phi,t} := \omega_\phi^\lambda := \tfrac{1}{(2\pi\mathbf{i})^{n-1}} \operatorname{Res}_{X_\phi^\lambda}\left(\tfrac{dx_1/x_1\wedge\cdots\wedge dx_n/x_n}{1-t\phi(\underline{x})}\right) \in \Omega^{n-1}(X_\phi^\lambda)$$

is of rank n, with Hodge numbers $\{h^{i,n-i-1}\}_{0\le i\le n-1}$ all 1. In this case we call ϕ *principal* (since $\mathcal{V}_\phi$ is a "principal VHS" [**Ro**]). Note that $\mathcal{V}_\phi$ is polarized by taking cup products on a (fiberwise) desingularization; write Q for this polarization.

Our $\mathcal{V}_\phi$ is thus a polarized VHS on $\mathbb{P}^1 \setminus \mathcal{D}_\phi$, where $\mathcal{D}_\phi$ denotes its singularity (discriminant) locus. In particular, since $X_{\phi,0} = \mathbb{D}_\Delta$, $\mathcal{V}_\phi$ has maximal unipotent monodromy at $t=0$, and so $0 \in \mathcal{D}_\phi$. We write $\mathcal{D}_\phi^\times := \mathcal{D}_\phi \cap \mathbb{G}_m$, $P_\phi(t) := \prod_{t_0\in\mathcal{D}_\phi^\times}\left(\frac{t}{t_0}-1\right)$, $d_\phi := |\mathcal{D}_\phi^\times| = \deg(P_\phi)$, and

$$\mathfrak{r}_\phi = \min\{|t_0| \mid t_0 \in \mathcal{D}_\phi^\times\}.$$

For ϕ of conifold type, the monodromy of $\mathcal{V}_\phi$ about each point of $\mathcal{D}_\phi^\times$ is given by a single Picard-Lefschetz transformation. Since the local exponents are then given by $\{0,1,\ldots,n-2\}\cup\{\frac{n-2}{2}\}$, an easy calculation[9] shows that the Picard-Fuchs operator for ω_ϕ takes the form

$$D_\phi(t) = P_\phi(t)\delta_t^n + \sum_{k=0}^{n-1} f_{\phi,k}(t)\delta_t^k \in \mathbb{C}[t,\delta_t].$$

For $n > 2$ we remark that ρ is not semistable over 0 (and possibly at other points of $\mathcal{D}_\phi$) without blowing up $\mathbb{P}_\Delta$ along the singularities of the base locus, but this won't be an issue for us.

The natural isomorphism $\mathcal{X}_\phi^\times \to \mathbb{G}_m^n$ provides n coordinates $x_i \in \mathcal{O}^*(\mathcal{X}_\phi^\times) \cong H^1_{\mathcal{M}}(\mathcal{X}_\phi^\times, \mathbb{Q}(1))$ whose cup product produces the *coordinate symbol*

$$\{\underline{x}\} = \{x_1,\ldots,x_n\} \in H^n_{\mathcal{M}}(\mathcal{X}_\phi^\times, \mathbb{Q}(n)).$$

We shall call ϕ *tempered* if this lifts to a motivic cohomology class

$$\Xi \in H^n_{\mathcal{M}}(\mathcal{X}_\phi^-, \mathbb{Q}(n)),$$

where $\mathcal{X}_\phi^- := \mathcal{X}_\phi \setminus X_\phi^\infty$. (See Theorem 4.2(a) for a down-to-earth temperedness criterion when $n \le 3$.) Writing T_{x_i} [resp. $T_{\{\underline{x}\}}$] for the analytic chain $x_i^{-1}(\mathbb{R}_-)$

---

[7]When ϕ is not Δ-regular, we also assume that no *non-generic* singularities of X_ϕ^λ along the base locus $X_\phi^\lambda \cap \mathbb{D}_\Delta$ occur for values $\lambda \in \mathcal{D}_\phi^\times$ (hence don't affect the local system $\mathbb{V}_\phi$), see below.

[8]For $n=1$, the generic $X_{\phi,t}$ is a pair of points, and H^0 [resp. H_0] means everywhere the augmentation cokernel [resp. kernel], i.e. reduced (co)homology. Involutivity and reflexivity imply that $x\phi(x)$ is a quadratic polynomial with two distinct roots.

[9]With no information on exponents, one would have $P_\phi(t)\delta_t^n + \sum_{k=0}^{n-1}\frac{F_k(t)}{P_\phi(t)^{n-k-1}}\delta_t^k$ ($F_k \in \mathbb{C}[t]$); the existence of holomorphic solutions with orders 0 thru $n-2$ about each root t_0 of P_ϕ forces $(t-t_0)^{n-k-1} \mid F_k$. Note that if n is even, the exponent $\frac{n-2}{2}$ is repeated.

[resp. $\cap_{i=1}^n x_i^{-1}(\mathbb{R}_-)$] and $U_\phi := \mathbb{P}^1 \setminus \rho(T_{\{\underline{x}\}})$, we term ϕ *strongly tempered* if the *higher normal function* defined by

$$\nu_{\phi,t} := \mathrm{AJ}^{n,n}_{X_{\phi,t}}(\Xi|_{X_{\phi,t}}) \in H^{n-1}(X_{\phi,t}, \mathbb{C}/\mathbb{Q}(n))$$

has a single-valued holomorphic family of lifts

$$\tilde{\nu}_{\phi,t} \in H^{n-1}(X_{\phi,t}, \mathbb{C})$$

over U_ϕ.[10] Here by a "lift", we simply mean that the natural map $H^{n-1}(X_{\phi,t}, \mathbb{C}) \twoheadrightarrow H^{n-1}(X_{\phi,t}, \mathbb{C}/\mathbb{Q}(n))$ maps $\tilde{\nu}_{\phi,t} \mapsto \nu_{\phi,t}$.

REMARK 3.1. As defined, ν_ϕ gives an element of the space $\mathrm{ANF}(\mathcal{H}_\phi(n))$ of admissible normal functions [**7K**, §5.2]. We shall only really care about the $\mathcal{V}_\phi$-component of ν_ϕ below, so one might as well regard it (by projection) as an element of $\mathrm{ANF}(\mathcal{V}_\phi(n))$. Note that any lift (such as $\tilde{\nu}_\phi$) of this component must have nontrivial monodromy on $\mathbb{P}^1 \setminus \mathcal{D}_\phi$ (as opposed to just U_ϕ) c: otherwise its topological invariant $[\nu_\phi] \in \mathrm{Hom}\left(\mathbb{Q}(0), H^n\left(\rho^{-1}(\mathbb{P}^1 \setminus \mathcal{D}_\phi), \mathbb{Q}(n)\right)\right)$ would vanish. Since the latter is computed by Ω_Ξ (see the proof of [**DK**, Cor. 4.6]), which restricts to the Haar form on $\mathcal{X}^*_\times \cong \mathbb{G}_m^n$, this is absurd.

Now either $\omega_{\phi,t}$ or $\tilde{\omega}_{\phi,t} := t\omega_{\phi,t}$ is a holomorphic section of the canonically extended Hodge bundle $\mathcal{F}_e := \mathcal{F}^{n-1}\mathcal{V}_{\phi,e}$. As residue forms, they are really most naturally regarded as elements of $H_{n-1}(X_{\phi,t}, \mathbb{C})$ for any $t \in \mathbb{P}^1$. If $t \notin \mathcal{D}_\phi$, then purity of $\mathcal{H}_\phi$ makes $H^{n-1} \to H_{n-1}$ an isomorphism,[11] allowing us to treat them as cohomology classes; but over all of $\mathbb{P}^1$ or U_ϕ, they only make sense as sections of H_{n-1}. Conveniently enough, this pairs with $H^{n-1}(X_{\phi,t}, \mathbb{C})$, allowing us to define

$$\tilde{V}_\phi(\lambda) := \langle \tilde{\nu}_\phi^\lambda, \tilde{\omega}_\phi^\lambda \rangle \in \mathcal{O}(U_\phi). \tag{3.1}$$

This extends to a multivalued-holomorphic function

$$V_\phi(\lambda) := \langle \nu_\phi^\lambda, \tilde{\omega}_\phi^\lambda \rangle \mod \tilde{\mathscr{P}}_\phi^{\mathbb{Q}(n)}$$

on $\mathbb{P}^1 \setminus \mathcal{D}_\phi$, defined modulo $\mathbb{Q}(n)$-periods of $\tilde{\omega}_\phi^\lambda$; we shall refer to both V and $\tilde{V}$ as the *truncated higher normal function* associated to Ξ and $\tilde{\omega}_\phi$. Note that $\tilde{V}_\phi$ *cannot* extend to an entire function, since $\tilde{\nu}_\phi$ has nontrivial monodromy on $\mathbb{P}^1 \setminus \mathcal{D}_\phi$ and $\tilde{\omega}_\phi$ has n independent periods (courtesy of the maximal unipotent monodromy).

We can now state the main result of this section:

THEOREM 3.2. *Let $\phi(x_1, \ldots, x_n)$ be an integral Laurent polynomial, such that $\phi(-\underline{x})$ has all positive coefficients, which is reflexive, involutive, principal, strongly tempered, of conifold type, and admits a facile r-cover. Assume that $d_\phi = 2$, $D_\phi \in \mathbb{Z}[t, \delta_t]$, and $\mathfrak{r}_\phi < e^{-n}$. Then $\tilde{V}_\phi(0) \notin \mathbb{Q}^\times$.*

---

[10] Note that if we assume this only over $U_\phi \setminus U_\phi \cap \mathcal{D}_\phi$, the lift extends to U_ϕ anyway: the single-valuedness of $\tilde{\nu}$ on a punctured disk about t_0 means ν has no singularity at the center, so that $\tilde{\nu}$ uniquely extends to the whole disk. The value $\tilde{\nu}(t_0)$ lies in $\ker(T - I)$ in the canonical extension $\mathcal{H}_\phi^{\lim}$, which contains the image of $H^{n-1}(X_{\phi,t_0})$; see [**7K**, §5].

[11] Here and below, we use the polarization Q to make this identification (up to twist).

3.1. Proof of Theorem 3.2.

Step 1: The power series. By involutivity, the four points of $\mathcal{D}_\phi$ have λ-values $0, \lambda_0, \pm\lambda_0^{-1}, \infty$, with $|\lambda_0| = \mathfrak{r}_\phi < 1 < |\lambda_0|^{-1}$. (The choice of $\pm$ here matches that in the involution $t \mapsto \pm t^{-1}$.) Moreover, the $\mathbb{Z}$-local system $\mathbb{V}_\phi$ underlying $\mathcal{V}_\phi$ has maximal unipotent monodromy at $\lambda = 0$, with (rank 1) invariant subsystem on the disk $D_{\mathfrak{r}_\phi}$ generated by a family of $(n-1)$-cycles φ_0^λ. Indeed, we may assume that $\varphi_0^\lambda = \mathcal{I}_* {}'\varphi_0^{\mathcal{I}(\lambda)}$ where $\mathrm{Tube}('\varphi_0) = \mathbb{T}_n := \{|x_1| = \cdots = |x_n| = 1\} \subset \mathbb{P}_\Delta \setminus X_\phi^{\mathcal{I}(\lambda)}$. As the Hodge bundle $\mathcal{F}_e$ has degree 1 [**GGK**, p. 504], and $\omega_\phi \in \Gamma(\mathbb{P}^1, \mathcal{F}_e)$ has a zero at ∞, we have $(\mathcal{I}^*\omega_\phi)/\omega_\phi = M\lambda^{-1}$ for some $M \in \mathbb{Q}^\times$. But then $(\mathcal{I}^*)^2\omega_\phi = \mathcal{I}^*\frac{M}{\lambda}\omega_\phi = \frac{M^2}{\lambda\mathcal{I}(\lambda)}\omega_\phi = \pm M^2\omega_\phi$, and since $(\mathcal{I}^*)^2$ acts on $\mathbb{V}_\phi$ (and $\mathbb{V}_\phi^{\mathbb{C}}$ is irreducible), we must have $M^2 \in \mathbb{Z}^\times = \{\pm 1\}$. This forces $M = \pm 1$ hence $\mathcal{I}^*\omega_\phi = \pm\tilde{\omega}_\phi$. The holomorphic period at $\lambda = 0$ is therefore

$$\begin{aligned} A_\phi(\lambda) &:= \int_{\varphi_0^\lambda} \tilde{\omega}_\phi^\lambda = \int_{'\varphi_0^{\mathcal{I}(\lambda)}} \mathcal{I}^*\tilde{\omega}_\phi^\lambda \\ &= \pm\int_{'\varphi_0^{\mathcal{I}(\lambda)}} \omega_\phi^{\mathcal{I}(\lambda)} = \pm\int_{'\varphi_{0,\pm\lambda}} \omega_{\phi,\pm\lambda} \\ &= \pm\int_{\mathbb{T}_n} \frac{dx_1/x_1 \wedge\cdots\wedge dx_n/x_n}{1 \mp \lambda\phi(\underline{x})}, \end{aligned}$$

and of course we may change the signs of φ_0 and λ if needed so that

$$A_\phi(\lambda) = \sum_{k\geq 0} [\phi^k]_{\underline{0}}\lambda^k =: \sum_{k\geq 0} a_k\lambda^k \tag{3.2}$$

(where $[\cdot]_{\underline{0}}$ denotes "constant term"), with $a_0 = 1$. By the fundamental result of [**DvdK**], we have $\mathfrak{r}_A := \left(\limsup_{k\to\infty} |a_k|^{1/k}\right)^{-1} \in \mathbb{C}^\times$; and since A_ϕ is a period of $(\mathbb{V}_\phi, \tilde{\omega}_\phi)$, $\mathfrak{r}_A$ must be the modulus of an element of $\mathcal{D}_\phi^\times$ (i.e., $\mathfrak{r}_\phi$ or $\mathfrak{r}_\phi^{-1}$). Since the $[\phi^k]_{\underline{0}}$ are all (positive) integers, $\mathfrak{r}_A = \mathfrak{r}_\phi$.

On the other hand, positivity of the coefficients of $\phi(-\underline{x})$ forces $\phi(T_{\{\underline{x}\}}) \subset [1, \infty]$. Now the global minimum of $\phi(T_{\{\underline{x}\}})$ necessarily occurs at a critical point of ϕ; more precisely, it may be regarded as the terminus of a Lefschetz thimble on the generator ∂T_Ξ of $\mathcal{V}_\phi^{0,\lim}/N\mathcal{V}_\phi^{0,\lim} \cong \mathrm{Gr}_0^W H_{n-1}(X_\phi^0)$, and thus is $\pm$ the λ-coordinate of a point of $\mathcal{D}_\phi^\times$. Clearly this must be the larger one, so that U_ϕ contains the disk $D_{\mathfrak{r}_\phi^{-1}}$ about $\lambda = 0$. Writing

$$\tilde{V}_\phi(\lambda) =: \sum_{k\geq 0} v_k\lambda^k,$$

it follows that $\mathfrak{r}_V := \left(\limsup_{k\to\infty} |v_k|^{1/k}\right)^{-1} = \mathfrak{r}_\phi^{-1}$.[12] Moreover, since $A(\lambda)$ is (up to scale) the *only* period of $\tilde{\omega}_\phi$ invariant about $\lambda = 0$, and it is *not* invariant about $\lambda = \lambda_0$, we conclude that $\tilde{V}_\phi(\lambda)$ [resp. $\tilde{\nu}_\phi^\lambda$] is the unique analytic continuation of $V_\phi(\lambda)$ [resp. holomorphic lift of ν_ϕ^λ] which is well-defined on U_ϕ.

In particular, $\tilde{V}_\phi(0) = v_0 \in \mathbb{C}$ is well-defined. Assuming henceforth that $v_0 \neq 0$ (recall we are proving that $v_0 \notin \mathbb{Q}^\times$), we may consider

$$B_\phi(\lambda) := -\tilde{V}_\phi(\lambda) + \tilde{V}_\phi(0)A_\phi(0) =: \sum_{k\geq 1} b_k\lambda^k.$$

---

[12]Recall that $\tilde{V}_\phi$ cannot be entire.

Obviously this has radius of convergence $\mathfrak{r}_B = \mathfrak{r}_A\ (< 1)$, while $v_k = v_0 a_k - b_k \to 0$ as $k \to 0$. Restricting if necessary to a subsequence $a_{k_j} \to \infty$, we therefore have

$$\lim_{j\to\infty} \frac{b_{k_j}}{a_{k_j}} = v_0.$$

Step 2: The inhomogeneous equation. Since the holomorphic period of $\tilde{\omega}$ about $\lambda = 0$ is obtained by substituting λ for t in the holomorphic period of ω about $t = 0$, the same goes for the (inhomogeneous) Picard-Fuchs equations: that is,

$$D_\phi(\lambda)\tilde{\mathscr{P}}_\phi^{\mathbb{C}} = 0.$$

Consequently, in $\mathcal{V}_\phi$ we have

$$P_\phi(\lambda)\nabla^n_{\delta_\lambda}[\tilde{\omega}_\phi] = -\sum_{k=0}^{n-1} f_{\phi,k}(\lambda)\nabla^k_{\delta_k}[\tilde{\omega}_\phi]; \tag{3.3}$$

and regarding $\tilde{\omega}_\phi$ as a section of $\mathcal{V}_{\phi,e}$, maximal unipotent monodromy at $\lambda = 0$ and $\mathrm{Res}_0\,\nabla = \frac{1}{2\pi\mathbf{i}}N$ imply the independence of the $\{\nabla^k_{\delta_k}[\tilde{\omega}_\phi]\}_{k=0}^{n-1}$ in the fiber $\mathcal{V}^0_{\phi,e}$. As $N^n = 0$, $\nabla^n_{\delta_\lambda}[\tilde{\omega}_\phi]$ vanishes in $\mathcal{V}^0_{\phi,e}$, and so all $f_{\phi,k}(0) = 0 \implies f_{\phi,k}(\lambda) =: \lambda g_{\phi,k}(\lambda)$. Moreover, since $D_\phi(\lambda) \circ \lambda^{-1}$ (like $D_\phi(t)$) kills $[\omega_\phi]$, we have

$$F(t) \circ D_\phi(t) = D_\phi(\lambda) \circ \lambda^{-1} = D_\phi(\pm t^{-1}) \circ t$$

for some function $F(t)$. Since by assumption $P_\phi(\lambda)$ is quadratic, a short computation shows that $F(t) = \pm\frac{1}{t}$ and that the $tg_{\phi,k}(t^{-1})$ must be polynomials, forcing the $g_{\phi,k}$ to be linear.

Next, define the *Yukawa coupling* by

$$Y(\lambda) := Q\left([\tilde{\omega}^\lambda_\phi], \nabla^{n-1}_{\delta_\lambda}[\tilde{\omega}^\lambda_\phi]\right) \in \mathbb{C}(\lambda)^\times. \tag{3.4}$$

Let w be a local coordinate at a point $p \in \mathcal{D}^\times_\phi$, about which monodromy is described by $\varphi \mapsto \varphi - \langle\varphi, \sigma\rangle\sigma$ for some vanishing cycle σ (since ϕ is of conifold type). Writing $[\tilde{\omega}^\lambda_\phi]$ with respect to a local basis $\{\sigma_0, \ldots, \sigma_{n-1}\}$ of $\mathbb{V}_\phi$ with $\sigma_1, \ldots, \sigma_{n-1}$ invariant about p and $\sigma = \sigma_0$ [resp. σ_1] for n odd [resp. even] (with $\sigma_0 \mapsto -\sigma_0$ resp. $\sigma_0 \mapsto \sigma_0 - \sigma_1$), the coefficients $u_i(w)$ of $\{\sigma_1, \ldots, \sigma_n\}$ are holomorphic and we have (near p) $u_0(w) \sim w^{\frac{n-2}{2}}$ [resp. $u_1(w) \sim u_0(w)\log(w) \sim w^{\frac{n-2}{2}}\log(w)$]. From this we compute $Y \sim w^{\frac{n-2}{2}-(n-1)} \times w^{\frac{n-2}{2}} = w^{-1}$, so that Y has (at worst) simple poles at $\mathcal{D}^\times_\phi$. Moreover, the pairing (3.4) makes sense at $\lambda = 0$ in $\mathcal{V}^0_{\phi,e}$, so that $Y(0) \neq \infty$; while $\mathcal{I}^*\tilde{\omega}^\lambda_\phi = \pm\lambda\tilde{\omega}^\lambda_\phi \implies Y(\mathcal{I}(\lambda)) = \pm\lambda^2 Y(\lambda) \implies \mathrm{ord}_\infty(Y) \geq 2$. Since Y is rational with only 2 simple poles, it can have only the double zero at ∞, and thus[13]

$$Y(\lambda) = \frac{Y(0)}{P_\phi(\lambda)}.$$

To evaluate $Y(0)$, and also anticipating Step 3, we extend φ^λ_0 to a basis $\{\varphi^\lambda_j\}_{j=0}^{n-1}$ of $(\mathbb{V}^{\lambda,\mathbb{Q}}_\phi)^\vee \cong H_{n-1}(X^\lambda_\phi, \mathbb{Q})$ on an angular sector of the punctured disk $D^\times_{\mathfrak{r}_\phi}$ satisfying $\varphi_{n-j-1} = N^j\varphi_{n-1}$ hence $\varphi_j \in W_{2(j-n+1)}$ ($N = \log(T)$ the monodromy logarithm at 0, and $W_\bullet = W(N)[n-1]_\bullet$). From $NQ = -QN$ and $N^n = 0$, the $Q(\varphi_j, \varphi_k)$

[13]One may also show that $g_{\phi,n-1}(\lambda) = \frac{n}{2}P'_\phi(\lambda)$, but we won't need this.

are zero for $j+k<n-1$, and the $(-1)^j Q(\varphi_j,\varphi_{n-j-1})$ equal a common constant $Q_0^{-1}\in\mathbb{Q}^\times$. We may then modify

$$\varphi_j \mapsto \hat{\varphi}_j = \varphi_j + \sum_{i<j}\alpha_{ij}\varphi_i \in W_{2(j-n+1)} \qquad (\alpha_{ij}\in\mathbb{Q})$$

so that $Q(\hat{\varphi}_i,\hat{\varphi}_{n-j-1}) = (-1)^i Q_0^{-1}\delta_{ij}$, with $N\hat{\varphi}_j = \hat{\varphi}_{j-1} + \sum_{i<j-1}\eta_{ij}\hat{\varphi}_i$ ($\eta_{ij}\in\mathbb{Q}$) and dual basis $\{\check{\varphi}_j\}$ of $\mathbb{V}_\phi^{\mathbb{Q}}$. (Note that $\check{\varphi}_j \in W_{2(n-j-1)}\mathbb{V}_\phi$, $-N\check{\varphi}_j = \check{\varphi}_{j+1} + \sum_{i>j+1}\eta_{ji}\check{\varphi}_i$, while $Q(\check{\varphi}_{n-i-1},\check{\varphi}_j) = (-1)^i Q_0\delta_{ij}$.) Writing locally

$$[\tilde{\omega}_\phi^\lambda] = \sum_{j=0}^{n-1}\left(\int_{\hat{\varphi}_j}\tilde{\omega}_\phi^\lambda\right)\check{\varphi}_j =: \sum_{j=0}^{n-1}\hat{\pi}_j(\lambda)\check{\varphi}_j$$

in $\mathcal{V}_\phi$, and $\pi_j(\lambda) := \int_{\varphi_j}\tilde{\omega}_\phi^\lambda$, we have

$$\pi_0(\lambda) = \hat{\pi}_0(\lambda) = A_\phi(\lambda) =: A_\phi^{(0)}(\lambda)$$

and

$$\hat{\pi}_j = \pi_j(\lambda) + \sum_{i<j}\alpha_{ij}\pi_i(\lambda);$$

while in accordance with the monodromy properties of φ_j,

$$\pi_j(\lambda) = \sum_{k=0}^{j}\frac{\log^k(\lambda)}{(2\pi\mathbf{i})^k k!}A_\phi^{(j-k)}(\lambda) \tag{3.5}$$

for some functions $A_\phi^{(\ell)}(\lambda)$ holomorphic on $D_{\mathfrak{r}_\phi}$. Clearly, the limits

$$\lim_{\lambda\to 0}\log^\ell(\lambda)\left(\delta_\lambda^{n-1}\hat{\pi}_j\right)(\lambda)$$

are zero for $j<n-1$, while we have $\lim_{\lambda\to 0}(2\pi\mathbf{i})^{n-1}\left(\delta_\lambda^{n-1}\hat{\pi}_{n-1}\right)(\lambda) = 1 = A_\phi^{(0)}(0)$; and so $Y(0) = (2\pi\mathbf{i})^{1-n}\times Q(\check{\varphi}_0,\check{\varphi}_{n-1}) = \frac{\pm Q_0}{(2\pi\mathbf{i})^{n-1}}$.

Now the fiberwise restrictions $\Xi^\lambda := \Xi|_{X_\phi^\lambda}$ ($\lambda\neq\infty$) have trivial class in $\mathrm{Hom}_{\mathrm{MHS}}\left(\mathbb{Q}(0), H^n(X_\phi^\lambda,\mathbb{Q}(n))\right)$, so that T_Ξ is a coboundary over any sufficiently small $B\subset\mathbb{A}_\lambda^1$. Since the regulator current R_Ξ has $dR_\Xi = \frac{dx_1}{x_1}\wedge\frac{dx_n}{x_n} - (2\pi\mathbf{i})^n\delta_{T_\Xi}$ on $\mathcal{X}_\phi^-$, writing $T_\Xi|_{\rho^{-1}(B)}\equiv\partial\Gamma_B \pmod{\partial\rho^{-1}(B)}$ yields a current $\tilde{R}_\Xi^B := R_\Xi|_{\rho^{-1}(B)} + (2\pi\mathbf{i})^n\delta_{\Gamma_B}$ with closed fiberwise pullbacks. This yields

$$\nabla_{\delta_\lambda}\nu_\phi^\lambda = \nabla_{\delta_\lambda}[\tilde{R}_{\Xi^\lambda}^B] = \left[\left.\left(\tfrac{dx_1}{x_1}\wedge\cdots\wedge\tfrac{dx_n}{x_n}\lrcorner\,\widetilde{\delta_\lambda}\right)\right|_{X_\phi^\lambda}\right] = -(2\pi\mathbf{i})^{n-1}[\omega_\phi^\lambda]. \tag{3.6}$$

Using this (and $Q(\omega,\nabla_{\delta_\lambda}^k\tilde{\omega}) = 0$ for $k<n-1$), one computes that $\delta_\lambda^k V(\lambda) = \delta_\lambda^k Q(\nu,\tilde{\omega}) = Q(\nu,\nabla_{\delta_\lambda}^k\tilde{\omega})$ for $k<n$ and $-(2\pi\mathbf{i})^{n-1}Q(\omega,\nabla_{\delta_\lambda}^{n-1}\tilde{\omega}) + Q(\nu,\nabla_{\delta_\lambda}^n\tilde{\omega})$ for $k=n$. Using (3.4), we have at once

$$D_\phi(\lambda)V_\phi(\lambda) = \pm(2\pi\mathbf{i})^{n-1}\lambda Y(\lambda)P_\phi(\lambda) = \pm Q_0\lambda. \tag{3.7}$$

Notice that while $V_\phi(\lambda)$ is multivalued, the ambiguities are killed by $D_\phi(\lambda)$. Moreover, $\tilde{V}_\phi(\lambda)$ satisfies the same equation, so that $D_\phi(\lambda)A_\phi(\lambda) = 0 \implies$

$$D_\phi(\lambda)B_\phi(\lambda) = \mp Q_0\lambda. \tag{3.8}$$

Step 3: Arithmetic of coefficients. By (3.2) and the integrality of ϕ, we have $a_m \in \mathbb{Z}$ ($\forall m$). Expressing b_m as $\frac{p_m}{q_m}$ with $p_m \in \mathbb{Z}$ and $q_m \in \mathbb{N}$ relatively prime, we claim that for some fixed $\varepsilon_B \in \mathbb{N}$,

$$q_m \mid \varepsilon_B (m!)^n \quad (\forall m). \tag{3.9}$$

Indeed, if we write $P_\phi(\lambda) = c_n' \lambda^2 + c_n'' \lambda + 1$ and $g_{\phi,j}(\lambda) = c_j' \lambda + c_j''$ (all $c_j', c_j'' \in \mathbb{Z}$; $c_n' = \pm 1$), substituting $\sum_{m \geq 1} b_m \lambda^m = B_\phi(\lambda)$ in (3.8) yields $b_1 = \pm Q_0 \in \mathbb{Q}$ and the recurrence

$$-m^n b_m = \left(\Sigma_{j=0}^n c_j'(m-2)^j\right) b_{m-2} + \left(\Sigma_{j=0}^n c_j''(m-1)^j\right) b_{m-1}.$$

Taking $\varepsilon_B = q_1$, this establishes (3.9).

We claim that, in addition (modifying ε_B if necessary),

$$q_m \mid \varepsilon_B L_{rm}^n \quad (\forall m). \tag{3.10}$$

To show this, we make use of the facile r-cover $\mathcal{J}$, which induces an isomorphism of VHS $\mathcal{J}^* \mathcal{V}_\phi \cong \mathcal{W}_\Phi$ hence of their extended Hodge bundles $\mathcal{O}_{\mathbb{P}^1_\xi}(r) \cong \mathcal{J}^* \mathcal{F}_{e,\phi}^{n-1} \cong \mathcal{F}_{e,\Phi}^{n-1}$ over $\mathbb{P}^1_\xi$, of which $\mathcal{J}^* \omega_{\phi,\xi^r}$ and

$$\varpi_\xi := (2\pi \mathbf{i})^{1-n} \operatorname{Res}_{X_\xi} \left(\tfrac{dy_1/y_1 \wedge \cdots \wedge dy_n/y_n}{\Phi(\xi, \underline{y})} \right)$$

are sections. Since $\gcd_{\underline{m} \in \mathcal{A}_\diamond} \{P_{\underline{m}}(\xi)\} = 1$ (see Defn. 2.5), ϖ_ξ has no zeroes over $\mathbb{A}^1_\xi$, so that both sections share the divisor $r[\infty]$. Since $(\mathcal{J}_\xi^{\pm 1})_*$ exchanges the generators $\mathcal{I}_*(\varphi_0)$ and γ_0 of the integral invariant cycles about $\xi = 0$ and $\lambda = 0$, and $\lim_{t \to 0} \int_{\mathcal{I}_*(\varphi_0)} \omega_{\phi,t} = 1 = \lim_{\xi \to 0} \int_{\gamma_0} \varpi_\xi$, we find that $\varpi_\xi = \mathcal{J}^* \omega_{\phi,\xi^r}$. Via $\mathcal{I}^* \tilde{\omega}_\phi^{\xi^r} = \omega_{\phi,\xi^r}$, Corollary 2.6, and (3.5), we therefore have ($0 \leq \ell \leq n-1$)

$$\hat{\pi}_\ell(\xi^r) = \int_{\varphi_\ell} \tilde{\omega}_\phi^{\xi^r} = \int_{\mathcal{J}_* \hat{\gamma}_\ell} \omega_{\phi,\xi^r} = \int_{\hat{\gamma}_\ell} \varpi_\xi$$

for some $\hat{\gamma}_\ell = \sum_{k \leq \ell} (2\pi \mathbf{i})^{k-\ell} \beta_{k\ell} \gamma_k \in W_{2(\ell-n+1)} \mathcal{W}_{\Phi,\mathbb{Q}}^\vee$ ($\beta_{k\ell} \in \mathbb{C}$; $\beta_{\ell\ell} = r^\ell$). That is,

$$\hat{\pi}_\ell(\xi^r) \;=\; \frac{1}{(2\pi \mathbf{i})^\ell} \sum_{\substack{m \geq 0, \\ 0 \leq j \leq k \leq \ell}} \frac{1}{j!} \mathfrak{f}_m^{(k-j)} \, \beta_{k\ell} \, \xi^m \log^j(\xi). \tag{3.11}$$

Returning to the λ-disk for a moment, we have $[\tilde{\omega}_\phi^0] \equiv Q_0 \hat{\varphi}_{n-1}$ in $H_{n-1}(X_\phi^0)$ so that $\tilde{\nu}_\phi^0 = Q_0^{-1} v_0 \check{\varphi}_{n-1}$ in $H^{n-1}(X_\phi^0)$. This fixes the constant of integration, so that (3.6) gives

$$\begin{aligned} \tilde{\nu}_\phi^\lambda - Q_0^{-1} v_0 \check{\varphi}_{n-1} &= \int_0 \left(\nabla_{\delta_\lambda} \nu_\phi^\lambda \right) \frac{d\lambda}{\lambda} \\ &= -(2\pi \mathbf{i})^{n-1} \int_0 [\omega_\phi^\lambda] \frac{d\lambda}{\lambda} \\ &= (2\pi \mathbf{i})^{n-1} \sum_{\ell=0}^{n-1} \int_0 \hat{\pi}_\ell(\lambda) d\lambda \; \check{\varphi}_\ell \end{aligned} \tag{3.12}$$

in $\mathcal{V}_\phi$. Since $Q(Q_0^{-1} v_0 \check{\varphi}_{n-1}, \tilde{\omega}_\phi^\lambda) = v_0 A_\phi(\lambda)$ and $Q(\tilde{\nu}_\phi^\lambda, \tilde{\omega}_\phi^\lambda) = \tilde{V}_\phi(\lambda)$, pairing (3.12) with $[\tilde{\omega}_\phi^\lambda] = \sum_{\ell'=0}^{n-1} \hat{\pi}_{\ell'}(\lambda) \hat{\varphi}_{\ell'}$ yields the key formula

$$-B_\phi(\lambda) = (-2\pi \mathbf{i})^{n-1} Q_0 \sum_{\ell=0}^{n-1} (-1)^\ell \hat{\pi}_{n-\ell-1}(\lambda) \int_0 \hat{\pi}_\ell(\lambda) d\lambda.$$

Substituting $\xi^r = \lambda$, this becomes

$$\pm \sum_{\mu \geq 1} b_\mu \xi^{r\mu} = (2\pi \mathbf{i})^{n-1} Q_0 \sum_{\ell=0}^{n-1} (-1)^\ell \hat{\pi}_{n-\ell-1}(\xi^r) \int_0 \hat{\pi}_\ell(\xi^r) r \xi^{r-1} d\xi,$$

and using $\int_0 \log^a(x) x^{b-1} dx = x^b \sum_{c=0}^{a} \frac{(-1)^{a-c} a!}{(a-c)! b^{c+1}} \log^{a-c}(x)$ together with (3.11),

$$= Q_0 r \sum_{m \geq 0} \xi^m \sum_{(*)_m} \frac{(-1)^{\ell+j'-i'}}{(j'-i')! j''!} \frac{\mathfrak{f}_{m''}^{(k''-j'')} \mathfrak{f}_{m'}^{(k'-j')} \beta_{k'',n-\ell-1} \beta_{k',\ell}}{(m'+r)^{i'+1}} \log^{j''+j'-i'}(\xi),$$

where the $\sum_{(*)_m}$ (finite for each m) is over $m' + m'' = m - r$, $0 \leq \ell \leq n-1$, $0 \leq j'' \leq k'' \leq n - \ell - 1$, and $0 \leq i' \leq j' \leq k' \leq \ell$. Since this plainly has to be a power series in ξ^r, the $\log^* \xi$ terms must cancel out (forcing $j'' = 0 = j' - i'$), leaving us with

$$= Q_0 r \sum_{m \geq 0} \xi^m \sum_{(*)_m} \frac{(-1)^\ell}{(m'+r)^{j'+1}} \mathfrak{f}_{m''}^{(k'')} \mathfrak{f}_{m'}^{(k'-j')} \beta_{k'',n-\ell-1} \beta_{k',\ell}. \tag{3.13}$$

Let $\{\sigma_1, \ldots, \sigma_d\}$ ($\sigma_1 = 1$) be a basis of the $\mathbb{Q}$-vector space $\mathscr{B}$ generated by all the products $\{\beta_{ij}\beta_{k\ell}\}$, and write $\beta_{ij}\beta_{k\ell} = \sum_{s=1}^d \mathfrak{q}_s^{ijk\ell} \sigma_s$. Since (3.13) $= \pm \sum b_\mu \xi^{r\mu}$ with $b_\mu \in \mathbb{Q}$ (and $\mathfrak{f}_b^{(a)} \in \mathbb{Q}$), the resulting "$\sigma_s$-components" of (3.13) vanish for $s > 1$, while

$$\pm b_\mu = Q_0 r \sum_{(*)_{r\mu}} \frac{(-1)^\ell}{(m'+r)^{j'+1}} \mathfrak{f}_{m''}^{(k'')} \mathfrak{f}_{m'}^{(k'-j')} \mathfrak{q}_1^{k'',n-\ell-1,k',\ell}. \tag{3.14}$$

Choose $\epsilon \in \mathbb{N}$ sufficiently large that all the $\{\epsilon Q_0 \mathfrak{q}_1^{ijk\ell}\}$ (a finite set) are integers. Now $m'+r \leq r\mu \implies \frac{L_{r\mu}^{j'+1}}{(m'+r)^{j'+1}} \in \mathbb{Z}$, while Corollary 2.6 $\implies \varepsilon \mathfrak{f}_{m''}^{(k'')} L_{r\mu}^{k''}, \varepsilon \mathfrak{f}_{m'}^{(k'-j')} L_{r\mu}^{k'-j'} \in \mathbb{Z}$. Since in each term of RHS(3.14) we have $k'' + k' - j' + j' - 1 = k'' + k' + 1 \leq (n - \ell - 1) + \ell + 1 = n$, multiplying the original ε_B by $\epsilon\varepsilon^2$ gives $\varepsilon_B b_\mu L_{r\mu}^n \in \mathbb{Z}$, as desired.

Step 4: Irrationality of v_0. Let $\Lambda_m := \gcd(m!, L_{rm})$, so that by Step 3 we have $\varepsilon_B \Lambda_m^n b_m =: B_m \in \mathbb{Z}$ ($\forall m$). Writing $\mathcal{P}_m := \{p \text{ prime} \mid p \leq m\}$, set

$$\pi(m) = |\mathcal{P}_m| \text{ and } \chi(m) := \sum_{p \in \mathcal{P}_m} \log(p).$$

Evidently

$$e^{\chi(m)} \leq \Lambda_m \leq \prod_{p \in \mathcal{P}_m} p^{\lfloor \log_p(rm) \rfloor} \leq (rm)^{\pi(m)},$$

hence

$$e^{\chi(m)/m} \leq \Lambda_m^{1/m} \leq (rm)^{\pi(m)/m}. \tag{3.15}$$

By the Prime Number Theorem and its proof, $\frac{\pi(m)}{m} \sim \frac{1}{\log(m)}$ and $\lim_{m\to\infty} \frac{\chi(m)}{m} = 1$. So the outer terms of (3.15) limit to e, and so does $\Lambda_m^{1/m}$.

Finally, suppose $v_0 = \frac{P}{Q}$ ($P \in \mathbb{Z}$, $Q \in \mathbb{Z} \setminus \{0\}$). Then

$$\begin{aligned}
&\limsup_{m\to\infty} |\varepsilon_B \Lambda_m^n a_m P - B_m Q|^{\frac{1}{m}} \\
&\quad = \left(\lim_{m\to\infty} \Lambda_m^{1/m}\right)^n \cdot \lim_{m\to\infty} |\varepsilon_B Q|^{\frac{1}{m}} \cdot \limsup_{m\to\infty} |a_m v_0 - b_m|^{\frac{1}{m}} \\
&\quad = e^n \cdot 1 \cdot \mathfrak{r}_\phi \qquad \text{(by Step 1)} \\
&\quad < 1 \qquad \text{(by assumption)},
\end{aligned}$$

and for some sufficiently large m we therefore have

$$0 < |\varepsilon_B \Lambda_m^n a_m P - B_m Q| < 1,$$

a contradiction. Q.E.D.

3.2. Casting a wider net. By [**7K**, Thm. 5.2], we can think of $\tilde{\nu}_\phi^0$ as computing the extension of $\mathbb{Q}(0)$ by $W_{-2n}H^{n-1}(X_\phi^0, \mathbb{Q}(n)) \cong \mathbb{Q}(n)$ associated to $\Xi|_{X_\phi^0} \in H^n_{\mathcal{M}}(X_\phi^0, \mathbb{Q}(0))$. Since $(\mathcal{I}^*\omega)|_{X_\phi^0}$ generates the top graded piece $\mathrm{Gr}_0^W H_{n-1}(X_\phi^0, \mathbb{Q})$, pairing with it sends the above $\mathbb{Q}(n)$ isomorphically to $(2\pi\mathbf{i})^n\mathbb{Q}$. So $\tilde{V}_\phi(0)$ realizes $\mathrm{AJ}\left(\Xi|_{X_\phi^0}\right) \in \mathbb{C}/(2\pi\mathbf{i})^n\mathbb{Q}$, which one interprets (arguing as in [op. cit.], at least for $n \leq 3$) as a Borel regulator value r_{Bor} for K_{2n-1} of the number field k required to resolve X_ϕ^0. Since this is just the field of definition of $\mathcal{I}$, the numbers $\tilde{V}_\phi(0)$ potentially appearing in Theorem 3.2 are limited (for $n \geq 2$) to $\zeta(n)$.

The first step in a generalization of this result would be to expand the notion of involutivity:

- drop the requirement $k = \mathbb{Q}$,
- replace $t \mapsto \pm t^{-1}$ by $t \mapsto \frac{1}{Ct}$ ($C \in \mathbb{Z}$), and
- allow $\mathcal{I}$ to be a correspondence inducing an isomorphism between the $\mathbb{Q}$-VHS $\mathcal{V}_\phi$ and its pullback.

Unfortunately, we have to pay for this expansion with a stronger bound:

PROPOSITION 3.3. *Let ϕ be as in Theorem 3.2, but with the weaker involutivity just described. Assume in addition that $\mathfrak{r}_\phi < \frac{e^{-n}}{|C|}$. Then $\tilde{V}_\phi(0) \notin \mathbb{Q}$.*

The proof is a straightforward, but tedious, generalization of the above. One nice formal consequence is that $\mathcal{I}^*\omega = \frac{c}{\lambda}\omega$ for $c \in \mathbb{C}^\times$ with $\frac{c^2}{C} = \pm 1$, so that $\tilde{V}_\phi(0) = \langle \tilde{\nu}_\phi^0, \tilde{\omega}_\phi^0 \rangle = c^{-1}\langle \tilde{\nu}_\phi^0, \mathcal{I}^*\omega_{\phi,0}\rangle = c^{-1} r_{\mathrm{Bor}}$. For example, if $n = 2$ and $\mathcal{I}$ is defined over $\mathbb{Q}(\mathbf{i})$, then one expects $r_{\mathrm{Bor}} \underset{\mathbb{Q}^\times}{\sim} \mathbf{i}G$, where $G = L(\chi_{-4}, 2)$ is Catalan's constant. But then, one also expects $c \underset{\mathbb{Q}^\times}{\sim} \mathbf{i}$, so that $\tilde{V}_\phi(0) \underset{\mathbb{Q}^\times}{\sim} G$—a not-insignificant "calibration", as irrationality of $\mathbf{i}G$ and of G are rather different things. Naturally, we don't have a proof of the latter, but we will briefly discuss some higher normal functions with G as a special value below.

REMARK 3.4. The function of $\mathcal{I}$ and $\mathcal{J}$ in the above proof is to match periods of a pullback of $\mathcal{V}_\phi$ with those of $\mathcal{V}_\phi$ resp. the facile family. If one has such a matching by other means, there is obviously no need for the maps of varieties.

Another natural way to relax the hypotheses is to permit the Newton polytope of ϕ to be non-reflexive, as long as $\underline{0}$ remains its unique interior integral point. This ensures that, while X_ϕ^λ may not be Calabi-Yau, $h^{n-1,0}(X_\phi^\lambda)$ remains 1. More significantly, one could abandon the "principality" constraint that Hodge numbers of $\mathcal{V}_\phi$ *all* equal 1, in order to make contact with results (such as [**Z2**]) involving linear forms in more than one zeta value; two likely sources of interesting examples will be discussed in §5.4.

4. Low dimension

In order to implement Theorem 3.2 in any specific cases, we must be able to check strong temperedness and compute $\tilde{V}_\phi(0)$. For the temperedness, we begin by examining the toric boundary structure more closely. (This part draws heavily on [**DK**, §§2-3].) Let ϕ and $\mathbb{P}_\Delta$ be as in the first paragraph of §3, with associated family $\mathcal{X}_\phi \underset{\rho}{\to} \mathbb{P}^1$ of anticanonical hypersurfaces.

Consider the toric variety $\hat{\mathbb{P}}_\Delta$ associated to Δ° without the triangulation, with canonical blow-down morphisms $\mathbb{P}_\Delta \underset{b}{\twoheadrightarrow} \hat{\mathbb{P}}_\Delta$, $\mathbb{D}_\Delta \underset{b}{\twoheadrightarrow} \hat{\mathbb{D}}_\Delta = \hat{\mathbb{P}}_\Delta \setminus \mathbb{G}_m^n$. Let $\sigma \subset \Delta$ be a codimension-j face with ($(j-1)$-dimensional) dual $\sigma^\circ \subset \Delta^\circ$, and $\mathbb{G}_m^{n-j} \cong \hat{\mathbb{D}}_\sigma^\times \subset \hat{\mathbb{D}}_\sigma \subset \hat{\mathbb{D}}_\Delta$ the corresponding (open resp. closed) codimension-j stratum. The ($(n-j+1)$-dimensional) strata $\mathbb{D}_\sigma^\alpha \subseteq b^{-1}(\hat{\mathbb{D}}_\sigma)$ correspond to the ($(j-i-1)$-dimensional) faces $\sigma_\alpha^\circ \subseteq \sigma^\circ$ in the triangulation of Δ°. A basis of the flag $\mathrm{ann}(\sigma^\circ) \subseteq \mathrm{ann}(\sigma_\alpha^\circ) \subseteq \mathbb{Z}^n$ produces $n-j+1$ toric coordinates $\{x_1^\sigma, \ldots, x_{n-j}^\sigma; y_1, \ldots, y_i\}$ on $\mathbb{D}_\sigma^{\alpha,\times}$, with b induced by forgetting the $\{y_i\}$. Writing $Y_\phi := X_\phi^\lambda \cap \mathbb{D}_\Delta \underset{b}{\twoheadrightarrow} \hat{Y}_\phi$, each[14]

$$Y_\sigma^\alpha := Y_\phi \cap \mathbb{D}_\sigma^\alpha = \cup_{i=1}^{\mu_\sigma} Y_\sigma^{\alpha,i} \underset{b}{\twoheadrightarrow} \cup_{i=1}^{\mu_\sigma} \hat{Y}_\sigma^i = \hat{Y}_\sigma =: \hat{Y}_\phi \cap \mathbb{D}_\sigma$$

is cut out of $\mathbb{D}_\sigma^\alpha$ resp. $\hat{\mathbb{D}}_\sigma$ by a "face polynomial" $\phi_\sigma(x_1^\sigma, \ldots, x_{n-j}^\sigma)$ given (up to a shift) by rewriting the terms $\{c\underline{x}^{\underline{m}} \mid \underline{m} \in \sigma\}$ of ϕ in terms of $\underline{x}^\sigma$.

A precondition for temperedness of ϕ (as defined above) is that the iterated residues of the coordinate symbol $\{\underline{x}\}|_{(X_\phi^\lambda)^\times}$ along strata Y_σ^α of $X_\phi^\lambda \setminus (X_\phi^\lambda)^\times$ all vanish, which is equivalent to the vanishing of the $\{\underline{x}^\sigma\}$ on $\hat{Y}_\sigma^\times$:

DEFINITION 4.1. ϕ is *weakly tempered*[15] if the symbols

$$\{x_1^\sigma, \ldots, x_{n-j}^\sigma\}|_{\hat{Y}_\sigma^\times} = 0 \in K_{n-j}^M(\bar{\mathbb{Q}}(\hat{Y}_\sigma^i)) \otimes \mathbb{Q} \tag{4.1}$$

for all $1 \le j \le n-1$, $\sigma \in \Delta(j)$, and $i = 1, \ldots, \mu_\sigma$.

This is a condition on the face polynomials $\phi_\sigma = \prod \phi_{\sigma,i}$. For example, (4.1) holds:

- for $j = n-1$ if all edge polynomials are cyclotomic; and
- for $j = n-2$ if (for all 2-faces) the $\phi_{\sigma,i}$ are *Steinberg* (i.e. $\phi_{\sigma,i}(x,y) = 0$ makes $\{x, y\} = 0$ in K_2).

We will say that ϕ_σ is $\mathbb{Q}$-*Steinberg* if its factors are Steinberg and defined over $\mathbb{Q}$.

For $n \ge 4$, it may not even be the case that weak temperedness implies the existence of fiberwise lifts $\Xi^\lambda \in \mathrm{CH}^n(\widetilde{X_\phi^\lambda}, n) \otimes \mathbb{Q}$ of $\{\underline{x}\}|_{X_\phi^{\lambda,\times}}$ to a desingularization. This is always true for $n \le 3$; for $n = 4, 5$ it holds if (for instance) all boundary strata are rational and defined over a totally real number field. (See [**DK**, §3] for

[14]The $\{\hat{Y}_\sigma, Y_\sigma^\alpha\}$ need not be irreducible, as $\lambda - \phi$ is not assumed Δ-regular for generic λ.

[15]Note that this is the definition of *temperedness* in [**DK**, §3]. Our use of "tempered" here correlates to the property of "$\{\underline{x}\}$ completes to a family of motivic cohomology classes" in [op. cit.].

other refinements.) For particular families one can certainly check temperedness in any dimension. However, in the remainder of this section, we prefer to restrict to the case $n \leq 3$.

Let $\mathcal{K}_\phi := \rho(T_{\{\underline{x}\}}) \subset \mathbb{P}^1$ and $U_\phi := \mathbb{P}^1 \setminus \mathcal{K}_\phi$ (as in §3). We are interested in conditions under which not only is ϕ strongly tempered, but where the (unique) single-valued lift $\tilde{\nu}_\phi^\lambda$ over U_ϕ is given by fiberwise restrictions of the regulator current

$$R_{\{\underline{x}\}} = \sum_{j=1}^{n} (-1)^{(n-1)(j-1)} (2\pi\mathbf{i})^{j-1} \log(x_j) \frac{dx_{j+1}}{x_{j+1}} \wedge \cdots \wedge \frac{dx_n}{x_n} \delta_{T_{x_1} \cap \cdots \cap T_{x_{j-1}}}$$

up to push-fowards of currents from Y_ϕ and coboundary currents. The biggest nuisance turns out to be the correction term $(2\pi\mathbf{i})^n \delta_{\Gamma^\lambda}$ added to R_{Ξ^λ} to produce a closed current; part of this correction is (the current of integration over) a chain on X_ϕ^λ bounding on an unknown $(n-2)$-cycle on Y_ϕ. In order that this not contribute to $\tilde{V}_\phi$, we have to assume that H_{n-2} of Y_ϕ (or part of it) vanishes.[16]

To make these conditions relatively weak, we define some "bad" subsets of the boundary. Let $\mathscr{I}_\phi \subset Y_\phi$ be the generic Δ-irregularity locus of $\lambda - \phi$—that is, the closure of the union of all singularities and nonreduced components of all $Y_\sigma^\times$ (computable by taking partials of ϕ_σ). Let $\mathscr{A} \subset \mathscr{I}$ be the locus of generic singularities of X_ϕ^λ ($\mathscr{A}$ is empty for $n = 1$ or 2). Denoting by $\mathbb{I}_\Delta$ the intersection with $\mathbb{D}_\Delta$ of the closure of the locus $\cup_{i=1}^n \{x_i = 1\}$ in $\mathbb{P}_\Delta$, we write $\mathscr{J}_\phi$ for the union of all components of Y_ϕ not contained in $\mathbb{I}_\Delta$, and (for $n = 3$) not of "Steinberg type" $\{x_{i_1} + x_{i_2} = 1,\ x_{i_3} = 0 \text{ or } \infty\}$.

THEOREM 4.2. *Let $\phi \in \mathbb{Z}[x_1^{\pm 1}, \ldots, x_n^{\pm 1}]$ be a reflexive Laurent polynomial for $n = 1$, 2, or 3.*

(a) *Assume that edge polynomials [resp. 2-face polynomials] of ϕ are cyclotomic [resp. $\mathbb{Q}$-Steinberg], and that $\mathscr{I} \subset \mathbb{I}_\Delta$ and $\mathscr{I} \cap \mathscr{J} \subset \mathscr{A}$; if $n = 3$, assume $\mathscr{A}$ consists only of A_1 singularities and lies in $\mathbb{I}_\Delta \cap \operatorname{sing}(\mathbb{D}_\Delta)$. Then ϕ is tempered.*

(b) *If $n = 2$, assume in addition that $\mathscr{J}$ is one point. If $n = 3$, assume in addition that $H_1(\mathscr{J} \setminus \mathscr{J} \cap \mathscr{A}) = \{0\}$. Then ϕ is strongly tempered, with single-valued truncated higher normal function*

$$\tilde{V}_\phi(\lambda) = \int_{X_\phi^\lambda} R\{\underline{x}\} \wedge \tilde{\omega}_\phi^\lambda \tag{4.2}$$

over U_ϕ.

SKETCH. (a) is a restatement of part of [**DK**, Thm. 3.8]. The strong-temperedness part of (b) recapitulates [**DK**, Prop. 4.12]. We now show that (4.2) is consistent with the lift constructed in the proof of [loc. cit.]. Assuming first that $\mathscr{A} = \emptyset$, we can extend $\{\underline{x}\}$ to a closed precycle on $Z^n(\mathcal{X}_\phi^- \setminus \mathscr{J} \times \mathbb{A}^1, n)$ by taking Zariski closure and (for $n = 3$) adding precycles of the form $\left(t, 1-t, 1-\frac{x_{i_1}}{t}\right)_{t \in \mathbb{P}^1}$ over the "Steinberg type" locus $\mathscr{S}$. As in [loc. cit.], one then has $\gamma \in Z^n(\mathcal{X}_\phi^- \setminus \mathscr{J} \times \mathbb{A}^1, n+1)$ and (closed) $\Xi \in Z^n(\mathcal{X}_\phi^-, n)$ with $\Xi|_{\mathcal{X}_\phi^- \setminus \mathscr{J} \times \mathbb{A}^1} = \{\underline{x}\} + \imath_*^{\mathscr{S}} Z + \partial\gamma$ $\implies R_\Xi|_{\mathcal{X}_\phi^- \setminus \mathscr{J} \times \mathbb{A}^1} = R_{\{\underline{x}\}} + \frac{1}{2\pi\mathbf{i}} dR_\gamma + \imath_*^{\mathscr{S}} R_Z - (2\pi\mathbf{i})^n \delta_{T_\gamma}$ and $T_\Xi \equiv \partial\overline{T_\gamma} + \tau$ mod

[16] To get a feel for this condition, and the proof that follows, one may consult [**BKV**, §4] where a similar argument is run on a concrete example.

$\rho^{-1}(\mathcal{K})$, with τ supported on $\mathscr{J} \times (\mathbb{P}^1, \mathcal{K})$. Arguing as in [loc. cit.], our hypotheses give that $\tau \underset{\text{hom}}{\equiv} 0$ on $\mathscr{J} \times (\mathbb{P}^1, \mathcal{K})$. Writing this as $\tau \equiv \partial C_{\mathscr{J}}$, the "lift" of R_{Ξ} given there is $R'_{\Xi} = R_{\Xi} + (2\pi\mathbf{i})^n \delta_{\overline{T_\gamma + C_{\mathscr{J}}}}$, so that

$$R'_{\Xi}|_{\mathcal{X}_\phi^- \setminus \mathscr{J} \times \mathbb{A}^1} = R_{\{\underline{x}\}} + \tfrac{1}{2\pi\mathbf{i}} dR_\gamma.$$

Writing $X^\lambda_{\phi,\epsilon}$ for X^λ_ϕ minus a small tubular neighborhood of $\mathscr{J}$, this leads to

$$\tilde{V}_\phi(\lambda) = \int_{X^\lambda_\phi} R'_{\Xi} \wedge \tilde{\omega}^\lambda_\phi = \lim_{\epsilon \to 0} \left\{ \int_{X^\lambda_{\phi,\epsilon}} R_{\{\underline{x}\}} \wedge \tilde{\omega} + \tfrac{1}{2\pi\mathbf{i}} \int_{\partial X^\lambda_{\phi,\epsilon}} R_\gamma \wedge \tilde{\omega} \right\}$$

hence (4.2), provided $\lim_{\epsilon\to 0} \int_{\partial X^\lambda_{\phi,\epsilon}} R_\gamma \wedge \tilde{\omega} = 0$. Viewing γ as a family of curves in $\square^{n+1}$ over $\mathcal{X}_\phi$, the $(n-2)$-current R_γ is the push-forward of $R_{\{\underline{z}\}}$ ($\underline{z}$ = coordinates on $\square^{n+1}$). If $n = 2$, and n is a local holomorphic coordinate about $\mathscr{J}$, one checks that R_γ is locally $\mathcal{O}(\log^2 u)$; clearly $\oint_{|u|=\epsilon} \log^2(u) du \to 0$. For $n = 3$ (with $|(u)_0| = \mathscr{J}$ locally), R_γ is the current of integration over a 3-chain times a locally-$\mathcal{O}(\log^2(u))$ function, with the same result. Finally, in the event that $\mathscr{A}$ is nonempty, and $\widetilde{\mathcal{X}_\phi^-}$ the blowup along $\mathscr{A} \times \mathbb{P}^1$ (with exceptional divisor $\mathcal{E}$, one replaces all complexes (of higher Chow cycles and currents) on $\mathcal{X}_\phi^-$ by cone (double-)complexes for the morphism $\mathcal{E} \longrightarrow \mathscr{A} \amalg \widetilde{\mathcal{X}_\phi^-}$. The assumption that $H_1(\mathscr{J} \setminus \mathscr{J} \cap \mathscr{A}) = \{0\}$ allows $C_{\mathscr{J}}$ to be drawn so as to avoid $\mathcal{E}$. □

This theorem is likely far from optimal, but suffices for the applications in the next section.

Turning to the computation of $\tilde{V}_\phi(0)$, we have the following specialization result:[17]

Corollary 4.3. *For ϕ as in Theorem 4.2(b), if $\phi(-\underline{x})$ has all positive coefficients, then*

$$\tilde{V}_\phi(0) = \int_\psi R_{\{\underline{x}\}}|_{X^0_\phi}$$

for any $(n-1)$-cycle $\psi \subset X^0_\phi \setminus \mathscr{J}$ representing $Q_0 \hat{\varphi}_{n-1} \in H_{n-1}(X^0_\phi)$, the class (from §3.1, Step 2) with intersection number ± 1 against φ_0.[18]

Sketch. The additional hypothesis on ϕ ensures that $\lambda = 0$ belongs to U_ϕ. Now $Z^0 = \mathrm{Res}_{X^0_\phi}(\mathcal{I}^*\Xi) = \mathcal{I}^* \mathrm{Res}_{X_{\phi,0}}(\Xi) = \mathcal{I}^* Z_0$ has $\Omega_{Z^0} = \mathcal{I}^*\Omega_{Z_0} = (2\pi\mathbf{i})^{n-1}\mathcal{I}^*\omega_0 = (2\pi\mathbf{i})^{n-1}\tilde{\omega}^0$, while T_{Z^0} is an $(n-1)$-cycle with $dR_{Z^0} = \Omega_{Z^0} - (2\pi\mathbf{i})^{n-1}\delta_{T_{Z^0}} \implies [T_{Z^0}] = [\tilde{\omega}^0] = Q_0\hat{\varphi}_{n-1}$ in homology. So for ψ as above, there exists an $(n-2)$-current $\mathcal{R}$ on X^0_ϕ with $d\mathcal{R} = \omega^0 - \delta_\psi$, with closed hence (by hypothesis on $\mathscr{J}$) exact restriction to $\mathscr{J}$. We may therefore assume that $\mathcal{R}$ (is of intersection type with respect to $\mathscr{J}$ and) pulls back to 0 on $\mathscr{J}$, so that $\lim_{\epsilon\to 0}\int_{\partial X^0_{\phi,\epsilon}} R_{\{\underline{x}\}} \wedge \mathcal{R} = 0$. It now follows that $\tilde{V}_\phi(0) = \lim_{\epsilon\to 0}\int_{X^0_{\phi,\epsilon}} R_{\{\underline{x}\}} \wedge \tilde{\omega}^0_\phi = \lim_{\epsilon\to 0}\int_{X^0_{\phi,\epsilon}} R_{\{\underline{x}\}} \wedge \delta_\psi$ as claimed. □

---

[17] This may be viewed an analogue of [**7K**, Cor. 5.3], but on the level of lifts (i.e. as an exact equality of complex numbers).

[18] One should think of ψ as a membrane "stretched once around" X^0_λ.

5. Examples and near-examples

Here we record some Laurent polynomials that satisfy the conditions of Theorems 3.2 and 4.2, as well as a few which stray close enough to warrant attention.

5.1. $n = 1$. Let $b \in \mathbb{Z}_{>0}$, $a = 2b + 1$, and set

$$\phi(x) := -x + a + \frac{1 - a^2}{4x}.$$

X_ϕ^λ is a pair of points $\{p_+^\lambda, p_-^\lambda\}$ which are distinct unless λ is a root of $P_a(\lambda) := \lambda^2 - 2a\lambda + 1$, and $\mathcal{X}_\phi$ has involution[19]

$$(x, \lambda) \longmapsto \left(\lambda^{-1}x + \tfrac{1}{2}(1 - \lambda^{-1})(a + 1), \lambda^{-1}\right).$$

We have $p_\pm^\lambda = \frac{1}{2}(a - \lambda \pm P_a(\lambda)^{1/2})$, and the 0-form

$$\tilde{\omega}_\phi^\lambda = \lambda^{-1}\omega_\phi^\lambda = \frac{\pm 1}{2\sqrt{P_a(\lambda)}} \quad \text{on} \quad p_\pm^\lambda$$

has period

$$A(\lambda) = \int_{p_+^\lambda - p_-^\lambda} \tilde{\omega}_\phi^\lambda = \frac{1}{\sqrt{P_a(\lambda)}}.$$

The regulator 0-current $R_\lambda = \log(p_\pm^\lambda)$ on $p_\pm^\lambda$, and so the higher normal function is

$$\tilde{V}_\phi(\lambda) = \frac{1}{\sqrt{P_a(\lambda)}} \log(p_+^\lambda / p_-^\lambda).$$

Since $\mathfrak{r}_\phi = a - \sqrt{a^2 - 1} < e^{-1}$ for all $b \geq 1$, we conclude by Theorem 3.2 that

$$\tilde{V}_\phi(0) = \log\left(\frac{b+1}{b}\right) \notin \mathbb{Q}.$$

5.2. $n = 2$. Let

$$\phi(x_1, x_2) := x_1^{-1}x_2^{-1}(1 - x_1)(1 - x_2)(1 - x_1 - x_2);$$

the picture

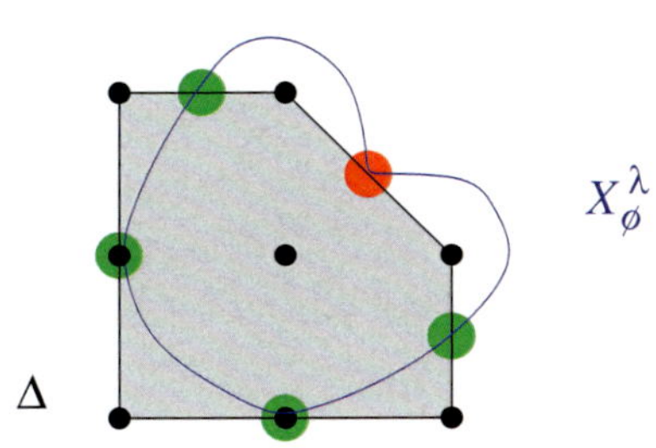

indicates the (reflexive) Newton polytope Δ and $X_\phi^\lambda \subset \mathbb{P}_\Delta$. The green [resp. red] dots represent $Y_\phi \setminus \mathscr{J}$ [resp. $\mathscr{J}$]; edge polynomials are $x^\sigma - 1$ or $(x^\sigma - 1)^2$, so ϕ is strongly tempered by Theorem 4.2(b).

The singular fibers of $\mathcal{X}_\phi$ are over 0, ∞, and $t_\pm = \frac{-11 \pm 5\sqrt{5}}{2}$, with $\mathfrak{r}_\phi = t_+ < e^{-2}$ (and $\mathcal{X}_{\phi, t_\pm}$ of Kodaira type I_1), while

$$D_\phi(t) = (t^2 + 11t - 1)\delta_t^2 + t(2t + 11)\delta_t + t(t + 3)$$

[19] of course, it is its own facile 1-cover!

is integral.[20] Involutivity is ensured by the (order 4) automorphism

$$\mathcal{I}:\ (x_1, x_2, t) \longmapsto \left(\tfrac{x_1}{x_1-1}, \tfrac{1-x_2}{1-x_1-x_2}, -\tfrac{1}{t}\right),$$

while the facile 2-cover given by

$$(1-\xi^2) + \xi y_1 + \xi^2 y_2 - \xi y_1 y_2 - \xi y_1^{-1} y_2^{-1} = 0$$

maps down by

$$\mathcal{J}:\ (y_1, y_2, \xi) \longmapsto \left(\tfrac{y_1}{\xi}, \tfrac{y_1-\xi}{y_1 y_2-\xi}, \xi^2\right).$$

To compute the special value $\tilde{V}_\phi(0)$ using Corollary 4.3, note that $X^0 = \{x_1 = 1\} \cup \{x_2 = 1\} \cup \{x_1 + x_2 = 1\}$, with ψ going "once around" the figure. Furthermore, $R_{\{\underline{x}\}} = \log(x_1)\frac{dx_1}{x_1} - 2\pi\mathbf{i}\log(x_2)\delta_{T_{x_1}}$ vanishes on the first two components. Parametrizing the remaining component of $-\psi$ by $[0,1] \ni s \mapsto (1-s, s)$, we conclude (by Theorem 3.2) that

$$\tilde{V}_\phi(0) = -\int_0^1 \log(1-s)\tfrac{ds}{s} = \mathrm{Li}_2(1) = \tfrac{\pi^2}{6} \notin \mathbb{Q}.$$

Remark 5.1. In view of the results of Zagier's search for recurrencies of Apéry type [**Za**], it seems likely that this ϕ is the unique example for $n = 2$ that satisfies the conditions of Proposition 3.3. One can match tempered Laurent polynomials (hence higher normal functions) to the sporadic examples of [op. cit.], but outside case "D" (just treated), both the bound on $\mathfrak{r}_\phi$ and involutivity fail. For instance, case "E" of [op. cit.] is $\phi(\underline{x}) = (x_1 + x_1^{-1})(x_2 + x_2^{-1}) + 4$; this has $\tilde{V}_\phi(0) = 2G$ (Catalan), but $\mathfrak{r}_\phi C = \frac{1}{8} \cdot 32 = 4$ (too big), and the Kodaira fiber types at λ_0 and C/λ_0 (or 0 and ∞) don't match, so that $\mathcal{V}_\phi \not\cong \mathcal{I}^*\mathcal{V}_\phi$. This non-involutivity is not a problem for the approach via modular forms, which gives a different means for obtaining period expansions about any cusp; we are trading off this advantage for (at least in principle) the ability to treat non-modular families in higher dimension.

Remark 5.2. One other "near-example" related to Catalan's constant arises from work of Zudilin [**Z3**], who found an Apéry-like recurrence with rational solutions a_m, b_m whose ratios b_m/a_m converge rapidly to G. With some work, one can write the generating series $\sum_{m\geq 0}(a_m G - b_m)\lambda^m$ as a normal function associated to a higher cycle on a family of open[21] genus-9 curves, which are branched 4:1 covers of the "baby Apéry" family of elliptic curves above! By construction, this has $V(0) = G$.

5.3. $n = 3$**.** The Newton polytope Δ of

$$\phi(x_1, x_2, x_3) = x_1^{-1}x_2^{-1}x_3^{-1}(x_1-1)(x_2-1)(x_3-1)(1 - x_1 - x_2 + x_1x_2 - x_1x_2x_3) \tag{5.1}$$

[20] The coefficients of the holomorphic solution are the "baby Apéry" sequence $a_m = \sum_{k=0}^m \binom{m}{k}^2\binom{m+k}{k} = 1, 3, 19, 147, \ldots$.

[21] The corresponding Laurent polynomial is neither reflexive nor tempered; two points are removed from each fiber.

and its dual are

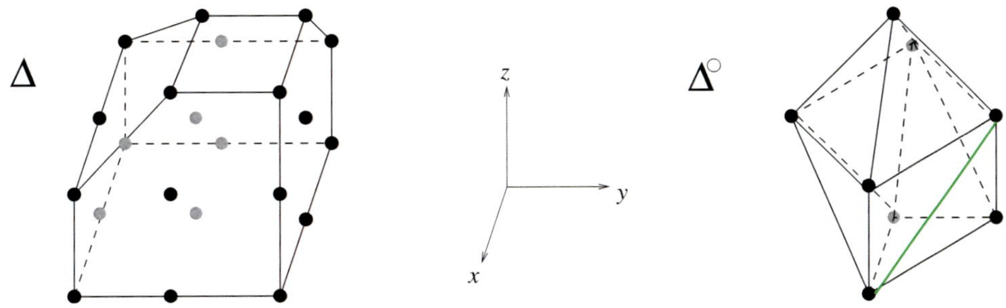

showing reflexivity; maximal triangulation adds the green edge. The edge and facet polynomials are products of $(x_i^\sigma - 1)$, $(x_i^\sigma - 1)^2$, and $(1 - x_1^\sigma \pm x_2^\sigma)$, and $\mathscr{A}$ (red), $\mathscr{J}$ (blue), and $\mathscr{I}$ (green) are as depicted:

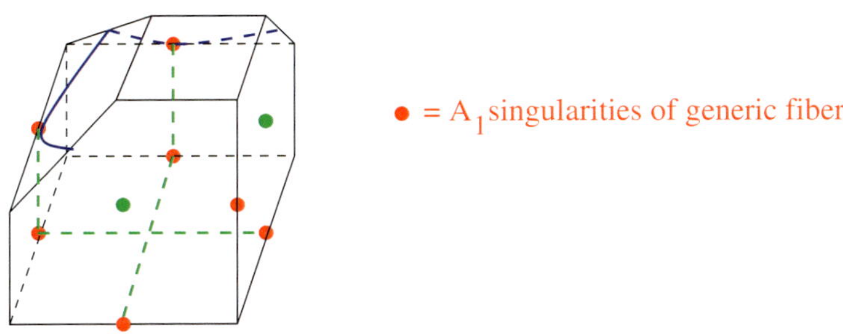

In particular, $\mathscr{J} \setminus \mathscr{J} \cap \mathscr{A}$ is two copies of $\mathbb{A}^1$ attached at a point, and we conclude by Theorem 4.2(b) that ϕ is strongly tempered.

Singular fibers are at 0, ∞, $t_\pm = (\sqrt{2} \pm 1)^4$, and 1; the last of these does not contribute to monodromy of $\mathbb{V}_\phi$, and so $d_\phi = 2$, while $\mathfrak{r}_\phi = t_- < e^{-3}$. To see that the generic Picard rank is 19, one can use a torically-induced elliptic fibration (cf. [**Ke1**, §2]). The Picard-Fuchs operator is

$$D_\phi(t) = (t^2 - 34t + 1)\delta_t^3 + 3t(t-17)\delta_t^2 + 3t(t-9)\delta_t + t(t-5),$$

and the $\{a_m\}$ the famous Apéry sequence $a_m = \sum_{k=0}^m \binom{m}{k}^2 \binom{m+k}{k}^2 = 1, 5, 73, 1445, \ldots$.

Changing coordinates by $\mathsf{X}_i = \frac{x_i}{x_i - 1}$ brings $1 - t\phi(\underline{x}) = 0$ into the form studied by Beukers and Peters [**BP**]. By the results of Peters and Stienstra [**PS**], $\mathcal{X}_{\phi,t}$ thus has a (facile) 2-cover by the *Fermi family*

$$\mathrm{X}_\xi := \left\{ \xi \sum_{i=1}^3 (y_i + y_i^{-1}) + 1 + \xi^2 = 0 \right\}.$$

It also has an involution, by

$$\mathcal{I} : (x_1, x_2, x_3, t) \longmapsto \left(\tfrac{x_3}{x_3 - 1}, \tfrac{(x_1-1)(x_2-1)}{1 - x_1 - x_2 + x_1 x_2 - x_1 x_2 x_3}, \tfrac{x_1}{x_1 - 1}, \tfrac{1}{t} \right).$$

The 2-current

$$R_{\{\underline{x}\}} = \log(x_1)\tfrac{dx_2}{x_2} \wedge \tfrac{dx_3}{x_3} + (2\pi\mathbf{i})\log(x_2)\tfrac{dx_3}{x_3}\delta_{T_{x_1}} + (2\pi\mathbf{i})^2 \log(x_3)\delta_{T_{x_1} \cap T_{x_2}}$$

vanishes on the components $\{x_i = 1\}$ of X_ϕ^0. The piece of ψ on the remaining component $x_3 = \frac{(1-x_1)(1-x_2)}{x_1 x_2}$ is parametrized by $0 \le r \le s \le 1 \longmapsto (1 -$

$r, s, \frac{r(1-s)}{s(1-r)}$), and so (invoking Cor. 4.3 and Thm. 3.2)

$$\begin{aligned}\tilde{V}_\phi(0) &= \int_{0\le r\le s\le 1} \log(1-r)\,\mathrm{dlog}(s)\wedge \mathrm{dlog}(\tfrac{r}{1-r})\\ &= -\int_0^1 \log(1-r)\log(r)\,\mathrm{dlog}(\tfrac{r}{1-r})\\ &= -2\int_0^1 \log(1-r)\log(r)\,\mathrm{dlog}(r)\\ &= 2\sum_{k\ge 1}\frac{1}{k}\int_0^1 r^{k-1}\log(r)dr \quad = \quad -2\zeta(3) \notin \mathbb{Q}.\end{aligned}$$

REMARK 5.3. In our normalization (see the definition of b_m in §3.1, Step 1) one has $\{b_m\} = 0, -12, -\frac{351}{2}, -\frac{62531}{18}, \ldots$, which is -2 times the usual second Apéry sequence. We note, however, that explicit knowledge of $\{a_m\}$ and $\{b_m\}$ is not needed for proving irrationality of $\zeta(3)$ via Theorem 3.2.

REMARK 5.4. There are at least three "near-examples" for $n = 3$, identified in [**dS**] (and closely related to [**Go**]), which satisfy all the criteria in Proposition 3.3 that we have checked, except for the bound: writing ϕ_{I} for (5.1), these are

$$\begin{aligned}\phi_{\mathrm{II}} &= (1-x_1-x_2-x_3)(1-x_1^{-1})(1-x_2^{-1})(1-x_3^{-1})\\ \phi_{\mathrm{III}} &= (x_1+x_2+x_3)(-1+x_1^{-1}+x_2^{-1}+x_3^{-1}-x_1^{-1}x_2^{-1}-x_1^{-1}x_3^{-1}-x_2^{-1}x_3^{-1})\\ \phi_{\mathrm{IV}} &= (1-x_1-x_2-x_3)(1-x_1^{-1}-x_2^{-1}-x_3^{-1}).\end{aligned}$$

They are reflexive, tempered, and involutive, with $C_{\mathrm{I}} = 1$, $C_{\mathrm{II}} = 16$, $C_{\mathrm{III}} = -27$, and $C_{\mathrm{IV}} = 64$; while $\tilde{V}(0) \underset{\mathbb{Q}^\times}{\sim} \zeta(3)$ except for ϕ_{III}, where $\tilde{V}(0) \underset{\mathbb{Q}^\times}{\sim} L(\chi_{-3}, 3) \underset{\mathbb{Q}^\times}{\sim} \pi^3/\sqrt{3}$. For $\mathfrak{r}_\phi|C|e^3$ we obtain ≈ 0.59, 13.78, 27.97, resp. 80.34, which satisfies the required bound (< 1) only in the first case.

5.4. Higher dimension? Here we propose two sources for examples with $n \ge 4$, if one is prepared to weaken the hypotheses as in the last paragraph of §3.2. In both cases, the Laurent polynomials considered, while not in general reflexive, all have Newton polytope $\Delta \subset [-1, 1]^n$ having the origin as unique integer interior point. Details, proofs, and further developments will appear elsewhere.

Define the *VZ polynomials* $\{\phi_n\}$ inductively by $\psi_1 = 1$,

$$\psi_n(x_1, \ldots, x_n) := x_1 \cdots x_n + (1-x_n)\psi_{n-1}(x_1, \ldots, x_{n-1}),$$

$$\phi_n(\underline{x}) := (1-x_1^{-1})\cdots(1-x_n^{-1})\psi_n(\underline{x});$$

they are obtained (by substituting $\mathsf{X}_i := \frac{x_i}{x_i - 1}$) from denominators of integrals first considered by Vasilyev [**Va**] and Zudilin [**Z2**] in their works on linear forms in zeta values. For $n = 2$ and 3, this recovers (up to inversion and permutation of coordinates) the Apéry polynomials above. For $n = 5$, we expect that ϕ_5 is strongly tempered, and note that Hodge numbers of $\mathcal{V}_\phi$ are $(1, 1, 2, 1, 1)$. This means there are *two* invariant periods $A(\lambda) = 1 + \sum_{m\ge 1} a_m\lambda^m$, $B(\lambda) = \lambda + \sum_{m\ge 2} b_m\lambda^m$ about the maximal unipotent monodromy point, as N has two primitive classes. Writing

$V(\lambda)$ for the higher normal function, one then expects $\mathfrak{r}_\phi < e^{-5}$, and

$$\begin{aligned} C(\lambda) &:= -V(\lambda) + A(\lambda)V(0) + B(\lambda)\left(-V(0)A'(0) + V'(0)\right) \\ &= \sum_{m\geq 2} c_m \lambda^m \end{aligned}$$

to satisfy an inhomogeneous equation as above. Combining this with Vasilyev's results, one would conclude that

$$V(\lambda) = \sum_{m\geq 0} \left(2a_m\zeta(5) + b_m\zeta(3) - c_m\right),$$

where $a_m, L_m^2 b_m, L_m^5 c_m \in \mathbb{Z}$, with the innocuous consequence that "at least one of $\zeta(3)$ and $\zeta(5)$ is irrational."

Recent work of F. Brown [**Br**] provides another source of interesting Laurent polynomials. Given a permutation $\pi \in \mathfrak{S}_{n+3}$, write formally

$$\theta_\pi(z_1, \ldots, z_{n+3}) := \prod_{i\in\mathbb{Z}/(n+3)\mathbb{Z}} (z_{\pi(i)} - z_{\pi(i+1)}),$$

$$x_j^\pi := -\operatorname{CR}\left(z_{\pi(1)}, z_{\pi(n+2-j)}, z_{\pi(n+3-j)}, z_{\pi(n+4-j)}\right),$$

where $j = 1, \ldots, n$, and

$$\Omega_\pi := \frac{dx_1^\pi}{x_1^\pi} \wedge \cdots \wedge \frac{dx_n^\pi}{x_n^\pi};$$

if $\pi = \mathrm{Id}$ then we drop the sub- and superscripts. Now let $\sigma \in \mathfrak{S}_{n+3}$ be a *convergent* permutation in the sense of [op. cit.]; namely, we assume that for any $i \in \mathbb{Z}/(n+3)\mathbb{Z}$ and $2 \leq k \leq n+1$, $\{\sigma(i), \ldots, \sigma(i+k)\}$ is not a consecutive sequence of integers mod $(n+3)$. It turns out that $\theta_\sigma(\underline{z})/\theta(\underline{z})$ can be written as a Laurent polynomial $\phi_\sigma(x_1, \ldots, x_n)$ (with Newton polytope Δ_σ), and the *basic cellular integrals* on $\mathcal{M}_{0,n}$ of [op. cit.] become the integrals

$$I_\sigma^{(k)} := \int_{T_{\{\underline{x}\}}} \phi_\sigma(\underline{x})^{-k}\Omega_\sigma$$

on $\mathbb{P}_{\Delta_\sigma}$. Defining $X_\sigma^\lambda \subset \mathbb{P}_{\Delta_\sigma}$ by $\lambda = \phi_\sigma(\underline{x})$, the generating series

$$\begin{aligned} V_\sigma(\lambda) &:= (2\pi\mathbf{i})^{1-n} \sum_{k\geq 0} I_\sigma^{(k)} \lambda^k \\ &= (2\pi\mathbf{i})^{1-n} \int_{T_{\{\underline{x}\}}} \frac{dx_1/x_1 \wedge \cdots \wedge dx_n/x_n}{\lambda - \phi_\sigma(\underline{x})} \end{aligned}$$

may be rewritten (using integration by parts) in the form

$$= \int_{X_\sigma^\lambda} R_{\{\underline{x}\}}|_{X_\sigma^\lambda} \wedge \tilde{\omega}^\lambda,$$

which is a truncated higher normal function under a strong temperedness hypothesis. Finally, involutivity may be arranged via the additional hypothesis that $\sigma^{-1} = \pi_1 \circ \sigma \circ \pi_2$, with π_1, π_2 belonging to the dihedral group D_{n+3}.

References

[AESZ] G. Almkvist, C. van Enckevort, D. van Straten, and W. Zudilin, *Tables of Calabi-Yau equations*, arXiv:math/0507430; database at http://cydb.mathematik.uni-mainz.de

[ASZ] G. Almkvist, D. van Straten, and W. Zudilin, *Apéry limits of differential equations of order 4 and 5*, Modular forms and string duality, Fields Inst. Commun., vol. 54, Amer. Math. Soc., Providence, RI, 2008, pp. 105–123, DOI 10.1090/fic/054/05. MR2454322

[AGM] P. S. Aspinwall, B. R. Greene, and D. R. Morrison, *The monomial-divisor mirror map*, Internat. Math. Res. Notices **12** (1993), 319–337, DOI 10.1155/S1073792893000376. MR1253648

[BR] K. Ball and T. Rivoal, *Irrationalité d'une infinité de valeurs de la fonction zêta aux entiers impairs* (French), Invent. Math. **146** (2001), no. 1, 193–207, DOI 10.1007/s002220100168. MR1859021

[Ba] V. V. Batyrev, *Dual polyhedra and mirror symmetry for Calabi-Yau hypersurfaces in toric varieties*, J. Algebraic Geom. **3** (1994), no. 3, 493–535. MR1269718

[Be] F. Beukers, *Irrationality of π^2, periods of an elliptic curve and $\Gamma_1(5)$*, Diophantine approximations and transcendental numbers (Luminy, 1982), Progr. Math., vol. 31, Birkhäuser, Boston, MA, 1983, pp. 47–66. MR702189

[BP] F. Beukers and C. A. M. Peters, *A family of $K3$ surfaces and $\zeta(3)$*, J. Reine Angew. Math. **351** (1984), 42–54, DOI 10.1515/crll.1984.351.42. MR749676

[BKV] S. Bloch, M. Kerr, and P. Vanhove, *A Feynman integral via higher normal functions*, Compos. Math. **151** (2015), no. 12, 2329–2375, DOI 10.1112/S0010437X15007472. MR3433889

[BV] S. Bloch and M. Vlasenko, *Gamma functions, monodromy and Frobenius constants*, Commun. Number Theory Phys. **15** (2021), no. 1, 91–147, DOI 10.4310/CNTP.2021.v15.n1.a3. MR4195428

[Br] F. Brown, *Irrationality proofs for zeta values, moduli spaces and dinner parties*, Mosc. J. Comb. Number Theory **6** (2016), no. 2-3, 102–165. MR3607782

[BZ] F. Brown and W. Zudilin, *On cellular rational approximations to $\zeta(5)$*, preprint, arXiv:2210.03391.

[Fano1] G. Brown, A Kasprzyk et al, "Graded Ring Database", at http://www.grdb.co.uk.

[CDT] F. Calegari, V. Dimitrov, and Y. Tang, *The linear independence of 1, $\zeta(2)$, and $L(2,\chi_{-3})$*, preprint, arXiv:2408.15403.

[Fano2] T. Coates, A. Corti, S. Galkin, V. Golyshev and A. Kasprzyk, "Fano varieties and extremal Laurent polynomials", research blog and database, http://coates.ma.ic.ac.uk/fanosearch/

[CK] D. A. Cox and S. Katz, *Mirror symmetry and algebraic geometry*, Mathematical Surveys and Monographs, vol. 68, American Mathematical Society, Providence, RI, 1999, DOI 10.1090/surv/068. MR1677117

[dS] G. Da Silva Jr., *On the arithmetic of Landau-Ginzburg model of a certain class of threefolds*, Commun. Number Theory Phys. **13** (2019), no. 1, 149–163, DOI 10.4310/CNTP.2019.v13.n1.a5. MR3951107

[7K] P. L. del Ángel R., C. Doran, M. Kerr, J. Lewis, J. Iyer, S. Müller-Stach, and D. Patel, *Specialization of cycles and the K-theory elevator*, Commun. Number Theory Phys. **13** (2019), no. 2, 299–349, DOI 10.4310/CNTP.2019.v13.n2.a2. MR3951112

[DK] C. F. Doran and M. Kerr, *Algebraic K-theory of toric hypersurfaces*, Commun. Number Theory Phys. **5** (2011), no. 2, 397–600, DOI 10.4310/CNTP.2011.v5.n2.a3. MR2851155

[DM] C. F. Doran and A. Malmendier, *Calabi-Yau manifolds realizing symplectically rigid monodromy tuples*, Adv. Theor. Math. Phys. **23** (2019), no. 5, 1271–1359, DOI 10.4310/atmp.2019.v23.n5.a3. MR4069107

[DvdK] J. J. Duistermaat and W. van der Kallen, *Constant terms in powers of a Laurent polynomial*, Indag. Math. (N.S.) **9** (1998), no. 2, 221–231, DOI 10.1016/S0019-3577(98)80020-7. MR1691479

[Ga] S. Galkin, *Apéry constants of homogeneous varieties*, arXiv:1604.04652.

[GGI] S. Galkin, V. Golyshev, and H. Iritani, *Gamma classes and quantum cohomology of Fano manifolds: gamma conjectures*, Duke Math. J. **165** (2016), no. 11, 2005–2077, DOI 10.1215/00127094-3476593. MR3536989

[GI] S. Galkin and H. Iritani, *Gamma conjecture via mirror symmetry*, Primitive forms and related subjects—Kavli IPMU 2014, Adv. Stud. Pure Math., vol. 83, Math. Soc. Japan, [Tokyo], [2019] ©2019, pp. 55–115. MR4384381

[GGK] M. Green, P. Griffiths, and M. Kerr, *Some enumerative global properties of variations of Hodge structures* (English, with English and Russian summaries), Mosc. Math. J. **9** (2009), no. 3, 469–530, back matter, DOI 10.17323/1609-4514-2009-9-3-469-530. MR2562791

[Go] V. Golyshev, *Deresonating a Tate period*, arXiv:0908.1458.

[GKS] V. Golyshev, M. Kerr, and T. Sasaki, *Apéry extensions*, J. Lond. Math. Soc. (2) **109** (2024), no. 1, Paper No. e12825, 38, DOI 10.1112/jlms.12825. MR4680214

[GZ] V. V. Golyshev and D. Zagir, *Proof of the gamma conjecture for Fano 3-folds with a Picard lattice of rank one* (Russian, with Russian summary), Izv. Ross. Akad. Nauk Ser. Mat. **80** (2016), no. 1, 27–54, DOI 10.4213/im8343; English transl., Izv. Math. **80** (2016), no. 1, 24–49. MR3462676

[HLY] S. Hosono, B. H. Lian, and S.-T. Yau, *GKZ-generalized hypergeometric systems in mirror symmetry of Calabi-Yau hypersurfaces*, Comm. Math. Phys. **182** (1996), no. 3, 535–577. MR1461942

[HLYZ] A. Huang, B. H. Lian, S.-T. Yau, and X. Zhu, *Chain integral solutions to tautological systems*, Math. Res. Lett. **23** (2016), no. 6, 1721–1736, DOI 10.4310/MRL.2016.v23.n6.a7. MR3621104

[Ir] H. Iritani, *Quantum cohomology and periods* (English, with English and French summaries), Ann. Inst. Fourier (Grenoble) **61** (2011), no. 7, 2909–2958, DOI 10.5802/aif.2798. MR3112512

[IKSY] K. Iwasaki, H. Kimura, S. Shimomura, and M. Yoshida, *From Gauss to Painlevé*, Aspects of Mathematics, E16, Friedr. Vieweg & Sohn, Braunschweig, 1991. A modern theory of special functions, DOI 10.1007/978-3-322-90163-7. MR1118604

[Ke1] M. Kerr, *K_1^{ind} of elliptically fibered $K3$ surfaces: a tale of two cycles*, Arithmetic and geometry of K3 surfaces and Calabi-Yau threefolds, Fields Inst. Commun., vol. 67, Springer, New York, 2013, pp. 387–409, DOI 10.1007/978-1-4614-6403-7_13. MR3156423

[Ke2] M. Kerr, *Unipotent extensions and differential equations (after Bloch-Vlasenko)*, Commun. Number Theory Phys. **16** (2022), no. 4, 801–849, DOI 10.4310/cntp.2022.v16.n4.a5. MR4503993

[KL] M. Kerr and J. D. Lewis, *The Abel-Jacobi map for higher Chow groups. II*, Invent. Math. **170** (2007), no. 2, 355–420, DOI 10.1007/s00222-007-0066-x. MR2342640

[KLM] M. Kerr, J. D. Lewis, and S. Müller-Stach, *The Abel-Jacobi map for higher Chow groups*, Compos. Math. **142** (2006), no. 2, 374–396, DOI 10.1112/S0010437X05001867. MR2218900

[LZ] B. H. Lian and M. Zhu, *On the hyperplane conjecture for periods of Calabi-Yau hypersurfaces in $\mathbf{P}^n$*, J. Differential Geom. **118** (2021), no. 1, 101–146, DOI 10.4310/jdg/1620272942. MR4255072

[Na] M. Nair, *On Chebyshev-type inequalities for primes*, Amer. Math. Monthly **89** (1982), no. 2, 126–129, DOI 10.2307/2320934. MR643279

[Pe] C. Peters, *Monodromy and Picard-Fuchs equations for families of $K3$-surfaces and elliptic curves*, Ann. Sci. École Norm. Sup. (4) **19** (1986), no. 4, 583–607. MR875089

[PS] C. Peters and J. Stienstra, *A pencil of $K3$-surfaces related to Apéry's recurrence for $\zeta(3)$ and Fermi surfaces for potential zero*, Arithmetic of complex manifolds (Erlangen, 1988), Lecture Notes in Math., vol. 1399, Springer, Berlin, 1989, pp. 110–127, DOI 10.1007/BFb0095972. MR1034260

[Ro] C. Robles, *Principal Hodge representations*, Hodge theory, complex geometry, and representation theory, Contemp. Math., vol. 608, Amer. Math. Soc., Providence, RI, 2014, pp. 259–283, DOI 10.1090/conm/608/12183. MR3220093

[vdP] A. van der Poorten, *A proof that Euler missed...Apéry's proof of the irrationality of $\zeta(3)$*, Math. Intelligencer **1** (1978/79), no. 4, 195–203, DOI 10.1007/BF03028234. An informal report. MR547748

[Va] D. Vasilyev, *On small linear forms for the values of the Riemann zeta-function at odd points*, Preprint no. 1 (558), Nat. Acad. Sci. Belarus, Institute Math., Minsk, 2001.

[Za] D. Zagier, *Integral solutions of Apéry-like recurrence equations*, Groups and symmetries, CRM Proc. Lecture Notes, vol. 47, Amer. Math. Soc., Providence, RI, 2009, pp. 349–366, DOI 10.1090/crmp/047/22. MR2500571

[Z1] V. V. Zudilin, *One of the numbers* $\zeta(5)$, $\zeta(7)$, $\zeta(9)$, $\zeta(11)$ *is irrational* (Russian), Uspekhi Mat. Nauk **56** (2001), no. 4(340), 149–150, DOI 10.1070/RM2001v056n04ABEH000427; English transl., Russian Math. Surveys **56** (2001), no. 4, 774–776. MR1861452

[Z2] W. Zudilin, *Well-poised hypergeometric service for Diophantine problems of zeta values* (English, with English and French summaries), J. Théor. Nombres Bordeaux **15** (2003), no. 2, 593–626, DOI 10.5802/jtnb.415. MR2140869

[Z3] W. Zudilin, *An Apéry-like difference equation for Catalan's constant*, Electron. J. Combin. **10** (2003), Research Paper 14, 10, DOI 10.37236/1707. MR1975764

Department of Mathematics, Campus Box 1146, Washington University in St. Louis, St. Louis, MO, 63130

Email address: matkerr@wustl.edu

Contemporary Mathematics
Volume **842**, 2026
https://doi.org/10.1090/conm/842/16858

On singular supports in mixed characteristic

Takeshi Saito

Abstract. We fix an excellent regular noetherian scheme S over $\mathbf{Z}_{(p)}$ satisfying a certain finiteness condition. For a constructible étale sheaf $\mathcal{F}$ on a regular scheme X of finite type over S, we introduce a variant of the singular support relatively to S and prove the existence of a saturated relative variant of the singular support by adopting the method of Beilinson (2016) in using the Radon transform. We may deduce the existence of the singular support itself, if we admit an expected property on the micro support of tensor product and if the scheme X is sufficiently ramified over the base S.

For a constructible étale sheaf on a smooth scheme over a field, Beilinson proved in [**2**] the existence of the singular support and its fundamental properties. A key tool used in the proof of the existence is the Radon transform studied in [**4**]. In mixed characteristic, the existence of singular support remains an open question although a framework to study the singular support was prepared in [**14**] using the Frobenius–Witt cotangent bundle FT^*X supported on the characteristic p fiber $X_{\mathbf{F}_p}$. We prove in Theorem 2.4.12 the existence of a saturated relative variant of the singular support in mixed characteristic by adopting the method of Beilinson.

Using Radon transform requires the introduction of two ingredients. The first is to work in a relative setting with a fixed base scheme S. The second is to work with pairs (h, f) of morphisms $h\colon W \to X$ and $f\colon W \to Y$ of schemes over S. The Radon transform is defined for a projective space $\mathbf{P} = \mathbf{P}_S$ over a fixed scheme S using the pair $(q, q^\vee)$ of the universal family $q^\vee\colon Q \to \mathbf{P}^\vee$ of hyperplanes in $\mathbf{P}$ and the canonical morphism $q\colon Q \to \mathbf{P}$. Thus, we need to work in a relative setting over S and also with pairs (h, f) of morphisms while the definition of singular support in [**14**] is formulated only with morphisms h without the second morphisms f.

The format of the definitions of various variants of singular supports is summarized as follows. First, we introduce a certain condition on morphisms of schemes with respect to a given closed conical subset C in the cotangent bundle and a corresponding condition with respect to an étale sheaf $\mathcal{F}$. Next, we define to say that $\mathcal{F}$ is micro supported on C if every morphism satisfying the condition for C satisfies the corresponding condition for $\mathcal{F}$. The singular support $SS\mathcal{F}$ is then defined to be the smallest C on which $\mathcal{F}$ is micro supported. The existence of singular support is non-trivial because the condition that $\mathcal{F}$ is micro supported on C and on C' does not a priori imply that $\mathcal{F}$ is micro supported on the intersection $C \cap C'$.

In the original definition by Beilinson in [**2**], he considers pairs (h, f) of morphisms and the condition on $\mathcal{F}$ is given by the local acyclicity of f relatively to

$h^*\mathcal{F}$. Since a naive imitation does not give a good definition of singular support in mixed characteristic, we consider in [**14**] morphisms h without f and the $\mathcal{F}$-transversality as a condition on h relatively to $\mathcal{F}$: A separated morphism h of finite type is $\mathcal{F}$-transversal if the canonical morphism

$$c_{\mathcal{F},h}\colon h^*\mathcal{F}\otimes h^!\Lambda\to h^!\mathcal{F}$$

is an isomorphism. Here and in the following the letter R indicating the derived functor is omitted.

We fix a base scheme S. To define in Definition 2.4.1 the S-singular support of a sheaf $\mathcal{F}$ on a regular scheme X of finite type over S along the format above, we introduce in Definition 2.2.1 a condition for a pair of morphisms $h\colon W\to X$ and $f\colon W\to Y$ over S to be $\mathcal{F}$-acyclic over S, by modifying the local acyclicity of f relatively to $h^*\mathcal{F}$ used by Beilinson in his definition of the singular support. For the modification, we use a criterion of the local acyclicity due to Braverman–Gaitsgory [**3**, Appendix Theorem B.2] recalled in Proposition 1.2.9: A morphism $f\colon X\to Y$ to a smooth scheme Y over a field k is locally acyclic relatively to $\mathcal{F}$ if and only if the morphism $(1,f)\colon X\to X\times_k Y$ is $\mathcal{F}\boxtimes\mathcal{G}$-transversal for every sheaf $\mathcal{G}$ on Y.

If S is a scheme over a field k and if $Y=Y_0\times_k S$ is the base change of a smooth scheme Y_0 over k, then we may consider the condition that $(h,f)\colon W\to X\times_S Y=X\times_k Y_0$ is $\mathcal{F}\boxtimes\mathcal{G}$-transversal for the pull-back $\mathcal{G}$ to Y of every sheaf $\mathcal{G}_0$ on Y_0. In mixed characteristic case where a base field cannot exist, we need to find a replacement of the condition for a sheaf $\mathcal{G}$ on a smooth scheme Y over S to 'come from Y_0 as pull-back'. We will formulate such a condition called the S-acyclicity in Definition 1.6.3 by using micro support defined in an absolute setting without a base scheme and use it to define the $\mathcal{F}$-acyclicity over S in Definition 2.2.1.

The $\mathcal{F}$-acyclicity over S enables us to define the S-singular support $SS_S\mathcal{F}$ using the format above and to ask its existence as a closed conical subset of the FW cotangent bundle FT^*X. A key reason why the use of Radon transform is effective in the proof of the existence of singular support is that the universal family Q of hyperplanes in $\mathbf{P}$ over S is canonically identified with the projective space bundle $\mathbf{P}(T^*\mathbf{P}/S)$ associated to the relative cotangent bundle $T^*\mathbf{P}/S$. To fill the gap between the FW cotangent bundle FT^*X and the relative cotangent bundle $T^*\mathbf{P}/S$, we need to introduce the S-saturation $SS^{\rm sat}_S\mathcal{F}$ defined as the smallest closed conical subset stable under the action of $FT^*S\times_S X$ and containing $SS_S\mathcal{F}$ as a subset.

With this variant, the method of Beilinson allows us to prove the existence of $SS^{\rm sat}_S\mathcal{F}$ in Theorem 2.4.12. If X is smooth over S, the S-saturated singular support $SS^{\rm sat}_S\mathcal{F}$ defines a closed conical subset of the relative cotangent bundle $T^*X/S|_{X_{\mathbf{F}_p}}$ restricted to the characteristic p fiber. In the case where $S=\operatorname{Spec}k$ for a perfect field k of characteristic $p>0$, we have $T^*X/S|_{X_{\mathbf{F}_p}}=T^*X$ and recover the singular support defined by Beilinson in [**2**]. If k is not perfect, ours is different from that in [**2**] as we see in Example after Definition 2.4.1. If we admit a certain property on micro support of tensor product, we would have $SS_S\mathcal{F}\subset SS\mathcal{F}\subset SS^{\rm sat}_S\mathcal{F}$ by Lemma 2.3.3. The inclusions may be strict in general. If we further assume that the morphism $FT^*S\times_S X\to FT^*X$ is 0, then we have $SS_S\mathcal{F}=SS^{\rm sat}_S\mathcal{F}$.

In Section 1, we recall the definition of singular support from [**14**] and give various complements. We recall the definition of $\mathcal{F}$-transversality and of the C-transversality and prepare their basic properties in Sections 1.1 and 1.3 respectively.

We study the relation between the $\mathcal{F}$-transversality and the local acyclicity in Section 1.2 and close the section with a proof of the criterion of the local acyclicity due to Braverman–Gaitsgory in Proposition 1.2.9. After recalling the definition of micro support in Section 1.4, we recall the definition of the singular support in Section 1.5 following the general format. We show in Corollary 1.5.4 that the existence of singular support is reduced to the case of projective spaces. Finally, we introduce the S-acyclicity as a preparation of the definition of the $\mathcal{F}$-acyclicity over S.

In Section 2, we introduce the S-singular support and prove the existence of its S-saturation. First, in Sections 2.1 and 2.2, we define the C-acyclicity over S and the $\mathcal{F}$-acyclicity over S and study their basic properties respectively. Following the general format, we define the notion of S-micro support in Section 2.3 and the S-singular support in Section 2.4 respectively. After recalling basic properties of the Legendre transform and the Radon transform, we prove the existence of S-saturated singular support in Theorem 2.4.12 adopting the method of Beilinson.

The research was partially supported by Kakenhi Grant-in-Aid (C) 24K06683. The author thanks greatly Daichi Takeuchi for proposing an essential improvement on the main result and correcting some errors in earlier versions. He also thanks an anonymous referee for extremely careful reading and for pointing out several inaccuracies and numerous typographical errors.

Let Λ be a finite field of characteristic ℓ. Schemes are over $\mathbf{Z}[1/\ell]$. An object of the derived category $D^+(X,\Lambda)$ of bounded below complexes on the étale site of a scheme X will be called a *sheaf* on X by abuse of terminology. If a sheaf belongs to the full sub category $D^b_c(X,\Lambda)$ consisting of bounded constructible complexes, we say it is *constructible*. A constructible sheaf $\mathcal{F}$ is said to be locally constant if every cohomology sheaf $\mathcal{H}^q\mathcal{F}$ is locally constant. We drop R in the notation for derived functors and Λ in $\otimes_\Lambda$.

1. Micro support

1.1. $\mathcal{F}$-transversality. Let $h\colon W \to X$ be a separated morphism of finite type of noetherian schemes and let $\mathcal{F}$ and $\mathcal{G}$ be sheaves on X. We define a canonical morphism

$$c_{h,\mathcal{F},\mathcal{G}}\colon h^*\mathcal{F}\otimes h^!\mathcal{G}\to h^!(\mathcal{F}\otimes\mathcal{G})$$

to be the adjoint of the composition

$$h_!(h^*\mathcal{F}\otimes h^!\mathcal{G}) \longrightarrow \mathcal{F}\otimes h_!h^!\mathcal{G} \xrightarrow{1_{\mathcal{F}}\otimes\mathrm{adj}} \mathcal{F}\otimes\mathcal{G}$$

where the first arrow is the isomorphism of the projection formula and the second is induced by the adjunction. There is another way to view the morphism $c_{h,\mathcal{F},\mathcal{G}}$ under the assumption that $\mathcal{F}$ is bounded. By taking the composition of the canonical morphism $h^!\mathcal{G}\to h^!\mathcal{H}om(\mathcal{F},\mathcal{F}\otimes\mathcal{G})$ with the inverse of the isomorphism $\mathcal{H}om(h^*\mathcal{F},h^!(\mathcal{F}\otimes\mathcal{G}))\to h^!\mathcal{H}om(\mathcal{F},\mathcal{F}\otimes\mathcal{G})$ of [**5**, Corollaire 3.1.12.2], [**7**, Theorem 8.4.7], we obtain a morphism $h^!\mathcal{G}\to\mathcal{H}om(h^*\mathcal{F},h^!(\mathcal{F}\otimes\mathcal{G}))$. Taking the adjoint, we obtain $c_{h,\mathcal{F},\mathcal{G}}\colon h^*\mathcal{F}\otimes h^!\mathcal{G}\to h^!(\mathcal{F}\otimes\mathcal{G})$.

For $\mathcal{G}=\Lambda$,

$$c_{h,\mathcal{F}}\colon h^*\mathcal{F}\otimes h^!\Lambda\to h^!\mathcal{F}$$

is defined as $c_{h,\mathcal{F},\Lambda}$.

LEMMA 1.1.1. *The following diagram is commutative:*

$$\begin{array}{ccc} h^*\mathcal{F}\otimes h^!\mathcal{G} & \xrightarrow{c_{h,\mathcal{F},\mathcal{G}}} & h^!(\mathcal{F}\otimes\mathcal{G}) \\ {\scriptstyle 1_{h^*\mathcal{F}}\otimes c_{h,\mathcal{G}}}\uparrow & & \uparrow{\scriptstyle c_{h,\mathcal{F}\otimes\mathcal{G}}} \\ h^*\mathcal{F}\otimes h^*\mathcal{G}\otimes h^!\Lambda & \xrightarrow{\mathrm{can}\otimes 1} & h^*(\mathcal{F}\otimes\mathcal{G})\otimes h^!\Lambda. \end{array}$$

PROOF. By taking the adjoint and applying the projection formula, it suffices to show that the diagram

$$\begin{array}{ccc} \mathcal{F}\otimes h_!h^!\mathcal{G} & \xrightarrow{1_{\mathcal{F}}\otimes\mathrm{adj}} & \mathcal{F}\otimes\mathcal{G} \\ {\scriptstyle 1_{\mathcal{F}}\otimes c_{h,\mathcal{G}}}\uparrow & & \uparrow{\scriptstyle 1_{\mathcal{F}}\otimes 1_{\mathcal{G}}\otimes\mathrm{adj}} \\ \mathcal{F}\otimes h_!(h^*\mathcal{G}\otimes h^!\Lambda) & \xrightarrow{1_{\mathcal{F}}\otimes\mathrm{proj}} & \mathcal{F}\otimes\mathcal{G}\otimes h_!h^!\Lambda \end{array}$$

is commutative. It suffices to show the commutativity before tensoring $\mathcal{F}$. Then, it follows from the definition of $c_{h,\mathcal{G}}$. □

We prove the transitivity of $c_{h,\mathcal{F}}$.

LEMMA 1.1.2. *Let $h\colon W\to X'$ and $g\colon X'\to X$ be separated morphisms of finite type of noetherian schemes. Then, the following diagram is commutative:*

(1.1)
$$\begin{array}{ccc} (gh)^*\mathcal{F}\otimes(gh)^!\Lambda & \xrightarrow{c_{gh,\mathcal{F}}} & (gh)^!\mathcal{F} \\ {\scriptstyle 1\otimes c_{h,g^!\Lambda}}\uparrow & \searrow{\scriptstyle c_{h,g^*\mathcal{F},g^!\Lambda}} & \uparrow{\scriptstyle h^!(c_{g,\mathcal{F}})} \\ h^*(g^*\mathcal{F}\otimes g^!\Lambda)\otimes h^!\Lambda & \xrightarrow{c_{h,g^*\mathcal{F}\otimes g^!\Lambda}} & h^!(g^*\mathcal{F}\otimes g^!\Lambda) \end{array}$$

PROOF. The lower left triangle is commutative by Lemma 1.1.1. To prove that the upper right triangle is commutative, we consider the diagram

$$\begin{array}{ccccc} (gh)_!((gh)^*\mathcal{F}\otimes(gh)^!\Lambda) & & \xrightarrow{\mathrm{proj}} & & \mathcal{F}\otimes(gh)_!(gh)^!\Lambda \\ \downarrow & & & & \downarrow \quad \searrow{\scriptstyle\mathrm{adj}} \\ g_!h_!(h^*g^*\mathcal{F}\otimes h^!g^!\Lambda) & \xrightarrow{\mathrm{proj}} & g_!(g^*\mathcal{F}\otimes h_!h^!g^!\Lambda) & \xrightarrow{\mathrm{proj}} & \mathcal{F}\otimes g_!h_!h^!g^!\Lambda \\ & \searrow & \downarrow{\scriptstyle\mathrm{adj}} & & \downarrow{\scriptstyle\mathrm{adj}} \\ & & g_!(g^*\mathcal{F}\otimes g^!\Lambda) & \xrightarrow{\mathrm{proj}} & \mathcal{F}\otimes g_!g^!\Lambda \xrightarrow{\mathrm{adj}} \mathcal{F}. \end{array}$$

The morphism $c_{gh,\mathcal{F}}$ is induced by the composition via the upper right, $c_{g,\mathcal{F}}$ is induced by the composition of the lower line and $c_{h,g^*\mathcal{F},g^!\Lambda}$ is induced by the lower left slant arrow without $g_!$. Hence, it is reduced to showing the commutativity of the diagram.

By the transitivity of the isomorphisms of the projection formula, the upper rectangle of the diagram is commutative. The lower square is commutative by the functoriality of the isomorphism of the projection formula and the right triangle is commutative by the transitivity of the adjunction. Hence the diagram is commutative as required. □

We prove the compatibility with base change. We consider a cartesian diagram

$$\begin{array}{ccc} W' & \xrightarrow{h'} & X' \\ {\scriptstyle g}\downarrow & & \downarrow{\scriptstyle f} \\ W & \xrightarrow{h} & X, \end{array} \tag{1.2}$$

of noetherian schemes where the horizontal arrows are separated morphisms of finite type. The base change morphism

$$g^*h^! \to h'^!f^* \tag{1.3}$$

[**5**, (3.1.14.2)] is defined as the adjoint of the composition $h'_!g^*h^! \to f^*h_!h^! \to f^*$ of the inverse of the isomorphism of proper base change theorem and the adjunction. As the adjoint of the isomorphism $f^*h_! \to h'_!g^*$ of proper base change theorem, we obtain an isomorphism

$$h^!f_* \to g_*h'^! \tag{1.4}$$

[**5**, Corollaire 3.1.12.3], [**7**, Proposition 8.4.9]. The morphism (1.4) is the adjoint of the composition of $g^*h^!f_* \to h'^!f^*f_* \to h'^!$ where the first arrow is induced by (1.3) and the second is induced by the adjunction.

LEMMA 1.1.3. *Let (1.2) be a cartesian diagram of noetherian schemes and assume that the horizontal arrows are separated morphisms of finite type.*

1. *For a sheaf $\mathcal{F}$ on X, the diagram*

$$\begin{array}{ccc} g^*(h^*\mathcal{F} \otimes h^!\Lambda) & \xrightarrow{g^*c_{h,\mathcal{F}}} & g^*h^!\mathcal{F} \\ {\scriptstyle \mathrm{can}\otimes(1.3)}\downarrow & & \downarrow{\scriptstyle (1.3)} \\ h'^*f^*\mathcal{F} \otimes h'^!\Lambda & \xrightarrow{c_{h',f^*\mathcal{F}}} & h'^!f^*\mathcal{F} \end{array} \tag{1.5}$$

is commutative.

2. *Let $\mathcal{F}'$ be a sheaf on X'. Then, the following diagram is commutative:*

$$\begin{array}{ccc} h^*f_*\mathcal{F}' \otimes h^!\Lambda & \xrightarrow{c_{h,f_*\mathcal{F}'}} & h^!f_*\mathcal{F}' \\ \downarrow & & \downarrow{\scriptstyle (1.4)} \\ g_*(h'^*\mathcal{F}' \otimes h'^!\Lambda) & \xrightarrow{g_*(c_{h',\mathcal{F}'})} & g_*h'^!\mathcal{F}' \end{array} \tag{1.6}$$

where the left vertical arrow is the adjoint of

$$g^*(h^*f_*\mathcal{F}' \otimes h^!\Lambda) = g^*h^*f_*\mathcal{F}' \otimes g^*h^!\Lambda \xrightarrow{\mathrm{can}\otimes 1}$$
$$h'^*f^*f_*\mathcal{F}' \otimes g^*h^!\Lambda \xrightarrow{h'^*\mathrm{adj}\otimes(1.3)} h'^*\mathcal{F}' \otimes h'^!\Lambda.$$

PROOF.

(1) By the definition of (1.3), it suffices to show that the diagram

$$\begin{array}{ccccc} f^*h_!(h^*\mathcal{F} \otimes h^!\Lambda) & \longrightarrow & f^*(\mathcal{F} \otimes h_!h^!\Lambda) & \xrightarrow{f^*(1\otimes\mathrm{adj})} & f^*\mathcal{F} \\ \downarrow & & \downarrow{\scriptstyle \mathrm{can}\otimes(1.3)} & & \| \\ h'_!(h'^*f^*\mathcal{F} \otimes h'^!\Lambda) & \longrightarrow & f^*\mathcal{F} \otimes h'_!h'^!\Lambda & \xrightarrow{1\otimes\mathrm{adj}} & f^*\mathcal{F} \end{array}$$

is commutative. The vertical arrows are induced by the base change morphisms. The left square is the compatibility of the projection formula with base change. By the definition of (1.3), the diagram

$$\begin{array}{ccc} h'_! g^* h^! & \longrightarrow & h'_! h'^! f^* \\ \downarrow & & \downarrow{\scriptstyle \mathrm{adj}} \\ f^* h_! h^! & \xrightarrow{\mathrm{adj}} & f^* \end{array}$$

is commutative. This implies the commutativity of the right square.

(2) By the commutative diagram (1.5) in 1 for $\mathcal{F} = f_*\mathcal{F}'$ and the functoriality of $c_{h',\mathcal{F}}$ for $f^*f_*\mathcal{F} \to \mathcal{F}$, we obtain a commutative diagram

$$\begin{array}{ccc} g^*(h^* f_*\mathcal{F}' \otimes h^!\Lambda) & \xrightarrow{g^* c_{h, f_*\mathcal{F}'}} & g^* h^! f_*\mathcal{F}' \\ \downarrow & & \downarrow \\ h'^*\mathcal{F}' \otimes h'^!\Lambda & \xrightarrow{c_{h',\mathcal{F}'}} & h'^!\mathcal{F}'. \end{array}$$

Taking the adjoint, we obtain the commutative diagram since (1.4) is the adjoint of the composition of the left vertical arrow.

□

DEFINITION 1.1.4. *Let $h\colon W \to X$ be a morphism of finite type of regular noetherian schemes.*

arabic*. *If, locally on W, h is the composition of a smooth morphism $P \to X$ of relative dimension n and a regular immersion $W \to P$ of codimension c, we say that W is of relative dimension $n - c$ over X.*

arabic*. *Let $g\colon X' \to X$ be a morphism of regular noetherian schemes and let $U \subset W' = W \times_X X'$ be an open subscheme. We say that h is transversal to g on U if U is regular and if the relative dimension of U over X' equals the relative dimension of W over X.*

LEMMA 1.1.5. *Let $h\colon W \to X$ be a separated morphism of finite type of regular noetherian schemes.*

arabic*. *Let d be the relative dimension of h. Then, there is a canonical isomorphism $h^!\Lambda \to \Lambda(d)[2d]$.*

arabic*. *Let $g\colon X' \to X$ be a morphism of regular noetherian schemes and let*

$$\begin{array}{ccc} W' & \xrightarrow{h'} & X' \\ {\scriptstyle g'}\downarrow & & \downarrow{\scriptstyle g} \\ W & \xrightarrow{h} & X \end{array}$$

be a cartesian diagram of noetherian schemes. Assume that h is transversal to g on an open subscheme $U' \subset W'$. Then the base change morphism $g'^* h^!\Lambda \to h'^!\Lambda$ (1.3) is an isomorphism on U'.

PROOF. Since the assertion is local on W, we may factor h into the composition of a smooth morphism and an immersion. The smooth case is the Poincaré duality [**5**, Théorème 3.2.5] and the immersion case follows from the purity [**8**, Exposé XVI, Théorème 3.1.1]. □

DEFINITION 1.1.6. *Let $h\colon W \to X$ be a separated morphism of finite type of noetherian schemes and let $\mathcal{F}$ be a sheaf on X.*

We say that h is $\mathcal{F}$-transversal if $c_{h,\mathcal{F}}\colon h^*\mathcal{F} \otimes h^!\Lambda \to h^!\mathcal{F}$ is an isomorphism.

LEMMA 1.1.7. *Let $h\colon W \to X$ be a separated morphism of finite type of noetherian schemes and let $\mathcal{F}$ be a sheaf on X.*

1. *If h is smooth, then h is $\mathcal{F}$-transversal.*
2. *If $\mathcal{F}' \to \mathcal{F} \to \mathcal{F}'' \to$ is a distinguished triangle of sheaves on X, and if h is transversal for two of $\mathcal{F}', \mathcal{F}$ and $\mathcal{F}''$. Then, h is transversal for the rest.*
3. *If $\mathcal{F}$ is locally constant, then every morphism h is $\mathcal{F}$-transversal.*
4. *Assume that h is a closed immersion and that $\mathcal{F}$ is a constructible sheaf supported on the image $h(W) \subset X$. If h is $\mathcal{F}$-transversal and if $h^!\Lambda$ is isomorphic to $\Lambda(-c)[-2c]$ for an integer $c \neq 0$, then $\mathcal{F} = 0$.*
5. *Assume that h is a closed immersion and let $j\colon U = X - W \to X$ be the open immersion of the complement. Then, the following conditions are equivalent:*
   (1) *h is $\mathcal{F}$-transversal.*
   (2) *The canonical morphism $\mathcal{F} \otimes j_*\Lambda \to j_*j^*\mathcal{F}$ is an isomorphism.*

We show a partial converse of 3 for constructible sheaves on regular schemes in Proposition 1.1.11 below.

PROOF. 1. Poincaré duality [**5**, Théorème 3.2.5].

2. It suffices to consider the morphism

$$\begin{array}{ccccccc} h^*\mathcal{F}' \otimes h^!\Lambda & \longrightarrow & h^*\mathcal{F} \otimes h^!\Lambda & \longrightarrow & h^*\mathcal{F}'' \otimes h^!\Lambda & \longrightarrow \\ {\scriptstyle c_{h,\mathcal{F}'}}\downarrow & & {\scriptstyle c_{h,\mathcal{F}}}\downarrow & & {\scriptstyle c_{h,\mathcal{F}''}}\downarrow & \\ h^!\mathcal{F}' & \longrightarrow & h^!\mathcal{F} & \longrightarrow & h^!\mathcal{F}'' & \longrightarrow \end{array}$$

of distinguished triangles.

3. Since the question is étale local, we may assume that $\mathcal{F}$ is constant and the assertion follows.

4. Since $\mathcal{F}$ is supported on $h(W)$, the morphism $h^!\mathcal{F} \to h^*\mathcal{F}$ is an isomorphism. Further if $h^*\mathcal{F}(-c)[-2c] \to h^!\mathcal{F}$ is an isomorphism for $c \neq 0$, we have $h^*\mathcal{F} = 0$ and $\mathcal{F} = 0$.

5. In the commutative diagram

$$\begin{array}{ccccccc} \mathcal{F} \otimes h_*h^!\Lambda & \longrightarrow & \mathcal{F} & \longrightarrow & \mathcal{F} \otimes j_*\Lambda & \longrightarrow \\ \downarrow & & \| & & \downarrow & \\ h_*h^!\mathcal{F} & \longrightarrow & \mathcal{F} & \longrightarrow & j_*j^*\mathcal{F} & \longrightarrow \end{array}$$

of distinguished triangles, the left vertical arrow is an isomorphism if and only if so is the right vertical arrow. By the projection formula, the left vertical arrow is identified with $h_*c_{h,\mathcal{F}}$ and the assertion follows.

□

We show the transitivity of the transversality.

LEMMA 1.1.8. *Let $h\colon W \to X'$ and $g\colon X' \to X$ be separated morphisms of finite type of noetherian schemes and let $\mathcal{F}$ be a sheaf on X. Assume that $g^!\Lambda$ is isomorphic to $\Lambda(d)[2d]$ for some integer d and that g is $\mathcal{F}$-transversal. Then, the following conditions are equivalent:*

(1) *gh is $\mathcal{F}$-transversal.*
(2) *h is $g^*\mathcal{F}$-transversal.*

The assumption that $g^!\Lambda$ is isomorphic to $\Lambda(d)[2d]$ is satisfied if g is smooth of relative dimension d.

PROOF. We consider the commutative diagram (1.1). Since $g^!\Lambda$ is locally constant, the left vertical arrow is an isomorphism by Lemma 1.1.7.3. The assumption that g is $\mathcal{F}$-transversal implies that the right vertical arrow is an isomorphism. Since $g^!\Lambda$ is isomorphic to $\Lambda(d)[2d]$, condition (2) is equivalent to the condition that the lower horizontal arrow is an isomorphism. Since condition (1) is equivalent to the condition that the upper horizontal arrow is an isomorphism, the assertion follows. □

LEMMA 1.1.9 ([**1**, Proposition 2.11]). *Let X be a scheme and $\mathcal{F}$ be a constructible sheaf. Let $x \to X$ be a geometric point. Then, the following conditions are equivalent:*

(1) *$\mathcal{F}$ is locally constant on a neighborhood of the image of x.*
(2) *For every geometric point t of the strict localization X_x, the morphism $\mathcal{F}_x \to \mathcal{F}_t$ of specialization is an isomorphism.*

PROOF. (1)⇒(2): Since we may replace X by an étale neighborhood of x, we may assume that $\mathcal{F}$ is constant. Then the morphism $\mathcal{F}_x \to \mathcal{F}_t$ is an isomorphism.

(2)⇒(1): By taking the cohomology sheaves, we may assume that $\mathcal{F}$ is concentrated in degree 0. Since $\mathcal{F}_x$ is finite, after replacing X by an étale neighborhood of x, we may assume that there is a morphism $\mathcal{C} \to \mathcal{F}$ from a constant sheaf such that $\mathcal{C}_x \to \mathcal{F}_x$ is an isomorphism. Then, for every geometric point t of the strict localization X_x, the morphism $\mathcal{C}_t \to \mathcal{F}_t$ is also an isomorphism. Since $\mathcal{F}$ and $\mathcal{C}$ are constructible, the subset Z of X consisting of the image of geometric points z such that $\mathcal{C}_z \to \mathcal{F}_z$ is an isomorphism is constructible as in the proof of [**1**, Proposition 2.11]. By this and (2), Z contains an open neighborhood U of the image of x and the restriction $\mathcal{F}|_U$ is constant.

□

LEMMA 1.1.10. *Let $X = \operatorname{Spec} \mathcal{O}_K$ for a henselian discrete valuation ring and $j\colon U = \operatorname{Spec} K \to X$ be the open immersion. Let $\mathcal{F}$ be a constructible sheaf on X and assume that for every integer q, the action of the inertia subgroup $I \subset G_K = \operatorname{Gal}(\overline{K}/K)$ on $\mathcal{H}^q\mathcal{F}_{\overline{K}}$ is trivial. Then the following conditions are equivalent:*

(1) *$\mathcal{F}$ is locally constant.*
(2) *The morphism $\mathcal{F} \otimes j_*\Lambda \to j_*j^*\mathcal{F}$ is an isomorphism.*

PROOF. Since the implication (1)⇒(2) is trivial, we show (2)⇒(1). Assume that $\mathcal{F}$ is acyclic outside $[a, b]$ and we show it by induction on b. Let $\overline{s}$ and $\overline{\eta}$ be the closed and generic geometric points of X and let I denote the inertia group. Then, the stalk $(R^1j_*j^*\mathcal{H}^q\mathcal{F})_{\overline{s}}$ is canonically isomorphic to the Tate twist $\mathcal{H}^q\mathcal{F}_{\overline{\eta}I}(-1)$ of the coinvariant with respect to the action of the inertia group I. Since the cohomological dimension of Rj_* is 1 and the inertia action is assumed to be trivial, we have an isomorphism $\mathcal{H}^b(\mathcal{F})_{\overline{s}}(-1) \to \mathcal{H}^b(\mathcal{F})_{\overline{\eta}}(-1)$. Thus $\mathcal{H}^b(\mathcal{F})$ is locally constant and the assertion follows by induction. □

PROPOSITION 1.1.11. *Let X be a regular excellent noetherian scheme and $Z \subset X$ be a closed subset. For a constructible sheaf $\mathcal{F}$ on X, the following conditions are equivalent:*

(1) *$\mathcal{F}$ is locally constant on a neighborhood of Z.*

(2) *Every separated morphism $h\colon W \to X$ of finite type of regular noetherian schemes is $\mathcal{F}$-transversal on a neighborhood of the inverse image of Z.*

PROOF. Since the implication (1)⇒(2) is trivial, we show (2)⇒(1). Let x be a geometric point of Z and let t be a geometric point of the strict localization X_x. By Lemma 1.1.9, it suffices to show that the morphism $\mathcal{F}_x \to \mathcal{F}_t$ of specialization is an isomorphism. We may assume that the image $t_0 \in X$ of t is different from the image x_0 of x. Let $T \subset X$ be the closure of t_0 regarded as a reduced closed subscheme and let $T' \to T$ be the blow-up at the closure S of x_0.

Let W be the normalization of T' in a finite extension of the function field such that the monodromy action on the stalk of the pull-back is trivial and D be the reduced inverse image of $S \subset T$. By replacing W by an open neighborhood of the generic point s of an irreducible component D, we may assume that W is regular and $D \subset W$ is a regular divisor. Then, the closed immersion $i\colon D \to W$ is $\mathcal{F}_W$-transversal for the pull-back $\mathcal{F}_W$ of $\mathcal{F}$ by Lemma 1.1.8. Hence for the open immersion $j\colon U = W - D \to W$, the morphism $\mathcal{F}_W \otimes Rj_*\Lambda \to Rj_*j^*\mathcal{F}_W$ is an isomorphism by Lemma 1.1.7.5. Thus the pull-back of $\mathcal{F}_W$ on the local ring $\operatorname{Spec} \mathcal{O}_{W,s}$ is locally constant by Lemma 1.1.10 and $\mathcal{F}_x \to \mathcal{F}_t$ is an isomorphism. □

Let $g\colon X' \to X$ be a morphism of noetherian schemes and $Z' \subset X'$ be a closed subset. We say that g is proper (resp. finite) on Z, if the restriction of g to Z is proper (resp. finite) with respect to a closed subscheme structure on Z.

LEMMA 1.1.12. *Let*

$$\begin{array}{ccc} W' & \xrightarrow{h'} & X' \\ {\scriptstyle g'}\downarrow & & \downarrow{\scriptstyle g} \\ W & \xrightarrow{h} & X \end{array}$$

be a cartesian diagram of separated morphisms of finite type of noetherian schemes and let $\mathcal{F}'$ be a constructible sheaf on X'. Assume that g is proper on the support of $\mathcal{F}$ and that the base change morphism $g'^*h^!\Lambda \to h'^!\Lambda$ (1.3) is an isomorphism. We consider the following conditions:

(1) *h is $g_*\mathcal{F}'$-transversal.*

(2) *h' is $\mathcal{F}'$-transversal.*

We have the implication (2)⇒(1). If g is finite on the support of $\mathcal{F}'$, then these conditions are equivalent.

PROOF. In the commutative diagram (1.6), the left vertical arrow is an isomorphism by the projection formula, the proper base change theorem and the assumption that $g'^*h^!\Lambda \to h'^!\Lambda$ (1.3) is an isomorphism. The right vertical arrow is an isomorphism by the proper base change theorem. If h' is $\mathcal{F}'$-transversal, then the lower horizontal arrow is an isomorphism and hence h is $g_*\mathcal{F}'$-transversal.

Conversely, if h is $g_*\mathcal{F}'$-transversal, the horizontal arrows are isomorphism. Further if g is finite on the support of $\mathcal{F}'$, then the condition that $g_*(c_{h',\mathcal{F}'})$ is an isomorphism implies that $c_{h',\mathcal{F}'}$ itself is an isomorphism. □

1.2. $\mathcal{F}$-transversality and local acyclicity.

DEFINITION 1.2.1 ([**6**, Definition 2.12]). *Let $f\colon X \to Y$ be a morphism and $\mathcal{F}$ be a sheaf on X.*

1. *We say that f is locally acyclic relatively to $\mathcal{F}$ or $\mathcal{F}$-acyclic for short if the following condition is satisfied:*

Let y be any geometric point of Y and let $t \to Y_{(y)}$ be any geometric point of the strict localization $Y_{(y)}$ of Y at y. Let $i\colon X_y = X \times_Y y \to X \times_Y Y_{(y)}$ and $j\colon X_t = X \times_Y t \to X \times_Y Y_{(y)}$ be the base changes of the canonical morphisms $y \to Y_{(y)} \leftarrow t$. Then, the morphism

$$\mathcal{F}|_{X_y} \to i^* j_* \mathcal{F}|_{X_t} \tag{1.7}$$

is an isomorphism.

Let Z be a closed subset of X, we say that f is $\mathcal{F}$-acyclic along Z if (1.7) is an isomorphism on the inverse image of Z.

2. *We say that f is universally locally acyclic if for any morphism $Y' \to Y$ and the pull-back $\mathcal{F}'$ of $\mathcal{F}$ on $X' = X \times_Y Y'$, the base change $f'\colon X' \to Y'$ is $\mathcal{F}'$-acyclic.*

We consider the cartesian diagram

$$\begin{array}{ccc} X \times_Y Y_{(y)} & \xleftarrow{\ j\ } & X_t \\ {\scriptstyle f_{(y)}}\big\downarrow & & \big\downarrow{\scriptstyle f_t} \\ Y_{(y)} & \xleftarrow{\ j_Y\ } & t. \end{array} \tag{1.8}$$

Then, the morphism (1.7) is the pull-back by $i\colon X_y \to X \times_Y Y_{(y)}$ of the morphism

$$\mathcal{F}|_{X \times_Y Y_{(y)}} \otimes f^*_{(y)} j_{Y*}\Lambda \to j_* j^* (\mathcal{F}|_{X \times_Y Y_{(y)}}) \tag{1.9}$$

defined as the adjoint of $j^*\mathcal{F}|_{X \times_Y Y_{(y)}} \otimes j^* f^*_{(y)} j_{Y*}\Lambda \to j^*\mathcal{F}|_{X \times_Y Y_{(y)}} \otimes f_t^* j_Y^* j_{Y*}\Lambda \xrightarrow{1\otimes\mathrm{adj}} j^*(\mathcal{F}|_{X \times_Y Y_{(y)}})$.

THEOREM 1.2.2 (cf. [**6**, Théorème 2.12]). *Let k be a field and X be a scheme of finite type over k. Then the morphism $X \to \operatorname{Spec} k$ is universally $\mathcal{F}$-acyclic for any sheaf $\mathcal{F}$ on X.*

PROOF. If $\mathcal{F}$ is constructible, then this is a special case of [**6**, Théorème 2.12]. The general case follows by taking the limit. □

LEMMA 1.2.3 (cf. [**7**, Proposition 7.6.3]). *In the definition of universal local acyclicity, it suffices to consider Y' smooth over Y.*

PROOF. Since the question is local on Y', we may assume that Y and Y' are affine and Y' is a closed subscheme of $\mathbf{A}^n_Y$. Since $Y' \to \mathbf{A}^n_Y$ is an immersion, the local acyclicity of $\mathbf{A}^n_X \to \mathbf{A}^n_Y$ implies the local acyclicity of $X' \to Y'$. □

PROPOSITION 1.2.4 (cf. [**7**, Theorem 7.6.9]). *Let $f\colon X \to Y$ be a morphism of schemes and $\mathcal{F}$ be a sheaf on X. The following conditions are equivalent:*

(1) *f is $\mathcal{F}$-acyclic.*

(2) *For every cartesian diagram*

$$\begin{array}{ccc} X & \xleftarrow{h} & W \\ {\scriptstyle f}\downarrow & & \downarrow{\scriptstyle g} \\ Y & \xleftarrow{i} & V \end{array} \tag{1.10}$$

such that i is an immersion and for every sheaf $\mathcal{G}$ on V, the morphism

$$\mathcal{F} \otimes f^* i_* \mathcal{G} \to h_*(h^* \mathcal{F} \otimes g^* \mathcal{G}) \tag{1.11}$$

is an isomorphism.

(2′) *For every cartesian diagram* (1.10) *such that i is an immersion and for every locally constant constructible sheaf $\mathcal{L}$ on V, the morphism*

$$\mathcal{F} \otimes f^* i_* \mathcal{L} \to h_*(h^* \mathcal{F} \otimes g^* \mathcal{L}) \tag{1.12}$$

is an isomorphism.

If Y is an excellent noetherian scheme, we may assume that V is regular in (2′).

The morphism (1.11) is defined as the adjoint of $1 \otimes \mathrm{adj}\colon h^*(\mathcal{F} \otimes f^* i_* \mathcal{G}) = h^* \mathcal{F} \otimes h^* f^* i_* \mathcal{G} = h^* \mathcal{F} \otimes g^* i^* i_* \mathcal{G} \to h^* \mathcal{F} \otimes g^* \mathcal{G}$.

PROOF. (1)⇒(2): The immersion i is the composition of an open immersion and a closed immersion. For an open immersion, the isomorphism (1.11) is [**10**, Proposition 2.10], [**7**, Lemma 7.6.7]. If i is a closed immersion, the isomorphism (1.11) follows from the projection formula and the proper base change theorem.

(2)⇒(2′) is clear.

(2′)⇒(1): Let y be a geometric point of Y and t be a geometric point of $Y_{(y)}$. We show that (1.7) is an isomorphism. Let v be the image of t by $Y_{(y)} \to Y$ and $V \subset Y$ be the closure of $\{v\}$. Then, there exists a projective system $p_\lambda\colon U_\lambda \to V_\lambda$ of finite étale morphisms of integral schemes and open immersions $j_\lambda\colon V_\lambda \to V$ and an isomorphism $\varinjlim (j_\lambda p_\lambda)_* \Lambda \to j_* \Lambda$. If Y is excellent, we may assume that V_λ are regular. Let $j\colon t \to V$ be the canonical morphism and consider the cartesian diagram

$$\begin{array}{ccccccc} X & \xleftarrow{h} & W & \xleftarrow{h_\lambda} & W \times_V V_\lambda & \xleftarrow{k_\lambda} & W \times_V U_\lambda \\ {\scriptstyle f}\downarrow & & \downarrow{\scriptstyle g} & & \downarrow{\scriptstyle g_\lambda} & & \downarrow \\ Y & \xleftarrow{i} & V & \xleftarrow{j_\lambda} & V_\lambda & \xleftarrow{p_\lambda} & U_\lambda \end{array}$$

extending (1.10). As the isomorphism (1.12) for the immersions $ij_\lambda\colon V_\lambda \to Y$ and the locally constant sheaves $p_{\lambda*}\Lambda$ on V_λ, we obtain an isomorphism

$$\mathcal{F} \otimes f^* i_* j_{\lambda*} p_{\lambda*} \Lambda \to (hh_\lambda)_*((hh_\lambda)^* \mathcal{F} \otimes g_\lambda^* p_{\lambda*} \Lambda) \tag{1.13}$$

on X. The target of (1.13) is identified with $(hh_\lambda k_\lambda)_*(hh_\lambda k_\lambda)^* \mathcal{F}$ by the projection formula.

We consider the cartesian diagram

$$\begin{array}{ccccccccc} Y & \xleftarrow{i} & V & \xleftarrow{j_\lambda} & V_\lambda & \xleftarrow{p_\lambda} & U_\lambda & \longleftarrow & t \\ \uparrow & & \uparrow & & \uparrow & & \uparrow & & \uparrow \\ Y_{(y)} & \xleftarrow{i_{(y)}} & V_{(y)} & \xleftarrow{j_{\lambda(y)}} & V_{\lambda(y)} & \xleftarrow{p_{\lambda(y)}} & U_{\lambda(y)} & \longleftarrow & t_{(y)} \end{array}$$

and let ${}_{(y)}$ denote the base change by $Y \leftarrow Y_{(y)}$. Then the pull-back of (1.13) to $X \times_Y Y_{(y)}$ gives an isomorphism

$$\mathcal{F}|_{X\times_Y Y_{(y)}} \otimes f_{(y)}^*(i_{(y)}p_{\lambda(y)}j_{\lambda(y)})_*\Lambda \to (hh_\lambda k_\lambda)_{(y)*}(hh_\lambda k_\lambda)_{(y)}^*\mathcal{F}|_{X\times_Y Y_{(y)}}. \tag{1.14}$$

Let $T_{\lambda(y)} \subset U_{\lambda(y)}$ be the connected component containing the image of $t \to U_{\lambda(y)}$ induced by $t \to Y_{(y)}$ and consider the cartesian diagram

$$\begin{array}{ccc} X \times_Y Y_{(y)} & \xleftarrow{h_{T\lambda(y)}} & X \times_Y T_{\lambda(y)} \\ {\scriptstyle f_{(y)}}\downarrow & & \downarrow{\scriptstyle f_{T_{\lambda(y)}}} \\ Y_{(y)} & \xleftarrow{j_{T\lambda(y)}} & T_{\lambda(y)} \end{array}$$

whose limit gives (1.8). Then, we obtain an isomorphism

$$\mathcal{F}|_{X\times_Y Y_{(y)}} \otimes f_{(y)}^* j_{T\lambda(y)*}\Lambda \to h_{T\lambda(y)*}h_{T\lambda(y)}{}^*\mathcal{F}|_{X\times_Y Y_{(y)}}. \tag{1.15}$$

as a direct summand of (1.14). Since $t = \varprojlim T_{\lambda(y)}$, by taking the limit of (1.15), we obtain the isomorphism (1.9). □

COROLLARY 1.2.5. *Let $f\colon X \to Y$ and $g\colon Y \to Z$ be morphisms of schemes and let $\mathcal{F}$ be a sheaf on X.*

1. *(*[**13**, Proposition 1.4], [**10**, Corollaire 2.7]*) Let $\mathcal{G}$ be a sheaf on Y and assume that f is $\mathcal{F}$-acyclic and that g is $\mathcal{G}$-acyclic. Then, gf is $\mathcal{F}\otimes f^*\mathcal{G}$-acyclic.*
2. *(*[**2**, Lemma 3.9 (i)]*) Assume that f is proper and that gf is $\mathcal{F}$-acyclic. Then, g is $f_*\mathcal{F}$-acyclic.*

PROOF. Let $i\colon V \to Z$ be an immersion and consider the cartesian diagram

$$\begin{array}{ccccc} W & \xrightarrow{f'} & W' & \xrightarrow{g'} & V \\ {\scriptstyle h}\downarrow & & {\scriptstyle h'}\downarrow & & \downarrow{\scriptstyle i} \\ X & \xrightarrow{f} & Y & \xrightarrow{g} & Z. \end{array}$$

1. For any sheaf $\mathcal{H}$ on V, we have a commutative diagram

$$\begin{array}{ccccc} \mathcal{F}\otimes f^*\mathcal{G}\otimes(gf)^*i_*\mathcal{H} & \longrightarrow & & \longrightarrow & h_*(h^*(\mathcal{F}\otimes f^*\mathcal{G})\otimes(g'f')^*\mathcal{H})) \\ \downarrow & & & & \downarrow \\ \mathcal{F}\otimes f^*(\mathcal{G}\otimes g^*i_*\mathcal{H}) & \longrightarrow & \mathcal{F}\otimes f^*(h'_*(h'^*\mathcal{G}\otimes g'^*\mathcal{H})) & \longrightarrow & h_*(h^*\mathcal{F}\otimes f'^*(h'^*\mathcal{G}\otimes g'^*\mathcal{H})). \end{array}$$

The lower arrows are isomorphisms by the assumption and Proposition 1.2.4 (1)⇒(2). The vertical arrows are the canonical isomorphisms. Hence the upper line is an isomorphism and the assertion follows by Proposition 1.2.4 (2)⇒(1).

2. For any sheaf $\mathcal{H}$ on V, we have a commutative diagram

$$\begin{array}{ccccc} f_*\mathcal{F}\otimes g^*i_*\mathcal{H} & \longrightarrow & & \longrightarrow & h'_*(h'^*f_*\mathcal{F}\otimes g'^*\mathcal{H}) \\ \downarrow & & & & \downarrow \\ f_*(\mathcal{F}\otimes(gf)^*i_*\mathcal{H}) & \longrightarrow & f_*(h_*(h^*\mathcal{F}\otimes(g'f')^*\mathcal{H})) & \longrightarrow & h'_*(f'_*(h^*\mathcal{F}\otimes(g'f')^*\mathcal{H})). \end{array}$$

The lower left horizontal arrow is an isomorphism by the assumption and Proposition 1.2.4 (1)⇒(2). The lower right horizontal arrow is the canonical isomorphism. The vertical arrows are the isomorphisms of the projection formula together with the proper base change theorem. Hence the upper line is an isomorphism and the assertion follows by Proposition 1.2.4 (2)⇒(1). □

COROLLARY 1.2.6. *Let*

$$\begin{array}{ccc} X & \xleftarrow{h} & W \\ {\scriptstyle f}\downarrow & & \downarrow \\ Y & \xleftarrow{i} & V \end{array}$$

be a cartesian diagram of schemes such that the vertical arrows are smooth and that the horizontal arrows are immersions. Let $\mathcal{F}$ *be a sheaf on* X *and* $\mathcal{G}$ *be a sheaf on* Y. *Assume that* $f\colon X \to Y$ *is* $\mathcal{F}$*-acyclic. If* $i\colon V \to Y$ *is* $\mathcal{G}$*-transversal, then* $h\colon W \to X$ *is* $\mathcal{F} \otimes f^*\mathcal{G}$*-transversal.*

In particular, for any immersion $i\colon V \to Y$, *the morphism* $h\colon W \to X$ *is* $\mathcal{F}$*-transversal.*

PROOF. Since the assertion is local on V, after replacing Y by an open neighborhood of V, we may assume that i is a closed immersion. Let $U_0 = Y - V$ be the complement and consider the cartesian diagram

$$\begin{array}{ccccc} W & \xrightarrow{h} & X & \xleftarrow{j} & U \\ {\scriptstyle g}\downarrow & & {\scriptstyle f}\downarrow & & \downarrow{\scriptstyle f_U} \\ V & \xrightarrow{i} & Y & \xleftarrow{j_0} & U_0. \end{array}$$

We consider the commutative diagram

$$\begin{array}{ccccc} \mathcal{F} \otimes f^*(\mathcal{G} \otimes j_{0*}\Lambda) & \longrightarrow & \mathcal{F} \otimes f^*j_{0*}j_0^*\mathcal{G} & \longrightarrow & j_*(j^*\mathcal{F} \otimes f_U^*j_0^*\mathcal{G}) \\ \downarrow & & & & \downarrow \\ \mathcal{F} \otimes f^*\mathcal{G} \otimes j_*\Lambda & & \longrightarrow & & j_*j^*(\mathcal{F} \otimes f^*\mathcal{G}) \end{array}$$

Since f is smooth, f is Λ-acyclic by the local acyclicity of smooth morphisms. Since f is $\mathcal{F}$-acyclic and Λ-acyclic, the upper right horizontal arrow and the left vertical arrow are isomorphisms by Proposition 1.2.4 (1)⇒(2). The right vertical arrow is a canonical isomorphism. Since i is $\mathcal{G}$-transversal, the upper left horizontal arrow is an isomorphism by Lemma 1.1.7.5 (1)⇒(2). Hence the lower horizontal arrow is an isomorphism and the assertion follows by Lemma 1.1.7.5 (2)⇒(1). □

COROLLARY 1.2.7 ([**11**, Proposition 2.2]). *Assume that* $X \to S$ *is universally* $\mathcal{F}$*-acyclic. Then, for any immersion* $i\colon V \to Y$ *of schemes over* S *and for any sheaf* $\mathcal{G}$ *on* V, *the morphism* $\mathcal{F} \boxtimes i_*\mathcal{G} \to (1 \times i)_*(\mathcal{F} \boxtimes \mathcal{G})$ *is an isomorphism.*

PROOF. We apply Proposition 1.2.4 (1)⇒(2) to the cartesian square

$$\begin{array}{ccc} X \times_S Y & \xleftarrow{1\times i} & X \times_S V \\ {\scriptstyle \mathrm{pr}_2}\downarrow & & \downarrow \\ Y & \xleftarrow{i} & V \end{array}$$

and $\mathrm{pr}_1^*\mathcal{F}$ on $X \times_S Y$ and $\mathcal{G}$ on V. Then since $\mathrm{pr}_2\colon X \times_S Y \to Y$ is $\mathrm{pr}_1^*\mathcal{F}$-acyclic, the morphism

$$\mathcal{F} \boxtimes i_*\mathcal{G} = \mathrm{pr}_1^*\mathcal{F} \otimes \mathrm{pr}_2^*i_*\mathcal{G} \to (1 \times i)_*((1 \times i)^*\mathrm{pr}_1^*\mathcal{F} \otimes \mathrm{pr}_2^*\mathcal{G}) = (1 \times i)_*(\mathcal{F} \boxtimes \mathcal{G})$$

is an isomorphism. □

Proposition 1.2.8 (cf. [**12**, Corollary 8.12] [**14**, Proposition 1.2.4])). *Let $f\colon X \to Y$ be a smooth morphism of excellent noetherian schemes and $\mathcal{F}$ be a sheaf on X. Let $Z \subset X$ be a closed subset. Assume that for every cartesian diagram*

$$\begin{array}{ccccc} X & \longleftarrow & X' & \stackrel{h'}{\longleftarrow} & W \\ {\scriptstyle f}\downarrow & & \downarrow & & \downarrow \\ Y & \longleftarrow & Y' & \longleftarrow & V \end{array} \tag{1.16}$$

where $V \to Y'$ is an immersion of regular schemes of finite type over Y, the morphism h' is $\mathcal{F}'$-transversal for the pull-back of $\mathcal{F}'$ on X' on a neighborhood of the inverse image of Z. Then f is $\mathcal{F}$-acyclic along Z.

The same proof as loc. cit. works with a form of alteration due to Gabber [**8**, Exposé VII, Théorème 1.1].

Proposition 1.2.9 (cf. [**3**, Appendix Theorem B.2]). *Let $f\colon X \to Y$ be a morphism of schemes over an excellent regular noetherian scheme S and let $\mathcal{F}$ be a sheaf on X. If $X \to S$ is universally $\mathcal{F}$-acyclic and if Y is smooth over S, then the following conditions are equivalent:*

(1) *f is $\mathcal{F}$-acyclic.*

(2) *For any constructible sheaf $\mathcal{G}$ on Y, the morphism $\gamma = (1_X, f)\colon X \to X \times_S Y$ is $\mathcal{F} \boxtimes \mathcal{G}$-transversal.*

The assumption that $X \to S$ is universally $\mathcal{F}$-acyclic is satisfied if $S = \operatorname{Spec} k$ for a field k by Theorem 1.2.2.

Proof. By devissage, the condition (2) is equivalent to the condition where $\mathcal{G}$ is restricted to $\mathcal{G} = i_*\mathcal{L}$ for immersions $i\colon V \to Y$ of regular subschemes and locally constant constructible sheaves $\mathcal{L}$ on V as in Proposition 1.2.4 (2′) since Y is excellent.

Let $h\colon W = X \times_Y V \to X$ be the base change of $i\colon V \to Y$ and consider the cartesian diagram

$$\begin{array}{ccccc} W & \xrightarrow{\gamma'} & X \times_S V & \xrightarrow{\mathrm{pr}_2} & V \\ {\scriptstyle h}\downarrow & & \downarrow{\scriptstyle 1\times i} & & \downarrow{\scriptstyle i} \\ X & \xrightarrow{\gamma} & X \times_S Y & \xrightarrow{\mathrm{pr}_2} & Y. \end{array} \tag{1.17}$$

The condition (2′) in Proposition 1.2.4 means that the morphism

$$\gamma^*(\mathcal{F} \boxtimes i_*\mathcal{L}) = \mathcal{F} \otimes f^* i_*\mathcal{L} \to h_*(h^*\mathcal{F} \otimes g^*\mathcal{L}) = h_*\gamma'^*(\mathcal{F} \boxtimes \mathcal{L})$$

is an isomorphism. Hence this for every V and $\mathcal{L}$ as above is equivalent to (1) by Proposition 1.2.4 (2′)⇔(1). Thus it suffices to show that the following conditions are equivalent:

(1′) For every immersion $i\colon V \to Y$ of regular subscheme and every locally constant constructible sheaf $\mathcal{L}$ on V, the morphism

$$\gamma^*(\mathcal{F} \boxtimes i_*\mathcal{L}) \to h_*\gamma'^*(\mathcal{F} \boxtimes \mathcal{L}) \tag{1.18}$$

is an isomorphism.

(2′) For every immersion $i\colon V \to Y$ of regular subscheme and every locally constant constructible sheaf $\mathcal{L}$ on V, the morphism

$$c_{\gamma,\mathcal{F}\boxtimes i_*\mathcal{L}}\colon \gamma^*(\mathcal{F}\boxtimes i_*\mathcal{L})\otimes\gamma^!\Lambda \to \gamma^!(\mathcal{F}\boxtimes i_*\mathcal{L}) \tag{1.19}$$

is an isomorphism.

We construct a commutative diagram

$$\begin{array}{ccc} \gamma^*(\mathcal{F}\boxtimes i_*\mathcal{L})\otimes\gamma^!\Lambda & \xrightarrow{c_{\gamma,\mathcal{F}\boxtimes i_*\mathcal{L}}} & \gamma^!(\mathcal{F}\boxtimes i_*\mathcal{L}) \\ \downarrow & & \downarrow \\ h_*(\gamma'^*(\mathcal{F}\boxtimes\mathcal{L})\otimes h^*\gamma^!\Lambda) & \longrightarrow & h_*\gamma'^!(\mathcal{F}\boxtimes\mathcal{L}) \end{array} \tag{1.20}$$

similarly as (1.6). The lower horizontal arrow is obtained by applying h_* to the composition

$$\gamma'^*(\mathcal{F}\boxtimes\mathcal{L})\otimes h^*\gamma^!\Lambda \to \gamma'^!(\mathcal{F}\boxtimes\mathcal{L}) \tag{1.21}$$

of $c_{\gamma',\mathcal{F}\boxtimes\mathcal{L}}$ with the morphism induced by $h^*\gamma^!\Lambda \to \gamma'^!\Lambda$ (1.3). The right vertical arrow is the composition of $\gamma^!(\mathcal{F}\boxtimes i_*\mathcal{L}) \to \gamma^!(1\times i)_*(\mathcal{F}\boxtimes\mathcal{L})$ and the base change isomorphism $\gamma^!(1\times i)_* \to h_*\gamma'^!$ (1.4). By the assumption that the morphism $X\to S$ is universally $\mathcal{F}$-acyclic and by Corollary 1.2.7, the morphism $\mathcal{F}\boxtimes i_*\mathcal{L}\to(1\times i)_*(\mathcal{F}\boxtimes\mathcal{L})$ is an isomorphism. Hence the right vertical arrow is an isomorphism. The left vertical arrow is the adjoint of the isomorphism $h^*\gamma^*(\mathcal{F}\boxtimes i_*\mathcal{L})\otimes h^*\gamma^!\Lambda\to\gamma'^*(\mathcal{F}\boxtimes\mathcal{L})\otimes h^*\gamma^!\Lambda$. By the assumption that $Y\to S$ is smooth, we have an isomorphism $\Lambda(-d)[-2d]\to\gamma^!\Lambda$ for the relative dimension d of Y over S. Hence the left vertical arrow is an isomorphism if and only if (1.18) is an isomorphism. By applying Lemma 1.1.3.2 to the left square of (1.17), we see that (1.20) is commutative.

We show that the following conditions are equivalent:

(1°) (1.21) and the left vertical arrow in (1.20) are isomorphisms.

(2°) (1.19) is an isomorphism,

(3°) All arrows in (1.20) are isomorphisms.

(1°)⇒(3°): If (1.21) is an isomorphism, then the lower horizontal arrow in (1.20) is an isomorphism. Further if the left vertical arrow in (1.20) is an isomorphism, (3°) is satisfied.

(2°)⇒(1°): If (1.19) is an isomorphism, then the composition of (1.20) via upper right is an isomorphism. Since h is an immersion, this implies that (1.21) is an isomorphism. The rest is similar to the proof of (1°)⇒(3°).

(3°)⇒(2°) is clear.

Thus it suffices to show that one of (1.21) and (1.19) is an isomorphism assuming (1′). First, we prove (1.21) in the case where i is an open immersion and $\mathcal{L}=\Lambda$. Since Y is smooth over S, the morphism γ is a section of a smooth morphism $\mathrm{pr}_1\colon X\times_S Y\to X$. Hence γ is $\mathcal{F}\boxtimes\Lambda$-transversal by Lemma 1.1.7.1 and Lemma 1.1.8. Further, its restriction $\gamma'\colon W\to X\times_S V$ is also $\mathcal{F}\boxtimes\Lambda$-transversal. This means that $\gamma'^*(\mathcal{F}\boxtimes\Lambda)\otimes\gamma'^!\Lambda\to\gamma'^!(\mathcal{F}\boxtimes\Lambda)$ (1.21) is an isomorphism and implies that γ is $\mathcal{F}\boxtimes i_*\Lambda$-transversal as we have shown above.

Next, we prove (1.19) in the case where i is a closed immersion and $\mathcal{L}=i^!\Lambda$. Note that if i is a regular immersion of codimension c, then $i^!\Lambda=\Lambda(-c)[-2c]$ is locally constant constructible by purity [**8**, Exposé XVI, Théorème 3.1.1]. Let $j\colon Y - V\to Y$ be the open immersion of the complement and consider the distinguished triangle $\mathcal{F}\boxtimes i_*i^!\Lambda\to\mathcal{F}\boxtimes\Lambda\to\mathcal{F}\boxtimes j_*\Lambda\to$. As we have already shown above, $\gamma\colon X\to X\times_S Y$ is transversal for the last two terms. Hence by Lemma

1.1.7.2, the morphism γ is also $\mathcal{F} \boxtimes i_* i^! \Lambda$-transversal and (1.19) for $\mathcal{L} = i^! \Lambda$ is an isomorphism.

We show (1.21) in the general case. Since the isomorphism (1.21) is an étale local condition on V and since $\mathcal{L}$ is assumed locally constant, it is reduced to the case where $i\colon V \to Y$ is a closed immersion by replacing Y by an open neighborhood of V and $\mathcal{L} = i^! \Lambda = \Lambda(-c)[-2c]$. □

1.3. C-transversality.

DEFINITION 1.3.1. *Let X be a scheme. Let V be a vector bundle on X and let $\mathbf{P}(V) = (V - X)/\mathbf{G}_m$ be the associated projective space bundle.*

1. *We say that a closed subset $C \subset V$ is conical if it is stable under the $\mathbf{G}_m$-action.*
2. *For a closed conical subset $C \subset V$, we call the intersection $C \cap X$ with the 0-section $X \subset V$ the base of C and the image $\mathbf{P}(C)$ of $C - C \cap X$ by the projection $V - X \to \mathbf{P}(V)$ the projectivization of C.*
3. *Let $f\colon V \to W$ be a linear morphism of vector bundles on X and $C \subset V$ be a closed conical subset. For $x \in X$, we say that f is C-transversal at x if the fiber at x of the intersection*

$$C \cap \operatorname{Ker}(f\colon V \to W) = C \times_W X$$

is a subset of $\{0\}$. We say that f is C-transversal if we have the inclusion everywhere on X.

The base $C \cap X \subset X$ and the projectivization $\mathbf{P}(C) \subset \mathbf{P}(V)$ are closed subsets. The base of C equals the image of C by the projection $V \to X$.

LEMMA 1.3.2. *Let C and C' be closed conical subsets of a vector bundle V. Then the following conditions are equivalent:*

(1) *$C \subset C'$.*

(2) *$C \cap X \subset C' \cap X$ and $\mathbf{P}(C) \subset \mathbf{P}(C')$.*

PROOF. (1)⇒(2) is clear. Since C is the union of $C \cap X$ and the inverse image of $\mathbf{P}(C)$ by $V - X \to \mathbf{P}(V)$, we have (2)⇒(1). □

LEMMA 1.3.3. *Let X be a scheme and $f\colon V \to W$ be a linear morphism of vector bundles on X. Let $C \subset V$ be a closed conical subset.*

1. *The subset $U = \{x \in X \mid f \text{ is } C\text{-transversal at } x\} \subset X$ is an open subset.*
2. *Endow $C \subset V$ with a closed subscheme structure stable under the action of $\mathbf{G}_m$. Then the following conditions are equivalent:*

(1) *f is C-transversal.*

(2) *The restriction of f on C is finite.*

(3) *The fiber product $C \times_W X$ is finite over X.*

If these equivalent conditions are satisfied, then the image $f(C) \subset W$ is a closed conical subset.

PROOF. 1. The intersection $C' = C \cap \operatorname{Ker}(f\colon V \to W)$ is a closed conical subset of V and its projectivization $\mathbf{P}(C')$ is a closed subset of the associated projective space bundle $\mathbf{P}(V)$. Since $U \subset X$ is the complement of the image of $\mathbf{P}(C') \subset \mathbf{P}(V)$, it is an open subset.

(1)⇔(3): Condition (1) means that the underlying set of $C \times_W X \subset V$ is a subset of the 0-section X. Since the ideal defining the closed immersion $X \to V$ is locally of finite type, this implies (3). Conversely, since C is stable under the $\mathbf{G}_m$-action, condition (3) implies that the underlying set of $C \times_W X \subset V$ is a subset of the 0-section X.

(2)⇔(3): Since the assertion is local on X, we may assume that $W = \operatorname{Spec} A$, $V = \operatorname{Spec} B$ are affine for graded rings $A = \bigoplus_{n \in \mathbf{N}} A_n, B = \bigoplus_{n \in \mathbf{N}} B_n$ and $C \subset V$ is defined by a graded ideal $J = \bigoplus_{n \in \mathbf{N}} J_n \subset B$. Hence, the assertion follows from Lemma 1.3.4 below.

□

LEMMA 1.3.4. *Let* $A = \bigoplus_{n \in \mathbf{N}} A_n$ *be a graded ring.*

1 *Let* $M = \bigoplus_{n \in \mathbf{N}} M_n$ *be a graded* A*-module and let* S *be a subset of* M *consisting of homogeneous elements. Then the following are equivalent:*

(1) *The* A*-module* M *is generated by* S.

(2) *For every* $n \geqq 0$, *the* $A/A_{\geqslant n}$*-module* $M/M_{\geqslant n}$ *is generated by the image of* S.

2 *Let* $A = \bigoplus_{n \in \mathbf{N}} A_n \to B = \bigoplus_{n \in \mathbf{N}} B_n$ *be a morphism of graded rings and* $J = \bigoplus_{n \in \mathbf{N}} J_n \subset B$ *be a graded ideal. Let* $S \subset B$ *be a set of homogeneous elements. Then the following conditions are equivalent:*

(1) *The* A*-module* B/J *is generated by the image of* S.

(2) *The* $A/A_{\geqslant 1}$*-module* $B/(J + A_{\geqslant 1}B)$ *is generated by the image of* S.

PROOF. 1. By replacing M by the quotient generated by S, we may assume that $S = \varnothing$. Then each condition means that $M = 0$.

(2)⇒(1): By induction on n, the $A/A_{\geqslant n}$-module $B/(J + A_{\geqslant n}B)$ is generated by the image of S. Since $A_{\geqslant n}B \subset B_{\geqslant n}$, the $A/A_{\geqslant n}$-module $B/(J + B_{\geqslant n})$ is also generated by the image of S. Hence it suffices to apply 1 to $M = B/J$.

(1)⇒(2) is clear.

□

Let p be a prime number and we fix a regular noetherian scheme S over $\mathbf{Z}_{(p)}$. satisfying the following finiteness condition:

(F) The reduced part of $S_{\mathbf{F}_p} = S \times_{\operatorname{Spec} \mathbf{Z}_{(p)}} \operatorname{Spec} \mathbf{F}_p$ is a scheme of finite type over a field of characteristic p with finite p-basis.

Let X be a regular scheme of finite type over S. Then the sheaf $F\Omega_X$ of Frobenius–Witt differentials is a locally free $\mathcal{O}_{X_{\mathbf{F}_p}}$-module of finite rank [**15**, Theorem 3.4] and define a vector bundle FT^*X on $X_{\mathbf{F}_p}$. For every point x of $X_{\mathbf{F}_p}$, we have a short exact sequence

$$0 \to F^*(\mathfrak{m}_x/\mathfrak{m}_x^2) \to F\Omega_{X,x} \otimes_{\mathcal{O}_{X,x}} k(x) \to F^*\Omega_{k(x)/\mathbf{F}_p} \to 0$$

where the functor F^* denotes the extension of scalars by the absolute Frobenius $k(x) \to k(x)$ [loc. cit. Proposition 2.4]. For a morphism $X \to Y$ of regular schemes of finite type over S, if $X \to Y$ is smooth, then $FT^*Y \times_Y X \to FT^*X$ is injective. Conversely, if $FT^*Y \times_Y X \to FT^*X$ is injective, then $X \to Y$ is smooth on a neighborhood of $X_{\mathbf{F}_p}$ by [loc. cit. Proposition 2.10]. Further, in this case, if $F^*T^*X/Y|_{X_{\mathbf{F}_p}}$ denote the Frobenius pull-back of the restriction to $X_{\mathbf{F}_p}$ of the relative cotangent bundle defined by $\Omega^1_{X/Y}$, we have a short exact sequence

$$0 \to FT^*Y \times_Y X \to FT^*X \to F^*T^*X/Y|_{X_{\mathbf{F}_p}} \to 0 \tag{1.22}$$

of vector bundles on $X_{\mathbf{F}_p}$.

Definition 1.3.5. *Let $h\colon W \to X$ be a morphism of regular schemes of finite type over S. Let $C \subset FT^*X$ be a closed conical subset and let h^*C denote the closed conical subset $C \times_X W \subset FT^*X \times_X W$.*

1. *We say that h is C-transversal if $FT^*X \times_X W \to FT^*W$ is h^*C-transversal.*
2. *If h is C-transversal, we define a closed conical subset $h^\circ C \subset FT^*W$ to be the image of h^*C by $FT^*X \times_X W \to FT^*W$.*

The terminology is explained by Lemma 1.3.7. If h is C-transversal and if $C' \subset C$ is another closed conical subset, then h is C'-transversal. If h is C-transversal, then $h^\circ C$ is a closed conical subset by Lemma 1.3.3.2.

Lemma 1.3.6. *Let $h\colon W \to X$ be a morphism of regular schemes of finite type over S. Let $C \subset FT^*X$ be a closed conical subset.*

1. *If h is smooth, then h is C-transversal.*
2. *Assume that C is a closed subset of the 0-section $FT^*_X X$. Then, h is C-transversal.*
3. *Assume that $C = FT^*X$ and that h is C-transversal. Then h is smooth on a neighborhood of $W_{\mathbf{F}_p}$.*

Proof. 1. Since $FT^*X \times_X W \to FT^*W$ is an injection, the intersection $h^*C \cap \operatorname{Ker}(FT^*X \times_X W \to FT^*W) \subset \operatorname{Ker}(FT^*X \times_X W \to FT^*W)$ is a subset of the 0-section.

2. If C is a subset of the 0-section $FT^*_X X$, then $h^*C \cap \operatorname{Ker}(FT^*X \times_X W \to FT^*W) \subset h^*C$ is a subset of the 0-section.

3. Since $FT^*X \times_X W \to FT^*W$ is an injection, h is smooth on a neighborhood of $W_{\mathbf{F}_p}$ by [**15**, Proposition 2.10]. □

Lemma 1.3.7. *Let $h\colon W \to X$ be a morphism of finite type of regular noetherian schemes. Let $Z \subset X$ be a regular closed subscheme and let $C \subset FT^*X$ be the image of the Frobenius pull-back of the conormal bundle $F^*T^*_Z X|_{Z_{\mathbf{F}_p}}$. Assume that X is excellent. Then, the following conditions are equivalent:*

(1) *h is C-transversal.*

(2) *h is transversal to the immersion $Z \to X$ on a neighborhood of $Z_{\mathbf{F}_p}$.*

Proof. Similarly as in the proof of Lemma 1.1.5, it is reduced to the case where h is a closed immersion. This case follows from the regularity criterion [**15**, Corollary 2.8]. □

Lemma 1.3.8. *Let $h\colon W \to X'$ and $g\colon X' \to X$ be morphisms of regular schemes of finite type over S and let C be a closed conical subset of FT^*X. Assume that g is C-transversal. Then, the following conditions are equivalent:*

(1) *gh is C-transversal.*

(2) *h is $g^\circ C$-transversal.*

Proof. Since $g^\circ C$ is defined as the image of g^*C and $g^*C \to g^\circ C$ is finite, condition (2) is equivalent to (1) $h^*g^*C \cap \operatorname{Ker}(FT^*X \times_X W \to FT^*W)$ is a subset of the 0-section. □

Definition 1.3.9. *Let $g\colon X \to Y$ be a morphism of regular schemes of finite type over S and let $C \subset FT^*X$ be a closed conical subset. Assume that g is proper* *on the base B of C. Then, we define a closed conical subset $g_\circ C \subset FT^*Y$ to be the image by $FT^*Y \times_Y X \to FT^*Y$ of the inverse image of C by $FT^*Y \times_Y X \to FT^*X$.*

LEMMA 1.3.10. *Let $g\colon X' \to X$ be a morphism of regular schemes of finite type over S and let $C' \subset FT^*X'$ be a closed conical subset. Assume that g is proper on the base B' of C'. Let $h\colon W \to X$ be a morphism of regular schemes of finite type over S and assume that h is $g_\circ C'$-transversal. Let $h'\colon W' \to X'$ be the base change of h. Assume that S is excellent.*

1. *The morphism h is transversal to g on a neighborhood of $h'^{-1}(B')$.*
2. *There exists a regular neighborhood $U' \subset W' = W \times_X X'$ of $h'^{-1}(B')$ on which $h'|_{U'}\colon U' \to X'$ is C'-transversal.*

PROOF. 1. Since $h\colon W \to X$ is C-transversal, on $h'^{-1}(B')$, the intersection of the kernel of $(FT^*X \times_X X' \to FT^*X') \times_{X'} W'$ with the kernel $(FT^*X \times_X W \to FT^*W) \times_W W'$ is a subset of the 0-section. Namely, the morphism $FT^*X \times_X W' \to FT^*X' \times_{X'} W' \oplus FT^*W \times_W W'$ is an injection on $h'^{-1}(B')$. This implies that W' is regular on a neighborhood U' of $h'^{-1}(B')$ by [**15**, Theorem 3.4] and $\operatorname{rank} FT^*U' - \operatorname{rank} FT^*X' = \operatorname{rank} FT^*W - \operatorname{rank} FT^*X$.

2. Since the question is local on W, we may decompose h as the composition of a smooth morphism and an immersion. Hence by Lemma 1.3.8, we may assume that h is an immersion. By 1, $g\colon X' \to X$ is transversal to the immersion $h\colon W' \to W$ on a neighborhood $U' \subset W'$ of $h'^{-1}(B')$. Then $\operatorname{Ker}(FT^*X' \times_{X'} U' \to FT^*U')$ is identified with $\operatorname{Ker}(FT^*X \times_X W \to FT^*W) \times_W U'$. Since $g_\circ C'$ is defined as the image of the intersection of the inverse image of C' with the image $\operatorname{Im}(FT^*X \times_X X' \to FT^*X')$ of the injection, the C'-transversality follows.

□

1.4. Micro support. We keep the notation that S is a regular noetherian scheme over $\mathbf{Z}_{(p)}$ satisfying the finiteness condition:
(F) The reduced part of $S_{\mathbf{F}_p}$ is of finite type over a field of characteristic p with finite p-basis.

DEFINITION 1.4.1 ([**14**, Definition 3.1.1]). *Let X be a regular scheme of finite type over S. Let $\mathcal{F}$ be a constructible sheaf on X and C be a closed conical subset of FT^*X. We say that $\mathcal{F}$ is micro supported on C if the following conditions (1) and (2) are satisfied:*

(1) *The intersection of the support of $\mathcal{F}$ and $X_{\mathbf{F}_p}$ is a subset of the base of C.*
(2) *Every C-transversal separated morphism $h\colon W \to X$ of regular schemes of finite type over S is $\mathcal{F}$-transversal on a neighborhood of $W_{\mathbf{F}_p}$.*

Note that the conditions (1) and (2) are on neighborhoods of $X_{\mathbf{F}_p}$ and of $W_{\mathbf{F}_p}$. This reflects the fact that the FW cotangent bundle FT^*X is a vector bundle on $X_{\mathbf{F}_p}$.

LEMMA 1.4.2. *Let X be a regular scheme of finite type over S and $\mathcal{F}$ be a constructible sheaf on X.*

1 *$\mathcal{F}$ is micro supported on FT^*X.*
2 *We consider the following conditions:*
(1) *$\mathcal{F}$ is locally constant on a neighborhood of $X_{\mathbf{F}_p}$.*
(2) *$\mathcal{F}$ is micro supported on the 0-section FT^*_XX.*
We have (1)⇒(2). Conversely if X is excellent, we have (2)⇒(1).

3 *The following conditions are equivalent:*
(1) $\mathcal{F} = 0$ *on a neighborhood of* $X_{\mathbf{F}_p}$.
(2) $\mathcal{F}$ *is micro supported on* $\varnothing$.

PROOF. 1. Let $h\colon W \to X$ be a separated morphism of regular schemes of finite type over S and assume that h is FT^*X-transversal. Then h is smooth on a neighborhood of $W_{\mathbf{F}_p}$ by Lemma 1.3.6.3 and is $\mathcal{F}$-transversal on a neighborhood of $W_{\mathbf{F}_p}$ by Lemma 1.1.7.1. Condition (1) in Definition 1.4.1 is clearly satisfied.

(1)⇒(2): If $\mathcal{F}$ is locally constant on a neighborhood of $X_{\mathbf{F}_p}$, every separated morphism $h\colon W \to X$ of regular schemes of finite type over S is $\mathcal{F}$-transversal on a neighborhood of $W_{\mathbf{F}_p}$ by Lemma 1.1.7.3. Condition (1) in Definition 1.4.1 is clearly satisfied.

(2)⇒(1): Assume that $\mathcal{F}$ is micro supported on the 0-section FT^*_XX. Then, every separated morphism $h\colon W \to X$ of regular schemes of finite type over S is FT^*_XX-transversal by Lemma 1.3.6.2 and hence is $\mathcal{F}$-transversal on a neighborhood of $W_{\mathbf{F}_p}$. Thus the assertion follows from Proposition 1.1.11.

2. By condition (1) in Definition 1.4.1, condition (2) is equivalent to that the support of $\mathcal{F}$ is a subset of $\varnothing$ on a neighborhood of $X_{\mathbf{F}_p}$. □

LEMMA 1.4.3. *Let* $h\colon W \to X$ *be a separated morphism of regular schemes of finite type over* S. *Let* $\mathcal{F}$ *be a constructible sheaf on* X *and* C *be a closed conical subset of* FT^*X. *If* $\mathcal{F}$ *is micro supported on* C *and if* $h\colon W \to X$ *is* C*-transversal, then* $h^*\mathcal{F}$ *is micro supported on* $h^\circ C$.

PROOF. It suffices to show that condition (2) in Definition 1.4.1 is satisfied. Since $h\colon W \to X$ is C-transversal, it is $\mathcal{F}$-transversal on a neighborhood of $W_{\mathbf{F}_p}$. Let $g\colon V \to W$ be a separated $h^\circ C$-transversal morphism of regular schemes of finite type over S. By Lemma 1.3.8 (2)⇒(1), $hg\colon V \to X$ is C-transversal and hence $\mathcal{F}$-transversal on a neighborhood of $V_{\mathbf{F}_p}$. By Lemma 1.1.8 (1)⇒(2), $g\colon V \to W$ is $h^*\mathcal{F}$-transversal on a neighborhood of $V_{\mathbf{F}_p}$ since the assumption is satisfied by purity [**8**, Exposé XVI, Théorème 3.1.1]. □

LEMMA 1.4.4. *Let* X *be a regular scheme of finite type over* S. *Let* $\mathcal{F}$ *be a constructible sheaf on* X *and* $C \subset FT^*X$ *be a closed conical subset.*

1. *Let* $X = \bigcup U_i$ *be an open covering of* X. *Then the following conditions are equivalent:*

(1) $\mathcal{F}$ *is micro supported on* C.
(2) $\mathcal{F}|_{U_i}$ *is micro supported on* $C|_{U_i}$ *for every* $i \in I$.

2. *Let* $U \subset X$ *be an open subscheme and let* $Z = X - U$ *be the complement. Let* $C' \subset FT^*U$ *be a closed subset. Assume that* $\mathcal{F}$ *is micro supported on* C *and that* $\mathcal{F}|_U$ *is micro supported on* C'. *Then,* $\mathcal{F}$ *is micro supported on the union* $C_1 = \overline{C'} \cup C|_Z$.

PROOF. (1)⇒(2): Since the open immersion $j\colon U \to X$ is C-transversal by Lemma 1.3.6.1, this follows from Lemma 1.4.3.

(2)⇒(1): Let $h\colon W \to X$ be a separated C-transversal morphism of regular schemes of finite type over S. Then for every i, the base change $h_i\colon W_i = W \times_X U_i \to U_i$ is $C|_{U_i}$-transversal. Since $\mathcal{F}|_{U_i}$ is micro supported on $C|_{U_i}$, h_i is $\mathcal{F}|_{U_i}$-transversal on a neighborhood of $W_{i,\mathbf{F}_p}$. Hence h is $\mathcal{F}$-transversal on a neighborhood of $W_{\mathbf{F}_p}$.

3. Let $h\colon W \to X$ be a separated C_1-transversal morphism of regular schemes of finite type over S. Then, the restriction $h_U\colon h^{-1}(U) \to U$ is C'-transversal and $h\colon W \to X$ is C-transversal on a neighborhood of $h^{-1}(Z)$ by Lemma 1.3.3.1. Hence by 1, h_U is $\mathcal{F}|_U$-transversal on a neighborhood of $h^{-1}(U)_{\mathbf{F}_p}$ and h is $\mathcal{F}$-transversal on a neighborhood of $h^{-1}(Z)_{\mathbf{F}_p}$. Since $W = h^{-1}(U) \cup h^{-1}(Z)$, h is $\mathcal{F}$-transversal on a neighborhood of $W_{\mathbf{F}_p}$.

□

LEMMA 1.4.5. *Let $g\colon X' \to X$ be a morphism of regular schemes of finite type over S and let $C' \subset FT^*X'$ be a closed conical subset. Assume that g is proper on the base B' of C'. Let $\mathcal{F}'$ be a constructible sheaf on X' and assume that $\mathcal{F}'$ is micro supported on $C' \subset FT^*X'$. Then, $g_*\mathcal{F}'$ is micro supported on $g_\circ C'$.*

PROOF. It suffices to show that condition (2) in Definition 1.4.1 is satisfied. Let $h\colon W \to X$ be a morphism of regular schemes of finite type over S and assume that h is $g_\circ C'$-transversal. By Lemma 1.3.10, the morphism h is transversal to g on a neighborhood $U' \subset W' = W \times_X X'$ of $h'^{-1}(B')$ and $h'\colon U' \to X'$ is C'-transversal. Since the assumption that the base change morphism $g'^*h^!\Lambda \to h'^!\Lambda$ (1.3) is an isomorphism on U' by Lemma 1.1.5.2, the morphism h is $g_*\mathcal{F}'$-transversal on a neighborhood of $W_{\mathbf{F}_p}$ by Lemma 1.1.12 (2)⇒(1). □

LEMMA 1.4.6. *Let $i\colon X \to P$ be a closed immersion of regular schemes of finite type over S and let $\mathcal{F}$ be a constructible sheaf on X. Let C_P be a closed conical subset of $FT^*P|_X \subset FT^*P$ and define $C \subset FT^*X$ to be the closure of the image of C_P by the surjection $FT^*P|_X \to FT^*X$. If $i_*\mathcal{F}$ is micro supported on C_P, then $\mathcal{F}$ is micro supported on C.*

PROOF. It suffices to show that condition (2) in Definition 1.4.1 is satisfied. Let $h\colon W \to X$ be a separated C-transversal morphism of regular schemes of finite type over S. We show that h is $\mathcal{F}$-transversal on a neighborhood of $W_{\mathbf{F}_p}$. Since the question is local on W, we may assume that there exists an immersion $W \to \mathbf{A}^n_X$. By replacing $W \to X \to P$ by $W \to \mathbf{A}^n_X \to \mathbf{A}^n_P$ and by Lemma 1.1.7.1, Lemma 1.1.8, Lemma 1.3.6.1 and Lemma 1.3.8, we may assume that h is an immersion.

Since the assertion is local on W, by taking a lifting of a local basis of the conormal bundle $T^*_W X$, we may extend the immersion $h\colon W \to X$ to an immersion $g\colon V \to P$ of regular closed subscheme transversal with the immersion $X \to P$. Then, the C-transversality of $h\colon W \to X$ implies the C_P-transversality of $V \to P$. Since $i_*\mathcal{F}$ is micro supported on C_P, this implies that $g\colon V \to P$ is $i_*\mathcal{F}$-transversal on a neighborhood of $V_{\mathbf{F}_p}$. Since the intersection $W = V \cap X$ is transversal, if $i'\colon W \to V$ denote the immersion, the morphism $i'^*g^!\Lambda \to h^!\Lambda$ is an isomorphism. Hence by Lemma 1.1.12 (1)⇒(2), the $i_*\mathcal{F}$-transversality of g implies the $\mathcal{F}$-transversality of h on a neighborhood of $W_{\mathbf{F}_p}$. Hence $\mathcal{F}$ is micro supported on C. □

The following lemma will be used in the reduction of the proof of the existence of singular support to the case of projective spaces, to produce enough sections étale locally.

LEMMA 1.4.7. *Let X be a regular scheme of finite type over S and $\mathcal{F}$ be a constructible sheaf on X. Let S' be a G-torsor over S for a finite group G. Let $g\colon X' = X \times_S S' \to X$ be the base change morphism and let $\mathcal{F}' = g^*\mathcal{F}$.*

1. *If $\mathcal{F}$ is micro supported on a closed conical subset $C \subset FT^*X$, then $\mathcal{F}'$ is micro supported on $C' = g^\circ C \subset FT^*X'$.*

2. *Assume that $\mathcal{F}'$ is micro supported on a closed conical subset $C' \subset FT^*X'$ and that C' is stable by the action of G. Then, $\mathcal{F}$ is micro supported on $C = g_\circ C' \subset FT^*X$.*

PROOF. It suffices to show that condition (2) in Definition 1.4.1 is satisfied.

1. Let $h: W \to X'$ be a separated morphism of regular schemes of finite type over S'. Then $h: W \to X'$ is a separated morphism of regular schemes of finite type over S and the condition is satisfied.

2. Let $h: W \to X$ be a separated morphism of regular schemes of finite type over S. Then the base change $h': W' \to X'$ is a separated morphism of regular schemes of finite type over S'. Since C' is assumed stable under the G-action, h is C-transversal if and only if h' is C'-transversal. Hence the condition is satisfied.

□

If k is a perfect field of characteristic $p > 0$ and X is a smooth scheme over k, the exact sequence (1.22) means a canonical isomorphism $FT^*X \to F^*T^*X$.

PROPOSITION 1.4.8. *Let k be a perfect field of characteristic $p > 0$ and let X be a smooth scheme over k. Let $\mathcal{F}$ be a constructible sheaf on X and $C \subset T^*X$ be a closed conical subset.*

1. *The following conditions are equivalent:*
*(1) $\mathcal{F}$ is micro supported on $C \subset T^*X$ in the sense of Beilinson [**2**].*
(2) $\mathcal{F}$ is micro supported on $F^*C \subset FT^*X$.

2. *Let $\mathcal{G}$ be another constructible sheaf and assume that $\mathcal{F}$ and $\mathcal{G}$ are micro supported on closed conical subsets C and on C' of T^*X respectively. If $C \cap C'$ is a subset of the 0-section, then $\mathcal{F} \otimes \mathcal{G}$ is micro supported on $C + C' \subset T^*X$.*

PROOF. 1. Since $F^*T^*X = T^*X \times_X X \to T^*X$ is a homeomorphism, the assertion follows from [**12**, Proposition 8.13].

2. The external tensor product $\mathcal{F} \boxtimes \mathcal{G} = \mathrm{pr}_1^*\mathcal{F} \otimes \mathrm{pr}_2^*\mathcal{G}$ on $X \times X$ is micro supported on $C \times C' = T^*X \times T^*X = T^*(X \times X)$ by [**13**, Theorem 2.2.1]. By the assumption that $C \cap C'$ is a subset of the 0-section, the diagonal morphism $\delta: X \to X \times X$ is $C \times C'$-transversal. Hence $\mathcal{F} \otimes \mathcal{G} = \delta^*(\mathcal{F} \boxtimes \mathcal{G})$ is micro supported on $\delta^\circ(C \times C') = C + C'$ by Lemma 1.4.3.

□

1.5. Singular support. We keep the notation that S is a regular scheme over $\mathbf{Z}_{(p)}$ satisfying the finiteness condition (F). For a constructible sheaf $\mathcal{F}$ on a regular scheme X of finite type over S, we call the smallest closed conical subset $C \subset FT^*X$ on which $\mathcal{F}$ is micro supported the singular support of $\mathcal{F}$ if it exists.

DEFINITION 1.5.1. *Let X be a regular scheme of finite type over S and let $\mathcal{F}$ be a constructible sheaf on X. We say that a closed conical subset $C \subset FT^*X$ is the singular support $SS\mathcal{F}$ of $\mathcal{F}$ if for any closed conical subset $C' \subset FT^*X$, the inclusion $C \subset C'$ is equivalent to the condition that $\mathcal{F}$ is micro supported on C'.*

We don't know the existence of singular support in general because the condition that $\mathcal{F}$ is micro supported on C and on C' does not a priori imply that $\mathcal{F}$ is micro supported on the intersection $C \cap C'$. Similarly as in the proof of [**2**, Theorem

1.3 (i)], the proof of the existence of singular support is reduced to the case where $X = \mathbf{P}^n_S$ as Corollary 1.5.4 below.

LEMMA 1.5.2. *Let X be a regular schemes of finite type over S and let $\mathcal{F}$ be a constructible sheaf on X.*

1. *Let $U \subset X$ be an open subscheme. Assume that $C \subset FT^*X$ is the singular support of $\mathcal{F}$. Then, $C|_U$ is the singular support of $\mathcal{F}|_U$.*
2. *Let (U_i) be an open covering of X and $C_i \subset FT^*U_i$ be the singular support of $\mathcal{F}|_{U_i}$. Then, $C = \bigcup_i C_i \subset FT^*X$ is the singular support of $\mathcal{F}$.*

PROOF. 1. The restriction $\mathcal{F}|_U$ is micro supported on $C|_U$ by Lemma 1.4.4.1. We show that $C|_U$ is the smallest. Let $C' \subset FT^*U$ be a closed conical subset on which $\mathcal{F}|_U$ is micro supported and $Z = X - U$ be the complement. Then $\mathcal{F}$ is micro supported on $\overline{C'} \cup C|_Z$ by Lemma 1.4.4.2. Since C is the smallest, we have $C \subset \overline{C'} \cup C|_Z$ and $C|_U \subset (\overline{C'} \cup C|_Z)|_U = C'$.

2. For every i, j, the restrictions $C_i|_{U_i \cap U_j}$ and $C_j|_{U_i \cap U_j}$ are the singular support of $\mathcal{F}|_{U_i \cap U_j}$ and are the same by 1. Hence the union $C = \bigcup_i C_i$ is a closed conical subset of FT^*X. Since $C_i = C|_{U_i}$, $\mathcal{F}$ is micro supported on C by Lemma 1.4.4.1.

We show that C is the smallest. Let $C' \subset FT^*X$ be a closed conical subset on which $\mathcal{F}$ is micro supported. Then, for each i, we have $C_i \subset C'|_{U_i}$. Hence we have $C \subset C'$.

□

PROPOSITION 1.5.3. *Let $i\colon X \to P$ be a closed immersion of regular schemes of finite type over S and let $\mathcal{F}$ be a constructible sheaf on X. Let $C_P \subset FT^*P$ be a closed conical subset and assume that C_P is the singular support of $i_*\mathcal{F}$. Then the following holds.*

1. *C_P is a subset of $FT^*P|_X$.*
2. *Define $C \subset FT^*X$ to be the closure of the image of C_P by the surjection $FT^*P|_X \to FT^*X$. Then C is the singular support $SS\mathcal{F}$.*

PROOF. 1. Let $U = P - X$ be the complement. Since $\mathcal{F}|_U = 0$ is micro supported on $\varnothing$ by Lemma 1.4.2.3, $\mathcal{F}$ is micro supported on $C_P|_X \subset FT^*P$ by Lemma 1.4.4.2. Since C_P is the smallest, we have $C_P = C_P|_X \subset FT^*P|_X$.

2. By Lemma 1.4.6, $\mathcal{F}$ is micro supported on C. We show that C is the smallest. Assume that $\mathcal{F}$ is micro supported on a closed conical subset $C' \subset FT^*X$. Then, since $i_*\mathcal{F}$ is micro supported on $i_\circ C'$ by Lemma 1.4.6.1, we have $C_P \subset i_\circ C'$. By taking the closure of the image by the surjection $FT^*P|_X \to FT^*X$, we obtain $C \subset C'$.

□

COROLLARY 1.5.4. *Assume that the singular support exists for every constructible sheaf on every projective space $\mathbf{P}$ over S. Then, the singular support $SS\mathcal{F}$ exists for every constructible sheaf $\mathcal{F}$ on every regular scheme X of finite type over S.*

PROOF. By Lemma 1.5.2, the assertion is local on X. Hence, we may assume that X is affine and is a closed subscheme of $\mathbf{A}^n_S$. By the assumption and Lemma 1.5.2, the assertion holds for $\mathbf{A}^n_S$. Hence the assertion follows from Proposition 1.5.3. □

1.6. S-acyclicity. We keep the notation that S is a regular noetherian scheme over $\mathbf{Z}_{(p)}$ satisfying the finiteness condition (F). We will introduce two notions the S-acyclicity (Definition 1.6.1 and 1.6.3) and the smoothness modulo S (Definition 1.6.6) on a scheme Y smooth over S as a preparation of the definitions (Definitions 2.1.1 and 2.2.1) and a property (Lemma 2.1.2.3) of acyclicity over S in the next chapter. If $Y = Y_0 \times_k S$ is a base change of a smooth scheme Y_0 over a perfect field k of characteristic $p > 0$, the two notions should be conditions on Y_0. Because we don't have such a situation in mixed characteristic, we formulate the conditions using the injection $FT^*S \times_S Y \to FT^*Y$.

DEFINITION 1.6.1. *Let $Y \to S$ be a smooth scheme over S. We say that a closed conical subset $C' \subset FT^*Y$ is S-acyclic if the inverse image of C' by the injection $FT^*S \times_S Y \to FT^*Y$ is a subset of the 0-section.*

We may define the S-acyclicity without assuming that Y is smooth over S. However, since the condition implies that Y is smooth on a neighborhood of the base of C' if S is assumed excellent, we impose the smoothness.

LEMMA 1.6.2. *Let $g\colon Y \to S$ be a smooth scheme over S and let $C' \subset FT^*Y$ be an S-acyclic closed conical subset. Let X be a regular scheme of finite type over S.*

1. *The second projection $\mathrm{pr}_2\colon X \times_S Y \to Y$ is C'-transversal and $\mathrm{pr}_2^\circ C' \subset FT^*(X \times_S Y)$ is X-acyclic.*
2. *Let $C \subset FT^*X$ be a closed conical subset. Then, the intersection $\mathrm{pr}_1^\circ C \cap \mathrm{pr}_2^\circ C' \subset FT^*(X \times_S Y)$ is a subset of the 0-section.*
3. *Let $C'_1 \subset FT^*Y$ be a closed conical subset satisfying $C' + (FT^*S \times_S Y) \subset C'_1$. Let $f\colon X \to Y$ be a morphism over S and assume that f is C'_1-transversal. Then, on a neighborhood of the inverse image $f^{-1}(B')$ of the base B' of C', the scheme X is smooth over S and $f^\circ C'$ is S-acyclic.*

Further if $S' \to S$ is a morphism of regular schemes of finite type, the base change $X' = X \times_S S' \to Y' = Y \times_S S'$ is $\mathrm{pr}_1^\circ C'$-transversal on a neighborhood of the inverse image of $f^{-1}(B')$.

PROOF. 1. Since $\mathrm{Ker}(FT^*Y \times_Y (X \times_S Y) \to FT^*(X \times_S Y))$ is a subset of

$$\begin{aligned}&\mathrm{Ker}(FT^*Y \times_Y (X \times_S Y) \to F^*T^*(X \times_S Y)/X|_{(X \times_S Y)_{\mathbf{F}_p}})\\ &= \mathrm{Ker}(FT^*Y \to F^*T^*Y/S|_{Y_{\mathbf{F}_p}}) \times_Y (X \times_S Y),\end{aligned}$$

the projection $\mathrm{pr}_2\colon X \times_S Y \to Y$ is C'-transversal.

Since $F^*T^*Y/S|_{Y_{\mathbf{F}_p}} \times_Y (X \times_S Y) \to F^*T^*(X \times_S Y)/X|_{(X \times_S Y)_{\mathbf{F}_p}}$ is an isomorphism, the intersection $\mathrm{pr}_2^\circ C' \cap \mathrm{Im}(FT^*X \times_X (X \times_S Y) \to FT^*(X \times_S Y))$ equals the image of $(C' \cap \mathrm{Im}(FT^*S \times_S Y \to FT^*Y)) \times_Y (X \times_S Y)$. Hence the S-acyclicity of C' implies the X-acyclicity of $\mathrm{pr}_2^\circ C'$.

2. The intersection $\mathrm{pr}_1^\circ C \cap \mathrm{pr}_2^\circ C' \subset \mathrm{pr}_2^\circ C' \cap \mathrm{Im}(FT^*X \times_X (X \times_S Y) \to FT^*(X \times_S Y))$ is a subset of the 0-section by 1.

3. Since f is C'_1-transversal and $(FT^*S \times_S Y)|_{B'} \subset C'_1$, the morphism $FT^*S \times_S X \to FT^*X$ is an injection on the inverse image $f^{-1}(B')$. Hence X is smooth over S on a neighborhood of $f^{-1}(B')$ by [**15**, Proposition 2.10].

To show the rest, we may assume X is smooth. Since f is C'_1-transversal, the intersection $f^\circ C' \cap \mathrm{Im}(FT^*S \times_S X \to FT^*X)$ is a subset of the image of

$f^*C' \cap \mathrm{Im}(FT^*S \times_S X \to FT^*Y \times_Y X)$. This is a subset of the 0-section since C' is assumed S-acyclic.

By 1, $\mathrm{pr}_1 \colon Y' \to Y$ is C'-transversal and $\mathrm{pr}_1^\circ C'$ is defined. Since $f^\circ C'$ is S-acyclic, $\mathrm{pr}_1 \colon X' \to X$ is $f^\circ C'$-transversal. Since $\mathrm{pr}_1 f' = f \circ \mathrm{pr}_1$, the base change $f' \colon X' \to Y'$ is $\mathrm{pr}_1^\circ C'$-transversal by Lemma 1.3.8.

□

DEFINITION 1.6.3. *Let Y be a smooth scheme of finite type over S. We say that a constructible sheaf $\mathcal{G}$ on Y is S-acyclic if there exists an S-acyclic closed conical subset $C' \subset FT^*Y$ on which $\mathcal{G}$ is micro supported.*

If $S = \mathrm{Spec}\, k$ for a perfect field of characteristic p, then every C' is S-acyclic and hence every $\mathcal{G}$ is S-acyclic. If $Y = S$ is excellent, the condition that $\mathcal{F}$ is S-acyclic means that $\mathcal{F}$ is locally constant on a neighborhood of $S_{\mathbf{F}_p}$ by Lemma 1.4.2.

LEMMA 1.6.4. *Assume that S is excellent. Let Y be a smooth scheme of finite type over S and let $\mathcal{G}$ be an S-acyclic constructible sheaf on Y. Then, $g \colon Y \to S$ is universally $\mathcal{G}$-acyclic along $Y_{\mathbf{F}_p}$.*

PROOF. Let $C' \subset FT^*Y$ be an S-acyclic closed conical subset on which $\mathcal{G}$ is micro supported. First, we show that it suffices to show that g is $\mathcal{G}$-acyclic at $Y_{\mathbf{F}_p}$. By Lemma 1.6.2.1, for every morphism $S' \to S$ of finite type of regular noetherian schemes, the base change $f \colon Y' = Y \times_S S' \to Y$ is C'-transversal and $f^\circ C'$ is S'-acyclic. Hence the pull-back $\mathcal{G}' = f^*\mathcal{G}$ on Y' is micro supported on $f^\circ C'$ by Lemma 1.4.3. Thus by Lemma 1.2.3, it suffices to show that g is $\mathcal{G}$-acyclic along $Y_{\mathbf{F}_p}$.

Let

$$\begin{array}{ccccc} Y & \xleftarrow{\ f\ } & Y' & \xleftarrow{\ h\ } & W \\ {\scriptstyle g}\big\downarrow & & \big\downarrow & & \big\downarrow \\ S & \longleftarrow & S' & \longleftarrow & V \end{array}$$

be a cartesian diagram where $V \to S'$ is an immersion of regular schemes of finite type over S as in (1.16) in Proposition 1.2.8. By Lemma 1.6.2.1, f and fh are C'-transversal and $f^\circ C'$ is S'-acyclic. Hence h is $f^\circ C'$-transversal by Lemma 1.3.8. Hence the pull-back $\mathcal{G}' = f^*\mathcal{G}$ on Y' is micro supported on $f^\circ C'$ by Lemma 1.4.3 and h is $\mathcal{G}'$-transversal on a neighborhood of $W_{\mathbf{F}_p}$. Thus by Proposition 1.2.8, f is $\mathcal{G}$-acyclic along $Y_{\mathbf{F}_p}$. □

LEMMA 1.6.5. 1. *Let $i \colon Z \to Y$ be a closed immersion of smooth schemes over S and let $\mathcal{G}$ be an S-acyclic constructible sheaf on Z. Then $i_*\mathcal{G}$ on Y is S-acyclic.*

2. *Let $f \colon Y \to Y'$ be a morphism of smooth schemes over S and $\mathcal{G}'$ be an S-acyclic constructible sheaf on Y' micro supported on an S-acyclic closed conical subset $C' \subset FT^*Y'$. Let $C'_1 \subset FT^*Y'$ be a closed conical subset satisfying $C' + (FT^*S \times_S Y') \subset C'_1$ and assume that f is C'_1-transversal. Then the pull-back $\mathcal{G} = f^*\mathcal{G}'$ on Y is S-acyclic.*

3. *Let $S' \to S$ be a separated morphism of finite type of regular noetherian schemes. Let $Y \to S$ be a smooth morphism and $h \colon Y' = Y \times_S S' \to Y$ be the base change. If a constructible sheaf $\mathcal{G}$ on Y is S-acyclic, then h is $\mathcal{G}$-transversal on a neighborhood of $Y'_{\mathbf{F}_p}$ and the pull-back $\mathcal{G}' = h^*\mathcal{G}$ on Y' is S'-acyclic.*

PROOF. 1. Let $C' \subset FT^*Z$ be an S-acyclic closed conical subset on which $\mathcal{G}$ is micro supported. Then $i_*\mathcal{G}$ is micro supported on $i_\circ C' \subset FT^*Y$ by Lemma 1.4.5 and $i_\circ C'$ is S-acyclic.

2. The pull-back $\mathcal{G}$ on Y of $\mathcal{G}'$ is micro supported on $C = g^\circ C' \subset FT^*Y$. Since C is S-acyclic by Lemma 1.6.2.3, $\mathcal{G}$ is S-acyclic.

3. Assume that $\mathcal{G}$ is micro supported on S-acyclic C'. By Lemma 1.6.2.1, $h\colon Y' \to Y$ is C'-transversal and $h^\circ C'$ is S'-acyclic. Hence h is $\mathcal{G}$-transversal and $\mathcal{G}' = h^*\mathcal{G}$ is micro supported on $h^\circ C'$ by Lemma 1.4.3. Hence $\mathcal{G}'$ is S'-acyclic.

□

DEFINITION 1.6.6. *Let $f\colon W \to Y$ be a morphism of regular scheme of finite type over S. Assume that Y is smooth over S. We say that f is smooth modulo S if we have an inclusion*

$$\mathrm{Ker}(FT^*Y \times_Y W \to FT^*W) \subset \mathrm{Im}(FT^*S \times_S W \to FT^*Y \times_Y W). \tag{1.23}$$

For a closed subset $Z \subset W$, we say that f is smooth modulo S on Z if we have an inclusion (1.23) *on the inverse image of Z.*

If S is smooth over a perfect field k of characteristic p and if $Y = Y_0 \times_k S$ for a scheme Y_0 smooth over k, then the smoothness of $W \to Y$ modulo S implies the smoothness of the composition $W \to Y_0$.

LEMMA 1.6.7. *Let $f\colon W \to Y$ be a morphism of regular schemes of finite type over S and assume that Y is smooth over S.*

1. *If $Y = S$, then every morphism $W \to S$ is smooth modulo S.*

2. *Let $g\colon Y' \to Y$ be a morphism of smooth schemes over S and set $W' = W \times_Y Y'$. If f is smooth modulo S and if S is excellent, then there exists a neighborhood $U' \subset W'$ of $W'_{\mathbf{F}_p}$ such that g is transversal to f on U' and the base change $f'\colon W' \to Y'$ of f is smooth modulo S on U'. In particular, the projection $\mathrm{pr}_2\colon W \times_S Y \to Y$ is smooth modulo S.*

3. *Let $S' \to S$ be a smooth morphism and let $f'\colon W \to Y' = Y \times_S S'$ be a morphism over S'. If f' is smooth modulo S', then $f = \mathrm{pr}_1 \circ f'$ is smooth modulo S.*

4. *Assume that S is smooth over a perfect field k of characteristic p and $Y = Y_0 \times_k S$ for a smooth scheme Y_0 over k. Then f is smooth modulo S if and only if the composition $\mathrm{pr}_1 \circ f\colon W \to Y_0$ is smooth. In particular if $S = \mathrm{Spec}\, k$, then f is smooth modulo k means that f is smooth.*

PROOF. 1. Clear from (1.23).

2. By replacing $W \to Y$ by $W \times_S Y' \to Y \times_S Y'$, we may assume that $Y' \to Y$ is a regular immersion of smooth schemes over S. Then, the morphism $T^*_{Y'}Y \to T^*Y/S$ is an injection. Hence the intersection $F^*T^*_{Y'}Y \cap \mathrm{Im}(FT^*S \times_S Y \to FT^*Y)$ is a subset of the 0-section. Thus the morphism $F^*T^*_{Y'}Y \times_Y W \to FT^*W$ is also an injection and W' is regular on a neighborhood of $W'_{\mathbf{F}_p}$ by [**15**, Theorem 3.4]. Further $\mathrm{Ker}(FT^*Y \times_Y W \to FT^*W) \to \mathrm{Ker}(FT^*Y' \times_{Y'} W' \to FT^*W')$ is an isomorphism and the assertion follows.

3. Since the inverse image of $\mathrm{Im}(FT^*S' \times_{S'} W \to FT^*Y' \times_{Y'} W)$ by $FT^*Y \times_Y W \to FT^*Y' \times_{Y'} W$ equals $\mathrm{Im}(FT^*S \times_S W \to FT^*Y \times_Y W)$, the assertion follows.

4. The assertion follows from the exact sequence $0 \to FT^*S \times_S W \to FT^*Y \times_Y W \to FT^*Y_0 \times_{Y_0} W \to 0$.

□

LEMMA 1.6.8. *Let $f\colon W \to Y$ be a morphism of regular schemes of finite type over S. Assume that Y is smooth over S and that f is smooth modulo S.*

1. *Let $C' \subset FT^*Y$ be a closed conical subset. If C' is S-acyclic, then f is C'-transversal.*
2. *Let $\mathcal{G}$ be an S-acyclic constructible sheaf on Y. Then f is $\mathcal{G}$-transversal on a neighborhood of $W_{\mathbf{F}_p}$.*

PROOF. 1. Clear from the inclusion (1.23).
2. Clear from 1.

□

LEMMA 1.6.9. *Let S be a smooth scheme over a perfect field k. Let Y be a smooth scheme of finite type over S and let $\mathcal{G}$ be a sheaf on Y micro supported on an S-acyclic $C' \subset FT^*Y$. Let X be a regular scheme of finite type over S and let $\mathcal{F}$ be a sheaf on X micro supported on $C \subset FT^*X$. Then, $\mathcal{F} \boxtimes \mathcal{G}$ on $X \times_S Y$ is micro supported on $\mathrm{pr}_1^\circ C + \mathrm{pr}_2^\circ C'$.*

PROOF. Since the first projection $\mathrm{pr}_1\colon X \times_S Y \to X$ is smooth, pr_1 is C-transversal and $\mathrm{pr}_1^*\mathcal{F}$ is micro supported on $\mathrm{pr}_1^\circ C$ by Lemma 1.4.3. Since the second projection $\mathrm{pr}_2\colon X \times_S Y \to Y$ is smooth modulo S by Lemma 1.6.7.2, pr_2 is C'-transversal by Lemma 1.6.8.1 and $\mathrm{pr}_2^*\mathcal{G}$ is micro supported on $\mathrm{pr}_2^\circ C'$ by Lemma 1.4.3. The intersection $\mathrm{pr}_1^\circ C \cap \mathrm{pr}_2^\circ C'$ is a subset of the 0-section by Lemma 1.6.2.2. Hence by Proposition 1.4.8, $\mathcal{F} \boxtimes \mathcal{G}$ is micro supported on $\mathrm{pr}_1^\circ C + \mathrm{pr}_2^\circ C'$. □

We prove a weaker assertion in general.

PROPOSITION 1.6.10. *Let $g\colon Y \to S$ be a smooth scheme over S and let $\mathcal{G}$ be an S-acyclic constructible sheaf on Y micro supported on an S-acyclic closed conical subset $C' \subset FT^*Y$. Assume that S is excellent. Let $C_1' \subset FT^*Y$ be a closed conical subset satisfying $C' + (FT^*S \times_S Y) \subset C_1'$. Then, for every constructible sheaf $\mathcal{H}$ on S, the tensor product $\mathcal{G} \otimes g^*\mathcal{H}$ is micro supported on C_1'.*

PROOF. Let

$$\begin{array}{ccc} W & \xrightarrow{f} & Y \\ & \searrow^{q} & \downarrow{\scriptstyle g} \\ & & S \end{array} \tag{1.24}$$

be a commutative diagram of separated morphism of regular schemes of finite type over S and assume that f is C_1'-transversal. We show that f is $\mathcal{G} \otimes g^*\mathcal{H}$-transversal on a neighborhood of $W_{\mathbf{F}_p}$. Since the base B' of C' contains the intersection of the support of $\mathcal{G}$ with $Y_{\mathbf{F}_p}$, we may assume that W is smooth over S by Lemma 1.6.2.3.

Since $f\colon W \to Y$ is C'-transversal, it is $\mathcal{G}$-transversal on a neighborhood of $W_{\mathbf{F}_p}$. Hence by Lemma 1.1.1, it suffices to show that the morphism $c_{f,\mathcal{G},g^*\mathcal{H}}\colon f^!\mathcal{G} \otimes q^*\mathcal{H} \to f^!(\mathcal{G} \otimes g^*\mathcal{H})$ is an isomorphism on a neighborhood of $W_{\mathbf{F}_p}$. By devissage, we may assume that there exists an immersion $j\colon V \to S$ of regular schemes, a locally constant constructible sheaf $\mathcal{L}$ on V and and $\mathcal{H} = j_*\mathcal{L}$.

Let

$$\begin{array}{ccc} W' & \xrightarrow{f'} & Y' \\ & \searrow^{q'} & \downarrow{\scriptstyle g'} \\ & & V \end{array}$$

be the base change of (1.24) by $j\colon V \to S$ and let $j_Y\colon Y' \to Y$ and $j_W\colon W' \to W$ be the induced morphisms. Since g is $\mathcal{G}$-acyclic by Lemma 1.6.4 along $Y_{\mathbf{F}_p}$, the canonical morphism $\mathcal{G} \otimes g^* j_* \mathcal{L} \to j_{Y*}(j_Y^* \mathcal{G} \otimes g'^* \mathcal{L})$ is an isomorphism on a neighborhood of $Y_{\mathbf{F}_p}$ by Proposition 1.2.4. Similarly, since $f^*\mathcal{G}$ is also S-acyclic by Lemma 1.6.5.2, the morphism $f^!\mathcal{G} \otimes q^* j_* \mathcal{L} \to j_{W*}(j_W^* f^! \mathcal{G} \otimes q'^* \mathcal{L})$ is an isomorphism on a neighborhood of $W_{\mathbf{F}_p}$. Thus, it suffices to show that the morphism $j_{W*}(j_W^* f^! \mathcal{G} \otimes q'^* \mathcal{L}) \to f^! j_{Y*}(j_Y^* \mathcal{G} \otimes g'^* \mathcal{L})$ is an isomorphism on a neighborhood of $W_{\mathbf{F}_p}$.

Since the morphism $f^! j_{Y*} \to j_{W*} f'^!$ (1.4) is an isomorphism, it suffices to show that $j_W^* f^! \mathcal{G} \otimes q'^* \mathcal{L} \to f'^!(j_Y^* \mathcal{G} \otimes g'^* \mathcal{L})$ is an isomorphism on a neighborhood of $W_{\mathbf{F}_p}$. Since $\mathcal{L}$ is locally constant, it is reduced to showing that $j_W^* f^! \mathcal{G} \to f'^! j_Y^* \mathcal{G}$ (1.3) is an isomorphism on a neighborhood of $W'_{\mathbf{F}_p}$. By Lemma 1.6.5.3 and Lemma 1.6.2.1, on a neighborhood of $Y'_{\mathbf{F}_p}$, j_Y is $\mathcal{G}$-transversal and $j_Y^* \mathcal{G}$ is V-acyclic and is micro supported on $j_Y^\circ C'$. By Lemma 1.6.2.3, f' is $j_Y^\circ C'$-transversal and is $j_Y^* \mathcal{G}$-transversal on a neighborhood of $W'_{\mathbf{F}_p}$. Since f is $\mathcal{G}$-transversal on a neighborhood of $W_{\mathbf{F}_p}$ and is transversal to j_Y, the morphism $j_W^* f^! \mathcal{G} \to f'^! j_Y^* \mathcal{G}$ is an isomorphism on a neighborhood of $W'_{\mathbf{F}_p}$ by Lemma 1.1.3.1. □

2. S-micro support

Let S be an excellent regular noetherian scheme over $\mathbf{Z}_{(p)}$ satisfying the finiteness condition:

(F) The reduced part of $S_{\mathbf{F}_p}$ is of finite type over a field of characteristic p with finite p-basis.

2.1. C-acyclicity over S.

DEFINITION 2.1.1. *Let $h\colon W \to X$ and $f\colon W \to Y$ be morphisms of regular schemes of finite type over S. Assume that Y is smooth over S. Let $C \subset FT^*X$ be a closed conical subset.*

1. *Let $w \in W_{\mathbf{F}_p}$. We say that (h, f) is C-acyclic over S at w if we have an inclusion*

(2.1)
$$\begin{aligned}(h^*C \times_W (FT^*Y \times_Y W)) &\cap \operatorname{Ker}((FT^*X \times_X W) \times_W (FT^*Y \times_Y W) \to FT^*W) \\ &\subset \operatorname{Ker}((FT^*X \times_X W) \times_W (FT^*Y \times_Y W) \to FT^*(X \times_S Y) \times_{X \times_S Y} W).\end{aligned}$$

on the fiber at w.

We say that (h, f) is C-acyclic over S if (h, f) is C-acyclic over S at every point of $W_{\mathbf{F}_p}$.

2. *We say* (h, f) *is universally* C*-acyclic over* S *if for every morphism* $g\colon Y' \to Y$ *of smooth schemes over* S *and the commutative diagram*

(2.2)
$$\begin{array}{ccccc} & & W' & \xrightarrow{f'} & Y' \\ & \swarrow^{h'} & \downarrow{\scriptstyle g'} & & \downarrow{\scriptstyle g} \\ X & \xleftarrow{h} & W & \xrightarrow{f} & Y \end{array}$$

with cartesian square, there exists a neighborhood $U' \subset W'$ *of the inverse image* $h'^{-1}(B)$ *of the base* B *of* C *such that* g *is transversal to* f *on* U' *and the pair* (h', f') *is* C*-acyclic on* U'.

Unlike the C-transversality, the C-acyclicity over S may not be an open condition in general. Since $(FT^*X \times_X W) \times_W (FT^*Y \times_Y W) \to FT^*W$ factors through $FT^*(X \times_S Y) \times_{X \times_S Y} W$, the inclusion (2.1) is equivalent to the inclusion

$(2.1)'$
$$\mathrm{Im}(h^*C \times_W (FT^*Y \times_Y W)) \cap \mathrm{Ker}(FT^*(X \times_S Y) \times_{X \times_S Y} W \to FT^*W) \subset 0\text{-section}$$

in $FT^*(X \times_S Y) \times_{X \times_S Y} W$. Using this interpretation, we will show in Lemma 2.1.9 that the C-acyclicity over S is an open condition for S-saturated C defined in Definition 2.1.8.1.

If $S = \mathrm{Spec}\, k$ for a perfect field k of characteristic p, then the C-acyclicity is the same as the C-transversality for a test pair defined by Beilinson [**2**, 1.2] by Lemma 2.1.2.4, Lemma 2.1.5 and Lemma 2.1.10. We show in Lemma 2.1.2.1 below that the C-acyclicity implies the universal C-acyclicity. If $C' \subset FT^*X$ is another closed conical subset satisfying $C \subset C' \subset C + FT^*S \times_S X$, then the C-acyclicity over S and the C'-acyclicity over S are equivalent to each other since the kernels in (2.1) are stable under the actions of $FT^*S \times_S W$.

LEMMA 2.1.2. *Let* $h\colon W \to X$ *and* $f\colon W \to Y$ *be morphisms of regular scheme of finite type over* S. *Let* $C \subset FT^*X$ *be a closed conical subset and let* $B \subset X$ *be the base of* C. *Assume that* Y *is smooth over* S *and that* (h, f) *is* C*-acyclic over* S.

1. *The pair* (h, f) *is universally* C*-acyclic over* S.
2. *On the inverse image* $h^{-1}(B) \subset W$, *the kernel* $\mathrm{Ker}(FT^*Y \times_Y W \to FT^*W)$ *is a subset of the image of* $\mathrm{Ker}(FT^*S \times_S W \to FT^*X \times_X W)$.
3. *The morphism* $f\colon W \to Y$ *is smooth modulo* S *on* $h^{-1}(B)$.
If X *is smooth over* S, *then* $W \to Y$ *is smooth on a neighborhood of* $h^{-1}(B)$.
4. *The morphism* h *is* C*-transversal. If* $q\colon W \to S$ *denotes the canonical morphism, then* (h, q) *is* C*-acyclic over* S.
5. *Let* $C' \subset FT^*Y$ *be a closed conical subset and assume that* C' *is* S*-acyclic. Then the morphism* $(h, f)\colon W \to X \times_S Y$ *is* $\mathrm{pr}_1^\circ C + \mathrm{pr}_2^\circ C'$*-transversal.*

PROOF. 2. The intersection

$$((FT^*_X X \times_X W) \times_W (FT^*Y \times_Y W)) \cap \mathrm{Ker}((FT^*X \times_X W) \times_W (FT^*Y \times_Y W) \to FT^*W)$$

is identified with $\mathrm{Ker}(FT^*Y \times_Y W \to FT^*W)$. In the commutative diagram

$$\begin{array}{ccccccccc} 0 & \longrightarrow & FT^*S \times_S W & \longrightarrow & FT^*Y \times_Y W & \longrightarrow & F^*T^*Y/S \times_Y W & \longrightarrow & 0 \\ & & \downarrow & & \downarrow & & \downarrow & & \\ 0 & \longrightarrow & FT^*X \times_X W & \longrightarrow & FT^*(X \times_S Y) \times_{X \times_S Y} W & \longrightarrow & (F^*T^*(X \times_S Y)/X) \times_{X \times_S Y} W & \longrightarrow & 0 \end{array}$$

of exact sequences, the right vertical arrow is an isomorphism. Hence the intersection

$$\begin{aligned}&((FT^*_X X \times_X W) \times_W (FT^*Y \times_Y W))\\ &\cap \operatorname{Ker}((FT^*X \times_X W) \times_W (FT^*Y \times_Y W) \to FT^*(X \times_S Y)_{X\times_S Y}W)\end{aligned}$$

is identified with the image of $\operatorname{Ker}(FT^*S \times_S W \to FT^*X \times_X W)$. Thus, we obtain the inclusion.

3. By 2, we have an inclusion (1.23) on the inverse image $h^{-1}(B)$.

If X is smooth over S, then $FT^*S \times_S X \to FT^*X$ is an injection. Hence by 2, $FT^*Y \times_Y W \to FT^*W$ is also an injection on $h^{-1}(B)$.

1. Let $g\colon Y' \to Y$ be a morphism of smooth schemes over S and consider the commutative diagram (2.2). By 3 and Lemma 1.6.7.2, the morphism g is transversal to f on a neighborhood $U' \subset W'$ of $h'^{-1}(B)$. To show that (h', f') is C-acyclic over S, it suffices to show that

$$\begin{aligned}&\operatorname{Ker}((FT^*X \times_X W) \times_W (FT^*Y \times_Y W) \to FT^*W) \times_W U'\\ &\to \operatorname{Ker}((FT^*X \times_X U') \times_{U'} (FT^*Y' \times_{Y'} U') \to FT^*U')\end{aligned}$$

is an isomorphism. By the factorization $Y' \to Y \times_S Y' \to Y$, it suffices to prove the cases where $Y' \to Y$ is smooth and $Y' \to Y$ is an immersion respectively. Since the morphism $f\colon W \to Y$ is transversal to $g\colon Y' \to Y$ in the second case, the assertion follows.

4. Assume that we have an inclusion (2.1). The intersection of (2.1) with the first factor $FT^*X \times_X W$ gives an inclusion

$$h^*C \cap \operatorname{Ker}(FT^*X \times_X W \to FT^*W) \subset \operatorname{Ker}(FT^*X \times_X W \to FT^*(X \times_S Y)_{X\times_S Y}W).$$

Since Y is smooth over S, the last morphism is an injection and hence h is C-transversal.

Since the morphism $FT^*S \times_S Y \to FT^*Y$ is an injection and the morphism $\operatorname{Ker}((FT^*X \times_X W) \times_W (FT^*S \times_S W) \to FT^*X \times_X W) \to \operatorname{Ker}((FT^*X \times_X W) \times_W (FT^*Y \times_Y W) \to FT^*(X \times_S Y) \times_{X\times_S Y} W)$ is an isomorphism, the C-transversality of (h, f) over S implies that of (h, q).

5. The assumption that C' is S-acyclic implies that that the intersection $(C \times C') \cap \operatorname{Ker}(FT^*X \times_S FT^*Y \to FT^*(X \times_S Y))$ is a subset of the 0-section and that $\mathrm{pr}_1^\circ C + \mathrm{pr}_2^\circ C' \subset T^*(X \times_S Y)$ is the image of $C \times C'$ and is a closed subset. If (h, f) is C-acyclic, then the intersection $(h^*C \times_W f^*C') \cap \operatorname{Ker}((FT^*X \times_X W) \times_W (FT^*Y \times_Y W) \to FT^*W)$ is a subset of $\operatorname{Ker}((FT^*X \times_X W) \times_W (FT^*Y \times_Y W) \to FT^*(X \times_S Y) \times_{X\times_S Y} W)$. Since $\mathrm{pr}_1^\circ C + \mathrm{pr}_2^\circ C'$ is the image of $h^*C \times_W f^*C'$ by the surjection $(FT^*X \times_X W) \times_W (FT^*Y \times_Y W) \to FT^*(X \times_S Y) \times_{X\times_S Y} W$, its intersection with the kernel of $FT^*(X \times_S Y) \times_{X\times_S Y} W \to FT^*W$ is a subset of the 0-section. □

LEMMA 2.1.3. *Let $f\colon X \to Y$ be morphisms of regular scheme of finite type over S. Let $C \subset FT^*X$ be a closed conical subset. Assume that Y is smooth over S. Then, the following conditions are equivalent:*

(1) *$(1, f)$ is C-acyclic over S.*

(2) *The inverse image of C by $FT^*Y \times_Y X \to FT^*X$ is a subset of the image of the injection $FT^*S \times_S X \to FT^*Y \times_Y X$.*

PROOF. Since the kernels $\mathrm{Ker}(FT^*X \times_X (FT^*Y \times_Y X) \to FT^*X)$ and $\mathrm{Ker}(FT^*X \times_X (FT^*Y \times_Y X) \to FT^*(X \times_S Y) \times_{X \times_S Y} X)$ are identified with the images of $FT^*Y \times_Y X$ and $FT^*S \times_S X$, the assertion follows. □

LEMMA 2.1.4. *Let $h\colon W \to X$ and $f\colon W \to Y$ be morphisms of regular schemes of finite type over S. Assume that Y is smooth over S.*

1. *For $C = FT^*X$, the following conditions are equivalent:*
(1) *The pair (h, f) is C-acyclic over S.*
(2) *The morphism $(h, f)\colon W \to X \times_S Y$ is smooth on a neighborhood of $W_{\mathbf{F}_p}$.*
2. *If C is the 0-section $FT^*_X X$, the following conditions are equivalent:*
(1) *The pair (h, f) is C-acyclic over S.*
(2) *The kernel $\mathrm{Ker}(FT^*Y \times_Y W \to FT^*W)$ is a subset of the image of $\mathrm{Ker}(FT^*S \times_S W \to FT^*X \times_X W)$ on a neighborhood of $W_{\mathbf{F}_p}$.*

PROOF.

1. The condition (1) holds if and only if $FT^*(X \times_S Y) \times_{X \times_S Y} W \to FT^*W$ is an injection. This is equivalent to (2) by [**15**, Proposition 2.8].
2. This is proved by the same argument as in the proof of Lemma 2.1.2.2.

□

LEMMA 2.1.5. *Let*

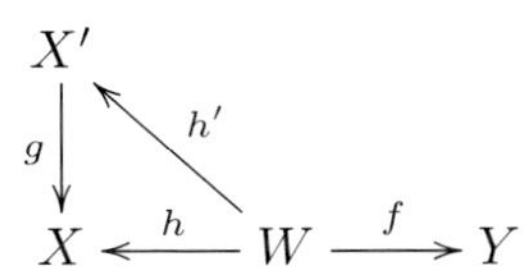

be a commutative diagram of regular schemes of finite type over S. Assume that Y is smooth over S. Let $C \subset FT^*X$ be a closed conical subset. Let $q\colon X' \to S$ be the canonical morphism and assume that (g, q) is C-acyclic over S. Then, the following conditions are equivalent:

(1) *(h, f) is C-acyclic over S.*
(2) *(h', f) is $g^\circ C$-acyclic over S.*

PROOF. (1)⇒(2): Since $\mathrm{Ker}((FT^*X \times_X W) \times_W (FT^*Y \times_Y W) \to FT^*W)$ equals the inverse image of $\mathrm{Ker}((FT^*X' \times_{X'} W) \times_W (FT^*Y \times_Y W) \to FT^*W)$, by taking the image of (2.1) by $(FT^*X \times_X W) \times_W (FT^*Y \times_Y W) \to (FT^*X' \times_{X'} W) \times_W (FT^*Y \times_Y W)$, we obtain an inclusion

(2.3)
$$
\begin{aligned}
(g^\circ C \times_W (FT^*Y \times_Y W)) \cap \mathrm{Ker}((FT^*X' \times_{X'} W) \times_W (FT^*Y \times_Y W) \to FT^*W)\\
\subset \mathrm{Ker}((FT^*X' \times_{X'} W) \times_W (FT^*Y \times_Y W) \to FT^*(X' \times_S Y) \times_{X' \times_S Y} W).
\end{aligned}
$$

(2)⇒(1): Since the kernel of $(FT^*X' \times_{X'} W) \times_W (FT^*Y \times_Y W) \to FT^*(X' \times_S Y) \times_{X' \times_S Y} W$ equals the image of $FT^*S \times_S W \to (FT^*X' \times_{X'} W) \times_W (FT^*Y \times_Y W)$, the intersection of its inverse image with the kernel of $(FT^*X \times_X W) \times_W (FT^*Y \times_Y W) \to FT^*W$ is a subset of the image of the kernel of $(FT^*X \times_X W) \times_W (FT^*S \times_S W) \to (FT^*X' \times_{X'} W)$. Hence the inclusion (2.3) implies that the left hand side of (2.1) is a subset of the intersection $(h^*C \times_W (FT^*S \times_S W)) \cap \mathrm{Ker}((FT^*X \times_X W) \times_W (FT^*S \times_S W) \to (FT^*X' \times_{X'} W))$. Hence by the assumption that (g, q) is C-acyclic over S, the assertion follows. □

LEMMA 2.1.6. *Let*

$$\begin{array}{ccccc} X & \xleftarrow{\ h\ } & W & \xrightarrow{\ f\ } & Y \\ & & & \searrow^{f'} & \downarrow{\scriptstyle g} \\ & & & & Y' \end{array}$$

be a commutative diagram of regular scheme of finite type over S. Assume that Y and Y' are smooth over S and that g is smooth. Let $C \subset FT^*X$ be a closed conical subset. Assume that (h, f) is C-acyclic over S. Then, (h, f') is C-acyclic over S.

PROOF. Since $FT^*Y' \times_{Y'} Y \to FT^*Y$ is an injection and the inverse image of the kernel $\mathrm{Ker}((FT^*X \times_X W) \times_W (FT^*Y \times_Y W) \to FT^*(X \times_S Y) \times_{X \times_S Y} W)$ by $(FT^*X \times_X W) \times_W (FT^*Y' \times_{Y'} W) \to (FT^*X \times_X W) \times_W (FT^*Y \times_Y W)$ is $\mathrm{Ker}((FT^*X \times_X W) \times_W (FT^*Y' \times_{Y'} W) \to FT^*(X \times_S Y') \times_{X \times_S Y'} W)$, the assertion follows. □

LEMMA 2.1.7. *Let $g\colon X' \to X$ be a morphism of regular schemes of finite type over S. Let $C' \subset FT^*X'$ be a closed conical subset and assume that g is proper on the base B' of C'. Let $C = g_\circ C'$. Let*

$$\begin{array}{ccccc} X' & \xleftarrow{\ h'\ } & W' & & \\ {\scriptstyle g}\downarrow & & {\scriptstyle g'}\downarrow & \searrow^{f'} & \\ X & \xleftarrow{\ h\ } & W & \xrightarrow{\ f\ } & Y \end{array}$$

be a commutative diagram of separated morphisms of schemes of finite type over S with cartesian square. Assume that W is regular and that Y is smooth over S. Assume that (h, f) is C-acyclic over S and that S is excellent.

1. *The morphism h is transversal to g on a neighborhood of $h'^{-1}(B')$.*
2. *There exists a regular neighborhood $U' \subset W'$ of $h'^{-1}(B')$ on which (h', f') is C'-acyclic over S.*

PROOF. 1. The morphism $h\colon W \to X$ is C-transversal by Lemma 2.1.2.4. Hence the assertion follows from Lemma 1.3.10.1.

2. By 1, there exists a regular neighborhood $U' \subset W'$ of $h'^{-1}(B')$ such that the sequence $0 \to FT^*X \times_X U' \to FT^*X' \times_{X'} U' \oplus FT^*W \times_W U' \to FT^*U'$ is exact. Then, the canonical morphism $\mathrm{Ker}((FT^*X \times_X U') \times_{U'} (FT^*Y \times_Y U') \to FT^*W \times_W U') \to \mathrm{Ker}((FT^*X' \times_{X'} U') \times_{U'} (FT^*Y \times_Y U') \to FT^*U')$ is an isomorphism. Since $\mathrm{Ker}((FT^*X \times_X U') \times_{U'} (FT^*Y \times_Y U') \to FT^*(X \times_S Y) \times_{X \times_S Y} U') \to \mathrm{Ker}((FT^*X' \times_{X'} U') \times_{U'} (FT^*Y \times_Y U') \to FT^*(X' \times_S Y) \times_{X' \times_S Y} U')$ is an isomorphism, the assertion follows from the definition of $C = g_\circ C'$. □

DEFINITION 2.1.8. *Let $C \subset FT^*X$ be a closed conical subset.*

1. *We say that C is S-saturated if C is stable under the action of $FT^*S \times_S X$.*
2. *Let $C^{\mathrm{sat}} \subset FT^*X$ denote the smallest S-saturated closed conical subset containing C as a subset.*

If the morphism $FT^*S \times_S X \to FT^*X$ is 0, then every closed conical subset $C \subset FT^*X$ is S-saturated and hence $C = C^{\mathrm{sat}}$. In particular, if $S = \mathrm{Spec}\, k$ for a perfect field k, every C is S-saturated. The saturation C^{sat} is given by the closure

$\overline{C + FT^*S \times_S X}$. If $C^{\rm sat} = C + FT^*S \times_S X$, the C-acyclicity over S and the $C^{\rm sat}$-acyclicity over S are equivalent to each other since the kernels in (2.1) are stable under the actions of $FT^*S \times_S W$.

LEMMA 2.1.9. *Let $h\colon W \to X$ and $f\colon W \to Y$ be morphisms of regular schemes of finite type over S and assume that Y is smooth over S. Let $C \subset FT^*X$ be an S-saturated closed conical subset. Then the subset $\{w_{\mathbf{F}_p} \in W \mid (h,f)$ is C-acyclic over S at $w\} \cup W_{\mathbf{Z}[\frac{1}{p}]} \subset W$ is an open subset.*

PROOF. Since C is S-saturated, by the split short exact sequence $0 \to FT^*S \times_S (X \times_S Y) \to FT^*X \times_S FT^*Y \to FT^*(X \times_S Y) \to 0$, the image $\mathrm{Im}(h^*C \times_W (FT^*Y \times_Y W)) \subset FT^*(X \times_S Y)$ is a closed subset. Hence the assertion follows from $(2.1)'$ and Lemma 1.3.3.1. □

LEMMA 2.1.10. *Let $h\colon W \to X$ be a morphism of regular schemes of finite type over S and $q\colon W \to S$ be the canonical morphism. Let $C \subset FT^*X$ be a closed conical subset. Assume that h is C-transversal and C is S-saturated. Then (h,q) is C-acyclic over S.*

PROOF. Assume that $(a,b) \in h^*C \times_W (FT^*S \times_S W)$ is in the kernel of $(FT^*X \times_X W) \times_W (FT^*S \times_S W) \to FT^*W$. Since C is S-stable, we have $a + b \in h^*C \subset FT^*X \times_X W$ and this is contained in the kernel of $FT^*X \times_X W \to FT^*W$. Since h is assumed C-transversal, we have $a + b = 0$ and hence (a,b) is in the kernel of $(FT^*X \times_X W) \times_W (FT^*S \times_S W) \to FT^*X \times_X W$. □

LEMMA 2.1.11. *Let*

$$\begin{array}{ccccccc} X & \xleftarrow{h} & W & \xrightarrow{f} & Y & \longrightarrow & S \\ & & & \searrow^{f_0} & \downarrow{\scriptstyle g} & & \downarrow \\ & & & & Y_0 & \longrightarrow & S_0 \end{array}$$

be a commutative diagram of regular noetherian schemes satisfying the finiteness condition (F) *with cartesian square. Assume that X, W and Y are of finite type over S and that S and Y_0 are smooth over S_0. We consider the following conditions:*

(1) *(h,f) is C-acyclic over S.*
(2) *(h,f_0) is C-acyclic over S_0.*

We have (1)⇒(2). Conversely, if C is S-saturated, we have (2)⇒(1).

PROOF. (1)⇒(2): Since $FT^*Y_0 \times_{Y_0} Y \to FT^*Y$ is an injection and since $X \times_{S_0} Y_0 = X \times_S Y$, the assertion follows.

(2)⇒(1): Let $q\colon W \to S$ be the canonical morphism. By Lemma 2.1.5, it suffices to show that (h,q) is C-acyclic over S and that $(1,f)$ is $h^\circ C$-acyclic over S. Since C is assumed S-saturated, (h,q) is C-acyclic by Lemma 2.1.2.4 and Lemma 2.1.10. Hence by replacing X and C by W and $h^\circ C$, we may assume $X = W$ and $h = 1_X$. By Lemma 2.1.3, it suffices to show that the condition (2) there for $X \to Y_0$ over S_0 implies that for $X \to Y$ over S. Since FT^*Y is generated by the images of FT^*Y_0 and FT^*S, this follows from the assumption that C is S-saturated. □

LEMMA 2.1.12. *Assume that X is smooth over S. Then, the mapping*

$$\{ \text{ closed conical subsets of } T^*X/S|_{X_{\mathbf{F}_p}}\} \to \{ \text{ closed conical subsets of } FT^*X\}$$

sending $\overline{C}$ to its inverse image is an injection and the image consists of saturated closed conical subsets.

PROOF. Since the base change $F^*(T^*X/S)|_{X_{\mathbf{F}_p}} \to T^*X/S|_{X_{\mathbf{F}_p}}$ is a homeomorphism, the assertion follows from the exact sequence

$$0 \to FT^*S \times_S X \to FT^*X \to F^*T^*X/S|_{X_{\mathbf{F}_p}} \to 0. \tag{1.22}$$

□

LEMMA 2.1.13. *Let $h\colon W \to X$ and $f\colon W \to Y$ be morphisms of smooth schemes over S. Let $C \subset FT^*X$ be an S-saturated closed conical subset and let $\overline{C} \subset T^*X/S$ be the corresponding closed conical subset.*

For $w \in W_{\mathbf{F}_p}$, the following conditions are equivalent:

(1) *(h,f) is C-acyclic over S at w.*

(2) *The fiber at w of the intersection*

$$((\overline{C}\times_X W)\times_W(T^*Y/S\times_Y W))\cap \mathrm{Ker}((T^*X/S\times_X W)\times_W(T^*Y/S\times_Y W)\to T^*W/S)$$

is a subset of $\{0\}$.

PROOF. The kernel $\mathrm{Ker}((FT^*X\times_X W)\times_W(FT^*Y\times_Y W)\to FT^*W)$ is the extension of $\mathrm{Ker}((F^*T^*X/S|_{X_{\mathbf{F}_p}}\times_X W)\times_W(F^*T^*Y/S|_{Y_{\mathbf{F}_p}}\times_Y W)\to F^*T^*W/S|_{W_{\mathbf{F}_p}})$ by the diagonal image of $FT^*S \times_S W$ and $T^*X/S \times_S T^*Y/S \to T^*(X\times_S Y)/S$ is an isomorphism. Hence the assertion follows. □

2.2. $\mathcal{F}$-acyclicity over S. Lemma 1.6.9 suggests the following definition.

DEFINITION 2.2.1. *Let $h\colon W \to X$ and $f\colon W \to Y$ be separated morphisms of regular schemes of finite type over S. Assume that Y is smooth over S. Let $\mathcal{F}$ be a constructible sheaf on X.*

1. *We say that (h,f) is $\mathcal{F}$-acyclic over S if the following condition is satisfied:*

For every S-acyclic constructible sheaf $\mathcal{G}$ on Y, the morphism $(h,f)\colon W \to X\times_S Y$ is $\mathcal{F}\boxtimes\mathcal{G}$-transversal on an open neighborhood of $W_{\mathbf{F}_p}$.

2. *Let $B \subset X$ be the support of $\mathcal{F}$. We say that (h,f) is universally $\mathcal{F}$-acyclic over S if the following condition is satisfied:*

For every morphism $g\colon Y' \to Y$ of smooth schemes over S and the commutative diagram

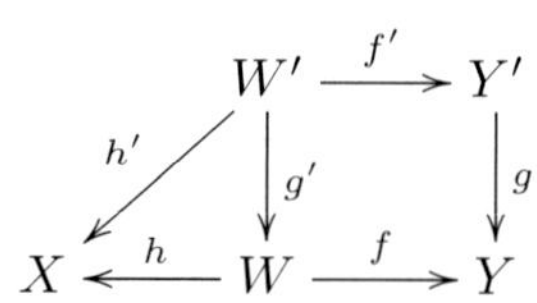

with cartesian square, there exists an open neighborhood $U' \subset W'$ of $h'^{-1}(B)_{\mathbf{F}_p}$ such that g is transversal to f on U' and the pair (h',f') is $\mathcal{F}$-acyclic over S on U'.

If $Y = S$, we say that h is universally $\mathcal{F}$-transversal over S.

In Definition 2.2.1.1, it is more natural to define a Zariski local condition on Y. However, since the C-acyclicity over S is a local condition on Y, this does not affect the definition of S-micro support in Definition 2.3.1. Thus, we adopt a simpler formulation here. If $Y = S$, the condition that (h,f) is $\mathcal{F}$-acyclic over S means that h is $\mathcal{F}$-transversal on a neighborhood of $W_{\mathbf{F}_p}$ since S-acyclicity of $\mathcal{G}$ on $Y = S$ means that $\mathcal{G}$ is locally constant on a neighborhood of $S_{\mathbf{F}_p}$ by Lemma 1.4.2.

In particular, for every $\mathcal{F}$ on X and the morphism $f: X \to S$, the pair of $(1_X, f)$ is $\mathcal{F}$-acyclic over S. The condition that $h: W \to X$ is universally $\mathcal{F}$-transversal over S means that for every smooth scheme Y over S and for every S-acyclic sheaf $\mathcal{G}$ on Y, the base change $h': W \times_S Y \to X \times_S Y$ is $\mathcal{F} \boxtimes \mathcal{G}$-transversal on a neighborhood of $(W \times_S Y)_{\mathbf{F}_p}$.

LEMMA 2.2.2. *Let $f: W \to Y$ be a smooth morphism of smooth schemes over S. Let $\mathcal{F}$ be a constructible sheaf on W. We consider the following conditions:*

(1) $(1_W, f)$ *is universally $\mathcal{F}$-acyclic over S.*
(2) f *is universally $\mathcal{F}$-acyclic.*
1. *We have* (2)⇒(1).
2. *If $S = \operatorname{Spec} k$ for a perfect field k of characteristic p, the conditions* (1) *and* (2) *are equivalent.*

The condition that f is smooth is satisfied on a neighborhood of the intersection of the support of $h^*\mathcal{F}$ with $W_{\mathbf{F}_p}$ if (h, f) is C-acyclic over S for a closed conical subset $C \subset FT^*X$ such that the intersection of the support of $\mathcal{F}$ with $X_{\mathbf{F}_p}$ is a subset of the base of C and if X is smooth over S by Lemma 2.1.2.3.

PROOF. 1. By replacing $W \to Y$ by the base change by a morphism $Y' \to Y$ of smooth schemes over Y, it suffices to show that $(1_W, f)$ is $\mathcal{F}$-acyclic over S. Let $\mathcal{G}$ be an S-acyclic constructible sheaf on Y. We show that $\gamma = (1, f): W \to W \times_S Y$ is $\mathcal{F} \boxtimes \mathcal{G}$-transversal. We consider the cartesian diagram

$$\begin{array}{ccc} W \times_S Y & \xleftarrow{\gamma} & W \\ {\scriptstyle f \times 1_Y}\downarrow & & \downarrow{\scriptstyle f} \\ Y \times_S Y & \xleftarrow{\delta} & Y. \end{array}$$

Since f is assumed universally $\mathcal{F}$-acyclic, the morphism $f \times 1_Y$ is $\mathcal{F} \boxtimes \Lambda$-acyclic. Since δ is a morphism of smooth schemes over Y, the diagonal δ is $\Lambda \boxtimes \mathcal{G}$-transversal. Since f is assumed smooth, by Corollary 1.2.6, the morphism γ is $\mathcal{F} \boxtimes \mathcal{G}$-transversal.

2. It suffices to show that f is $\mathcal{F}$-acyclic. Since every constructible sheaf on Y is S-acyclic, the assertion follows from Proposition 1.2.9.

□

LEMMA 2.2.3. *Let X be a regular scheme of finite type over S and $\mathcal{F}$ be a constructible sheaf on X. Let $h: W \to X$ be an open immersion and define $f: W \to Y = \mathbf{A}^1_S$ to be the composition of the canonical morphism $W \to S$ with the* 0-*section $i: S \to Y$. If (h, f) is $\mathcal{F}$-acyclic over S, on a neighborhood of $W_{\mathbf{F}_p}$, we have $h^*\mathcal{F} = 0$.*

PROOF. The constructible sheaf $\mathcal{G} = i_*\Lambda_S$ on Y is S-acyclic by Lemma 1.6.5.1. Hence the 0-section $i_W: W \to \mathbf{A}^1_X = X \times_S Y$ is $\mathcal{F} \boxtimes i_*\Lambda_S$-transversal on a neighborhood of $W_{\mathbf{F}_p}$. By Lemma 1.1.7.4, we have $\mathcal{F} = 0$ on a neighborhood of $W_{\mathbf{F}_p}$. □

LEMMA 2.2.4. *Let*

$$\begin{array}{ccccc} X' & & & & \\ {\scriptstyle g}\downarrow & \nwarrow^{h'} & & & \\ X & \xleftarrow{h} & W & \xrightarrow{f} & Y \end{array}$$

be a commutative diagram of separated morphisms regular schemes of finite type over S. Assume that Y is smooth over S. Let $\mathcal{F}$ be a constructible sheaf on X. Assume that g is universally $\mathcal{F}$-transversal. Then, the following conditions are equivalent:

(1) *(h, f) is $\mathcal{F}$-acyclic over S.*
(2) *(h', f) is $g^*\mathcal{F}$-acyclic over S.*

PROOF. Let $\mathcal{G}$ be an S-acyclic constructible sheaf on Y. Since g is universally $\mathcal{F}$-transversal, the morphism $g \times 1_Y : X' \times_S Y \to X \times_S Y$ is $\mathcal{F} \boxtimes \mathcal{G}$-transversal on a neighborhood of $(X' \times_S Y)_{\mathbf{F}_p}$. Since Y is smooth over S, $(g \times 1)^! \Lambda$ is isomorphic to $\Lambda(d)[2d]$ for the relative dimension d of g by Lemma 1.1.5.1. Hence by Lemma 1.1.8, $W \to X \times_S Y$ is $\mathcal{F} \boxtimes \mathcal{G}$-transversal on a neighborhood of $W_{\mathbf{F}_p}$ if and only if $W \to X' \times_S Y$ is $g^*\mathcal{F} \boxtimes \mathcal{G}$-transversal on a neighborhood of $W_{\mathbf{F}_p}$. □

LEMMA 2.2.5. *Let*

$$\begin{array}{ccccc} X & \xleftarrow{h} & W & \xrightarrow{f} & Y \\ & & & \searrow^{f'} & \downarrow g \\ & & & & Y' \end{array}$$

be a commutative diagram of separated morphisms of regular scheme of finite type over S. Assume that Y is smooth over S. Let $\mathcal{F}$ be a constructible sheaf on X. Assume that (h, f) is $\mathcal{F}$-acyclic over S and let $g \colon Y \to Y'$ be a smooth morphism of smooth schemes over S. Then (h, f') is $\mathcal{F}$-acyclic over S.

PROOF. Let $\mathcal{G}'$ be an S-acyclic constructible sheaf on Y'. Then, the pull-back $\mathcal{G}$ on Y of $\mathcal{G}'$ is S-acyclic by Lemma 1.6.5.2. Hence $(h, f) \colon W \to X \times_S Y$ is $\mathcal{F} \boxtimes \mathcal{G}$-transversal on a neighborhood of $W_{\mathbf{F}_p}$. Since $X \times_S Y \to X \times_S Y'$ is smooth, $(h, f') \colon W \to X \times_S Y'$ is also $\mathcal{F} \boxtimes \mathcal{G}'$-transversal on a neighborhood of $W_{\mathbf{F}_p}$ by Lemma 1.1.7.1 and Lemma 1.1.8. □

LEMMA 2.2.6. *Let $h \colon W \to X$ and $f \colon W \to Y$ be separated morphisms of regular schemes of finite type over S. Let*

$$\begin{array}{ccccc} X' & \xleftarrow{h'} & W' & & \\ \downarrow g & & \downarrow g' & \searrow^{f'} & \\ X & \xleftarrow{h} & W & \xrightarrow{f} & Y \end{array}$$

be a commutative diagram of schemes of finite type over S with cartesian square. Assume that X, W and X' are regular. Let $\mathcal{F}'$ be a constructible sheaf $\mathcal{F}'$ on X' and $B' \subset X'$ be the support. Assume that g is proper on B' and let $\mathcal{F} = g_*\mathcal{F}'$. Assume further that h is transversal to g on a neighborhood of $h'^{-1}(B')_{\mathbf{F}_p} \subset W'$. We consider the following conditions:

(1) *(h, f) is $\mathcal{F}$-acyclic over S.*
(2) *(h', f') is $\mathcal{F}'$-acyclic over S,*

We have the implication (2)⇒(1). If g is finite on B, then these conditions are equivalent.

PROOF. Consider the cartesian diagram

$$\begin{array}{ccc} W' & \xrightarrow{h'_Y} & X'\times_S Y\\ {\scriptstyle g'}\big\downarrow & & \big\downarrow{\scriptstyle g\times 1_Y}\\ W & \xrightarrow{h_Y} & X\times_S Y. \end{array}$$

Since h is transversal to g on a neighborhood U' of $h'^{-1}(B')_{\mathbf{F}_p} \subset W'$, the base change h_Y is transversal to $g \times 1_Y$ on U'. Hence the base change morphism $g'^* h^!_Y \Lambda \to h'^!_Y \Lambda$ is an isomorphism by Lemma 1.1.5.2. Let $\mathcal{G}$ be an S-acyclic constructible sheaf on Y. Since $g \times 1_Y$ is proper on the support of $\mathcal{F}' \boxtimes \mathcal{G}$, the $\mathcal{F}' \boxtimes \mathcal{G}$-transversality on a neighborhood of $W'_{\mathbf{F}_p}$ implies the $g_*\mathcal{F}' \boxtimes \mathcal{G}$-transversality on a neighborhood of $W_{\mathbf{F}_p}$ by Lemma 1.1.12 (2)⇒(1). Hence we have (2)⇒(1). Conversely, if g is finite on B, we have (1)⇒(2) by Lemma 1.1.12 (1)⇒(2). □

LEMMA 2.2.7. *Let*

$$\begin{array}{ccccccc} X & \xleftarrow{h} & W & \xrightarrow{f} & Y & \longrightarrow & S\\ & & & \searrow^{f_0} & \big\downarrow{\scriptstyle g} & & \big\downarrow\\ & & & & Y_0 & \longrightarrow & S_0 \end{array}$$

be a commutative diagram of separated morphisms of regular noetherian schemes satisfying the finiteness condition (F) *with cartesian square. Assume that X, W and Y are of finite type over S and that S and Y_0 are smooth over S_0. Let $\mathcal{F}$ be a constructible sheaf on X. If (h, f) is $\mathcal{F}$-acyclic over S, then (h, f_0) is $\mathcal{F}$-acyclic over S_0.*

PROOF. For an S_0-acyclic constructible sheaf $\mathcal{G}_0$ on Y_0, its pull-back $\mathcal{G}$ on Y is S-acyclic by Lemma 1.6.5.3. Hence if (h, f) is $\mathcal{F}$-acyclic over S, then the morphism $W \to X \times_S Y = X \times_{S_0} Y_0$ is $\mathcal{F} \boxtimes \mathcal{G} = \mathcal{F} \boxtimes \mathcal{G}_0$-transversal on a neighborhood of $W_{\mathbf{F}_p}$. □

2.3. *S*-micro support. Let S be a regular excellent noetherian scheme over $\mathbf{Z}_{(p)}$ satisfying the finiteness condition (F).

DEFINITION 2.3.1. *Let X be a regular scheme of finite type over S and let $\mathcal{F}$ be a constructible sheaf on X. Let $C \subset FT^*X$ be a closed conical subset. We say that $\mathcal{F}$ is S-micro supported on C if the following condition is satisfied:*

For every pair of separated morphisms $h\colon W \to X$ and $f\colon W \to Y$ of regular schemes of finite type over S such that Y is smooth over S, the C-acyclicity of (h, f) over S implies the $\mathcal{F}$-acyclicity of (h, f) over S.

We may replace the $\mathcal{F}$-acyclicity over S in the definition by the universal $\mathcal{F}$-acyclicity over S by Lemma 2.3.2 below. If $S = \operatorname{Spec} k$ for a perfect field of characteristic p, then being S-micro supported is equivalent to being micro supported by [**12**, Proposition 8.13].

LEMMA 2.3.2. *Let X be a regular scheme of finite type over S and let $\mathcal{F}$ be a constructible sheaf on X. Let $C \subset FT^*X$ be a closed conical subset and assume that $\mathcal{F}$ is S-micro supported on C.*

1. *The intersection of $X_{\mathbf{F}_p}$ with the support of $\mathcal{F}$ is a subset of the base B of C.*

2. *Let $h\colon W\to X$ and $f\colon W\to Y$ be morphisms of regular schemes of finite type over S. If (h,f) is C-acyclic over S, then (h,f) is universally $\mathcal{F}$-acyclic over S and h is universally $\mathcal{F}$-transversal over S.*

PROOF. 1. Let $W=X-B$ be the complement. Let $h\colon W\to X$ be the immersion and $f\colon W\to Y=\mathbf{A}^1_S$ be the composition of the canonical morphism $W\to S$ with the 0-section $i\colon S\to Y$ as in Lemma 2.2.3. Then, (h,f) is C-acyclic over S and hence $\mathcal{F}$-acyclic over S. Thus, we have $h^*\mathcal{F}=0$ on a neighborhood of $W_{\mathbf{F}_p}$ by Lemma 2.2.3.

2. Assume that (h,f) is C-acyclic over S. By 1 and Lemma 2.1.2.1, on a neighborhood of the intersection of $W_{\mathbf{F}_p}$ with the support of the pull-back of $\mathcal{F}$, (h,f) is universally C-acyclic over S and hence (h,f) is universally $\mathcal{F}$-acyclic over S.

Let $q\colon W\to S$ be the canonical morphism. Then, (h,q) is C-acyclic over S by Lemma 2.1.2.4. Hence (h,q) is universally $\mathcal{F}$-acyclic over S. This means that the morphism h is universally $\mathcal{F}$-transversal over S.

□

LEMMA 2.3.3. *Let X be a regular scheme of finite type over S. Let $\mathcal{F}$ be a constructible sheaf on X and $C\subset FT^*X$ be a closed conical subset.*

1. *If $\mathcal{F}$ is S-micro supported on C and if C is S-saturated, then $\mathcal{F}$ is micro supported on C.*

2. *Assume that $\mathcal{F}$ is micro supported on C. Assume further that the following condition is satisfied:*
(T) *For every smooth scheme Y over S, for every S-acyclic $C'\subset FT^*Y$ closed conical subset and for every constructible sheaf $\mathcal{G}$ on Y micro supported on C', the external tensor product $\mathcal{F}\boxtimes\mathcal{G}$ is micro supported on $\mathrm{pr}_1^\circ C+\mathrm{pr}_2^\circ C'$.*
Then $\mathcal{F}$ is S-micro supported on C.

3. *Let $S\to S_0$ be a smooth morphism of regular noetherian schemes over $\mathbf{Z}_{(p)}$. If $\mathcal{F}$ is S-micro supported on C and if C is S-saturated, then $\mathcal{F}$ is S_0-micro supported on C.*

The condition (T) in Lemma 2.3.3.2 is satisfied if S is smooth over a perfect field k of characteristic $p>0$ by By Lemma 1.6.9.

PROOF. 1. Let $h\colon W\to X$ be a separated morphism of regular schemes of finite type over S. Assume that h is C-transversal and let $q\colon W\to S$ be the canonical morphism. Then, (h,q) is C-acyclic over S by Lemma 2.1.10. Hence $h\colon W\to X=X\times_S S$ is $\mathcal{F}$-transversal on a neighborhood of $W_{\mathbf{F}_p}$.

2. Let $h\colon W\to X$ and $f\colon W\to Y$ be morphisms over S such that (h,f) is C-acyclic over S. Let $\mathcal{G}$ be an S-acyclic constructible sheaf on Y. Assume that $\mathcal{G}$ is micro supported on an S-acyclic closed conical subset $C'\subset FT^*Y$. By assumption, the tensor product $\mathcal{F}\boxtimes\mathcal{G}$ on $X\times_S Y$ is micro supported on $\mathrm{pr}_1^\circ C+\mathrm{pr}_2^\circ C'$. By Lemma 2.1.2.5, the morphism $h\colon W\to X\times_S Y$ is $\mathrm{pr}_1^\circ C+\mathrm{pr}_2^\circ C'$-transversal. Hence h is $\mathcal{F}\boxtimes\mathcal{G}$-transversal on a neighborhood of $W_{\mathbf{F}_p}$.

3. Let $h\colon W\to X$ and $f_0\colon W\to Y_0$ be morphisms over S_0. Assume that $Y_0\to S_0$ is smooth and that (h,f_0) is C-acyclic over S_0. Consider W as a scheme over S by the composition $W\to X\to S$ and let $f\colon W\to Y=Y_0\times_{S_0}S$ be the induced morphism over S. Then, (h,f) is C-acyclic over S by Lemma 2.1.11 and hence (h,f) is universally $\mathcal{F}$-acyclic over S. By Lemma 2.2.7, (h,f_0) is universally $\mathcal{F}$-acyclic over S_0. □

LEMMA 2.3.4. *Let X be a regular scheme of finite type over S.*

1. *Every constructible sheaf $\mathcal{F}$ on X is S-micro supported on FT^*X.*

2. *Every locally constant constructible sheaf $\mathcal{F}$ on X is S-micro supported on the 0-section FT^*_XX.*

3. *Let $\mathcal{F}$ be a constructible sheaf on X. Assume $\mathcal{F}$ is S-micro supported on the 0-section and that $FT^*S \times_S X \to FT^*X$ is the 0-morphism. Then $\mathcal{F}$ is locally constant on a neighborhood of $X_{\mathbf{F}_p}$.*

4. *Assume $X = S$. Then every constructible sheaf on S is S-micro supported on the 0-section $FT^*_SS \subset FT^*S$.*

Lemma 2.3.4.3 and 4. imply that the notion of S-micro support can be a reasonable one only for S-saturated conical closed subset.

PROOF. 1. Assume that (h, f) is FT^*X-acyclic over S. Then, by Lemma 2.1.4.1, $(h, f): W \to X \times_S Y$ is smooth on a neighborhood of $W_{\mathbf{F}_p}$. Hence for every constructible sheaf $\mathcal{F}$ on X and every constructible sheaf $\mathcal{G}$ on Y', the pair (h, f) is $\mathcal{F} \boxtimes \mathcal{G}$-transversal on a neighborhood of $W_{\mathbf{F}_p}$ by Lemma 1.1.7.1.

2. We may assume $\mathcal{F} = \Lambda$. Assume that (h, f) is FT^*_XX-acyclic over S. Then the base change $f: W \to Y$ is smooth modulo S by Lemma 2.1.2.3 and $\mathrm{pr}_2: X \times_S Y \to Y$ is smooth modulo S by Lemma 1.6.7.2. Hence for every S-acyclic sheaf $\mathcal{G}$ on Y, on a neighborhood of $W_{\mathbf{F}_p}$, the morphism f is $\mathcal{G}$-transversal and $\mathrm{pr}_2: X \times_S Y \to Y$ is also $\mathcal{G}$-transversal by Lemma 1.6.8.2. Hence $W \to X \times_S Y$ is $\Lambda \boxtimes \mathcal{G}$-transversal on a neighborhood of $W_{\mathbf{F}_p}$ by Lemma 1.1.8.

3. Assume that $FT^*S \times_S X \to FT^*X$ is the 0-morphism. By Lemma 2.1.4.2, for every morphism $h: W \to X$ of regular schemes of finite type and the canonical morphism $q: W \to S$, the pair (h, q) is C-acyclic over S for the 0-section $C = FT^*_XX$. Hence $h: W \to X = X \times_S S$ is $\mathcal{F} = \mathcal{F} \boxtimes \Lambda$-transversal on a neighborhood of $W_{\mathbf{F}_p}$. This implies that $\mathcal{F}$ is locally constant on a neighborhood of $X_{\mathbf{F}_p}$ by Proposition 1.1.11.

4. For $C = FT^*_SS$, we have $C \subset FT^*S \subset C + FT^*S$. Hence the C-acyclicity over S is equivalent to the FT^*S-acyclicity over S. Thus it follows from 1. □

LEMMA 2.3.5.

1. *Let $g: X' \to X$ be a separated morphism of regular schemes of finite type over S and $q: X' \to S$ be the canonical morphism. Let $\mathcal{F}$ be a constructible sheaf on X. If $\mathcal{F}$ is S-micro supported on C and if (g, q) is C-acyclic over S, then $g^*\mathcal{F}$ is S-micro supported on $g^\circ C$.*

2. *Let $g: X' \to X$ be a morphism of regular schemes of finite type over S. Assume that g is proper on the support of $C' \subset FT^*X'$. Let $\mathcal{F}'$ be a constructible sheaf on X'. If $\mathcal{F}'$ is S-micro supported on C', then $Rg_*\mathcal{F}'$ is S-micro supported on $C = g_\circ C'$.*

PROOF. 1. Let $h: W \to X'$ and $f: W \to Y$ be separated morphisms of regular schemes of finite type over S and assume that Y is smooth over S. Assume that (h, f) is $g^\circ C$-acyclic over S. Then by Lemma 2.1.5, (gh, f) is C-acyclic over S and is $\mathcal{F}$-acyclic over S. Since (g, q) is C-acyclic over S, the morphism g is universally $\mathcal{F}$-transversal over S by Lemma 2.3.2.2. Hence (h, f) is $g^*\mathcal{F}$-acyclic over S by Lemma 2.2.4.

2. Let $h: W \to X$ and $f: W \to Y$ be separated morphisms of regular schemes of finite type over S such that Y is smooth over S and (h, f) is C-acyclic over S.

Let

$$\begin{array}{ccc} X' & \xleftarrow{h'} & W' \\ {\scriptstyle g}\downarrow & & \downarrow{\scriptstyle g'} \\ X & \xleftarrow{h} & W \end{array}$$

be a cartesian diagram. Then, by Lemma 2.1.7, there exists an open regular neighborhood $U' \subset W' = W \times_X X'$ of the inverse image of B' where h is transversal to g and (h', fg') is C'-acyclic over S. Since B' contains the intersection of the support of $\mathcal{F}'$ and $X'_{\mathbf{F}_p}$, (h', fg') is $\mathcal{F}'$-acyclic over S on a neighborhood of $U'_{\mathbf{F}_p}$. Hence the assertion follows from Lemma 2.2.6.

□

LEMMA 2.3.6. *Let $i\colon X \to P$ be a closed immersion of regular schemes of finite type over S and let $\mathcal{F}$ be a constructible sheaf on X. Let C_P be a closed conical subset of $FT^*P|_X \subset FT^*P$ and define $C \subset FT^*X$ to be the closure of the image of C_P by the surjection $FT^*P|_X \to FT^*X$. If $i_*\mathcal{F}$ is S-micro supported on C_P, then $\mathcal{F}$ is S-micro supported on C.*

PROOF. Let $h\colon W \to X$ and $f\colon W \to Y$ be separated morphisms of regular schemes of finite type over S and assume that (h, f) is C-acyclic over S. We show that (h, f) is $\mathcal{F}$-acyclic over S. By the same argument as in the proof of Lemma 1.4.6, we may assume that h is an immersion. Then $(h, f)\colon W \to X \times_S Y$ is also an immersion.

Since the assertion is local on W, by taking a lifting of a local basis of the conormal bundle $T^*_W(X \times_S Y)$, we may extend the immersion (h, f) to an immersion $(g, q)\colon V \to P \times_S Y$ of regular closed subscheme transversal with the immersion $X \times_S Y \to P \times_S Y$. Then, the C-acyclicity over S of (h, f) implies the C_P-acyclicity over S of (g, q). Since $i_*\mathcal{F}$ is S-micro supported on C_P, this implies that (g, q) is $i_*\mathcal{F}$-acyclic over S. Since the intersection $W = V \cap (X \times_S Y)$ is transversal, if $i'\colon W \to V$ denote the immersion, the morphism $i'^* g^! \Lambda \to h^! \Lambda$ is an isomorphism. Hence by Lemma 2.2.6 (1)⇒(2), the $i_*\mathcal{F}$-acyclicity over S of (g, q) implies the $\mathcal{F}$-acyclicity over S of (h, f). Hence $\mathcal{F}$ is S-micro supported on C. □

LEMMA 2.3.7. *Let X be a regular schemes of finite type over S. Let C be a closed conical subset of FT^*X and $\mathcal{F}$ be a constructible sheaf on X.*

1. *Let $X = \bigcup U_i$ be an open covering of X. Then the following conditions are equivalent:*
   (1) *$\mathcal{F}$ is S-micro supported on C.*
   (2) *$\mathcal{F}|_{U_i}$ is S-micro supported on $C|_{U_i}$ for every $i \in I$.*
2. *Let $U \subset X$ be an open subscheme and let $Z = X - U$ be the complement. Let $C' \subset FT^*U$ be a closed subset. Assume that $\mathcal{F}$ is S-micro supported on C and that $\mathcal{F}|_U$ is S-micro supported on C'. Assume further that C and C' are S-saturated. Then, $\mathcal{F}$ is S-micro supported on the union $C_1 = \overline{C'} \cup C|_Z$.*

PROOF. The same argument as in the proof of Lemma 1.4.4 works using Lemma 2.1.9. □

For an S-saturated closed conical subset on a smooth scheme over S, the S-micro supported condition can be rephrased as follows.

LEMMA 2.3.8. *Let X be a smooth scheme over S and $\mathcal{F}$ be a constructible sheaf on X. Let $C \subset FT^*X$ be an S-saturated closed conical subset and let $\overline{C} \subset T^*X/S|_{X_{\mathbf{F}_p}}$ be the corresponding closed conical subset. Then, the following conditions are equivalent:*

(1) *$\mathcal{F}$ is S-micro supported on C.*

(2) *For every pair of morphisms $h\colon W \to X$ and $f\colon W \to Y$ of smooth separated schemes of finite type over S such that the morphism $(T^*X/S \times_X W) \times_W (T^*Y/S \times_Y W) \to T^*W/S$ is $h^*\overline{C} \times_W (T^*Y/S \times_Y W)|_{W_{\mathbf{F}_p}}$-transversal, the pair (h, f) is $\mathcal{F}$-acyclic over S.*

PROOF. By Lemma 2.1.2.3, it suffices to consider pairs (h, f) such that W is smooth over S. Hence the assertion follows from Lemma 2.1.13. □

LEMMA 2.3.9. *Let k be a field of characteristic $p > 0$ and X be a smooth scheme over $S = \operatorname{Spec} k$. Let $\mathcal{F}$ be a constructible sheaf on X. Let $C \subset FT^*X$ be an S-saturated closed conical subset and $\overline{C} \subset T^*X/k$ be the corresponding closed conical subset. If $\mathcal{F}$ is micro supported on $\overline{C} \subset T^*X/k$ in the sense of* [**2**], *then $\mathcal{F}$ is S-micro supported on C.*

PROOF. By Lemma 2.3.8, the assertion follows from Lemma 2.2.2. □

The converse of Lemma 2.3.9 does not hold if k is imperfect. For a concrete example, see Remark after Definition 2.4.1.

2.4. S-singular support. Let S be a regular excellent noetherian scheme over $\mathbf{Z}_{(p)}$ satisfying the finiteness condition (F).

DEFINITION 2.4.1. *Let X be a regular scheme of finite type over S and let $\mathcal{F}$ be a constructible sheaf on X.*

1. *We say that a closed conical subset (resp. an S-saturated closed conical subset) $C \subset FT^*X$ is the S-singular support $SS_S\mathcal{F}$ (resp. the S-saturated singular support $SS_S^{\rm sat}\mathcal{F}$) of $\mathcal{F}$ if for every closed conical subset (resp. every S-saturated closed conical subset) $C' \subset FT^*X$, the inclusion $C \subset C'$ is equivalent to the condition that $\mathcal{F}$ is S-micro supported on C'.*

2. *If X is smooth over S, we call the closed conical subset $\overline{SS}_S\mathcal{F} \subset T^*X/S|_{X_{\mathbf{F}_p}}$ corresponding to $SS_S^{\rm sat}\mathcal{F} \subset FT^*X$ the S-relative singular support.*

If $S = \operatorname{Spec} k$ for a perfect field of characteristic p, the S-singular support and the S-saturated singular support is the same as the singular support. If k' is not perfect, the k'-relative singular support may be different from that defined by Beilinson [**2**] or by Hu–Yang [**9**]. By [**2**, Theorem 1.4 (iii)], the singular support defined by Beilinson commutes with base change. In this sense, Beilinson's singular support is essentially an invariant for smooth schemes over a perfect field. On the other hand, the relative singular support defined in Definition 2.4.1 does not commute with base change in general.

Example. Let k be a perfect field and $k' = k(y)$ be the rational function field with one variable. Let $\mathcal{F}$ be the sheaf on $X = \mathbf{A}^1_{k'} = \operatorname{Spec} k'[x]$ defined by $t^p - t = y/x^p$. Then, the singular support $SS\mathcal{F} \subset FT^*X$ is the union of the 0-section and the line spanned by dy at the origin, the k'-singular support $SS_{k'}\mathcal{F} \subset FT^*X$ and the k'-relative singular support $\overline{SS}_{k'}\mathcal{F} \subset T^*X/k'$ are the 0-sections. On the other hand, the singular support defined by Beilinson in [**2**] is the union of the 0-section and the line spanned by dx at the origin.

We don't know the existence of $SS_S\mathcal{F}$ in general since the condition that $\mathcal{F}$ is S-micro supported on C and C' does not a priori imply that $\mathcal{F}$ is S-micro supported on $C \cap C'$.

LEMMA 2.4.2. *Assume that $SS\mathcal{F}$, $SS_S\mathcal{F}$ and $SS_S^{\rm sat}\mathcal{F}$ exist.*

1. *We have*

$$SS_S\mathcal{F} \subset SS_S^{\rm sat}\mathcal{F}$$

and the equality holds if the morphism $FT^*S \times_S X \to FT^*X$ is 0.

2. *We have*

$$SS\mathcal{F} \subset SS_S^{\rm sat}\mathcal{F}.$$

3. *If the condition* (T) *in Lemma* 2.3.3.*2 is satisfied, we have*

$$SS_S\mathcal{F} \subset SS\mathcal{F}.$$

In particular, this inclusion is satisfied if S is a smooth scheme over a perfect field k of characteristic $p > 0$.

PROOF. 1. Clear.

2. This follows from Lemma 2.3.3.1.

3. The inclusion follows from Lemma 2.3.3.2. In the last case, condition (T) is satisfied by Lemma 1.6.9.

□

We reduce the proof of the existence of $SS_S^{\rm sat}\mathcal{F}$ on a regular scheme X of finite type over S to the case where X is a projective space $\mathbf{P}_S$ over S as in Corollary 1.5.4.

LEMMA 2.4.3. *Let X be a regular scheme of finite type over S and let $\mathcal{F}$ be a constructible sheaf on X.*

1. *Let $U \subset X$ be an open subscheme. Assume that $C \subset FT^*X$ is the S-saturated singular support of $\mathcal{F}$. Then, $C|_U$ is the S-saturated singular support of $\mathcal{F}|_U$.*

2. *Let (U_i) be an open covering of X and $C_i \subset FT^*U_i$ be the S-saturated singular support of $\mathcal{F}|_{U_i}$. Then, $C = \bigcup_i C_i \subset FT^*X$ is the S-saturated singular support of $\mathcal{F}$.*

PROOF. The same argument as in the proof of Lemma 1.5.2 works using Lemma 2.3.7. □

PROPOSITION 2.4.4. *Let $i\colon X \to P$ be a closed immersion of regular schemes of finite type over S and let $\mathcal{F}$ be a constructible sheaf on X. Let $C_P \subset FT^*P$ be a closed conical subset and assume that C_P is the S-saturated singular support of $i_*\mathcal{F}$. Then the following holds.*

1. *C_P is a subset of $FT^*P|_X$.*

2. *Define $C \subset FT^*X$ to be the closure of the image of C_P by the surjection $FT^*P|_X \to FT^*X$. Then C is the S-saturated singular support $SS_S^{\rm sat}\mathcal{F}$.*

PROOF. 1. The same argument as in the proof of Proposition 1.5.3 works using Lemma 2.3.7.

2. By Lemma 2.3.6, $\mathcal{F}$ is S-micro supported on C. Since C_P is S-saturated, C is also S-saturated. We show that C is the smallest. Assume that $\mathcal{F}$ is S-micro supported on an S-saturated closed conical subset $C' \subset FT^*X$. Then, since $i_*\mathcal{F}$ is

S-micro supported on $i_\circ C'$ by Lemma 2.3.5.2, we have $C_P \subset i_\circ C'$. Thus the image of C_P by the surjection $FT^*P|_X \to FT^*X$ is a subset of C' and we obtain $C \subset C'$. □

We will prove the existence of $SS_S^{\rm sat}\mathcal{F}$ for the projective space $\mathbf{P}_S$ over S using the Radon transform, following Beilinson's argument in [**2**]. Let S be a scheme and $V = \mathbf{A}_S^{n+1}$ be the vector bundle of rank $n+1$. Let $\mathbf{P} = \mathbf{P}_S = \mathbf{P}(V)$ be the projective space bundle of dimension n parametrizing lines in V. The dual projective space $\mathbf{P}^\vee = \mathbf{P}(V^\vee)$ is the moduli space of hyperplanes in $\mathbf{P}$.

By the exact sequence $0 \to \Omega^1_{\mathbf{P}/S}(1) \to \mathcal{O}_{\mathbf{P}} \otimes V^\vee \to \mathcal{O}_{\mathbf{P}}(1) \to 0$ of locally free $\mathcal{O}_{\mathbf{P}}$-modules, we define a closed subscheme $Q = \mathbf{P}(T^*\mathbf{P}/S) \subset \mathbf{P} \times_S \mathbf{P}(V^\vee) = \mathbf{P} \times_S \mathbf{P}^\vee$ of codimension 1. This equals the universal family of hyperplanes since it is defined by the tautological section $\Gamma(\mathbf{P}\times_S\mathbf{P}^\vee, \mathcal{O}(1)\boxtimes\mathcal{O}(1)) = V^\vee \otimes V$ corresponding to the identity $1 \in \operatorname{End}(V)$. Let $q\colon Q \to \mathbf{P}$ and $q^\vee\colon Q \to \mathbf{P}^\vee$ be the restrictions of the projections $\mathbf{P} \times_S \mathbf{P}^\vee \to \mathbf{P}$ and $\mathbf{P} \times_S \mathbf{P}^\vee \to \mathbf{P}^\vee$. By symmetry, $Q \subset \mathbf{P} \times_S \mathbf{P}^\vee$ is identified with $\mathbf{P}(T^*\mathbf{P}^\vee/S)$.

The conormal bundle $L_Q = T^*_Q(\mathbf{P} \times_S \mathbf{P}^\vee) \subset (T^*\mathbf{P}/S \times_S T^*\mathbf{P}^\vee/S)|_Q$ is a line bundle. Since $1 \in \operatorname{End}(V) = V^\vee \otimes V$ regarded as a global section of $\mathcal{O}(1) \boxtimes \mathcal{O}(1)$ is the bilinear form defining $Q \subset \mathbf{P} \times_S \mathbf{P}^\vee$, the morphism $N_{Q/(\mathbf{P}\times_S\mathbf{P}^\vee)} \to \Omega^1_{(\mathbf{P}\times_S\mathbf{P}^\vee)/\mathbf{P}^\vee} \otimes_{\mathcal{O}_{\mathbf{P}\times_S\mathbf{P}^\vee}} \mathcal{O}_Q = \Omega^1_{\mathbf{P}/S} \otimes_{\mathcal{O}_{\mathbf{P}}} \mathcal{O}_Q$ defines a tautological sub invertible sheaf on $Q = \mathbf{P}(T^*\mathbf{P}/S)$. In other words, the tautological sub line bundle $L \subset T^*\mathbf{P}/S \times_{\mathbf{P}} Q$ is the image of L_Q by the first projection $\mathrm{pr}_1\colon (T^*\mathbf{P}/S \times_S T^*\mathbf{P}^\vee/S)|_Q \to T^*\mathbf{P}/S \times_{\mathbf{P}} Q$. By symmetry, the image by the second projection equals the tautological sub line bundle $L^\vee$ on $Q = \mathbf{P}(T^*\mathbf{P}^\vee/S)$.

Since the conormal bundle L_Q is the kernel of the surjection $(T^*\mathbf{P}/S \times_S T^*\mathbf{P}^\vee/S)|_Q \to T^*Q$, the intersection $q^\circ T^*\mathbf{P}/S \cap q^{\vee\circ}T^*\mathbf{P}^\vee/S \subset T^*Q$ equals the image of the tautological bundle $L \subset T^*\mathbf{P}/S \times_{\mathbf{P}} Q$. By symmetry, the intersection also equals the image of the tautological bundle $L^\vee \subset T^*\mathbf{P}^\vee/S \times_{\mathbf{P}^\vee} Q$.

Let $C \subset T^*\mathbf{P}/S$ denote a closed conical subset. We define the Legendre transform $C^\vee \subset T^*\mathbf{P}^\vee/S$ to be $q^\vee_\circ q^\circ C$. We consider projectivizations $\mathbf{P}(C) \subset \mathbf{P}(T^*\mathbf{P}/S)$ and $\mathbf{P}(C^\vee) \subset \mathbf{P}(T^*\mathbf{P}^\vee/S)$ as closed subsets of Q. Let $C^+ \subset T^*\mathbf{P}/S$ denote the union of C and the 0-section.

In the following, for a smooth scheme X over S and a closed conical subset $C \subset T^*X/S$, we say that a constructible sheaf $\mathcal{F}$ on X is micro supported on C, if it is so on the corresponding S-saturated closed conical subset of FT^*X, and similarly for other properties by abuse of terminology.

PROPOSITION 2.4.5. *Let $C \subset T^*\mathbf{P}/S$ be a closed conical subset. Let $E = \mathbf{P}(C) \subset Q = \mathbf{P}(T^*\mathbf{P}/S)$ be the projectivization. Let $L_Q = T^*_Q(\mathbf{P}\times_S\mathbf{P}^\vee) \subset (T^*\mathbf{P}\times_S T^*\mathbf{P}^\vee)|_Q$ be the conormal line bundle.*

1. *The projectivization $E = \mathbf{P}(C) \subset Q$ is the complement of the largest open subset where $(q, q^\vee)$ is C-acyclic.*
2. *The Legendre transform $C^\vee$ equals the union of the image of $L|_E \subset q^\circ T^*\mathbf{P} \cap q^{\vee\circ}T^*\mathbf{P}^\vee \subset T^*Q$ and its base. We have $\mathbf{P}(C) = \mathbf{P}(C^\vee)$.*
3. *We have $C^{\vee\vee} \subset C^+$.*

PROOF. 1. The kernel $\operatorname{Ker}((T^*\mathbf{P}/S \times_S T^*\mathbf{P}^\vee/S)|_Q \to T^*Q)$ equals the conormal bundle L_Q and the first projection induces an isomorphism $L_Q \to L \subset T^*\mathbf{P}/S \times_{\mathbf{P}} Q$ to the tautological bundle. By this isomorphism, the intersection

$(C \times_S T^*\mathbf{P}^\vee/S)|_Q \cap L_Q$ is identified with $q^*C \cap L$. Hence $(q, q^\vee)$ is C-acyclic on $U \subset Q$ if and only if the restriction $(q^*C \cap L)|_U$ is a subset of the 0-section. Since the projectivization $E = \mathbf{P}(C) \subset \mathbf{P}(T^*\mathbf{P}/S)$ equals $\mathbf{P}(q^*C \cap L) \subset \mathbf{P}(L) = Q$, the assertion follows.

2. The Legendre transform $C^\vee$ is the image of the intersection

$$q^\circ C \cap q^{\vee\circ} T^*\mathbf{P}^\vee/S \subset T^*Q$$

by $q^{\vee\circ}T^*\mathbf{P}^\vee/S \to T^*\mathbf{P}^\vee/S$. We identify the intersection $q^\circ T^*\mathbf{P}/S \cap q^{\vee\circ}T^*\mathbf{P}^\vee/S \subset T^*Q$ with the tautological line bundle $L \subset T^*\mathbf{P}/S \times_{\mathbf{P}} Q$. Then, the intersection $q^\circ C \cap q^{\vee\circ}T^*\mathbf{P}^\vee/S \subset T^*Q$ is identified with $q^*C \cap L$. Since the projectivization $E = \mathbf{P}(C) \subset \mathbf{P}(T^*\mathbf{P}/S)$ equals $\mathbf{P}(q^*C \cap L) \subset \mathbf{P}(L) = Q$, the closed conical subset $q^*C \cap L$ equals $L|_E$ up to the base.

Since $L|_E$ is identified with $L^\vee|_E$ inside T^*Q, we have $\mathbf{P}(C^\vee) = \mathbf{P}(L^\vee|_E) = E \subset \mathbf{P}(T^*\mathbf{P}^\vee/S) = Q$.

3. By 2 and symmetry, we have $\mathbf{P}(C) = \mathbf{P}(C^\vee) = \mathbf{P}(C^{\vee\vee})$. Hence we have $C^{\vee\vee} \subset C^+$.

□

LEMMA 2.4.6. *Let $C \subset T^*\mathbf{P}/S$ be a closed conical subset and $E = \mathbf{P}(C) \subset Q = \mathbf{P}(T^*\mathbf{P}/S)$ be its projectivization. Let*

$$\begin{array}{ccccc}
\mathbf{P} & \xleftarrow{\ q\ } & Q & \xleftarrow{\ h'\ } & Q_W & & \\
 & & \downarrow{\scriptstyle q^\vee} & & \downarrow{\scriptstyle q^\vee_W} & \searrow{\scriptstyle f'} & \\
 & & \mathbf{P}^\vee & \xleftarrow{\ h\ } & W & \xrightarrow{\ f\ } & Y
\end{array}$$

be a commutative diagram of smooth schemes over S with cartesian square. Let $C^{\vee+} = C^\vee \cup T^*_{\mathbf{P}^\vee}\mathbf{P}^\vee/S \subset T^*\mathbf{P}^\vee/S$ be the union of the Legendre transform with the 0-section and suppose that (h, f) is $C^{\vee+}$-acyclic.

1. *The morphism f is smooth on an open neighborhood of $W_{\mathbf{F}_p}$ and the pair (qh', f') is C-acyclic on a neighborhood of $(Q_W - E_W)_{\mathbf{F}_p}$.*
2. *On a neighborhood of $E_{W,\mathbf{F}_p}$, the pair (qh', f') is $T^*\mathbf{P}/S$-acyclic.*

PROOF. 1. Since (h, f) is $C^{\vee+}$-acyclic over S and $C^{\vee+}$ contains the 0-section, the morphism $f\colon W \to Y$ is smooth. By Proposition 2.4.5.1, $(q, q^\vee)$ is C-acyclic over S outside E. Hence $(qh', q^\vee_W)$ is C-acyclic over S outside E_W by Lemma 2.1.2.1 and (qh', f') is C-acyclic over S outside E_W by Lemma 2.1.6.

2. By the description of $C^\vee$ in Proposition 2.4.5.2 and by the open condition Lemma 2.1.9, the $C^\vee$-acyclicity over S of (h, f) implies the $T^*\mathbf{P}$-acyclicity over S of (qh', f') on a neighborhood U of $E_W \subset Q_W$.

□

We define the naive Radon transform $R\mathcal{F}$ to be $q^\vee_* q^*\mathcal{F}$ and the naive inverse Radon transform $R^\vee\mathcal{G}$ to be $q_*q^{\vee*}\mathcal{G}$. Since we consider only the naive Radon transform, we drop the adjective 'naive' in the sequel. We will refine in Proposition 2.4.10, after studying the difference between $R^\vee R\mathcal{F}$ and $\mathcal{F}$, the following elementary property.

LEMMA 2.4.7. *Assume that $\mathcal{F}$ is S-micro supported on C.*

1. *The Radon transform $R\mathcal{F}$ is S-micro supported on $C^\vee$.*
2. *$R^\vee R\mathcal{F}$ is S-micro supported on C^+.*

PROOF. 1. By Lemma 2.3.5, the Radon transform $R\mathcal{F} = q^\vee_* q^* \mathcal{F}$ is S-micro supported on the Legendre transform $C^\vee = q^\vee_\circ q^\circ C$.

2. By 1, $R^\vee R\mathcal{F}$ is S-micro supported on $C^{\vee\vee} \subset C^+$.

□

LEMMA 2.4.8. *We consider the commutative diagram*

$$\begin{array}{ccccc} Q \times_{\mathbf{P}^\vee} Q & \xrightarrow{\ i\ } & \mathbf{P} \times_S \mathbf{P}^\vee \times_S \mathbf{P} & & \\ & \searrow^{q \times q} & \downarrow \mathrm{pr}_{13} & & \\ & & \mathbf{P} \times_S \mathbf{P} & \xleftarrow{\ \delta_{\mathbf{P}}\ } & \mathbf{P} \end{array}$$

where $\delta_{\mathbf{P}} \colon \mathbf{P} \to \mathbf{P} \times_S \mathbf{P}$ is the diagonal immersion.

1. *The closed immersion $i = ((q, q^\vee), (q^\vee, q))$ induces isomorphisms*

(2.4)
$$R^s(q \times q)_* \Lambda_{Q \times_{\mathbf{P}^\vee} Q} \to \begin{cases} \Lambda(-t)[-2t] & \text{if } s = 2t \text{ and } 0 \leqslant t \leqslant n-2, \\ \delta_{\mathbf{P}*}\Lambda(-(n-1))[-2(n-1)] & \text{if } s = 2(n-1), \\ 0 & \text{if otherwise.} \end{cases}$$

2. *Let $p \colon \mathbf{P} \to S$ and $p^\vee \colon \mathbf{P}^\vee \to S$ denote the projections. Then, we have a distinguished triangle*

(2.5) $$\to \tau_{\leqq 2(n-2)}(p \times p)^* p^\vee_* \Lambda \to (q \times q)_* \Lambda_{Q \times_{\mathbf{P}^\vee} Q} \to \delta_{\mathbf{P}*}\Lambda_{\mathbf{P}}(n-1)[2(n-1)] \to .$$

PROOF. 1. The immersion i induces a morphism

(2.6) $$(p \times p)^* p^\vee_* \Lambda = \mathrm{pr}_{13*}\Lambda \to (q \times q)_* \Lambda_{Q \times_{\mathbf{P}^\vee} Q}$$

and we have isomorphisms $R^s p^\vee_* \Lambda \to \Lambda(-t)[-2t]$ for $s = 2t, 0 \leqslant t \leqslant n$ and $R^s p^\vee_* \Lambda = 0$ otherwise. The restriction of the closed immersion $i \colon Q \times_{\mathbf{P}^\vee} Q \to \mathbf{P} \times_S \mathbf{P}^\vee \times_S \mathbf{P}$ on the diagonal $\mathbf{P} \subset \mathbf{P} \times_S \mathbf{P}$ is the sub $\mathbf{P}^{n-1}$-bundle $Q \subset \mathbf{P} \times_S \mathbf{P}^\vee$. On the complement $\mathbf{P} \times_S \mathbf{P} - \mathbf{P}$, it is a sub $\mathbf{P}^{n-2}$-bundle. Hence (2.6) induces an isomorphism of cohomology sheaves except for degree s and induces an isomorphism $\delta_{\mathbf{P}*} p^* R^{2(n-1)} p^\vee_* \Lambda \to R^{2(n-1)}(q \times q)_* \Lambda_{Q \times_{\mathbf{P}^\vee} Q}$.

2. This follows from the isomorphisms (2.4).

□

Next, we consider the diagram

(2.7)
$$\begin{array}{ccccc} \mathbf{P} & \xleftarrow{\ \mathrm{pr}_1\ } & \mathbf{P} \times_S \mathbf{P} & \xleftarrow{\ q \times q\ } & Q \times_{\mathbf{P}^\vee} Q \\ & & \downarrow \mathrm{pr}_2 & & \\ & & \mathbf{P}. & & \end{array}$$

PROPOSITION 2.4.9. *We have a canonical isomorphism*

(2.8) $$R^\vee R\mathcal{F} \to R\mathrm{pr}_{2*}\big(\mathrm{pr}_1^* \mathcal{F} \otimes R(q \times q)_* \Lambda_{Q \times_{\mathbf{P}^\vee} Q}\big).$$

1. *The isomorphism (2.8) induces a distinguished triangle*

$$\to \bigoplus_{q=0}^{n-2} p^* p_* \mathcal{F}(-q)[-2q] \to R^\vee R\mathcal{F} \to \mathcal{F}(-(n-1))[-2(n-1)] \to .$$

PROOF. 1. By the cartesian diagram

$$\begin{array}{ccccc}
\mathbf{P} & \xleftarrow{q} & Q & \xleftarrow{\mathrm{pr}_1} & Q \times_{\mathbf{P}^\vee} Q \\
 & & {\scriptstyle q^\vee}\downarrow & & \downarrow{\scriptstyle \mathrm{pr}_2} \\
 & & \mathbf{P}^\vee & \xleftarrow{q^\vee} & Q \\
 & & & & \downarrow{\scriptstyle q} \\
 & & & & \mathbf{P},
\end{array}$$

we have $R^\vee R\mathcal{F} = Rq_* q^{\vee *} Rq^\vee_* q^* \mathcal{F}$. By the proper base change theorem, we have a canonical isomorphism $Rq_* q^{\vee *} Rq^\vee_* q^* \mathcal{F} \to R(q \circ \mathrm{pr}_2)_* (q \circ \mathrm{pr}_1)^* \mathcal{F}$. In the notation of (2.7), the latter is identified with $R(\mathrm{pr}_2 \circ (q \times q))_* (\mathrm{pr}_1 \circ (q \times q))^* \mathcal{F}$. This is identified with $R\mathrm{pr}_{2*}(\mathrm{pr}_{1*}\mathcal{F} \otimes R(q \times q)_* \Lambda_{Q \times_{\mathbf{P}^\vee} Q})$ by the projection formula.

2. This follows from the isomorphism (2.8) and the distinguished triangle (2.5). □

PROPOSITION 2.4.10. *For a constructible sheaf $\mathcal{F}$ on $\mathbf{P}$ and a closed conical subset $C \subset T^*\mathbf{P}$, we have implications* (1)⇒(2)⇒(3)⇒(4).

(1) *$\mathcal{F}$ is S-micro supported on C.*
(2) *The pair $(q, q^\vee)$ is universally $\mathcal{F}$-acyclic over S outside $E = \mathbf{P}(C)$.*
(3) *The Radon transform $R\mathcal{F}$ is S-micro supported on $C^{\vee+}$.*
(4) *$\mathcal{F}$ is S-micro supported on C^+.*

PROOF. (1)⇒(2): The pair $(q, q^\vee)$ of $q\colon Q \to \mathbf{P}$ and $q^\vee : Q \to \mathbf{P}^\vee$ is C-acyclic over S outside $E = \mathbf{P}(C)$ by Proposition 2.4.5.1. Hence (1) implies that $(q, q^\vee)$ is universally $\mathcal{F}$-acyclic over S on the complement $Q - E$ by Lemma 2.3.2.2.

(2)⇒(3): Assume that a pair of morphisms $h\colon W \to \mathbf{P}^\vee, f\colon W \to Y$ is $C^{\vee+}$-acyclic over S and show that (h, f) is $R\mathcal{F}$-acyclic over S. We consider the commutative diagram

$$\begin{array}{ccccccc}
\mathbf{P} & \xleftarrow{q} & Q & \xleftarrow{h'} & Q_W & & \\
 & & {\scriptstyle q^\vee}\downarrow & & {\scriptstyle q^\vee_W}\downarrow & \searrow{\scriptstyle f'} & \\
 & & \mathbf{P}^\vee & \xleftarrow{h} & W & \xrightarrow{f} & Y
\end{array}$$

with cartesian square as in Lemma 2.4.6. Since $q^\vee$ is proper, it suffices to show that (h', f') is $q^*\mathcal{F}$-acyclic over S by Lemma 2.2.6. Since q is smooth, it suffices to show that (qh', f') is $\mathcal{F}$-acyclic over S by Lemma 2.2.4 (1)⇒(2).

By (2), $(qh', q^\vee_W)$ is $\mathcal{F}$-acyclic over S on a neighborhood of $(Q_W - E_W)_{\mathbf{F}_p}$. By Lemma 2.4.6.1, the morphism $f\colon W \to Y$ is smooth on an open neighborhood of $W_{\mathbf{F}_p}$. Hence (qh', f') is $\mathcal{F}$-acyclic over S on a neighborhood of $(Q_W - E_W)_{\mathbf{F}_p}$ by Lemma 2.2.5.

By Lemma 2.4.6.2, the pair (qh', f') is $T^*\mathbf{P}$-acyclic over S on a neighborhood U of $E_W \subset Q_W$. Since $\mathcal{F}$ is S-micro supported on $FT^*\mathbf{P}$ by Lemma 2.3.4.1, (qh', f') is $\mathcal{F}$-acyclic over S on a neighborhood of $E_{W,\mathbf{F}_p}$. Thus (qh', f') is $\mathcal{F}$-acyclic over S on a neighborhood of $Q_{W,\mathbf{F}_p}$ as required.

(3)⇒(4) By (3) and (1)⇒(3), $R^\vee R\mathcal{F}$ is S-micro supported on $(C^{\vee+})^{\vee+} = C^+$. Since $p^*p_*\mathcal{F}$ is S-micro supported on the $FT^*S \times_S \mathbf{P}$, by the distinguished triangle in Proposition 2.4.9.2, $\mathcal{F}$ is also S-micro supported on C^+. □

COROLLARY 2.4.11. *Let $\mathcal{F}$ be a constructible sheaf on $\mathbf{P}$. Let $E \subset Q = \mathbf{P}(T^*\mathbf{P})$ be the complement of the largest open subset on which $(q, q^\vee)$ is universally $\mathcal{F}$-acyclic over S. Then the closed conical subset $C \subset FT^*\mathbf{P}$ corresponding to the base $B = \operatorname{supp} \mathcal{F} \cap \mathbf{P}_{\mathbf{F}_p}$ and $E \cap Q_{\mathbf{F}_p} \subset \mathbf{P}(T^*\mathbf{P}/S)_{\mathbf{F}_p}$ is the S-saturated singular support $SS_S^{\mathrm{sat}}\mathcal{F}$ of $\mathcal{F}$.*

PROOF. By Proposition 2.4.10 (2)⇒(4), $\mathcal{F}$ is S-micro supported on C^+. Hence $\mathcal{F}$ is S-micro supported on $C = C^+|_B$ by Lemma 2.3.7.2.

Assume that $\mathcal{F}$ is S-micro supported on $C' \subset T^*\mathbf{P}/S_{\mathbf{F}_p}$. Then, by Proposition 2.4.10 (1)⇒(2), we have $\mathbf{P}(C') \supset E = \mathbf{P}(C)$ since E is the smallest. Since the base of C' contains $B = \operatorname{supp} \mathcal{F}$ as a subset, we have $C' \supset C$ by Lemma 1.3.2. □

THEOREM 2.4.12. *Let S be an excellent regular scheme over $\mathbf{Z}_{(p)}$ such that the reduced part of $S_{\mathbf{F}_p}$ is of finite type over a field k such that $[k : k^p]$ is finite. Let X be a regular scheme of finite type over S and $\mathcal{F}$ be a constructible sheaf on X. Then, the S-saturated singular support $SS_S^{\mathrm{sat}}\mathcal{F}$ exists.*

PROOF. By the same argument as the proof of Corollary 1.5.4 using Lemma 2.4.3 and Proposition 2.4.4 in place of Lemma 1.5.2 and Proposition 1.5.3, it suffices to prove the case where $X = \mathbf{P}_S$. This case is proved in Corollary 2.4.11. □

By the same argument as in Lemma 2.4.2, we have the following.

COROLLARY 2.4.13. *Assume further that the morphism $FT^*S \times_S X \to FT^*X$ is 0. Then the S-singular support $SS_S\mathcal{F}$ exists and equals $SS_S^{\mathrm{sat}}\mathcal{F}$. If we further assume that the condition* (T) *in Lemma 2.3.3.2 is satisfied, then $SS\mathcal{F}$ exists and equals $SS_S\mathcal{F} = SS_S^{\mathrm{sat}}\mathcal{F}$.*

References

[1] M. Artin, *Faisceaux constructibles, cohomologie d'une courbe algébrique*, SGA4 Exposé IX. Théorie des Topos et Cohomologie Étale des Schémas, Lecture Notes in Mathematics, vol. 305, pp. 1–42 (1973)

[2] A. Beilinson, *Constructible sheaves are holonomic*, Selecta Math. (N.S.) **22** (2016), no. 4, 1797–1819, DOI 10.1007/s00029-016-0260-z. MR3573946

[3] A. Braverman and D. Gaitsgory, *Geometric Eisenstein series*, Invent. Math. **150** (2002), no. 2, 287–384, DOI 10.1007/s00222-002-0237-8. MR1933587

[4] J.-L. Brylinski, *Transformations canoniques, dualité projective, théorie de Lefschetz, transformations de Fourier et sommes trigonométriques* (French, with English summary), Astérisque **140-141** (1986), 3–134, 251. Géométrie et analyse microlocales. MR864073

[5] P. Deligne, *La formule de dualité globale, Théorie des topos et cohomologie étale des schémas*, SGA 4 Exposé XVIII. Springer Lecture Notes in Math. 305, 480–587 (1972)

[6] P. Deligne, *Théorèmes de finitude en cohomologie ℓ-adique* (French), Cohomologie étale, Lecture Notes in Math., vol. 569, Springer, Berlin, 1977, pp. 233–261. MR3727439

[7] L. Fu, *Etale cohomology theory*, Nankai Tracts in Mathematics, vol. 13, World Scientific Publishing Co. Pte. Ltd., Hackensack, NJ, 2011, DOI 10.1142/9789814307734. MR2791606

[8] L. Illusie, Y. Laszlo, and F. Orgogozo, *Introduction* (French), Astérisque **363-364** (2014), xiii–xix. Travaux de Gabber sur l'uniformisation locale et la cohomologie étale des schémas quasi-excellents. MR3329769

[9] H. Hu and E. Yang, *Relative singular support and the semi-continuity of characteristic cycles for étale sheaves*, Selecta Math. (N.S.) **24** (2018), no. 3, 2235–2273, DOI 10.1007/s00029-017-0355-1. MR3816504

[10] L. Illusie, *Appendice à Théorème de finitude en cohomologie ℓ-adique*, Cohomologie étale SGA $4\frac{1}{2}$. Springer Lecture Notes in Mathematics, vol. 569, pp. 252–261 (1977)

[11] Q. Lu and W. Zheng, *Categorical traces and a relative Lefschetz-Verdier formula*, Forum Math. Sigma **10** (2022), Paper No. e10, 24, DOI 10.1017/fms.2022.2. MR4377268

[12] T. Saito, *The characteristic cycle and the singular support of a constructible sheaf*, Invent. Math. **207** (2017), no. 2, 597–695, DOI 10.1007/s00222-016-0675-3. MR3595935

[13] T. Saito, *Characteristic cycle of the external product of constructible sheaves*, Manuscripta Math. **154** (2017), no. 1-2, 1–12, DOI 10.1007/s00229-016-0913-4. MR3682201

[14] T. Saito, *Cotangent bundle and microsupports in the mixed characteristic case*, Algebra Number Theory **16** (2022), no. 2, 335–368, DOI 10.2140/ant.2022.16.335. MR4412576

[15] T. Saito, *Frobenius-Witt differentials and regularity*, Algebra Number Theory **16** (2022), no. 2, 369–391, DOI 10.2140/ant.2022.16.369. MR4412577

Contemporary Mathematics
Volume **842**, 2026
https://doi.org/10.1090/conm/842/16859

Picard-Fuchs equations of twisted differential forms associated to Feynman integrals

Pierre Vanhove

ABSTRACT. Dimensionally or analytically regulated Feynman integrals lead to relative twisted period integrals. We present a recent extension of the Griffiths-Dwork pole reduction algorithm for deriving the D-module of differential operators acting on the twisted differential forms from Feynman integrals.

We illustrate the application of this algorithm by providing twisted Picard-Fuchs operators for hypergeometric, elliptic and Calabi-Yau differential motives arising from families of Feynman integrals.

Part 1. Feynman integrals in physics

Feynman integrals are a cornerstone in understanding fundamental interactions and the elementary building blocks of nature. Scattering amplitudes are used in particle physics to compare theoretical predictions with experimental measurements in particle colliders (see [**1**–**3**] for instance), more recently to gravitational wave physics [**4**–**6**], or the evaluation of the correlation functions of quantum fields at the end of inflation as they provide tools for analysing the formation of structure in the Universe [**7**, **8**]. Their accurate calculation, whether analytically or numerically, is needed for precision physics, but this remains a significant hurdle.

There are growing evidences [**9**–**26**] that Feynman integrals needed for precision physics correspond to (relative) period integrals of (singular) Calabi–Yau geometries.

Given a family of Feynman integrals attached to a Feynman graph, it is desirable to answer the questions:

(1) What is the class of functions to which belongs a given Feynman integral.
(2) Determine the complete set of partial differential operators acting on a given family Feynman integrals.

Feynman integrals satisfy several remarkable important properties:

(1) They are D-finite functions, that is they satisfy a differentiable module of partial differential equations with respect to their parameters [**27**, **28**].

2020 *Mathematics Subject Classification.* Primary 54C40, 14E20; Secondary 46E25, 20C20.
Key words and phrases. Twisted cohomology; Feynman integral; relative periods; variation of mixed Hodge structures.
This text is a contribution to the proceedings of the conference Regulators V, 3-13 juin 2024 Department of Mathematics, University of Pisa, Italy.

(2) Feynman integrals are relative period integrals of a variation of mixed Hodge structures [**9**, **29**].

These two properties set the question of analysing the nature of Feynman integral in a clear Hodge theoretical framework. This approach has been developed in [**9**–**11**, **17**, **29**–**37**].

We recall, in section 2, how Feynman integrals and motives are attached to a graph, and we make explicit the appearance of twisted differential in section. In section 3, we explain the extension of the Griffths-Dwork reduction for the specific twisted Feynman differentials detailled in [**38**]. In section 4, we make explicit twisted differential operators from Feynman integrals: hypergeometric differential operator in section 4.1, hyperelliptic and elliptic differential operators in section 4.2, and Calabi–Yau differential operators in section 4.3.

Part 2. Hodge structures of Feynman integrals

2.1. Feynman Graph polynomials

DEFINITION 2.1. A *Feynman* graph Γ is a finite collection of vertices $V(\Gamma)$, edges $E(\Gamma)$, and half-edges $H(\Gamma)$ satisfying the usual definitions; edges are adjacent to two vertices, and half-edges are adjacent to a single vertex, and allowing multiple edges between pairs of vertices.

We let $e(\Gamma) = |E(\Gamma)|$ the number of edges. To each edge of Γ we attach a mass variable $m_e \in \mathbb{R}$ and to each half-edge we attach a momentum vector $p_h \in \mathbb{R}^{1,D-1}$ in the D-dimensional Minkowski space equipped with a metric of signature $(1, D-1)$. To each half edge of Γ attach a vector $p_h \in \mathbb{C}^D$ subject to the so-called momentum conservation relation

$$\sum_{h \in H(\Gamma)} p_h = 0. \tag{2.1}$$

We assume that each vertex of Γ has a single outgoing half-edge. Therefore, one may view Γ as a graph in the usual sense, allowing multiple edges between vertices. For physical processes these vectors belong to of the D-dimensional Minkowski space $\mathbb{R}^{1,D-1}$. To simplify notation, we view momenta as being attached to vertices, and write p_v instead of p_h. Furthermore, we consider only the completely massive case with $m_e^2 > 0$ and all external vectors are of non-zero norm $p_v \cdot p_v \neq 0$. We take m_e, p_v as having complex values. The analytic properties of the Feynman integrals are studied by using analytic continuation in the multi-dimensional complex plane spanned by the number of independent scalar products $p_i \cdot p_j$ with $1 \leq i, j \leq |H(\Gamma)|$ and the masses m_i^2 with $1 \leq i \leq e(\Gamma)$.

We associate to the graph Γ two polynomials which are defined as follows [**39**, **40**]. Let $\{x_e \mid e \in e(\Gamma)\}$ be variables attached to all edges of Γ. A spanning tree of Γ is a subgraph T of Γ which contains all vertices of Γ, and so that $b_1(\mathsf{T}) = 0$ and $b_0(\mathsf{T}) = 1$. For each spanning tree T of Γ we attach the monomial $x^{\mathsf{T}} = \prod_{e \notin \mathsf{T}} x_e$. The *first Symanzik polynomial* is the polynomial

$$\mathscr{U} = \sum_{\substack{\text{Spanning}\\ \text{trees of } \Gamma}} x^{\mathsf{T}}. \tag{2.2}$$

A spanning k-forest of Γ is a subgraph F of Γ containing all vertices of Γ and so that $h_1(\mathsf{F}) = 0$ and $h_0(\mathsf{F}) = k$. We attach the polynomial $x^{\mathsf{F}} = \prod_{e \notin \mathsf{F}} x_i$ to each spanning 2-forest. A 2-forest is a disjoint union of two sub-trees $\mathsf{F} = \mathsf{T}_1 \cup \mathsf{T}_2$, and we define $s_{\mathsf{F}} = \sum_{(v_1,v_2)\in\mathsf{F}=\mathsf{T}_1\cup\mathsf{T}_2} p_{v_1}\cdot p_{v_2}$ Where the $\cdot$-product is the scalar product on $\mathbb{C}^D$. Then

(2.3)

$$\mathscr{V}(\vec{s},\vec{m};D) = \sum_{\substack{\text{Spanning}\\ \text{2-forests of }\Gamma}} s_{\mathsf{F}} x^{\mathsf{F}}, \quad \mathscr{F}(\vec{s},\vec{m};D) = \mathscr{U}\times\left(\sum_{e\in E(\Gamma)} m_e^2 x_e\right) - \mathscr{V}(\vec{s},\vec{m};D).$$

The polynomial $\mathscr{F}(\vec{s},\vec{m};D)$ is called the *second Symanzik polynomial* of Γ, depends the mass parameters and kinematic invariants, respectively:

$$\vec{m} := \left\{m_1^2,\dots,m_{e(\Gamma)}^2\right\} \in \mathbb{R}_{>0}^{e(\Gamma)}, \qquad \vec{s} = \{p_i\cdot p_j, i,j\in v(\Gamma)\}. \tag{2.4}$$

When $|v(\Gamma)| > D$, not all the scalar products are independent, and the number of independent variables satisfies certain Gram determinant conditions [**41**]. The discriminant locus of $\mathscr{F}$ depends on the linear relations between the scalar products in $\vec{s}$. This polynomial is a homogeneous polynomial of degree $L+1$ in the variables x_e for $e\in e(\Gamma)$, where $L = b_1(\Gamma)$. This L is often called the *loop order* of Γ. Henceforward, we will write instead $\mathscr{F}$ to simplify our notation.

2.2. Feynman integrals in parametric representation

Using the two polynomials associated to a graph Γ, we define a family of Feynman integrals [**40, 42**]

$$I_\Gamma(\underline{z};D,\underline{\nu}) = \int_{[0,+\infty[^{e(\Gamma)}} \frac{\mathscr{U}^{\nu-(l+1)\frac{D}{2}}}{\mathscr{F}^{\nu-l\frac{D}{2}}}\,\delta\left(\sum_{i=1}^{e(\Gamma)} x_i - 1\right)\prod_{i=1}^{e(\Gamma)} x_i^{\nu_i-1}dx_i. \tag{2.5}$$

where we have set $\underline{\nu} := (\nu_1,\dots,\nu_{e(\Gamma)})$ and collected the kinematic factors (the internal masses m_i and the independent scalar products between the external momenta) into $\underline{z} = (\vec{s},\vec{m})$.

Since the coordinate scaling $(x_1,\dots,x_{e(\Gamma)}) \to \lambda(x_1,\cdots,x_{e(\Gamma)})$ leaves invariant the integrand and the domain of integration, we can rewrite this integral as

$$I_\Gamma(\underline{z};D,\underline{\nu}) = \int_{\Delta_{e(\Gamma)}} \Omega_\Gamma^{D,\underline{\nu}} \tag{2.6}$$

with

$$\Omega_\Gamma^{D,\underline{\nu}} := \frac{\mathscr{U}^{\nu-(l+1)\frac{D}{2}}}{\mathscr{F}^{\nu-l\frac{D}{2}}}\prod_{e\in e(\Gamma)} x_e^{\nu_e-1}\,\Omega_0 \tag{2.7}$$

with the differential $e(\Gamma)-1$-form

$$\Omega_0 := \sum_{j=1}^{e(\Gamma)} (-1)^{j-1} x_j\, dx_1\wedge\cdots\wedge\widehat{dx_j}\wedge\cdots\wedge dx_{e(\Gamma)}, \tag{2.8}$$

where $\widehat{dx_j}$ means that dx_j is omitted. The domain of integration is defined as

$$\Delta_{e(\Gamma)} := \left\{[x_1,\dots,x_{e(\Gamma)}]\in\mathbb{P}^{e(\Gamma)-1} | x_i\in\mathbb{R}, x_i\geq 0\right\}. \tag{2.9}$$

2.3. Mixed Hodge structures for Feynman graph integrals

We define the vanishing loci for the Symanzik polynomials attached to the graph Γ:

(2.10)

$$X_{\Gamma;D} = \{\mathscr{F}(\vec{s},\vec{m};D) = 0 | x_i \in \mathbb{P}^{e(\Gamma)-1}(\mathbb{R})\}; \qquad Y_\Gamma = \{\mathscr{U} = 0 | x_i \in \mathbb{P}^{e(\Gamma)-1}(\mathbb{R})\}.$$

Notice that the vanishing locus for $\mathscr{F}$ depends on the space-time dimension D through the linear relations between the external momenta.

The integrand of the Feynman integral (2.6) is a differential form representing a class of $\mathrm{H}^{e(\Gamma)-1}(\mathbb{P}^{e(\Gamma)-1} - Z_{\Gamma;D})$ where $Z_{\Gamma;D}$ is the singular locus of the integrand. We see that if $e(\Gamma) - \frac{(L+1)D}{2} < 0$, $Z_{\Gamma;D} = Y_\Gamma$ and that if and $e(\Gamma) - \frac{LD}{2} > 0$ then $Z_{\Gamma;D} = X_{\Gamma;D}$. If neither of these inequalities is satisfied, then $Z_{\Gamma;D} = X_{\Gamma;D} \cup Y_\Gamma$.

Although the integrand $\Omega_\Gamma^{D,\underline{\nu}}$ is a closed form such that $\eta \in \mathrm{H}^{e(\Gamma)-1}(\mathbb{P}^{e(\Gamma)-1} - Z_{\Gamma;D})$, in general the domain $\Delta_{e(\Gamma)}$ has a boundary and therefore its homology class is not in $\mathrm{H}_{e(\Gamma)-1}(\mathbb{P}^{e(\Gamma)-1} - Z_{\Gamma;D})$. This difficulty is resolved by considering the relative cohomology [**29**, **33**].

We need to consider a blow-up in $\mathbb{P}^{e(\Gamma)-1}$ of linear space $f : \mathscr{P} \to \mathbb{P}^{e(\Gamma)-1}$, such that all the vertices of $\Delta_{e(\Gamma)}$ lie in $\mathscr{P}\backslash\mathscr{X}$ where $\mathscr{X}$ is the strict transform of $Z_{\Gamma;D}$. Let $\mathscr{B}$ be the total inverse image of the coordinate simplex $\{x_1 x_2 \cdots x_{e(\Gamma)} = 0 | [x_1, \ldots, x_{e(\Gamma)}] \in \mathbb{P}^{e(\Gamma)-1}\}$.

As been explained by Bloch, Esnault and Kreimer in [**29**] all of this lead to the mixed Hodge structure associated to the Feynman graph

$$M(\Gamma) := H^{e(\Gamma)-1}(\mathscr{P}\backslash\mathscr{X}, \mathscr{B}\backslash\mathscr{B} \cap \mathscr{X}; \mathbb{Q}). \tag{2.11}$$

2.4. Regulated Feynman integrals and twisted differential forms

The Feynman integral defined in (2.6) is a function of the parameters D and $\underline{\nu}$. For integer values of D and the powers ν_i the integral can be divergent. There are the ultaviolet or infrared divergences which have a special meaning in quantum field theory. We refer to e.g. [**43**] for a physics based discussion.

One can prove that the Feynman integral is a meromorphic function of $(D, \underline{\nu})$ in $\mathbb{C}^{e(\Gamma)+1}$, with simple poles located on affine hyperplanes defined by linear equations $c_0 D + \sum_{r=1}^{e(\Gamma)} c_r \nu_r$ with integer coefficients $(c_0, c_1, \ldots, c_{e(\Gamma)}) \in \mathbb{Z}^{e(\Gamma)+1}$. One can show that there is an open subset of $(D, \nu_1, \ldots, \nu_{e(\Gamma)}) \in \mathbb{C}^{e(\Gamma)+1}$ where the integral converges. The (unique) value of the Feynman integral is defined by analytic continuation. We refer to [**44**] for a throughouly discussion.

Using these properties there are two commonly used regularisation in physics:

(1) The dimensional regularisation where the analytic continuation is done in the spacetime D with fixed values for the indices $\underline{\nu}$ taken to be integers.
(2) The analytic continuation with a fixed integer value for the spacetime dimension D, but the analytic continuation is done with respect to (a subset) of the indices $\underline{\nu}$.

In the following we combine both of these regulators, and we introduce the following notations

$$I_\Gamma^{\epsilon,\kappa}(\underline{z}) := I_\Gamma(\underline{z}; 2\delta - 2\epsilon, \mathsf{v}_1 + \mu_1\kappa, \ldots, \mathsf{v}_{e(\Gamma)} + \mu_{e(\Gamma)}\kappa) \tag{2.12}$$

with δ an integer spacetime dimension of interest (say for instance $\delta = 2$), and $(\mathsf{v}_1, \ldots, \mathsf{v}_{e(\Gamma)}, \mu_1, \ldots, \mu_{e(\Gamma)}) \in \mathbb{Z}^{2e(\Gamma)}$, and ϵ and κ are positive real numbers. We set

$$I_\Gamma^{\epsilon,\kappa}(\underline{z}) = \int_{\Delta_{e(\Gamma)}} \Omega_\Gamma^{\epsilon,\kappa}, \tag{2.13}$$

with

$$\Omega_\Gamma^{\epsilon,\kappa} = \omega_\Gamma^{\mathrm{Rat}} \times \left(\frac{\mathscr{U}^{L+1}}{\mathscr{F}^L}\right)^\epsilon \prod_{i=1}^{e(\Gamma)} \left(\frac{x_i \mathscr{U}}{\mathscr{F}}\right)^{\mu_i\kappa} dx_1 \cdots dx_{e(\Gamma)}, \tag{2.14}$$

with the rational function

$$\omega_\Gamma^{\mathrm{Rat}} = \frac{\mathscr{U}^{\nu_1+\cdots+\nu_{e(\Gamma)}-(L+1)\delta}}{\mathscr{F}^{\nu_1+\cdots+\nu_{e(\Gamma)}-L\delta}} \prod_{i=1}^{e(\Gamma)} x_i^{\nu_i-1} \tag{2.15}$$

We make a few remarks:

(1) When $\epsilon = \kappa = 0$ we have that $\Omega_\Gamma^{0,0} = \omega_\Gamma^{\mathrm{Rat}}$ is a rational differential form.

(2) We remark that $\mathscr{U}^{L+1}/\mathscr{F}^L$ and $x_i\mathscr{U}/\mathscr{F}$ are degree zero rational functions in $\mathbb{P}^{e(\Gamma)-1}$, therefore when $\epsilon \neq 0$ or $\kappa \neq 0$ we have well-defined twisted differential forms.

(3) The twists in (2.14) does not introduce new poles. Therefore, the twisted differential has the *same* singular locus as the rational differential form (2.15).

Feynman integrals naturally lead to twisted differential forms of the kind studied in [**45**–**48**], which has been considered in recent applications, e.g. [**49**–**51**], for reducing the families of Feynman integrals attached to a given graph onto a basis of integrals.

In the present text will discuss a different approach aimed to derive the differential operators acting on the Feynman integrals [**38**]. Our approach uses that the twist for Feynman integral is a rational function build from the graph polynomials.

Part 3. D-modules for twisted differential forms

An important property of Feynman integral is that they are holonomic functions, which means that they satisfy finite order differential equation when differentiating with respect to their physical parameters $\underline{z} := \{\vec{s}, \vec{m}\}$.

Let us consider r parameters from the set of internal masses and independent kinematics, $\underline{z} := \{z_1, \ldots, z_r\} \in \vec{m} \cup \vec{s}$. We seek differential operators annihilating the twisted differential form $\Omega_\Gamma^{\epsilon,\kappa}$ in (2.14) in cohomology

$$\left(\sum_{a_1=0}^{o_1} \cdots \sum_{a_r=0}^{o_r} c_{a_1,\ldots,a_r}(\vec{m}, \vec{s}, \epsilon, \kappa) \left(\frac{\partial}{\partial z_1}\right)^{a_1} \cdots \left(\frac{\partial}{\partial z_r}\right)^{a_r}\right) \Omega_\Gamma^{\epsilon,\kappa} = d\beta_\Gamma^{\epsilon,\kappa}, \tag{3.1}$$

where $c_{a_1,\ldots,a_r}(\vec{m}, \vec{s}, \epsilon, \kappa)$ are rational functions of the physical parameters, but they are independent of the edge variables $x_1, \ldots, x_{e(\Gamma)}$, and $o_1, \ldots, o_r$ are some positive integers. The inhomogeneous term $d\beta_\Gamma^{\epsilon,\kappa}$ is a total derivative in x_i's where the only allowed poles are those already present in $\Omega_\Gamma^{\epsilon,\kappa}$ [**52**]. Because the domain of integration (2.9) of the Feynman integral does not depend on the physical parameters, we then deduce

$$\left(\sum_{a_1=0}^{o_1} \cdots \sum_{a_r=0}^{o_r} c_{a_1,\ldots,a_r}(\vec{m}, \vec{s}, \epsilon, \kappa) \left(\frac{\partial}{\partial z_1}\right)^{a_1} \cdots \left(\frac{\partial}{\partial z_r}\right)^{a_r}\right) I_\Gamma^{\epsilon,\kappa} = \mathscr{S}_\Gamma^{\epsilon,\kappa}, \tag{3.2}$$

where $\mathscr{S}_\Gamma^{\epsilon,\kappa}$ is an inhomogeneous term obtained by integrating $d\beta_\Gamma^{\epsilon,\kappa}$ over the boundary of the positive orthant (2.9). This is a non-trivial task because one needs to blow-up the intersections between the graph hypersurface and the domain of integration, so the integral is well-defined [**9, 11, 29, 53**]. For instance, section 3.2 of [**11**] gives a detailed derivation of the inhomogeneous term for the two-loop sunset integral along these lines.

If the integration is done over a cycle $\mathcal{C}$, like the one defined by the torus $\mathcal{C}_{\max} := \{|x_1| = \cdots = |x_{e(\Gamma)}| = 1\}$, the resulting integral is annihilated by the action of the differential operator [**54**]

$$(3.3)\qquad \left(\sum_{a_1=0}^{o_1}\cdots\sum_{a_r=0}^{o_r} c_{a_1,\dots,a_r}(\vec{m},\vec{s},\epsilon,\kappa)\left(\frac{\partial}{\partial z_1}\right)^{a_1}\cdots\left(\frac{\partial}{\partial z_r}\right)^{a_r}\right)\int_{\mathcal{C}_{\max}}\Omega_\Gamma^{\epsilon,\kappa}=0.$$

The ideal generated by these differential operators is a differential module (or D-module). Thus, the differential equations we are seeking can be obtained by deriving annihilators of $\Omega_\Gamma^{\epsilon,\kappa}$, i.e., partial differential operators that annihilate the integrand by acting on the physical parameter and the edge variables. An example of a system of partial differential equations for Feynman integrals is the Gel'fand-Kapranov-Zelevinskiĭ (GKZ) system, which provides a D-module of differential operators acting on the toric generalisation of the Feynman integral [**15, 54–61**]. Because the graph polynomials in the expression for $\Omega_\Gamma^{\epsilon,\kappa}$ in (2.14) are not generic polynomials the differential module acting on a given Feynman integral is obtained after restricting the GKZ D-module which is a higher non-trivial task [**55, 62–64**] and its systematic implementation is still an open problem. In the following sections we present an algorithmic procedure to derive the differential equations based on an extension of the Griffiths-Dwork reduction for twisted differential forms.

A motivation is to have an algorithm that applies to a large class of analytic regularised Feynman integrals which is missing for the commonly used programs in theoretical physics. This way we can analyse how the twists parameters ϵ, from space-time dimension, and κ, from the analytic regulator, deform the minimal order of the differential operators.

3.1. Variation of mixed Hodge structure and ODEs

If $\mathcal{H}_\mathbb{Q}$ is the local system underlying a variation of mixed Hodge structure over a 1-dimensional base M, and $\boldsymbol{s}$ is a meromorphic section of $\mathcal{H}_\mathbb{Q}\otimes\mathcal{O}_M$ then there is a minimal differential equation $\mathscr{L}_{\boldsymbol{s}}$ annihilating the period functions attached to $\boldsymbol{s}$.

Selecting a parameter t amongst the physical parameters $\vec{m}\cup\vec{s}$ we consider a pencil of graph $\mathscr{F}_\Gamma(t)$ and the differential form

$$(3.4)\qquad \Omega_\Gamma^{D,\underline{\nu}}(t)=\frac{\mathscr{U}^{\sum_{i=1}^{e(\Gamma)}\nu_i-(L+1)D/2}}{(\mathscr{F}_\Gamma(t))^{\sum_{i=1}^{e(\Gamma)}\nu_i-LD/2}}\prod_{i=1}^{e(\Gamma)}x_i^{\nu_i-1}\,\Omega_0,$$

determines a section of $\mathcal{H}_{\Gamma;D}\otimes\mathcal{O}_M$ where $\mathcal{H}_{\Gamma;D}$ is a variation of mixed Hodge structure over an open subset M of $\mathbb{A}_t^1$.

DEFINITION 3.1. Let $\mathscr{L}_\Gamma^D$ denote the minimal differential operator in $\mathbb{C}[M]\langle\partial_t\rangle$ which annihilates the form $\Omega_\Gamma^{D,\underline{\nu}}(t)$ in $\mathcal{H}_{\Gamma;D}\otimes\mathcal{O}_M$.

We recall how an ordinary differential equation is associated with a variation of mixed Hodge structure along with a holomorphic section of the underlying local system.

DEFINITION 3.2. A ($\mathbb{Q}$-)variation of mixed Hodge structure of weight n consists of several pieces of data

(1) A $\mathbb{Q}$-local system $\mathcal{H}_{\mathbb{Q}}$ over a complex manifold M,
(2) An increasing weight filtration by $\mathbb{Q}$-local systems $\mathcal{W}_0 \subseteq \mathcal{W}_1 \subseteq \cdots \subseteq \mathcal{W}_{2n} = \mathcal{H}_{\mathbb{Q}}$,
(3) A decreasing Hodge filtration $\mathcal{F}^n \subseteq \mathcal{F}^{n-1} \subseteq \cdots \subseteq \mathcal{F}^0 = \mathcal{H}_{\mathbb{C}} = \mathcal{H}_{\mathbb{Q}} \otimes_{\underline{\mathbb{Q}}_M} \underline{\mathbb{C}}_M$,
(4) A flat connection $\nabla : \mathcal{H}_{\mathbb{C}} \otimes \mathcal{O}_M \to \mathcal{H}_{\mathbb{C}} \otimes \Omega^1_M$ so that $\nabla(\mathcal{F}^i) \subseteq \mathcal{F}^{i-1}$,

so that on each fibre $\mathcal{H}_{\mathbb{Q}}$, the data $(\mathcal{H}_{\mathbb{Q}}, \mathcal{F}^{\bullet}_t, \mathcal{W}_{\bullet})$ is a mixed Hodge structure.

Given a local section $\boldsymbol{s}$ of $\mathcal{H}_{\mathbb{C}} \otimes \mathcal{O}_M$, and a local parameter t on M, we can construct local (or multivalued) period functions

$$\boldsymbol{\pi_s}(t) = \langle \boldsymbol{s}, \gamma_t \rangle \tag{3.5}$$

for a flat section γ_t of $\mathcal{H}^{\vee}_{\mathbb{Q}}$. We will often take $\boldsymbol{s} = \Omega^{D;\underline{\nu}}_{\Gamma}(t)$ and let $\mathcal{H}^{\vee}_{\mathbb{Q}}$ is the homology bundle underlying the family of varieties $\mathbb{P}^{e(\Gamma)-1} - X_{\Gamma;D}(t)$, in which case the pairing is integration.

Given a variation of mixed Hodge structure, $(\mathcal{H}_{\mathbb{Q}}, \mathcal{W}_{\bullet}, \mathcal{F}^{\bullet})$ over $M \subseteq \mathbb{A}^1$ with Gauss–Manin connection ∇, we have differential operators

$$\nabla_{\partial_t} : \mathcal{H} \otimes \mathcal{O}_M \to \mathcal{H} \otimes \mathcal{O}_M, [\omega] \mapsto \nabla([\omega])(\partial_t) \tag{3.6}$$

where ∂_t denotes the vector field corresponding to a choice of variable t. The pairing satisfies

$$\frac{d}{dt}\langle \boldsymbol{s}, \gamma_t \rangle = \langle \nabla_{\partial_t}(\omega), \gamma_t \rangle. \tag{3.7}$$

Consequently, there is a minimal collection of elements $\{f_0(t), \ldots, f_n(t)\}$ in the $\mathbb{C}(t)$-vector space $\Gamma(\mathcal{H} \otimes \mathcal{O}_M) \otimes \mathbb{C}(t)$ so that

$$\left[f_n(t)\nabla^n_{\partial_t} + f_{n-1}(t)\nabla^{n-1}_{\partial_t} + \cdots + f_1(t)\nabla_{\partial_t} + f_0(t)\right] \boldsymbol{s} = 0 \tag{3.8}$$

and thus there is a linear differential operator

$$\mathscr{L}_{\boldsymbol{s}} = f_n(t)\frac{d^n}{dt^n} + f_{n-1}(t)\frac{d^{(n-1)}}{dt^{(n-1)}} + \cdots + f_1(t)\frac{d}{dt} + f_0(t) \tag{3.9}$$

whose solutions are the period functions $\boldsymbol{\pi_s}(t)$. The local system $\mathcal{H}^{\vee}_{\mathbb{Q}}$ is equipped with a weight filtration $\mathcal{W}^*_{\bullet}$ dual to the weight filtration on $\mathcal{H}_{\Gamma;D}(t)$ determined by $\mathcal{W}^*_i = (\mathcal{W}_{-i-1})^{\vee}$. The pairing (3.5) induces a map from $\mathcal{H}^{\vee}_{\mathbb{Q}}$ to $\mathcal{O}_M$ whose image is $\mathcal{S}o\ell(\mathscr{L}_{\boldsymbol{s}})$, the local system of solutions of $\mathscr{L}_{\boldsymbol{s}}$. Therefore, $\mathcal{W}^*_i$ induces a filtration on $\mathcal{S}o\ell(\mathscr{L}_{\boldsymbol{s}})$.

LEMMA 3.3 (section 3.1 of [**37**]). *The local system $\mathcal{S}o\ell(\mathscr{L}_{\boldsymbol{s}})$ is a quotient of the dual local system $\mathcal{H}^{\vee}_{\mathbb{Q}}$ by a sub-local system $\mathbb{K}_{\boldsymbol{s}}$. If $\boldsymbol{s} \in \mathcal{W}_i \otimes \mathcal{O}_M$ then $\mathcal{W}^*_i \subseteq \mathbb{K}_{\boldsymbol{s}}$.*

We summarize the results of [**37**]

(1) The local systems $\mathcal{S}o\ell(\mathscr{L}_{\Gamma;D})$ are quotients of $\mathcal{H}^{\vee}_{\Gamma;D}$.
(2) The filtration induced by $\mathcal{W}^*_{\bullet}$ corresponds to a factorisation of $\mathscr{L}_{\Gamma;D}$, however there may be factorisations of $\mathscr{L}_{\Gamma;D}$ which do not correspond to $\mathcal{W}^*_{\bullet}$.
(3) The monodromy representation of $\mathcal{S}o\ell(\mathscr{L}_{\Gamma;D})$ is upper triangular with diagonal blocks equal to the monodromy representations of the factors of $\mathscr{L}_{\Gamma;D}$.

3.2. Griffiths-Dwork reduction for twisted differential forms

We present the Griffiths-Dwork reduction for twisted differential forms applied to the case of the differential form $\Omega_\Gamma^{\epsilon,\kappa}$ defined in (2.14).

Choosing r variables amongst the kinematic parameters $\underline{z} := \{z_1, \dots, z_r\} \in \vec{m} \cup \vec{s}$, the differentiation of $\Omega_\Gamma^{\epsilon,\kappa}$ leads to

(3.10)
$$\sum_{\substack{\mathsf{a}=a_1+\cdots+a_r\\ a_i\geq 0}} c_{\underline{a}}(\vec{m},\vec{s};\epsilon,\kappa)\left(\frac{\partial}{\partial z_1}\right)^{a_1}\cdots\left(\frac{\partial}{\partial z_r}\right)^{a_r}\Omega_\Gamma^{\epsilon,\kappa} = \sum_{\substack{\mathsf{a}=a_1+\cdots+a_r\\ a_i\geq 0}} \frac{c_{\underline{a}}(\vec{m},\vec{s};\epsilon,\kappa)P^{\underline{a}}(\underline{x})}{\mathscr{F}^{\mathsf{a}}}\Omega_\Gamma^{\epsilon,\kappa},$$

where $\underline{a} = (a_1, \dots, a_r) \in \mathbb{N}^r$ and $P^{\underline{a}}(\underline{x})$ is a homogeneous polynomial of degree $(L+1)(a_1 + \cdots + a_r)$ in the edge variables $\underline{x}$. The sum is over the differential operators of order $a_1 \geq 0, \dots, a_r \geq 0$ and fixed total order $\mathsf{a} := a_1 + \cdots + a_r$. The pole order in the second Symanzik polynomial $\mathscr{F}$ has increased by a.

We present an extension of the Griffths pole reduction [**65, 66**] adapted to include the twist factor in $\Omega_\Gamma^{\epsilon,\kappa}$:

Step 1: Reduction of polynomial $P^{\underline{a}}(\underline{x})$ in the Jacobian ideal of $\mathscr{F}$, $\mathrm{Jac}(\mathscr{F}_\Gamma) := \langle \vec{\nabla}\mathscr{F}(\underline{x})\rangle$

(3.11)
$$P^{\underline{a}}(\underline{x}) = \vec{C}^{\underline{a}}(\underline{x}) \cdot \vec{\nabla}\mathscr{F},$$

with $\vec{\nabla}\mathscr{F} := \left(\partial_{x_1}\mathscr{F}(\underline{x}), \dots, \partial_{x_{e(\Gamma)}}\mathscr{F}(\underline{x})\right)$.

Step 2: Reduction of $\vec{C}^{\underline{a}}(\underline{x})$ in the Jacobian ideal of $\mathscr{U}$, $\mathrm{Jac}(\mathscr{F}_\Gamma) := \langle\vec{\nabla}\mathscr{U}(\underline{x})\rangle$

(3.12)
$$\vec{C}^{\underline{a}}(\underline{x}) \cdot \vec{\nabla}\mathscr{U} = c^{\underline{a}}(\underline{x})\,\mathscr{U},$$

with $\vec{\nabla}\mathscr{U} := \left(\partial_{x_1}\mathscr{U}(\underline{x}), \dots, \partial_{x_{e(\Gamma)}}\mathscr{U}(\underline{x})\right)$.

Step 3: Thanks to steps 1 and 2 the differential form

(3.13)
$$\beta^{\underline{a}} = \sum_{1\leq i<j\leq e(\Gamma)} \frac{x_i C_j^{\underline{a}}(\underline{x}) - x_j C_i^{\underline{a}}(\underline{x})}{\mathscr{F}^{\mathsf{a}-1}}\Omega_\Gamma^{\epsilon,\kappa}\, dx_1 \wedge \cdots \wedge \widehat{dx_i} \wedge \cdots \wedge \widehat{dx_j} \wedge \cdots \wedge dx_{e(\Gamma)}.$$

satisfies the property

(3.14)
$$d\beta_\Gamma^{\underline{a}} = (\mathsf{a}-1)\frac{P^{\underline{a}}(\underline{x})}{\mathscr{F}^{\mathsf{a}}}\Omega_\Gamma^{\epsilon,\kappa} + \frac{\vec{\nabla}\cdot\vec{C}^{\underline{a}}(\underline{x}) + \lambda_U\, c^{\underline{a}}(\underline{x})}{\mathscr{F}^{\mathsf{a}-1}}\Omega_\Gamma^{\epsilon,\kappa}.$$

where $\lambda_U = e(\Gamma) - (L+1)(\delta - \epsilon) + \kappa\sum_{i=1}^{e(\Gamma)}\mu_i$ is the power of $\mathscr{U}$ in $\Omega_\Gamma^{\epsilon,\kappa}$.

Therefore, for a given set of derivatives, we have performed the pole reduction

(3.15)
$$\sum_{\substack{\mathsf{a}=a_1+\cdots+a_r\\ a_i\geq 0}} c_{\underline{a}}(\vec{m},\vec{s};\epsilon,\kappa)\left(\frac{\partial}{\partial z_1}\right)^{a_1}\cdots\left(\frac{\partial}{\partial z_r}\right)^{a_r}\Omega_\Gamma^{\epsilon,\kappa}$$
$$= \sum_{\substack{\mathsf{a}=a_1+\cdots+a_r\\ a_i\geq 0}} \frac{\vec{\nabla}\cdot\vec{C}^{\underline{a}}(\underline{x}) + \lambda_U\, c^{\underline{a}}(\underline{x})}{(\mathsf{a}-1)\mathscr{F}^{\mathsf{a}-1}}\Omega_\Gamma^{\epsilon,\kappa} + d\beta_\Gamma^{\epsilon,\kappa},$$

Iterating this procedure gives the partial differential equation (3.1). We refer to [**38**] for a proof of this reduction and details.

REMARK 3.4. We remark that this way of solving the linear system includes implicitly the freedom given by the syzygies of $\mathrm{Jac}(\mathscr{F})$ and $\mathrm{Jac}(\mathscr{U})$ since they belong to the kernel of the linear systems from (3.11) and (3.12) respectively. It

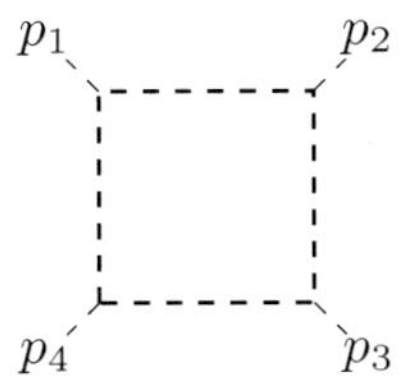

FIGURE 1. The box graph with massless external and internal states.

was noticed in [**52**], that in the rational case, only the first order syzygies are needed to take into account the non-isolated singularities of Feynman integrals.

Part 4. Differential operators for various graphs

In this section we present some differential equations acting on the regulated Feynman integrals. The twist does not change the singular locus of the integrand. Consequently, the real singularities of the associated differential operator are the same as when there are no twist ($\epsilon = \kappa = 0$). We illustrate the effects of the twist one some classes of period integrals.

4.1. Hypergeometric differential operators: the massless box graph

For the box graph in Fig. 1, we have four massless momenta p_i with $1 \leq i \leq 4$ such that $p_1 + \cdots + p_4 = 0$ and $p_1^2 = \cdots = p_4^2 = 0$. After scaling the integral and setting $X = \frac{p_1 \cdot p_4}{p_1 \cdot p_2}$ the graph polynomials are given by

$$\mathscr{U}_\square = x_1 + \cdots + x_4, \qquad \mathscr{F}_\square(X) = x_2 x_4 + X x_1 x_3. \tag{4.1}$$

The dimensionally regulated Feynman integral in $D = 4 - 2\epsilon$ is given by the twisted differential

$$I_\square^{\epsilon,0}(X) = \int_{\Delta_4} \left(\frac{\mathscr{U}_\square^2}{\mathscr{F}_\square(X)} \right)^\epsilon \frac{\Omega_0}{\mathscr{F}_\square(X)^2}, \tag{4.2}$$

with $\Omega_0 = x_1 dx_2 \wedge dx_3 \wedge dx_4 - x_2 dx_1 \wedge dx_3 \wedge dx_4 + x_3 dx_1 \wedge dx_2 \wedge dx_4 - x_4 dx_1 \wedge dx_2 \wedge dx_3$, and $\Delta_4 = \{x_i \geq 0,\ 1 \leq i \leq 4\}$. The application of the algorithm described in section 3.2 gives the differential equation

$$\mathscr{L}_\square^\epsilon I_\square^{\epsilon,0}(X) = \frac{(\epsilon+1)\Gamma(-\epsilon-1)^2}{\Gamma(-2\epsilon)} \left(1 + X^{-\epsilon-1}\right), \tag{4.3}$$

with the differential operator

$$\mathscr{L}_\square^\epsilon = (X+1)X\frac{d}{dX} + 1 + X + \epsilon. \tag{4.4}$$

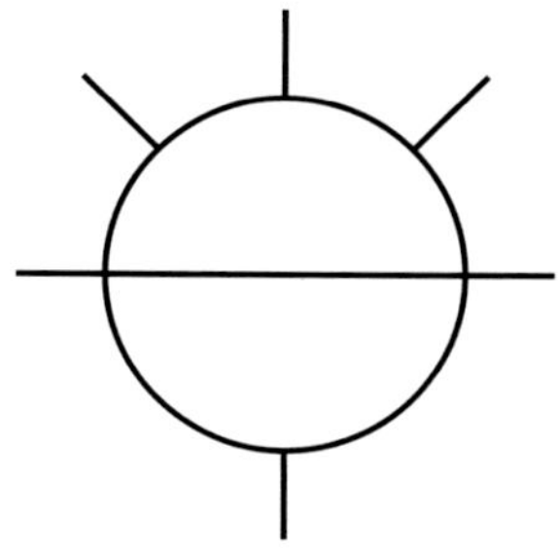

FIGURE 2. A two-loop graphs with $a = 4$, $b = 1$ and $c = 2$.

The Feynman integral integrates to hypergeometric functions, with the ϵ expansion as polylogarithms because we have a variation of mixed Tate motives [**67**, **68**]

$$I_{\square}^{\epsilon,0}(X) = \frac{4}{\epsilon^2 X} - \frac{4+2\ln(X)}{X\epsilon} + \frac{2\ln(X) - \frac{5\pi^2}{3} + 4}{X} \tag{4.5}$$
$$+ \frac{\epsilon}{X}\Big(2\,\mathrm{Li}_3(-X) - 2\ln(X)\,\mathrm{Li}_2(-X) - \ln(X)^2\ln(X+1) + \frac{\ln(X)^3}{3} + \frac{4\ln(X)\pi^2}{3}$$
$$- 2\ln(X) - 10\zeta(3) + \frac{5\pi^2}{3} - 4 - \pi^2\ln(X+1)\Big) + O(\epsilon^2),$$

where $\mathrm{Li}_r(X) = \sum_{n\geq 1} X^n/n^r$ are the polylogarithms.

4.2. Hyperelliptic differential operators: Planar two-loop graphs

Two-loop graphs can be labelled by the number of edges (a, b, c) on each cycle, in figure 2 we have represented a graph with $a = 4$, $b = 1$, $c = 2$. Planar graphs are graph for which least one edge number is equal to 1, and their graph polynomials are given by

$$\mathscr{U}_{(a,1,c)} = \left(z + \sum_{i=1}^{c} x_i\right)\left(\sum_{i=1}^{a} y_i\right) + z\left(\sum_{i=1}^{c} x_i\right), \tag{4.6}$$

$$\mathscr{V}_{(a,1,c);D} = z\left(\sum_{i=1}^{c}\sum_{j=1}^{a} r_{ij}^2 x_i y_j\right) + \left(z + \sum_{i=1}^{a} y_i\right)\left(\sum_{1\leq i<j\leq c} p_{ij}^2 x_i x_j\right) \tag{4.7}$$

$$+ \left(z + \sum_{i=1}^{c} x_i\right)\left(\sum_{1\leq i<j\leq a} q_{ij}^2 y_i y_j\right), \tag{4.8}$$

$$\mathscr{F}_{(a,1,c);D} = \mathscr{U}_{(a,1,c)}\left(\sum_{i=1}^{c} m_{i+a}^2 x_i + \sum_{i=1}^{a} m_i^2 y_i + m_{a+c+1}^2 z\right) - \mathscr{V}_{(a,1,c);D}.$$

We define the class of mixed Hodge structures (MHS) than can arise from the two-loop Feynman integrals:

DEFINITION 4.1 (MHS for planar two-loop graphs).

(1) Let $\mathbf{MHS}_{\mathbb{Q}}$ denote the Abelian category of $\mathbb{Q}$-mixed Hodge structures.
(2) The largest extension-closed subcategory of $\mathbf{MHS}_{\mathbb{Q}}$ containing the Tate twists of $\mathrm{H}^1(C;\mathbb{Q})$ for every hyperelliptic curve C is called $\mathbf{MHS}_{\mathbb{Q}}^{\mathrm{hyp}}$.

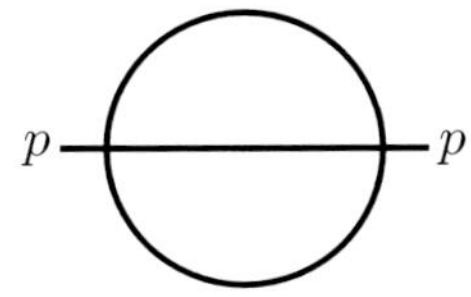

FIGURE 3. Two-loop sunset with $a = b = c = 1$

(3) The largest extension-closed subcategory of $\mathbf{MHS}_{\mathbb{Q}}$ containing the Tate twists of $\mathrm{H}^1(E;\mathbb{Q})$ for every elliptic curve E is called $\mathbf{MHS}_{\mathbb{Q}}^{\mathrm{ell}}$.

The main theorem of [**37**] states for generic values of $(a,1,c)$ and the spacetime dimension D the Feynman integral attached the planar two-loop graphs

$$I^{D,\underline{\nu}}_{(a,1,c)} = \int_{\Delta_{a+1+c}} \frac{\mathscr{U}^{\sum_{i=1}^{a+1+c}\nu_i - \frac{3D}{2}}}{\mathscr{F}^{\sum_{i=1}^{a+1+c}\nu_i - D}} \prod_{i=1}^{a} x_i^{\nu_i} \prod_{i=1}^{c} y_i^{\nu_{a+i}} z^{\nu_{a+c+1}} \Omega_0. \tag{4.9}$$

is a (relative) period integrals of $\mathbf{MHS}_{\mathbb{Q}}^{\mathrm{hyp}}$ because the singular locus is determined by the vanishing locus of $\mathscr{F}_{(a,1,c);D}$

THEOREM 4.2 (DHV [**37**]). *For any values of a, c, the cohomology groups of $X_{(a,1,c);D} = \{\mathscr{F}_{(a,1,c);D} = 0 | (x_i, y_i, z) \in \mathbb{P}^{a+c}\}$ are contained in $\mathbf{MHS}_{\mathbb{Q}}^{\mathrm{hyp}}$.*

Depending on the value of parameters D, a, c and the kinematic invariants p_{ij}^2, q_{ij}^2, r_{ij}^2, and the internal masses m_i^2, the singularities of the integrand change and Feynman integral can be become a period integral of $\mathbf{MHS}_{\mathbb{Q}}^{\mathrm{ell}}$ or just $\mathbf{MHS}_{\mathbb{Q}}$.

4.2.1. The elliptic curve: the two-loop sunset graph. We present the case of the sunset integral attached to the two-loop graph with $a = b = c = 1$.

The graph polynomials are, in the notations introduced above,

$$\mathscr{U}_{(1,1,1)} = x_1 z + x_1 y_1 + y_1 z, \tag{4.10}$$
$$\mathscr{V}_{(1,1,1);D} = p^2 x_1 y_1 z, \tag{4.11}$$
$$\mathscr{F}_{(1,1,1);D} = \mathscr{U}_{(1,1,1)} \times \left(m_1^2 x_1 + m_2^2 y_2 + m_3^2 z\right) - \mathscr{V}_{(1,1,1);D}.$$

and the Feynman integral for $D = 2 - 2\epsilon$ and $\nu_1 = \nu_2 = \nu_3 = 1$ reads

$$I^{\epsilon}_{(1,1,1)} = \int_{\Delta_3} \left(\frac{\mathscr{U}^3_{(1,1,1)}}{\mathscr{F}^2_{(1,1,1)}}\right)^{\epsilon} \frac{z dx_1 \wedge dy_1 - y_1 dx_1 \wedge dz + x_1 dy_1 \wedge dz}{\mathscr{F}_{(1,1,1)}}. \tag{4.12}$$

For $\epsilon = 0$ it is shown in [**11, 31**] that the integral is a regulator period integral of the $\mathbf{MHS}_{\mathbb{Q}}^{\mathrm{ell}}$ associated to the elliptic curve defined by $\mathscr{F}_{(1,1,1)} = 0$ in $\mathbb{P}^2$.

The equal-mass case: When all the mass parameters are the same $m_1 = m_2 = m_3$ and setting $t = p^2/m_1^2$, the sunset Feynman satisfies the differential equation

$$\left(\mathscr{L}^{(2)}_{(1,1,1)} + \epsilon \mathscr{L}^{(1)}_{(1,1,1)} + \epsilon^2 \mathscr{L}^{(0)}_{(1,1,1)}\right) I^{\epsilon}_{(1,1,1)} = -6 \frac{\Gamma(1+\epsilon)^2}{\Gamma(1+2\epsilon)} \tag{4.13}$$

with the differential operators $\mathscr{L}^{(r)}_{\ominus(3)}$ of order r

$$\mathscr{L}^{(2)}_{(1,1,1)} = \frac{d}{dt}\left(t(t-1)(t-9)\frac{d}{dt}\right) + (t-3), \tag{4.14}$$

$$\mathscr{L}^{(1)}_{(1,1,1)} = (3t^2-10t-9)\frac{d}{dt} + 3t - 5, \tag{4.15}$$

$$\mathscr{L}^{(0)}_{(1,1,1)} = 2(t+1). \tag{4.16}$$

The $\epsilon = 0$ piece is the Picard-Fuchs operator associated to the modular curve $X_1(6)$ defined by $(x_1y_1 + x_1y + y_1z)(x_1 + y_1 + z) = tx_1y_1z$ [**31**]. The twist induces the differential $\mathscr{L}^1_{(1,1,1)}$ and $\mathscr{L}^0_{(1,1,1)}$ without affecting the real singularities of the total differential operator.

The different mass case: For the non-equal-mass case $m_1 \neq m_2 \neq m_3$ the order of the differential equation is four with the following ϵ expansion [**38, 69, 70**]

$$\underbrace{\left(\mathscr{L}^{(4)}_0 + \sum_{r=0}^{4} \epsilon^{1+r}\hat{\mathscr{L}}^{(4-r)}_{r+1}\right)}_{=:\mathscr{L}^{\epsilon}_{(1,1,1)}} I^{\epsilon}_{(1,1,1)} = \mathscr{S}^{\epsilon}_{(1,1,1)}. \tag{4.17}$$

where the differential operators $\mathscr{L}^{(r)}_s$ are of order r. The $\epsilon = 0$ piece is a fourth order differential operator that factorizes

$$\mathscr{L}^{(4)}_0 = \mathscr{L}^{(1)}_a \circ \mathscr{L}^{(1)}_b \circ \mathscr{L}^{3-\text{mass}}_{\ominus(3)} \tag{4.18}$$

where $\mathscr{L}^{(1)}_a$ and $\mathscr{L}^{(1)}_b$ are order one differential operators. The differential operator $\mathscr{L}^{3-\text{mass}}_{\ominus(3)}$ is the Picard-Fuchs operator for the three masses two-loop sunset integral in two dimensions determined by the sunset elliptic curve $(x_1y_1+x_1z+y_1z)(m_1^2x_1+m_2^2y_1+m_3^2z) = p^2x_1y_1z$ [**11**].

The differential operators $\hat{\mathscr{L}}^{(4-r)}_{r+1}$ are irreducible differential operators of order $4-r$, therefore the differential operator (4.17) is irreducible for generic values of ϵ.

The factorisation of the differential operator (4.18) is understood from the fact that the Feynman integral are (relative) periods of the motive $\mathbf{MHS}^{\text{ell}}_{\mathbb{Q}}$ attached to the sunset elliptic curve. As shown in [**37**] such factorisation appears for the $(a, 1, c)$ graphs:

THEOREM 4.3 (Factorisation of differential operators). *For any a, c, the operator $\mathscr{L}_{(a,1,c);D}$ admits a factorisation*

$$\mathscr{L}_1\mathscr{L}_2\dots\mathscr{L}_k$$

where $\mathcal{S}o\ell(\mathscr{L}_i)$ is either:

(a) *a local system with finite order monodromy or*
(b) *a subquotient of the local system underlying a family of hyperelliptic curves over a Zariski open subset of $\mathbb{A}^1$.*

In particular, if a or c is ≤ 2 then the monodromy representation of $\mathcal{S}o\ell(\mathscr{L}_i)$ is either

(a) *finite, or*
(b) *a finite index subgroup of* $\mathrm{SL}_2(\mathbb{Z})$.

The twist in the differential form (4.12) induces an ϵ deformation of the Picard-Fuchs operator $\mathscr{L}^{\epsilon}_{(1,1,1)}$. This does not affect the real singularities of the differential operator because the ϵ factor in (4.12) does not change the nature of the singular locus which is still given by the same elliptic curve as in the $\epsilon = 0$ case. Therefore, the ϵ deformation only affects the local monodromies and the apparent singularities of the differential operator, as can be seen from the coefficient of the highest order term

(4.19)
$$\mathscr{L}^{\epsilon}_{(1,1,1)}\Big|_{(d/dt)^4} = (p^2)^3 \prod_{i=1}^{4}(p^2-\mu_i^2)\Big(-(2\epsilon+5)\,(p^2)^2 - 2\left(m_1^2+m_2^2+m_3^2\right)(1+2\epsilon)\,p^2 + (7+6\epsilon)\prod_{i=1}^{4}\mu_i\Big),$$

where $\mu_i = \{m_1+m_2+m_3, -m_1+m_2+m_3, m_1-m_2+m_3, m_1+m_2-m_3\}$ are the thresholds.

The action of $\mathscr{L}^{\epsilon}_{(1,1,1)}$ leads to the inhomogeneous differential equation (4.17), with an inhomogeneous term given by

(4.20)
$$\mathscr{S}_{\ominus}(\vec{m},t,\epsilon) = \frac{c_{23}(t,\epsilon)\Gamma(\epsilon+1)^2}{(m_2m_3)^{2\epsilon}\Gamma(1+2\epsilon)} + \frac{c_{13}(t,\epsilon)\Gamma(\epsilon+1)^2}{(m_1m_3)^{2\epsilon}\Gamma(1+2\epsilon)} + \frac{c_{12}(t,\epsilon)\Gamma(\epsilon+1)^2}{(m_1m_2)^{2\epsilon}\Gamma(1+2\epsilon)},$$

obtained from the integration of the exact differential $d\beta^{\epsilon}_{(1,1,1)}$ in (3.1). The coefficients $c_{12}(t,\epsilon)$, $c_{13}(t,\epsilon)$ and $c_{23}(t,\epsilon)$ are polynomials of degree 4 in t and degree 2 in ϵ, respectively which expressions are given on the `SageMath` worksheet Sunset-Twoloop-3mass-Epsilon.ipynb. Expanding in powers of ϵ, we have

$$\mathscr{S}_{\ominus}(\vec{m},t,\epsilon) = \mathscr{S}^0_{\ominus}(\vec{m},t) + \left(c_0^{(1)}(\vec{m}) + \sum_{i=1}^{3} c_i^{(1)}(\vec{m})\log(m_i)\right)\epsilon + O(\epsilon^2) \tag{4.21}$$

with the leading term given by

$$\mathscr{S}^0_{\ominus}(\vec{m},t) = 60t^4 + 56\left(m_1^2+m_2^2+m_3^2\right)t^3 - 308\prod_{i=1}^{4}\mu_i. \tag{4.22}$$

For $\epsilon = 0$ the two-loop sunset integral satisfies the differential equation [**11, 71**]

$$\mathscr{L}^{3-\text{mass}}_{\ominus} f^{(0)}_{\ominus}(t) = s_0(\vec{m},t) + \sum_{i=1}^{3} s_i(\vec{m},t)\log(m_i^2). \tag{4.23}$$

It can be checked that

$$\mathscr{S}^0_{\ominus}(\vec{m},t) = \mathscr{L}^{(1)}_1\mathscr{L}^{(2)}_1\mathscr{L}^{3-\text{mass}}_{\ominus} f^{(0)}_{\ominus}(t), \tag{4.24}$$

showing that the structure of the inhomogeneous term is compatible with the factorisation of the $\epsilon = 0$ piece of the differential operator in (4.17).

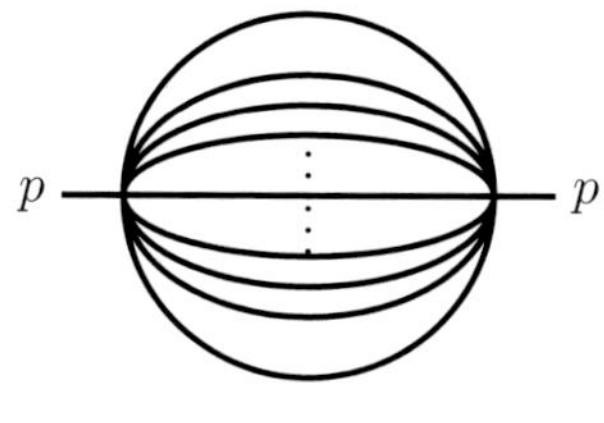

FIGURE 4. Multi-loop sunset with n edges

4.3. Calabi–Yau differential operators: sunset multiloop graphs

We now turn to the $n-1$-loop sunset integral in $D=2-2\epsilon$ dimensions attached to the graph in fig. 4 which reads

$$I^{\epsilon}_{\ominus(n)}(p^2,\vec{m},t)=\int_{\Delta_n}\left(\frac{\mathscr{U}^n_{\ominus(n)}}{\mathscr{F}^{n-1}_{\ominus(n)}}\right)^{\epsilon}\frac{\sum_{i=1}^n(-1)^{i-1}\bigwedge_{\substack{j=1\\ j\neq i}}^n dx_j}{\mathscr{F}_{\ominus(n)}}, \tag{4.25}$$

with the domain of integration $\Delta_n=\{x_i\geq 0, 1\leq i\leq n\}$ and the graph polynomials

$$\mathscr{U}_{\ominus(n)}=x_1\cdots x_n\sum_{i=1}^n\frac{1}{x_i}, \tag{4.26}$$

$$\mathscr{F}_{\ominus(n)}=\mathscr{U}_{\ominus(n)}\times\sum_{i=1}^n m_i^2x_i-p^2x_1\cdots x_n. \tag{4.27}$$

Notice that $\mathscr{U}^n_{\ominus(n)}/\mathscr{F}^{n-1}_{\ominus(n)}$ is a homogeneous rational function of degree 0 in $(x_1,\dots,x_n)$. As usual the differential form is defined in the complement of the vanishing locus of the denominator in $X_n=\{\mathscr{F}_{\ominus(n)}=0\}$. The Feynman integral being a (relative) period of a Calabi–Yau manifold of complex dimension $n-2$ defined by the equation $\mathscr{F}_{\ominus(n)}=0$ [**10, 11, 13, 16, 17, 19, 31, 72**].

The equal-mass case: For the equal-mass case $m_1=\cdots=m_n$ and $p^2=t\,m_n^2$ in (4.26) the sunset Feynman integral satisfies the differential equation

$$\underbrace{\left(\sum_{r=0}^{n-1}\epsilon^r\mathscr{L}^{(n-1-r)}_{\ominus(n)}\right)}_{=:\mathscr{L}^{\epsilon}_{\ominus(n)}}I^{\epsilon}_{\ominus}(t)=-n!\frac{\Gamma(1+\epsilon)^{n-1}}{\Gamma(1+(n-1)\epsilon)}. \tag{4.28}$$

Like the case $n=3$ described above the term of order ϵ^r is a differential operator $\mathscr{L}^{(r)}_{\ominus(n)}$ of order r in t. The coefficient of ϵ^0 is the differential operator of order $n-1$ derived in [**36**] (see as well [**16, 73–77**]).

The ϵ deformation does not change the real singularities of the differential operators because the twist in (4.25) does not introduce new singularities.

This is seen as well on the form of the ϵ-deformed Picard-Fuchs operator for the $n=4$ case with $m_1=m_2=m_3=m_4$

(4.29)
$$\begin{aligned}\mathscr{L}^{\epsilon}_{\ominus(4)} &= -(t-16)(t-4)t^2\left(\frac{d}{dt}\right)^3 - 6(t^3-15t^2+32t)\left(\frac{d}{dt}\right)^2 - (7t^2-68t+64)\left(\frac{d}{dt}\right)\\ &\quad - t+4+\epsilon\left(-6(t-10)t^2\left(\frac{d}{dt}\right)^2 - 6(3t-20)t\left(\frac{d}{dt}\right)+18-6t\right)\\ &\quad + \epsilon^2\left(-(11t^2-28t-64)\left(\frac{d}{dt}\right) - 11t+14\right) + \epsilon^3\left(-6t-12\right),\end{aligned}$$

where the $\mathscr{L}^{\epsilon}_{\ominus(4)}|_{\epsilon=0}$ operator is the Picard-Fuchs operator for the $K3$ surface with Picard number 19 [**10**].

Different mass case: We represent the result for the four different masses $m_1 \neq m_2 \neq m_3 \neq m_4$ for the case $n=4$. In that case the singular locus is a $K3$ surface of Picard number 16 [**52**]. The ϵ-deformed differential operator has order 11 and has the ϵ expansion

$$\mathscr{L}^{\epsilon}_{\ominus(4)} = \sum_{r=0}^{16} \epsilon^r \mathscr{L}^{(11)}_r + \sum_{r=0}^{11} \epsilon^{16+r} \mathscr{L}^{(11-r)}_{16+r}, \tag{4.30}$$

where the differential operators $\mathscr{L}^{(r)}_s$ are of order r. The order 11 part of this deformed Picard-Fuchs operator has degree 16 in ϵ.

The order $\epsilon=0$ operator factorises as

$$\mathscr{L}^{(11)}_0 = \mathscr{L}^{(1)}_{a_1} \circ \cdots \circ \mathscr{L}^{(1)}_{a_5} \circ \mathscr{L}^{4-\text{mass}}_{\ominus(4)}, \tag{4.31}$$

where $\mathscr{L}^{(1)}_{a_1}, \ldots, \mathscr{L}^{(1)}_{a_5}$ are first order operators and $\mathscr{L}^{4-\text{mass}}_{\ominus(4)}$ is the sixth order differential operator for the three-loop sunset integral with the all different mass configurations given in section 4.3 of [**52**].

The coefficient of the highest order term $(d/dt)^{11}$ is given by

$$\begin{aligned}\mathscr{L}^{\epsilon}_{\ominus(4)}\Big|_{(d/dt)^{11}} &= t^{11}\left(t-(m_1+m_2-m_3-m_4)^2\right)\\ &\times\left(t-(m_1-m_2+m_3-m_4)^2\right)\left(t-(m_1+m_2+m_3-m_4)^2\right)\\ &\times\left(t-(m_1-m_2-m_3+m_4)^2\right)\left(t-(m_1+m_2-m_3+m_4)^2\right)\\ &\times\left(t-(m_1-m_2+m_3+m_4)^2\right)\left(t-(-m_1+m_2+m_3+m_4)^2\right)\\ &\times\left(t-(m_1+m_2+m_3+m_4)^2\right) q^{[1111]}(t,\epsilon).\end{aligned} \tag{4.32}$$

The ϵ dependence appears only in the apparent singularities determined by the polynomial $q^{[1111]}(t,\epsilon)$ of degree 17 in t and 16 in ϵ. The polynomial is given in the only worksheet Sunset-Threeloop-Epsilon.ipynb.

4.4. Discussion

We have presented a generalisation of the Griffiths-Dwork reduction for deriving differential operators acting on Feynman integrals in dimensional or analytic regularisation. The algorithm makes a special use of the fact that the twist from the regularisations is the power of a degree zero homogeneous rational function build from the graph polynomials.

The algorithm gives the minimal order (non-factorisable) D-module of differential operators acting on regulated Feynman integrals. At each derivative order the procedure consists of solving the linear systems from the reductions with the respect the Jacobian ideal of the graph polynomials $\mathscr{F}$ in (3.11) and $\mathscr{U}$ in (3.12) in order to determine the coefficients $c_{\underline{a}}(\underline{z})$ and the inhomogeneous term $\beta_\Gamma^{\underline{a}}$ in (3.1).

Because the twisted differential $\Omega_\Gamma^{\epsilon,\kappa}$ has the same singularities as $\Omega_\Gamma^{0,0}$, the regularisation parameters ϵ or κ do not affect the discriminant locus but only the local monodromies. This reflects on the fact that these parameters only affect the apparent singularities of the differential operators.

With this algorithm we can derive a Gröbner basis of partial differential operators in some multiple scale cases. The differential operators produced by the algorithm of this paper might arise as specialisation of the system of partial differential operators obtained by GKZ approach. The restriction of the GKZ D-module is a difficult open problem, which we leave for further investigations.

Acknowledgments

I would like to thank Spencer Bloch, Matt Kerr, Leonardo de la Cruz, Pierre Lairez, Eric Pichon-Pharabod, Andrew Harder and Charles Doran for very enjoyable collaboration on the topics reported on this text.

I would like to thank the organisers of the Regulator V conference (3-13 June 2024 in Pisa, Italy) for the opportunity to present these results.

The author was supported in part by French National Agency for Research grant "Observables" (ANR-24-CE31-7996). This research was supported by the Munich Institute for Astro-, Particle and BioPhysics (MIAPbP) which is funded by the Deutsche Forschungsgemeinschaft (DFG, German Research Foundation) under Germany's Excellence Strategy - EXC-2094-390783311.

References

[1] J. R. Andersen, J. Bellm, J. Bendavid, N. Berger, D. Bhatia, B. Biedermann, S. Bräuer, D. Britzger, A. G. Buckley and R. Camacho, *et al.* "Les Houches 2017: Physics at TeV Colliders Standard Model Working Group Report," `arXiv:1803.07977 [hep-ph]`.

[2] S. Abreu, R. Britto, and C. Duhr, *Chapter 3: Mathematical structures in Feynman integrals*, J. Phys. A **55** (2022), no. 44, Paper No. 443004, 56. MR4516657

[3] N. Craig, C. Csáki, A. X. El-Khadra, Z. Bern, R. Boughezal, S. Catterall, Z. Davoudi, A. de Gouvêa, P. Draper and P. J. Fox, *et al.* "Snowmass Theory Frontier Report," `arXiv:2211.05772 [hep-ph]`.

[4] N. E. J. Bjerrum-Bohr, P. H. Damgaard, L. Planté, and P. Vanhove, *Chapter 13: Post-Minkowskian expansion from scattering amplitudes*, J. Phys. A **55** (2022), no. 44, Paper No. 443014, 33. MR4516667

[5] D. A. Kosower, R. Monteiro, and D. O'Connell, *Chapter 14: Classical gravity from scattering amplitudes*, J. Phys. A **55** (2022), no. 44, Paper No. 443015, 43. MR4516668

[6] N. E. J. Bjerrum-Bohr, L. Planté, and P. Vanhove, *Effective field theory and applications*, Handbook of quantum gravity, Springer, Singapore, [2024] ©2024, pp. 85–124, DOI 10.1007/978-981-99-7681-2_3. MR4877856

[7] D. Baumann, D. Green, A. Joyce, E. Pajer, G. L. Pimentel, C. Sleight and M. Taronna, "Snowmass White Paper: The Cosmological Bootstrap," `arXiv:2203.08121 [hep-th]`.

[8] P. Benincasa, *Amplitudes meet cosmology: a (scalar) primer*, Internat. J. Modern Phys. A **37** (2022), no. 26, Paper No. 2230010, 138, DOI 10.1142/S0217751X22300101. MR4514306

[9] F. C. S. Brown, "On the periods of some Feynman integrals," `arXiv:0910.0114 [math.AG]`.

[10] S. Bloch, M. Kerr, and P. Vanhove, *A Feynman integral via higher normal functions*, Compos. Math. **151** (2015), no. 12, 2329–2375, DOI 10.1112/S0010437X15007472. MR3433889
[11] S. Bloch, M. Kerr, and P. Vanhove, *Local mirror symmetry and the sunset Feynman integral*, Adv. Theor. Math. Phys. **21** (2017), no. 6, 1373–1454, DOI 10.4310/ATMP.2017.v21.n6.a1. MR3780269
[12] J. L. Bourjaily, Y.-H. He, A. J. McLeod, M. von Hippel, and M. Wilhelm, *Traintracks through Calabi-Yau manifolds: scattering amplitudes beyond elliptic polylogarithms*, Phys. Rev. Lett. **121** (2018), no. 7, 071603, 6, DOI 10.1103/PhysRevLett.121.071603. MR3849504
[13] J. L. Bourjaily, A. J. McLeod, C. Vergu, M. Volk, M. von Hippel, and M. Wilhelm, *Embedding Feynman integral (Calabi-Yau) geometries in weighted projective space*, J. High Energy Phys. **1** (2020), 078, 39, DOI 10.1007/jhep01(2020)078. MR4088241
[14] J. L. Bourjaily, A. J. McLeod, M. von Hippel and M. Wilhelm, "Bounded Collection of Feynman Integral Calabi-Yau Geometries," Phys. Rev. Lett. **122** (2019) no.3, 031601 `arXiv:1810.07689 [hep-th]`.
[15] A. Klemm, C. Nega, and R. Safari, *The l-loop banana amplitude from GKZ systems and relative Calabi-Yau periods*, J. High Energy Phys. **4** (2020), 088, 39, DOI 10.1007/jhep04(2020)088. MR4096961
[16] K. Bönisch, F. Fischbach, A. Klemm, C. Nega, and R. Safari, *Analytic structure of all loop banana integrals*, J. High Energy Phys. **5** (2021), Paper No. 066, 40. MR4575347
[17] K. Bönisch, C. Duhr, F. Fischbach, A. Klemm, and C. Nega, *Feynman integrals in dimensional regularization and extensions of Calabi-Yau motives*, J. High Energy Phys. **9** (2022), Paper No. 15, 116, DOI 10.1007/jhep09(2022)156. MR4490507
[18] J. L. Bourjaily, J. Broedel, E. Chaubey, C. Duhr, H. Frellesvig, M. Hidding, R. Marzucca, A. J. McLeod, M. Spradlin and L. Tancredi, *et al.* "Functions Beyond Multiple Polylogarithms for Precision Collider Physics," Contribution to Snowmass 2021 `arXiv:2203.07088 [hep-ph]`.
[19] A. Forum and M. von Hippel, "A symbol and coaction for higher-loop sunrise integrals," SciPost Phys. Core **6** (2023), 050 `arXiv:2209.03922 [hep-th]`.
[20] C. Duhr, A. Klemm, F. Loebbert, C. Nega, and F. Porkert, *Yangian-invariant fishnet integrals in two dimensions as volumes of Calabi-Yau varieties*, Phys. Rev. Lett. **130** (2023), no. 4, Paper No. 041602, 7, DOI 10.1103/physrevlett.130.041602. MR4549331
[21] H. Frellesvig, R. Morales, and M. Wilhelm, *Calabi-Yau meets gravity: a Calabi-Yau threefold at fifth post-Minkowskian order*, Phys. Rev. Lett. **132** (2024), no. 20, Paper No. 201602, 8, DOI 10.1103/physrevlett.132.201602. MR4751500
[22] S. Pögel, X. Wang and S. Weinzierl, "Feynman integrals, geometries and differential equations," PoS **RADCOR2023** (2024), 007 `arXiv:2309.07531 [hep-th]`.
[23] A. Klemm, C. Nega, B. Sauer, and J. Plefka, *Calabi-Yau periods for black hole scattering in classical general relativity*, Phys. Rev. D **109** (2024), no. 12, Paper No. 124046, 12, DOI 10.1103/physrevd.109.124046. MR4779715
[24] M. Driesse, G. U. Jakobsen, A. Klemm, G. Mogull, C. Nega, J. Plefka, B. Sauer and J. Usovitsch, "Emergence of Calabi–Yau manifolds in high-precision black-hole scattering," Nature **641** (2025) no.8063, 603–607 `arXiv:2411.11846 [hep-th]`.
[25] H. Frellesvig, R. Morales, and M. Wilhelm, *Calabi-Yau meets gravity: a Calabi-Yau threefold at fifth post-Minkowskian order*, Phys. Rev. Lett. **132** (2024), no. 20, Paper No. 201602, 8, DOI 10.1103/physrevlett.132.201602. MR4751500
[26] C. Duhr, S. Maggio, C. Nega, B. Sauer, L. Tancredi, and F. J. Wagner, *Aspects of canonical differential equations for Calabi-Yau geometries and beyond*, J. High Energy Phys. **6** (2025), Paper No. 128, 90, DOI 10.1007/jhep06(2025)128. MR4923641
[27] A. V. Smirnov and A. V. Petukhov, *The number of master integrals is finite*, Lett. Math. Phys. **97** (2011), no. 1, 37–44, DOI 10.1007/s11005-010-0450-0. MR2802315
[28] R. N. Lee and A. A. Pomeransky, *Critical points and number of master integrals*, J. High Energy Phys. **11** (2013), 165, front matter+16, DOI 10.1007/JHEP11(2013)165. MR3132172
[29] S. Bloch, H. Esnault, and D. Kreimer, *On motives associated to graph polynomials*, Comm. Math. Phys. **267** (2006), no. 1, 181–225, DOI 10.1007/s00220-006-0040-2. MR2238909
[30] S. Bloch, *Motives associated to sums of graphs*, The geometry of algebraic cycles, Clay Math. Proc., vol. 9, Amer. Math. Soc., Providence, RI, 2010, pp. 137–143. MR2648670
[31] S. Bloch and P. Vanhove, *The elliptic dilogarithm for the sunset graph*, J. Number Theory **148** (2015), 328–364, DOI 10.1016/j.jnt.2014.09.032. MR3283183

[32] S. Weinzierl, *Periods and Hodge structures in perturbative quantum field theory*, Feynman amplitudes, periods and motives, Contemp. Math., vol. 648, Amer. Math. Soc., Providence, RI, 2015, pp. 249–259, DOI 10.1090/conm/648/13005. MR3415417

[33] F. Brown, *Notes on motivic periods*, Commun. Number Theory Phys. **11** (2017), no. 3, 557–655, DOI 10.4310/CNTP.2017.v11.n3.a2. MR3713352

[34] F. Brown and O. Schnetz, *Single-valued multiple polylogarithms and a proof of the zig-zag conjecture*, J. Number Theory **148** (2015), 478–506, DOI 10.1016/j.jnt.2014.09.007. MR3283191

[35] M. Marcolli and G. Tabuada, *Feynman quadrics-motive of the massive sunset graph*, J. Number Theory **195** (2019), 159–183, DOI 10.1016/j.jnt.2018.06.001. MR3867438

[36] P. Vanhove, *The physics and the mixed Hodge structure of Feynman integrals*, String-Math 2013, Proc. Sympos. Pure Math., vol. 88, Amer. Math. Soc., Providence, RI, 2014, pp. 161–194, DOI 10.1090/pspum/088/01455. MR3330287

[37] C. F. Doran, A. Harder, P. Vanhove, and E. Pichon-Pharabod, *Motivic geometry of two-loop Feynman integrals*, Q. J. Math. **75** (2024), no. 3, 901–967, DOI 10.1093/qmath/haae015. MR4788720

[38] L. de la Cruz and P. Vanhove, *Algorithm for differential equations for Feynman integrals in general dimensions*, Lett. Math. Phys. **114** (2024), no. 3, Paper No. 89, 45, DOI 10.1007/s11005-024-01832-w. MR4765319

[39] Noboru Nakanishi, *Graph theory and Feynman integrals*, volume 11, Routledge, 1971.

[40] S. Weinzierl, *Feynman integrals—a comprehensive treatment for students and researchers*, Unitext for Physics, Springer, Cham, [2022] ©2022, DOI 10.1007/978-3-030-99558-4. MR4461090

[41] V. E. Asribekov, *Choice of invariant variables and analytical properties of many-point functions*, Phys. Lett. **2** (1962), 284–286. MR143515

[42] C. Itzykson and J. B. Zuber, *Quantum field theory*, International Series in Pure and Applied Physics, McGraw-Hill International Book Co., New York, 1980. MR585517

[43] M. E. Peskin and D. V. Schroeder, *An introduction to quantum field theory*, Addison-Wesley Publishing Company, Advanced Book Program, Reading, MA, 1995. Edited and with a foreword by David Pines. MR1402248

[44] E. R. Speer, *Generalized Feynman amplitudes*, Annals of Mathematics Studies, No. 62, Princeton University Press, Princeton, NJ; University of Tokyo Press, Tokyo, 1969. MR241077

[45] K. Aomoto, *Les équations aux différences linéaires et les intégrales des fonctions multiformes* (French), J. Fac. Sci. Univ. Tokyo Sect. IA Math. **22** (1975), no. 3, 271–297. MR508176

[46] K. Aomoto, *On vanishing of cohomology attached to certain many valued meromorphic functions*, J. Math. Soc. Japan **27** (1975), 248–255, DOI 10.2969/jmsj/02720248. MR440065

[47] K. Aomoto, *Configurations and invariant Gauss-Manin connections of integrals. I*, Tokyo J. Math. **5** (1982), no. 2, 249–287, DOI 10.3836/tjm/1270214894. MR688126

[48] K. Aomoto and M. Kita, *Theory of hypergeometric functions*, Springer Monographs in Mathematics, Springer-Verlag, Tokyo, 2011. With an appendix by Toshitake Kohno; Translated from the Japanese by Kenji Iohara, DOI 10.1007/978-4-431-53938-4. MR2799182

[49] S. Mizera, *Scattering amplitudes from intersection theory*, Phys. Rev. Lett. **120** (2018), no. 14, 141602, 6, DOI 10.1103/PhysRevLett.120.141602. MR3788412

[50] H. Frellesvig, F. Gasparotto, M. K. Mandal, P. Mastrolia, L. Mattiazzi, and S. Mizera, *Vector space of Feynman integrals and multivariate intersection numbers*, Phys. Rev. Lett. **123** (2019), no. 20, 201602, 7, DOI 10.1103/PhysRevLett.123.201602. MR4038991

[51] S. Mizera, "Status of Intersection Theory and Feynman Integrals," PoS **MA2019** (2019), 016 `arXiv:2002.10476 [hep-th]`.

[52] P. Lairez and P. Vanhove, *Algorithms for minimal Picard-Fuchs operators of Feynman integrals*, Lett. Math. Phys. **113** (2023), no. 2, Paper No. 37, 45, DOI 10.1007/s11005-023-01661-3. MR4569063

[53] S. Müller-Stach, S. Weinzierl, and R. Zayadeh, *Picard-Fuchs equations for Feynman integrals*, Comm. Math. Phys. **326** (2014), no. 1, 237–249, DOI 10.1007/s00220-013-1838-3. MR3162491

[54] P. Vanhove, *Feynman integrals, toric geometry and mirror symmetry*, Elliptic integrals, elliptic functions and modular forms in quantum field theory, Texts Monogr. Symbol. Comput., Springer, Cham, 2019, pp. 415–458. MR3889566

[55] L. de la Cruz, *Feynman integrals as A-hypergeometric functions*, J. High Energy Phys. **12** (2019), 123, 44, DOI 10.1007/jhep12(2019)123. MR4061234

[56] R. P. Klausen, *Hypergeometric series representations of Feynman integrals by GKZ hypergeometric systems*, J. High Energy Phys. **4** (2020), 121, 41, DOI 10.1007/jhep04(2020)121. MR4096928

[57] T.-F. Feng, C.-H. Chang, J.-B. Chen, and H.-B. Zhang, *GKZ-hypergeometric systems for Feynman integrals*, Nuclear Phys. B **953** (2020), 114952, 25, DOI 10.1016/j.nuclphysb.2020.114952. MR4065464

[58] B. Ananthanarayan, S. Banik, S. Bera and S. Datta, "FeynGKZ: A Mathematica package for solving Feynman integrals using GKZ hypergeometric systems," Comput. Phys. Commun. **287** (2023), 108699 [arXiv:2211.01285 [hep-th]].

[59] D. Agostini, C. Fevola, A.-L. Sattelberger, and S. Telen, *Vector spaces of generalized Euler integrals*, Commun. Number Theory Phys. **18** (2024), no. 2, 327–370, DOI 10.4310/cntp.2024.v18.n2.a2. With an appendix by Saiei-Jaeyeong Matsubara-Heo. MR4776160

[60] S. J. Matsubara-Heo, S. Mizera and S. Telen, "Four lectures on Euler integrals," SciPost Phys. Lect. Notes **75** (2023), 1 [arXiv:2306.13578 [math-ph]].

[61] H. J. Munch, "Feynman Integral Relations from GKZ Hypergeometric Systems," PoS **LL2022** (2022), 042 [arXiv:2207.09780 [hep-th]].

[62] R. P. Klausen, *Kinematic singularities of Feynman integrals and principal A-determinants*, J. High Energy Phys. **2** (2022), Paper No. 004, 44, DOI 10.1007/jhep02(2022)004. MR4407462

[63] V. Chestnov, S. J. Matsubara-Heo, H. J. Munch, and N. Takayama, *Restrictions of Pfaffian systems for Feynman integrals*, J. High Energy Phys. **11** (2023), Paper No. 202, 80, DOI 10.1007/jhep11(2023)202. MR4672770

[64] C. Dlapa, M. Helmer, G. Papathanasiou, and F. Tellander, *Symbol alphabets from the Landau singular locus*, J. High Energy Phys. **10** (2023), Paper No. 161, 56, DOI 10.1007/jhep10(2023)161. MR4662858

[65] Griffiths, P.A.: The Residue Calculus and Some Transcendental Results in Algebraic Geometry, I. Presented at the (1966)

[66] Griffiths, P.A.: "The Residue Calculus And Some Transcendental Results In Algebraic Geometry, II". Proceedings of the National Academy of Sciences. 55, 1392–1395 (1966).

[67] A. B. Goncharov, "Multiple polylogarithms and mixed Tate motives," [arXiv:math/0103059 [math.AG]].

[68] F. Brown, *Single-valued motivic periods and multiple zeta values*, Forum Math. Sigma **2** (2014), Paper No. e25, 37, DOI 10.1017/fms.2014.18. MR3271288

[69] E. Remiddi and L. Tancredi, *Schouten identities for Feynman graph amplitudes; the master integrals for the two-loop massive sunrise graph*, Nuclear Phys. B **880** (2014), 343–377, DOI 10.1016/j.nuclphysb.2014.01.009. MR3165370

[70] E. Remiddi and L. Tancredi, *Differential equations and dispersion relations for Feynman amplitudes. The two-loop massive sunrise and the kite integral*, Nuclear Phys. B **907** (2016), 400–444, DOI 10.1016/j.nuclphysb.2016.04.013. MR3498437

[71] L. Adams, C. Bogner, and S. Weinzierl, *The two-loop sunrise graph with arbitrary masses*, J. Math. Phys. **54** (2013), no. 5, 052303, 18, DOI 10.1063/1.4804996. MR3098926

[72] P. Candelas, X. de la Ossa, P. Kuusela, and J. McGovern, *Mirror symmetry for five-parameter Hulek-Verrill manifolds*, SciPost Phys. **15** (2023), no. 4, Paper No. 144, 144. MR4659102

[73] S. Pögel, X. Wang, and S. Weinzierl, *The three-loop equal-mass banana integral in ε-factorised form with meromorphic modular forms*, J. High Energy Phys. **9** (2022), Paper No. 62, 28, DOI 10.1007/jhep09(2022)062. MR4478462

[74] S. Pögel, X. Wang, and S. Weinzierl, *Taming Calabi-Yau Feynman integrals: the four-loop equal-mass banana integral*, Phys. Rev. Lett. **130** (2023), no. 10, Paper No. 101601, 7, DOI 10.1103/physrevlett.130.101601. MR4564810

[75] S. Pögel, X. Wang, and S. Weinzierl, *Bananas of equal mass: any loop, any order in the dimensional regularisation parameter*, J. High Energy Phys. **4** (2023), Paper No. 117, 33, DOI 10.1007/jhep04(2023)117. MR4581096

[76] V. Mishnyakov, A. Morozov, and P. Suprun, *Position space equations for banana Feynman diagrams*, Nuclear Phys. B **992** (2023), Paper No. 116245, 35, DOI 10.1016/j.nuclphysb.2023.116245. MR4595948

[77] V. Mishnyakov, A. Morozov, and M. Reva, *On factorization hierarchy of equations for banana Feynman integrals*, Nuclear Phys. B **1010** (2025), Paper No. 116746, 12, DOI 10.1016/j.nuclphysb.2024.116746. MR4832161

INSTITUT DE PHYSIQUE THÉORIQUE, UNIVERSITÉ PARIS-SACLAY, CEA, CNRS, F-91191 GIF-SUR-YVETTE CEDEX, FRANCE

Selected Published Titles in This Series

842 **Federico Binda, José Ignacio Burgos Gil, James Lewis, Deepam Patel, and Gregory Pearlstein, Editors,** Regulators V, 2026

839 **Sangita Jha, William Ott, and Mrinal Kanti Roychowdhury, Editors,** Current Trends in Dynamical Systems, Fractal Geometry, and Their Applications, 2026

838 **Julia E. Bergner, Joachim Kock, and Maru Sarazola, Editors,** Higher Segal Spaces and Applications, 2026

835 **Mee Seong Im and Tony Shaska, Editors,** Artificial Intelligence and Mathematics Research, 2026

834 **Mee Seong Im, Hanna E. Makaruk, Bach Nguyen, and Robert Owczarek, Editors,** Inverse Problems, Geometric and Topological Aspects of Representation Theory, and Low-Dimensional Topology, 2026

832 **Nils Bruin, David Kohel, and Chloe Martindale, Editors,** Arithmetic, Geometry, Cryptography and Coding Theory, 2026

831 **Irena Swanson and Kei-ichi Watanabe, Editors,** Lectures on Tight Closure and Its Applications, 2025

830 **O. Bezushchak, Yu. Drozd, P. Etingof, V. Futorny, R. Grigorchuk, and E. Zelmanov, Editors,** Modern Algebra Volume 2, 2025

829 **O. Bezushchak, Yu. Drozd, P. Etingof, V. Futorny, R. Grigorchuk, and E. Zelmanov, Editors,** Modern Algebra Volume 1, 2025

828 **Hafedh Herichi, Maria Rosaria Lancia, Therese Basa Landry, Anna Rozanova-Pierrat, and Steffen Winter, Editors,** Fractal Geometry in Pure and Applied Mathematics, 2025

827 **Carmen Caprau, J. Scott Carter, Neslihan Gügümcü, and Sam Nelson, Editors,** Algebraic Structures in Knot Theory, 2025

826 **Ahmed Laghribi and André Leroy, Editors,** Algebraic Structures and Applications, 2025

825 **Saurabh Verma, María A. Navascués, and Amit Priyadarshi, Editors,** Advances in Dimension Theory, Fractal Functions and Measures, 2025

824 **Hanna E. Makaruk and Robert M. Owczarek, Editors,** Inverse Problems: In Memory of Professor Zbigniew Oziewicz, 2025

823 **Mikhail Bershtein, Anton Dzhamay, and Andrei Okounkov, Editors,** The Versatility of Integrability, 2025

822 **Ahmad Barhoumi, Roozbeh Gharakhloo, and Andrei Martinez-Finkelshtein, Editors,** Recent Developments in Orthogonal Polynomials, 2025

821 **Bogdan D. Suceavă, Adara M. Blaga, Marian Ioan Munteanu, Cezar Oniciuc, Shoo Seto, Gabriel-Eduard Vîlcu, and Lihan Wang, Editors,** Recent Advances in Differential Geometry and Related Areas, 2025

820 **Sudeb Mitra, Rakesh Kumar Parmar, Saminathan Ponnusamy, and Toshiyuki Sugawa, Editors,** Geometric Function Theory and Related Topics, 2025

819 **Howard S. Cohl, Roberto S. Costas-Santos, and Robert S. Maier, Editors,** Applications and q-Extensions of Hypergeometric Functions, 2025

818 **Howard S. Cohl, Roberto S. Costas-Santos, and Robert S. Maier, Editors,** Classical Hypergeometric Functions and Generalizations, 2025

817 **Mikhail Khovanov, Joshua Sussan, and Anton Zeitlin, Editors,** From Representation Theory to Mathematical Physics and Back, 2025

816 **Luca F. Di Cerbo and Laurenţiu G. Maxim, Editors,** Geometry and Topology of Aspherical Manifolds, 2025

815 **Daniel Orr and Joshua Jeishing Wen, Editors,** Macdonald Theory and Beyond, 2025

For a complete list of titles in this series, visit the
AMS Bookstore at **www.ams.org/bookstore/conmseries/**.